High-Pressure Shock Compress
Condensed Matter

Detonation of a 500-ton TNT hemisphere. Photo courtesy of DRES (Defence Research Establishment Suffield, Medicine Hat, Alberta, Canada).

Jonas A. Zukas William P. Walters
Editors

Explosive Effects and Applications

With 160 Illustrations

Jonas A. Zukas
Computational Mechanics Associates
P.O. Box 11314
Baltimore, MD 21239
USA

William P. Walters
Army Research Laboratory
Aberdeen Proving Ground
Aberdeen, MD 21005
USA

Editors-in-Chief:
Lee Davison
39 Cañoncito Vista Road
Tijeras, NM 87059
USA
leedavison@aol.com

Yasuyuki Horie
Los Alamos National Laboratory
Los Alamos, NM 87545
USA
horie@lanl.gov

Library of Congress Cataloging-in-Publication Data
Explosive effects and applications / edited by Jonas A. Zukas,
William P. Walters.
p. cm. — (High pressure shock compression of condensed matter)
Includes bibliographical references.
ISBN 978-0-387-95558-2 ISBN 978-1-4612-0589-0 (eBook)
DOI 10.1007/978-1-4612-0589-0
1. Blast effect. 2. Shock waves. 3. Condensed matter.
I. Zukas, Jonas A. II. Walters, W.P. (William P.), 1943–
III. Series.
TA 654.7.E95 1997
662'.2—DC21 97-5779

ISBN 978-0-387-95558-2 Printed on acid-free paper.

9 8 7 6 5 4 3

springeronline.com

Preface

Our principal motivation in putting together this book, a project started in earnest in early 1996, was to have a book on energetic materials that was accessible to a newcomer to the field. To this end, we asked all contributors to emphasize basic principles and to provide a broad overview of their fields of specialization. In addition, we requested that they summarize the state of the art in their areas as well. By and large, we believe this goal has been met.

At the time this book was started, many of the classic books dealing with energetic materials were either "out of stock" or "out of print." These are publishing industry jargon for "If you don't already have it, you're not going to get it from us." Left were but a handful of books for specialists in the field. Fortunately, many of the authors and contributors to these extinct classics were active as teachers and lecturers in university classrooms and short courses. Indeed, many of the chapters in this book had their origins as course notes for short courses dealing with detonation physics and explosive technology. All of this was good material, aimed at an audience assumed to have no prior experience with energetic materials, but buried in short course notes that are not readily accessible. They accepted enthusiastically the invitation to prepare chapters covering their work and generously share a lifetime of expertise with the reader.

The book is divided into ten chapters. The first two provide a broad overview of the nature of explosives and their many applications. Most people consider explosives in a military context, and so will be surprised to find that the civilian use of explosives, particularly in rock blasting and demolitions, has always outweighed their use by the military. This has held true even in times of war. Since explosives release their energy over an extremely short period of time, the presence of shock waves is inevitable in all applications. This topic, as well as a description of explosive behavior under shock loading conditions, is considered in Chapter 3. Fundamentals of detonation theory are covered in Chapter 4 and the study of explosive initiation is continued in some detail in Chapter 6. Chapter 5 deals with the chemistry of explosives. In order to be useful in industry, the energy generated by an explosive, or explosive output, must be harnessed in a meaningful fashion. Chapter 7 covers one such application, the Gurney model, in great detail. Hazards and safety considerations are dealt with in Chapters 8 and 9. Chapter 10 considers one of many possible applications, the safe demolition of structures through the use of explosives. The astute reader will note some duplication between chapters. This is intentional. The editors felt that repetition of the basics from different points of view would be beneficial to understanding the subject matter and well worth the slight increase in the length of the book.

The material in this book has been prepared to be readily accessible to anyone with a first degree in science or engineering. Familiarity with the fundamental principles of thermodynamics and organic chemistry that are typically presented at an undergraduate level is particularly helpful.

All contributors to the book are internationally recognized experts in their respective fields. We thank them for the time they have taken to prepare and preserve this material. We are also grateful to the many reviewers whose advice contributed to the quality of the book.

This book is primarily meant for self-study by scientists and engineers who need an introduction to energetic materials and references for further study. With the addition of some supplemental material and class exercises, it can also serve as an advanced undergraduate or graduate text on explosive effects.

The Aberdeen Diner, where Bill Walters and I have spent more time than either of us care to remember, has as its motto:

"If you like the food, tell your friends. If not, tell us."

We encourage readers of the book to share their suggestions for improvement. Comments such as "great book" or "this is the worst piece of trash I've ever read" are useful in that they convey minimal information on how the book has been received. We would urge you though, having been moved enough by the book to write in the first place, to be just a tad more specific. Praise is delightful, but a detailed description of the faults of the book is invaluable. If there are enough of them, there may even be a "new and improved" version of this book. After all, why should the manufacturers of detergents have all the fun!

Baltimore, Maryland — Jonas A. Zukas
Aberdeen, Maryland — William P. Walters

Contents

Preface v
List of Editors and Contributors xv

CHAPTER 1
Introduction to Explosives 1
William C. Davis

1.1. History 1
1.2. Nomenclature 5
1.3. Blasting 6
1.4. Military Uses 9
1.4.1. Blast Waves 10
1.4.2. Fragments 13
1.5. Jet Penetrators 15
1.6. Reactive Armor 16
1.7. Explosive Welding 17
1.8. Wave Shaping and Lenses 18
1.9. Conclusions 20
1.10. Problems (Hints and Solutions) 20
References 21

CHAPTER 2
Explosives Development and Fundamentals of Explosives Technology 23
Peter R. Lee

2.1. Introduction 23
2.2. Nomenclature 24
2.2.1. Units of Measurement 24
2.2.2. Definitions of Physical, Chemical, and Materials Properties of Explosives 26
2.3. The Nature of Explosions 26
2.3.1. Physical Explosions 26
2.3.2. Chemical Explosions 27
2.3.3. Nuclear Explosions 27
2.4. What Are Explosives? 27

2.5. A Short History of Explosives 29
2.5.1. Greek Fire 29
2.5.2. Gunpowder, or Black Powder 29
2.5.3. Brief Chronology of the Discovery and Development of the Commoner High Explosives and Propellants 30
2.5.3.1. Nitroglycerine 30
2.5.3.2. Nitrocellulose 33
2.5.3.3. Gun Propellant Developments 36
2.5.3.4. Picric Acid (2,4,6-Trinitrophenol) 37
2.5.3.5. Tetryl (2,4,6-Trinitrophenylmethylnitramine) 38
2.5.3.6. TNT (Trinitrotoluene) 39
2.5.3.7. RDX (Cyclotrimethylenetrinitramine) 40
2.5.3.8. HMX (Cyclotetramethylenetetranitramine) 41
2.5.3.9. PETN (Pentaerithrytoltetranitrate) 41
2.5.3.10. HNS (Hexanitrostilbene) 41
2.5.3.11. TATB (Triaminotrinitrobenzene) 42
2.5.3.12. HBN (Hexanitrobenzene) 42
2.5.3.13. Commercial Explosives 42
2.5.3.14. Future Explosives Development 43

CHAPTER 3
Shock Waves; Rarefaction Waves; Equations of State 47
William C. Davis

3.1. Introduction 47
3.2. Notation and Units 48
3.3. List of Symbols 49
3.4. Shock Waves 50
3.5. Rarefaction Waves 55
3.6. Reference Frames 61
3.7. Sharp Shocks and Diffuse Rarefactions 63
3.8. Transmission and Reflection of Waves at Interfaces 64
3.9. The Shock Tube 72
3.10. Detonation 74
3.11. Phase Changes 77
3.12. Hydrodynamics and Thermodynamics 80
3.13. Equations of State 82
3.13.1. Ideal Gas and Polytropic Gas 83
3.13.2. Abel 85
3.13.3. Inverse Power Potential 86
3.13.4. Expansion of Equations of State in Powers of v 86
3.13.5. Tait 87
3.13.6. BKW 88
3.13.7. Intermolecular Potentials 88
3.13.8. JWL 89
3.13.9. Linear U–u 91
3.13.10. Walsh Mirror Image 93
3.13.11. Hayes 94
3.13.12. Davis 95
3.13.13. Williamsburg 95

3.13.14. Summary 96
3.14. Equations of State for Mixtures 96
3.15. The Adiabatic Gamma, the Grüneisen Gamma, and the Fundamental Derivative 100
Hugoniot Curve Data [Fritz (1996b)] 105
3.16. Problems (Hints and Solutions) 105
References 112

CHAPTER 4
Introduction to Detonation Physics 115
Paul W. Cooper

4.1. Nomenclature 115
4.2. The Simple Model 116
4.3. The Jump Equations 119
4.4. The Detonation Product P–V Isentrope 120
4.5. Detonation Velocity and Density 121
4.6. The C–J State 127
4.7. The Detonation Product P–u Hugoniot 128
4.8. Detonation Velocity and Charge Diameter 130
4.9. Conclusion 133
Bibliography 134
References 134

CHAPTER 5
The Chemistry of Explosives 137
Jimmie C. Oxley

5.1. Background 137
5.2. Conventional Explosives 140
5.3. Nitrate Esters 141
5.3.1. Nitrate Ester Formulations 144
5.4. Nitroarenes 146
5.4.1. Thermally Stable Nitroarenes 151
5.5. Nitroalkanes 154
5.6. Nitramines 156
5.6.1. Nitramine Composites 159
5.7. Heterocyclic Explosives 160
5.8. Energetic Salts 162
5.9. Composite Explosives 165
5.10. Liquid Oxidizers and Explosives 166
5.11. Unconventional Explosives 168
References 171

CHAPTER 6
Theories and Techniques of Initiation 173
Peter R. Lee

6.1. Introduction 173
6.2. Nomenclature 173

6.3. Initiation Theories 174
6.4. Thermal Explosion Theory 176
6.4.1. Semenov Theory 176
6.4.1.1. Critical Conditions 180
6.4.2. Frank-Kamenetskii Theory 182
6.5. Elementary Detonation Theory 189
6.5.1. Shock Initiation 192
6.5.2. The Hydrodynamic Theory of Detonation 194
6.5.2.1. Steady Detonation 194
6.5.2.2. Transient Detonation Waves 201
6.6. Relationships Between Thermal and Shock Initiation Theories 203
6.7. Initiation Mechanisms 205
6.7.1. Initiation by Heat 207
6.7.2. Friction or Stabbing 208
6.7.3. Flash or Flame 208
6.7.4. Percussion 208
6.7.5. Electrical 209
6.7.6. Coherent Light 212
6.8. Initiation Trains 214
6.9. Conclusions 217
References 217

CHAPTER 7
The Gurney Model for Explosive Output for Driving Metal 221
James E. Kennedy

7.1. Introduction 221
7.2. Nomenclature 223
7.3. Results of the Gurney Model 224
7.3.1. Terminal Velocity Formulas for Symmetric and Asymmetric Configurations 225
7.3.2. Gurney Equations 227
7.3.3. Imploding Geometries 228
7.3.4. Impulse Estimation 231
7.3.4.1. Specific Impulse of Unconfined Surface Charges 231
7.3.4.2. Impulse Increase by Tamping 232
7.3.5. Gurney Energy of Explosives 232
7.4. Applications of Gurney Analysis 235
7.4.1. Tamping Effectiveness 236
7.4.2. Direction of Metal Projection 236
7.5. Extensions of Gurney Analysis 238
7.5.1. Estimation of Gurney Velocity from Chemistry, Density, and Detonation Parameters 238
7.5.2. Effects of Gaps 239
7.5.3. Acceleration Solutions 239
7.5.3.1. Jones Analysis for Slab Geometries 239
7.5.3.2. Chanteret Analysis for Symmetric Slabs and Exploding Cylinders 242
7.5.3.3. Flis Analysis by Lagrange's Principle 243

7.5.4. Electrical Gurney Energy 245
7.5.5. Laser Ablation Gurney Energy 245
7.6. Limitations and Corrections in Gurney Analysis 246
7.6.1. Comparison with Gas-Dynamic Solution for Open Sandwich 246
7.6.2. Correction for Side Losses 247
7.6.3. Scaling 249
7.7. Combination of Gurney with Other Physics 249
7.7.1. Inelastic Collision Momentum Transfer 250
7.7.2. Detonation Transfer by Flyer Plate Impact 251
7.7.2.1. Shock-Initiation Criteria 251
7.7.2.2. Analysis of Impact Interaction 251
7.7.2.3. Gurney Analysis to Evaluate Initiation Criterion 252
Acknowledgment 254
References 255

CHAPTER 8
Hazard Assessment of Explosives and Propellants 259
Peter R. Lee

8.1. Introduction 259
8.1.1. Aims 259
8.1.2. Nomenclature 260
8.1.3. Background 261
8.2. Basic Precepts of Hazard Testing 263
8.2.1. Absolute and Relative Sensitivities 263
8.2.2. Differences Between Hazard and Reliability Testing 265
8.2.3. Analysis of Test Results 267
8.2.3.1. The Bruceton Staircase Technique 267
8.2.3.2. Method of Minimum Contradictoriness 269
8.2.3.3. The Role of Judgment in Hazard Testing 270
8.3. Sensitiveness, Sensitivity, and Explosiveness 271
8.4. Stages in Hazard Assessment 275
8.4.1. Powder Tests 276
8.4.2. Impact Tests 277
8.4.3. Design Details of Some Powder Impact Test Machines 278
8.4.3.1. US Bureau of Mines, Pittsburgh 279
8.4.3.2. US Naval Ordnance Laboratory, White Oak, MD: Laboratory Scale Test 280
8.4.3.3. Los Alamos Laboratory Scale Impact Test and Navy Weapon Center, China Lake 281
8.4.3.4. Picatinny Arsenal Laboratory Scale Test 281
8.4.3.5. Bureau of Mines Test 281
8.4.3.6. Rotter Test 282
8.4.3.7. Bureau of Explosives (NY) Laboratory Scale Tests 283
8.4.3.8. German BAM Technique 283
8.4.3.9. Test Results 284
8.4.4. Powder Friction Tests 286
8.4.4.1. UK Mallet Friction Test 286
8.4.4.2. US Bureau of Mines Pendulum Friction Test 287

8.4.4.3. US Navy Weapons Center, Friction Pendulum Test 288
8.4.4.4. German BAM Friction Test 288
8.4.4.5. UK Rotary Friction Test 288
8.4.4.6. Test Results 289
8.4.5. Other Powder Sensitiveness Tests 290
8.5. Charge Hazard Tests 290
8.5.1. Shock Initiation Tests 291
8.5.1.1. Uninstrumented Gap Tests 292
8.5.1.2. Instrumented Gap Tests 296
8.5.1.3. Gap Test Results 298
8.5.1.4. Other Shock Initiation Tests 298
8.5.2. Charge Impact Tests 302
8.5.2.1. Oblique Impact Test 302
8.5.2.2. LANL Oblique Impact Test 303
8.5.2.3. LLNL–Pantex Skid Test 304
8.5.2.4. Oblique Impact (Skid) Test Results 304
8.5.3. High-Speed Impact Test: The Susan Test 305
8.5.3.1. Results of Susan Test Firings 307
8.5.4. Intrusion Tests 309
8.5.4.1. LANL Spigot Intrusion Test 309
8.5.4.2. AWE Spigot Test 310
8.5.4.3. US Navy Spigot Test 311
8.5.4.4. Spigot Test Results 311
8.5.5. Thermal Hazard Tests 311
8.5.5.1. Thermal Hazard Testing 312
8.6. Electrostatic Sensitiveness 319
8.6.1. Powder Tests 320
8.7. Assessment of the Results of the Tests on Energetic Materials 321
8.8. "Insensitive" High Explosives and Propellants 322
8.9. System or Weapon Tests for IM 326
8.9.1. Insensitive Munitions Thermal Tests 328
8.9.1.1. Slow Cook-Off Tests 328
8.9.1.2. Fast Cook-Off Test (Fuel Fire Test) 329
8.9.1.3. Results of Cook-Off Tests 330
8.9.2. Impact Tests 331
8.9.2.1. Bullet and Fragment Attack Tests 331
8.9.2.2. Shaped-Charge Jet Impact Tests 333
8.9.2.3. Sympathetic Detonation Test 335
8.9.3. Procedures for Insensitive Munitions Testing 335
8.9.4. Research Required to Improve Understanding of Insensitive Munitions and the Results of Insensitive Munitions Testing 336
Acknowledgment 337
References 337

CHAPTER 9
Safe Handling of Explosives 341
Jimmie C. Oxley

9.1. Explosive Safety 341
9.2. Sensitivity Testing 343
9.2.1. Impact 343

9.2.2. Friction 346
9.2.3. ESD 346
9.3. Thermal Stability 346
9.3.1 Comparative Thermal Stabilities 347
9.3.2 Quantitative Thermal Stabilities 349
9.3.3 Summary of Thermal Analytical Tools 354
9.3.4. Time-to-Explosion 356
9.4. Case History of an Ammonium Nitrate (AN) Emulsion Accident 358
9.4.1. Thermal Stability of AN Formulations by DSC 359
9.4.2. Thermal Stability of AN Formulations by Isothermal Techniques 361
9.4.3. Thermal Stability of AN with Additives 366
9.4.4. Verification of Small-Scale AN Kinetics by Larger-Scale Tests 370
9.4.4.1. One-Liter Cook-Off 371
9.4.4.2. Sealed Cook-Off 372
9.4.5. Visual Observations 374
9.4.6. Summary 374
9.5. Thermal Stability of Organic Explosives 375
9.6. Toxicity of Explosives 375
References 378

CHAPTER 10
Demolitions 381
Chris A. Weickert

10.1. Introduction 381
10.2. Explosively Formed Projectiles 386
10.3. Wall Breaching 391
10.4. Bridge Demolition 399
10.5. Explosive Ordnance Disposal 413
References 420

Index 425

List of Editors and Contributors

Editors

William P. Walters
US Army Research Laboratory, Aberdeen Proving Ground, Aberdeen, Maryland, USA.

Jonas A. Zukas
Computational Mechanics Associates, Inc., Baltimore, Maryland, USA.

Contributors

Paul W. Cooper
Consultant, formerly Sandia National Laboratories, Albuquerque, New Mexico, USA.

William C. Davis
Energetic Dynamics Los Alamos, Los Alamos, New Mexico, USA.

James E. Kennedy
Los Alamos National Laboratory, Los Alamos, New Mexico, USA.

Peter R. Lee
Peter Lee Consulting Co., Ltd., Tunbridge Wells, Kent, England.

Jimmie C. Oxley
Gordon Research Conferences, University of Rhode Island, Kingston, Rhode Island, USA.

Chris A. Weickert
Defence Research Establishment Suffield, Medicine Hat, Alberta, Canada.

CHAPTER 1

Introduction to Explosives

William C. Davis

Explosions are violent, sudden, noisy, and startling. The human brain resists the thought that they are proper subjects for calm contemplation and detailed physical modeling. An engineer or scientist who begins to use explosives must subdue his or her startle reflex and aversion to loud noises, and consider exactly how explosives work. To begin, let us survey the vast range of uses for explosives, and the great variety in chemicals that explode.

1.1. History

Black powder has been known for centuries and used in small quantities for mining after about A.D. 1650, and in larger quantities after 1800. Black powder is extremely dangerous to use because it is so easily ignited by any spark. Its action is unpredictable because the burning rate and the pressure developed depend on the strength of the rock confining it. The economic surge of the industrial revolution demanded ore mining in hard rock, cutting defiles through hills for railroads and highways, and excavating and leveling of older structures before construction of huge factories. Black powder could not meet the needs.

Black powder is called a low explosive. It contains both the fuel and the oxydizer in the mixture, and is what is now called an energetic material. When it is ignited it burns, and the rate of burning increases as pressure develops from the release of gases, increasing the pressure still more. The maximum pressure is determined by the strength of the surroundings, the bore hole, for example. High explosives detonate rather than burn, and in a detonation the confinement that leads to high temperatures and very rapid reaction is provided by the explosive itself. The reaction is so rapid that the expansion, spreading in a wave propagating at the local speed of sound, is not fast enough to reduce the pressure appreciably, and the reaction is inertially confined by the explosive mass. Modern explosives are high explosives.

Detonating explosives have been known at least as long ago as the fifteenth century, when the alchemist Blasius Valentius produced "fulminating gold," which was used by wandering entertainers to amaze the public. Other high explosives were made over the years. From about 1830 on, systematic and purposeful studies produced many new explosives. The revolution in chemistry started in 1834 when Mitscherlich nitrated benzene to nitrobenzene, and then in 1842 Zinin reduced nitrobenzene to aniline. The demand for dyes fueled the search for new organic compounds, and many of these were either explosives or closely related. It is now thought that there are about 20,000 known molecules that are explosives, and perhaps 600 of those have been studied enough that they could be used or have been used. In addition, there are many mixtures of fuels and oxidizers that are explosives. Particularly influential for the development of explosives was the work of Sobrero (1847), and his nitroglycerin became the basis for the kieselguhr dynamite, a mixture of nitroglycerin and diatomaceous earth, and gelatin dynamite, a mixture of nitroglycerin and guncotton. Dynamite and gelatin dynamite were inventions of Alfred Nobel about 1867 and 1875, and they furnished the explosives needed for rapid expansion of the world economy in the last half of the nineteenth century. Nobel also invented the blasting cap in 1863, a detonator which will reliably initiate these explosives. The cap was similar to caps filled with potassium chlorate made by Forsyth in 1805, and with mercury fulminate by Ballot and Egg in 1815, for the initiation of black powder, later using the safety fuse invented by Bickford in 1831. Part of the fortune these inventions brought to Nobel funded the much-coveted Nobel prize. There is no prize for work on explosives.

Techniques for using explosives for mining developed rapidly. A remarkable change occurred in the early 1950s when technological advance and economic demand again coincided to revolutionize rock blasting. Tungsten carbide cutting edges for drills became available, thus increasing the lifetime of drill bits by a factor of hundreds. The small drill holes that had been used for dynamite, typically 50 mm in diameter, were replaced by large holes, 200–250 mm in diameter, which had suddenly become easily obtainable. At the same time, after a disastrous explosion in Texas City of a ship loaded with fertilizer, it became obvious that an inexpensive explosive could be made from ammonium nitrate and fuel oil. ANFO, as it is usually called from the initials of its constituents, had been patented in Sweden in 1867, but it had been used very little because it worked poorly in the small boreholes available then. At this same time, the easily worked ore deposits were nearly depleted. Low-grade ores required processing huge quantities of material, and hard-rock ores required huge quantities of explosive. Also, coal mining changed from underground mining to open-pit mining, and it became necessary to remove large amounts of the overburden. Now ANFO and other ammonium-nitrate-based explosives account for most of the explosive used in the world today. Many other explosives are used, in smaller quantities, for various applications.

Low explosives, propellants, and pyrotechnics have been used from ancient times in warfare. Military uses of high explosives came as soon as the blasting cap was available. The siege of Paris in 1871 spawned the first large research and development program in military explosives. Under Berthelot, the Parisians tried to find ways to keep the Germans from taking the city. The effort failed, but Berthelot became the giant upon whose shoulders the scientific study of high explosives now stands.

Many of the advances in the field of high explosives have come from military research. Applications span a tremendous range. As an example, bombs designed to be dropped from aircraft are classified according to filler explosive, chemical, incendiary, pyrotechnic, and inert, and according to use as armor-piercing, semi-armor-piercing, general purpose or demolition, light case, fragmentation, depth, gas, smoke, incendiary, photographic flash, target identification, leaflet, practice, and dummy. Most of these use explosives, if not for their main effect, then for some purpose in the operating mechanism. They range in mass from about 1 kg up to 20,000 kg. Many other weapons systems have similar variety and size ranges. Several hundred military explosive compositions are commonly used to satisfy all the special requirements for many kinds of weapons.

One particular military development is worth singling out for discussion. During World War II, the atomic bomb was developed at Los Alamos, and since then, several countries have produced their own bombs. Nuclear weapons use explosives for the precise, dynamic assembly of the fissionable components. Fissionable materials are among the rarest and most expensive materials used in the world; no one could afford to waste them by imprecise assembly. For blasting, the cost of the explosive has to be kept to the absolute minimum; for nuclear weapons, precision is a necessity and price is relatively unimportant. The new requirement for precision brought about a revolution in the explosive industry. All the attributes of high technology—precision, reproducibility, long shelf-life, and tailoring to infinitesimal gradations—came to the explosives industry.

Although nuclear weapons provided the initial requirement for precision explosives, other new uses appeared. Designers in many fields now think of explosives as power sources with extremely high-power densities that are extraordinarily safe, reliable, reproducible, and, best of all, incredibly inexpensive. The traditional applications for explosives are excavating, mining, tunneling, underwater excavation, and smooth blasting for underground construction. Explosives are cheaper and much safer than any available alternative method.

Demolition is often seen in the newspapers or on television in spectacular photographs of old hotels becoming just memories and a pile of rubble. Explosives cut the support beams and columns in an exactly timed sequence so that all the parts of the structure fall in predetermined spots. In metal working, explosives are used for forming, welding of otherwise unweldable combinations, cladding of sheets and tubes, and hardening of metal surfaces. Very high-velocity jets or sheets of metal can be produced by explosives, and these will penetrate or cut deeply into or through other materials. They are widely used by military forces to defeat armor, by rocket designers to separate rocket stages, and by oil well drillers to perforate pipe and rock. In steel furnaces they are used to release the molten metal. Scientists studying the Earth's magnetic field use explosives to inject ionizable barium at high altitudes, where the high-speed particles overcome the pull of the Earth's gravity, so that the geomagnetic field lines can be mapped. Still other applications are explosive rivets, motors, switches, valves, and bolts.

Less well known are explosively driven magnetic field compressors that can produce magnetic fields up to 10 megagauss, so strong that the field pressure would cause any static system to fly apart, and magnetic compression generators that produce enormous pulses of electricity. Intense light pulses from large surfaces with equivalent temperatures up to 35,000 K or more can be obtained from shock waves in rare gases driven by high explosive. These light pulses have been used for optical pumping of iodine compounds in lasers that emit very high energy directed beams.

Diamonds can be made from graphite, metal and ceramic powders can be sintered, submicrometer-sized powders can be made, and otherwise impossible chemical reactions can be forced to proceed at local hot spots in granular or polycrystalline matter, all by means of explosives as the power source. New applications are being found, some of them quite unusual. Recently there even were reports of small charges being use for blasting stones inside the human kidney.

The science of high explosives developed as the industry grew. The great physical chemist Berthelot (1871, 1883, 1892) contributed much to the understanding of what makes a good explosive, and, especially with the collaboration of Paul Vielle, to the science of how detonation propagates. Shock waves were described by Riemann in 1860, and again by Rankine in 1870, and by 1890 it was established that detonations were shock waves supported by the chemical energy release. Chapman (1899) showed how to calculate detonation velocities using the theory and measured thermochemical parameters, demonstrating accurate agreement with the extensive measurements of detonation speeds of gases collected by Dixon. By 1900 the basic theory was well established. Before 1920 Becker used realistic equations of state to calculate the detonation of solid explosives. In the early 1940s, Zeldovich, von Neumann, and Doering, working independently, developed the theory of the chemical reaction zone, and about 1960, Wood and Salzburg extended the theory to complex reaction zones. Excellent reviews of the early work are to be found in Manson (1986, 1987). Later work is reviewed in Fickett and Davis (1979).

Experiments with detonating explosives demanded ingenious new instrumentation. Berthelot and Vielle invented the bomb calorimeter to get accurate heats of formation. Time measurement to a few microseconds was extremely difficult in the period from 1860 to 1920, but the experimenters found new ways to do it. Some descriptions of apparatus, with references, are given by Manson (1987). Many others are found in Berthelot (1883), and in the original papers.

The demand for explosives taxed the chemical industry, particularly when new explosives were required for military use. The invention of a new explosive is a necessary but not sufficient condition for its use. It must be produced in quantity, economically and safely. A review of the role of chemical engineering in the production of these explosives is given by Ayerst and McLaren (1980). Davis (1941, 1943) describes many explosives, their properties, and how they are made. Accidents, causing many deaths, both in manufacture and in shipping, were a major problem in the development of the industry. Medard (1989) gives an analysis of how accidents occur, the important chemistry of explosions, and details for individual explosives. A review of early accidents, up to about 1927, is given by Assheton (1930) in a compilation made to determine safe distances for the storage of explosives. Some later accidents are described by Bailey and Murray (1989). Other accident descriptions are given by Nelson (1985) and Robinson (1944). A study of accidents is not just morbid preoccupation, but really necessary to understand what kind of things can happen so they can be prevented.

1.2. Nomenclature

a	Sound speed.
f	Dimensionless function for scaled pressure.
g	Dimensionless function for scaled impulse.
g	Acceleration of gravity.
i	Impulse.
p	Pressure.
r	Distance.
t	Time.
x	Distance.
A	Area.
B	Burden in bench blasting.
C_D	Drag coefficient.
D	Diameter.
D_f	Detonation velocity of fast explosive.
D_s	Detonation velocity of slow explosive.
E	Explosive energy.
F	Force.
L	Characteristic length for fragments.
M	Mass.
Q	Quantity (mass) of explosive for bench blasting.
R	Distance.
V	Velocity.

W Weight of explosive.

ρ_a Air density.
ρ_m Metal density.

1.3. Blasting

About five billion pounds of explosive are used in the United States every year. Most of that explosive is used to break rock, for many different purposes. The largest consumer of explosives is the coal mining industry, and most of the coal is used to produce electricity. The use of explosives and their application to blasting are extensively regulated by federal, state, and local laws and codes. Training and experience are needed before an individual may begin a career as a blaster. Here, we will outline the general principles of blasting and provide some references. This section must not be considered sufficient training to set oneself up as a blaster.

Rock blasting is a fully-developed engineering specialty with a long history. The goal is to break solid rock into pieces that can be moved, either for further processing or just to get them out of the way. Pieces of rock are of acceptable size if they can be handled with the available machinery, but many other constraints must be satisfied. The pieces of rock must not fly too far or they may cause injury or damage. Noise and ground vibration must be kept to acceptable levels. Blasters must comply with all applicable laws. Within these constraints and those of physical laws and rock properties, the cost of the operation must be minimized.

Because rock is much stronger in compression than in tension, efficient coupling of the explosive energy to the rock requires that the explosive be placed inside the rock, in a drill hole or a tunnel or other cavity. Then the explosive gases that are formed at high pressure when the explosive is detonated can do work on the rock and cause it to fracture and move. Properly done, blasting will break 1000 kg to 20,000 kg of rock for each kilogram of explosive used, with 5000 kg/kg being a typical ratio.

To begin, consider a uniform sphere of rock with explosive packed into a spherical cavity at its center. At an instant of time, the explosive reacts uniformly and instantaneously, changing into a sphere of high-pressure gas. A spherical shock wave expands outward as the gas pressure forces the rock near it to move. While the cavity expands, the gas pressure decreases and the rock near it decelerates. Deceleration follows the shock wave, with maximum rock velocity at the shock front, and slower velocities behind. When the shock wave reaches the free surface of the rock sphere, all the rock is moving out radially, with that near the surface moving faster than that toward the center. Mentally picture any spherical shell of rock: as it expands, it must stretch. Rock is not strong in tension and eventually it will break. Radial cracks are thus formed. Now, picture rock along a radius; an outer piece is moving faster than the one behind, and the rock must stretch along the radial line. This stretching also causes it to break at spherical surfaces. In this way, the sphere of rock is broken into fragments. The whole process is very complex. The waves are not strong enough to break the rock immediately except very near the borehole, and the continued motion is vital for the desired result. Stressing the rock requires energy proportional to the mass of rock, and therefore there will be a requirement for explosive mass proportional to the cube of a linear dimension.

Suppose that the fracture pattern is independent of the size of the sphere. That is, suppose that the number of fragments and their shapes are the same when a large sphere is blasted as they are when a small sphere is blasted. If this is true, then the mass of rock broken is proportional to the cube of the radius, whereas the surface area created is proportional to the square of the radius. Creating a new surface requires energy from the explosive. Therefore, the ratio of the weight of broken rock to the weight of explosive required to do the job increases with the size of the sphere. Although our supposition that the fracture pattern is independent of size is not exactly true, it is true that the larger the blast, the larger the ratio of rock weight to explosive weight. Thus, for quantity of rock broken, large blasts are more efficient in their consumption of explosive than smaller ones. To create new area, there will be a term in the required mass of explosive proportional to the square of a linear dimension.

On the other hand, large blasts result in larger fragments. At some point, the fragment size becomes unacceptable because the fragments are too large to be moved economically. Also, large blasts cause more noise and vibration, and they throw fragments farther. When all important factors are taken into account, there is an optimum size.

For very large blasts, another scaling becomes important. Broken rock occupies more space than consolidated rock, so when it is blasted the center of gravity is raised, thus requiring energy. This energy is proportional to the mass times the vertical change in the center of mass; thus it is proportional to the fourth power of the linear dimension. The principal term in determining the mass of explosive to use is the one proportional to the cube of a linear dimension, and next in importance is the one proportional to the square of the linear dimension. The term proportional to the fourth power of the linear dimension is a small correction.

In practice, the blasting method most like the oversimplified case of the sphere discussed above is "coyote" blasting, or concentrated charge blasting. A tunnel is driven into a cliff side, with a shorter tunnel at right angles to the main one at the end. The end of the short tunnel is loaded with explosive, and the whole tunnel is packed with rock, or "stemmed." The right-angle turn near the end prevents the explosion from pushing the stemming straight back out the tunnel. Often two charges are used. The tunnel makes a "T" and the charges are in the two ends of the cross bar. Sometimes more branch tunnels are used. This sort of arrangement is useful, for example, for blasting limestone for a remote cement plant, where noise and vibration, as well as flying rock, are acceptable. Limestone is often bedded or layered so that it breaks up into pieces of reasonable size even when the overall scale is very large. Coyote blasting is most suitable to rock formations that fragment by displacement only, breaking up simply when the rock is moved.

Most blasting is done in holes drilled into the rock. The drill holes may be from 15 mm to 300 mm (even occasionally 450 mm) in diameter, and they may be up to about 200 (or sometimes 1000) diameters long. Often holes are drilled vertically in a row parallel to the edge of a bench in an open-pit mine, and several rows may be used in a single blast. (An open-pit mine is a pit, with the ground stepped down into the pit. Each step is called a bench.) The distance from the holes to the edge is called the burden and is the length scale for the blast. The diameter of the holes is chosen to accommodate the required amount of explosive; typically the burden is 30 to 50 times the hole diameter. The spacing of the holes along the edge of the bench is usually about 1.25 times the burden. Breaking the rock loose at the bottom requires more explosive than shattering it up higher in the bench.

A higher-strength explosive may be used at the bottom of the hole, and a weaker one in the rest of it. At the top, the length left uncharged with explosive is stemmed with packed dry sand. The row of holes is often fired with delays between them. This practice reduces the ground vibration, allowing the firing of large, economical charges near populated areas, and it also improves the fragmentation. Blasting caps with delays built into them are available, so they may all be triggered by a single electrical impulse, but they fire after well-defined known delays that may be chosen for each hole. Delays are also important when more than one row of holes is fired, so that the outer layer can move away before the next layer is fired.

The mining engineer's problem is to find the combination of bench height, burden, drill-hole diameter, hole spacing, explosives, and rock-handling machinery that is the most economical at a particular place and time. Hundreds of possible combinations exist, and some of the costs are very difficult to estimate ahead of time. In the competitive mining industry, small differences are important.

To illustrate some of the details of rock blasting, consider the very simple arrangement with a single drill hole where the bench height equals the burden. To loosen the rock, the explosive is placed as a concentrated charge at the bottom of the hole. Langefors and Kihlstrom (1963) calculate the total quantity of explosive by

$$Q = 0.07B^2 + 0.4B^3 + 0.004B^4,$$

where B is the burden in meters and Q is the quantity of explosive in kilograms. The first term is the explosive required to create surface and to satisfy other dissipative processes. The second term is the principal term that relates the weight of explosive and the weight of rock. The third term, usually very small, provides the energy for the swell of the rock and the consequent lifting of the center of mass. Suppose the burden is 3 m. Then $Q = 11.75$ kg. This blast will loosen a triangular prism of rock, where the horizontal triangle has a base of about 6 m and a height of 3 m, and the vertical dimension of the prism is 3 m. The volume of rock is then 27 m^3, and the consumption of explosive is $11.75/27 = 0.435$ kg/m^3. If the rock density is 2700 kg/m^3, the weight of rock broken is 72,900 kg, and the ratio of rock weight to explosive weight is $72{,}900/11.75 = 6204$. To break rock down to the grade level, the hole should be drilled about 0.3 of the burden dimension below grade, so some of the explosive will be in the length of hole below grade. ANFO, an explosive made of ammonium nitrate and fuel oil, a sensible choice for this purpose, comes in bulk in bags, and can be poured into the hole. The blasting cap is inserted into a small booster charge, and the assembly is lowered into the hole by the wires. The ANFO is poured in on top of the booster. The final average density of the ANFO will be about 0.8 g/cm^3. A little calculation shows that a hole 75 mm (or 3 in.) in diameter will accommodate the explosive in 3.3 m length. The hole above the explosive is filled with tamped dry sand, and care is taken not to damage the cap wires.

In addition to rock blasting, there are many uses for explosives in construction, agriculture, and demolition. Small jobs in these fields are often attempted by persons with inadequate training or experience and occasionally the jobs end as disasters. Fortunately, texts on many technical levels discuss these problems. There is no excuse for throwing a stump up so it comes down through the roof of the house.

Explosives manufacturers publish handbooks with explicit directions for the use of their explosives, and they also provide engineering assistance when it is needed. A modern textbook is by Persson, Holmberg, and Lee (1993). Also of interest is Langefors and Kihlstrom (1963). A somewhat different approach is found in Hino (1959), which describes Japanese practice.

1.4. Military Uses

The governments of modern nations maintain military forces whose duties include preparing for and sometimes conducting operations designed to kill people and destroy property. Explosives are widely used for military purposes, and many devices have been developed for these purposes. Money for military research has been available for many years. It is interesting to note that Aristotle, Leonardo da Vinci, LaGrange, and many others had their work supported by governments for military applications. McShane, Kelly, and Reno (1953) contains a long appendix presenting the history of military support through the ages in interesting detail.

Among the explosive weapons that man uses to take life and destroy property are bombs dropped by aircraft, artillery shells, mines, and torpedoes. High-explosive bombs are metal cases filled with high explosive. Where the blast wave from the explosive is the main agent of destruction, the bomb case is light, just enough to protect the explosive, and the explosive weight may be as much as 80% of the total bomb weight. Where fragments of metal and the blast are both important, the case is made thicker, and in a "general purpose" bomb the explosive weight is about 50% of the total. Where the bomb must penetrate armor or concrete, the metal case is very heavy and the explosive weight is 10% to 20% of the total weight.

Artillery shells have metal cases filled with explosive, and the explosive weight is 10% to 20% of the total weight. These systems must withstand the enormous accelerations they experience in guns, up to about 30,000 g. One very serious problem in designing shells is to prevent initiation of the explosive during the acceleration. An explosion of this sort is called an "in-bore premature" explosion. It usually kills the gun crew, and it causes psychological problems for other crews.

The most complex design problem for military explosive devices is that of the fusing system, which includes timing devices, impact detectors, radar-ranging fuses, self-destruct devices, altitude detectors, etc., and the detonators and boosters that follow them. These problems and their solutions are discussed in considerable detail in the Picatinny Arsenal *Encyclopedia of Explosives and Related Items* [Fedoroff (1962)].

1.4.1. Blast Waves

Blast damage by explosives occurs when the air driven by the explosive (and sometimes the detonation products) moving at high speed, engulfs an object and accelerates it. To understand how this process works, a person or an object like an automobile can be modeled as a rigid body with friction between it and the Earth. Model the blast wave as a steep pressure rise, the shock wave, followed by an exponential decay of pressure. An actual blast wave has a negative-pressure region behind it, where the pressure is less than atmospheric pressure, neglected in this simple model. The negative-pressure region usually does much less damage than the positive part. The blast wave interacts with the object, and the object experiences a force proportional to the square of the air velocity, the object area, and the drag coefficient. If the force is larger than the friction force between the object and the earth, the object will move; if it is less, the object will remain stationary. Thus there is a minimum wave pressure that will cause the object to move. If it does move, it will accelerate as long as the force is larger than the frictional force.

When the force is very large compared with the frictional force, the final velocity will be proportional to the impulse delivered. The physical law is that the momentum change is equal to the time integral of the net force. The criterion for damage to the object is either the total distance it moves or the maximum velocity it attains. In either case, there is a minimum impulse or a minimum pressure that causes motion that exceeds the damage limit. In the impulse–pressure plane, a curve that resembles a rectangular hyperbola divides the plane into two regions. Below and to the left of the curve, no damage is done by the blast wave. Above and to the right of the curve, the object is damaged by the wave.

Another class of object that might be damaged by a blast wave is one that is fixed and that deflects elastically and, finally, plastically when it is stressed by the blast wave. An example might be a beam or a column in a structure, or a plate in a ship, or the nose cone of a rocket reentry vehicle, or a human eardrum. Model these as a rigid body attached to the Earth or some other massive body by a spring. In any real system the motion will be damped, so the model should also have a dashpot in parallel with the spring. The damping makes little difference for blast loading. The criterion for damage is the deflection of the system. When the total duration of the pressure wave is short, compared with the vibration time of the system, the loading is called impulsive loading, and the momentum is transferred to the body before it moves appreciably. Thus there is a minimum momentum that produces a damaging deflection. When the loading pulse is long compared with the natural vibration time of the system, the deflection depends only on the peak pressure, so there is a minimum pressure to produce damage. The impulse–pressure plane is separated by a curve like a rectangular hyperbola into two regions of damage and no damage.

The impulse–pressure plane is the useful one for considering blast damage. Note that a real structure might have several different parts with different damage criteria, and the impulse–pressure plane for the structure might have several dividing curves in it. The envelope of these curves is then the curve dividing the plane into undamaged–damaged regions. Except for detail near the focus, the dividing curve will still look much like a rectangular hyperbola.

These concepts of blast wave damage have been developed and applied by Baker, Cox, Westine, Kulesz, and Strehlow (1983). They give many examples, tables, and photographs. Anyone concerned with blast damage, whether from military or civil explosions, needs to study their book.

The military organizations that use bombs and shells have thoroughly studied damage and damage mechanisms. Many of their findings have been described, although some are in classified reports. Before a damage study is begun, a considerable research effort sifting through old reports should be done as the problem may have already been studied.

Measurements of the pressure and impulse in blast waves generated by explosives apply directly only for the size of explosive charge used and the particular distance of the measuring devices from the charge. They can, however be scaled so they can be used to apply more generally. Consider an explosive sphere of radius r at a distance R from a measuring device. If r and R are both increased in the same ratio, so that R/r remains constant, the maximum pressure will remain the same, and the time will increase in proportion to R or r. This scaling may seem intuitively obvious; it can be shown rigorously, as well. This scaling has been widely used, but its disadvantage is that the explosive must always be the same in every respect. The radius r for an explosive of fixed properties will always be proportional to the cube root of its weight, so that root $W^{1/3}$ is often used instead of r. This scaling has the disadvantage that it is dimensional, but it is used more widely than R/r. Researchers have often defined equivalent weight ratios for various explosives and have used the scaled weights to determine approximate pressures versus scaled distances for all explosives. In most experimental measurements the ambient air pressure is approximately the sea-level standard, although pressure and impulse are often needed for estimating damage at high altitude where the pressure is less. Another scaling used takes the atmospheric pressure into account. The energy of the explosive charge is used instead of its equivalent weight. In this system, the cube root of the ratio of energy to initial atmospheric pressure, which has the dimensions of length, is used as the length scale. For military explosives the scale is 70 or 80 times the charge radius when the initial atmospheric pressure is the standard sea-level value, and the scale increases at lower pressures. Using this scaling, the measured pressures can all be reduced to a single curve, with the scaled pressure is a function of the scaled distance, where

$$\frac{p}{p_0} = f\left[\frac{R}{(E/p_0)^{1/3}}\right],$$

where p is the measured pressure, p_0 is the initial atmospheric pressure, R is the distance from the charge to the measuring device, E is the energy of the charge, and f is the measured function for any choice of conditions. Some published plots of f have a large scatter in the points. The scatter is partially caused by a failure to appreciate how strongly the pressures depend on the charge shape. Later data taken using perfect explosive spheres look much better.

The impulse is the integral of pressure with respect to time. It has the units of pressure times time. To scale it a characteristic time is required. The characteristic distance is $(E/p_0)^{1/3}$, and the important velocity is the initial velocity of sound. Thus the characteristic time is $(E/p_0)^{1/3}/a_0$, where a_0 is the initial sound velocity. The pressure useful for scaling is the initial atmospheric pressure, so finally

$$\frac{i}{p_0[(E/p_0)^{1/3}/a_0]} = g\left[\frac{R}{(E/p_0)^{1/3}}\right],$$

where i is the impulse per unit area, and g is the scaled (dimensionless) impulse, a function of the scaled distance.

These scalings allow us to determine pressure and impulse for any set of conditions from existing measurements. Given the distance, energy, initial pressure, and initial sound velocity, calculate the argument of f and g. Look on the graph for the scaled pressure and impulse. Multiply the scaled pressure by p_0, and the scaled impulse by $p_0[(E/p_0)^{1/3}/a_0]$, to get the physical values needed. These scalings and experimental tests of the scalings are discussed by Baker et al. (1983) in their book.

The Earth's atmosphere is not really homogeneous, and pressure waves are focused unpredictably by the air. Just as we can sometimes hear voices clearly from a distant point, and another time we cannot make them out at all, so explosive wave propagation varies with conditions. The best we can hope for is an average, and we hope that conditions for the experiment in question are close to the average. Factors of 3 or more away from the average seem to be common in the results. Also, the data are for carefully made and carefully initiated spheres of explosive. A flat side gives a much greater blast wave, and the edge of a cube gives a very weak one. We can only locate the appropriate best approximation in the pressure–impulse plane, and see where it is relative to the damage threshold curve. If the point is far into the damage region, the structure or object will be damaged, and if it is far into the region for no damage, no damage will be done. If it is within a factor of 5 or so of the curve, the outcome is uncertain. All these factors and more are discussed with examples in the book by Baker et al. (1983).

1.4.2. Fragments

In addition to the blast wave, explosions throw fragments and missiles that cause damage. These can be divided into primary fragments, which are metal shards that were in contact with the explosive (for instance, part of the bomb case) and secondary fragments or objects that happened to be nearby and were accelerated by the blast wave. Both kinds cause damage, and perhaps secondary fragments are at least as important as primaries. Here, the discussion is limited to primary fragments only because they are easier to think about.

In the blasting process the mass of inert rock is usually 5000 times the explosive mass. In bombs and shells, the ratio is between 2 and 10, so even though steel is much stronger than rock these devices are heavily overloaded and will throw fragments large distances. A metal case in contact with explosive is usually broken into chunky fragments with dimension in one direction no more than a few times those in the other directions. The original thickness of the case is the primary scale for the fragments. Their width may be three times the thickness and their length ten times the thickness. A very large bomb may have fragments as large as a few hundred grams. The initial velocity for a fragment may be as high as 3 km/s or 4 km/s.

In general physics classes, every student determines the range of a projectile in a gravitational field without air resistance. The range is maximum when the projectile is shot at a 45° angle, and the range is V^2/g. If a fragment has an initial velocity $V = 3130$ m/s, and $g = 9.8$ m/s^2, the maximum range is quickly found to be 1,000,000 m. Real fragments do not travel that far because air drag forces are much larger than gravitational forces for fast-moving fragments. In fact, the drag forces are so much more important than the gravitational forces that, by neglecting gravity the range of a projectile can be approximated, and the equations are easier to solve. Newton's second law, that the mass times the acceleration equals the force, can be written

$$M\frac{dV}{dt} = F.$$

For a fragment with frontal area A, thickness in the direction of motion h, and density ρ_m, the mass is $M = \rho_m Ah$. We assume the drag force to be proportional to the square of the velocity and denote it as

$$F = -\tfrac{1}{2}C_D\rho_a AV^2,$$

where C_D is the drag coefficient, assumed to be independent of velocity, A is the frontal area, ρ_a is the density of the air, and V is the velocity of the fragment. The second law then becomes

$$\rho_m Ah\frac{dV}{dt} + \tfrac{1}{2}C_D A\rho_a V^2 = 0.$$

If the characteristic length is defined as

$$L = \frac{h\rho_m}{\tfrac{1}{2}C_D\rho_a}$$

the equation becomes

$$L\frac{dV}{dt} + V^2 = 0$$

and the solution is

$$V = \frac{LV_0}{V_0 t + L},$$

where V_0 is the initial velocity. Then the velocity equation can be integrated as

$$V = \frac{dx}{dt}$$

to find

$$V = V_0 \exp\left(-\frac{x}{L}\right).$$

This means that the velocity has fallen to less than 2% of its original value when $x = 4L$, and to $\frac{2}{3}$% when $x = 5L$. At these speeds little damage is done. Because C_D is a number between 1 and 2 for ordinary fragments, approximate values for L can be found by taking $C_D = 2$ so it cancels the factor $\frac{1}{2}$. Notice that, unfortunately, it is the largest fragments that have the largest value for L, and are the ones that go the farthest.

The complete equations with gravity and lift included cannot be solved analytically in closed form, but they are easily solved to any required accuracy with a computer. Baker et al. (1983) have solved the equations numerically with various ratios of lift-to-drag coefficients, and they show graphs of maximum range. Baker et al. (1983) also discuss damage criteria for fragments when they hit buildings, people, aircraft, etc.

PROBLEM 1. An experimental explosive system has in it a steel plate 1 in. thick, in contact with the explosive. Make the assumption that $C_D = 2$, and use the facts that the density of air at sea level is about 1.3 kg/m^3 and the density of steel is about 7800 kg/m^3, to calculate the characteristic length L for the fragments. To be safe, an area with radius at least $5L$ must be cleared of people, structures, and vehicles; calculate that area and its perimeter. How would the area and perimeter change if the plate could be changed to 0.25 in. thickness?

Secondary fragments, those nearby pieces that were accelerated by the blast, are as important as the primary fragments. They are much more difficult to describe. They are discussed by Baker et al. (1983, Chap. 6), along with primary fragments.

1.5. Jet Penetrators

Jet penetrators are explosively driven devices that emit a thin stream of metal at high velocity, and this stream penetrates large distances into any target that it is directed against. They are widely used by military forces against armor, but they are also used for penetrating the casing in oil wells, and tapping the bottom of steel furnaces so the molten metal can run out. They deserve a special section here because they illustrate so well some of the basic principles of sophisticated explosive systems.

Jet penetrators are made in several different geometries. The simplest to understand is a cylinder of explosive with a detonator at the center of one end, and a metal-lined hemispherical or conical cavity at the other end. The wave spreading from the detonator is approximately spherical, so it reaches the center of the liner first and starts it moving. Other zones of the liner are set into motion at successive times. The central part is driven by explosive surrounded by more explosive, and the pressure there stays high for the longest time. Nearer the edge, there is less surrounding explosive, so the pressure falls sooner, and the final velocity of metal near the edge is less than that of metal near the center. Thus the liner is turned inside out, and then, because the metal at the center started first and is going faster, it continues to stretch out into a long tube. The metal from the outer parts of the liner has a component of velocity in toward the center, and it is also pulled in by the tensile strength of the metal. If all conditions are met, the metal continues to stretch into a very long, straight piece.

When the long, straight jet contacts its target, it creates a very high pressure that pushes target material out of the way and starts a hole into the target. If the jet is straight, successive parts of the jet continue this process and make the hole deeper. A good jet charge, with the initial diameter of the explosive cylinder equal to D, will make a hole about $D/6$ in diameter and about $5D$ to $10D$ deep in steel.

A metal jet's tip velocity might equal 10 km/s and the tail velocity about 2 km/s, with the jet diameter about $D/25$, and the length about $8D$. If the cylinder of explosive has length $2D$, density 1800 kg/m^3, and specific energy 5 MJ/kg, and the density of the metal is 9000 kg/m^3, the kinetic energy of the jet metal is about one-eighth that of the total explosive energy. In this way, the explosive energy is concentrated so that it acts on a very small area of the target, only 1/625 of its own area.

PROBLEM 2. How far must the jet described in the preceding paragraph travel to stretch out to a length of $8D$? If its tip reaches the target at that distance, how long is it before the jet is completely consumed if it penetrates its own length in the target? Assume that the slowest particle of the jet tail maintains constant velocity while the jet is penetrating. Compute the kinetic energy of the jet and its ratio to the total energy of the explosive.

The penetration depth of the target by the jet depends on the jet length. As long as the jet continues to stretch and to travel straight, the penetration will increase. If the jet breaks up into small particles, the penetration stops increasing. Therefore, it is very important to choose the metal for its ductility in these special conditions.

The simple theory developed by Birkhoff, MacDougall, Pugh, and Taylor, (1948), is easy to understand and instructive. Modern references are Walters and Zukas (1989) and Carleone (1993). A history of the shaped charge effect is given by Kennedy (1983).

Much theoretical and computational work has been done recently on the behavior of astrophysical jets of gas moving at extraordinary velocities over fantastic distances. Conditions for these jets are very different from those for metal jets driven by explosive, but some of the investigations have explored the phenomena over the full range of the important variables. Their results may be worth studying to see whether any new insights on the behavior of jets can be found. A good introduction, with references, is given by Norman and Winkler (1985). Another discussion of the work is given by Smarr, Norman, and Winkler (1984).

1.6. Reactive Armor

One of the defenses developed to prevent jet charges from penetrating armor, particularly on tanks, is called reactive armor. One of the early antijet developments added a skirt to the tank so that the jet charge detonated when it hit the skirt, thus causing the jet to travel a relatively long distance before it hit the main armor of the tank. Any imperfection of the jet is magnified by the long travel. Reactive armor skirting is three layers: two sheets of metal with explosive between them. The jet strikes the skirt, usually not at normal incidence, penetrates the first metal sheet, and then initiates detonation in the explosive layer. Finally, the jet penetrates the final layer of the skirt and travels to the main body of the tank.

The outer plate of the skirt is accelerated outward by the explosive, and it moves across the jet, presenting new metal to the jet as it goes, using up much more of the jet than it would if the plate did not move. The outer plate then acts on the jet as if the plate were much thicker than it is.

The inner plate also moves across the jet and enhances the effect as though the plate were thicker. The plate's main action, however, is more subtle. The interaction of the jet and the plate is in an unstable regime, and the jet is drastically perturbed into an oscillatory mode. The jet breaks up into small particles unaligned with each other. The spread-out particles are unable to penetrate tank armor.

The action of reactive armor is described by Mayseless, Erlich, Falcovitz, and Rosenberg (1984).

1.7. Explosive Welding

To form good weld joints, two conditions must be met. The metal surfaces must be clean and uncontaminated by oxide or other layers, and they must be brought into intimate contact. In ordinary fusion welding, the metals are melted and contaminants are allowed to float away from the joint region. In pressure welding, any layer of contaminants is broken up by pressure, allowing intimate contact of clean metals. These processes are not effective if the metals have widely differing melting points or plastic flow strengths.

Ordnance specialists noted many years ago that sometimes projectiles or fragments bonded to their targets. Much later exploitation of this observation developed into explosive welding, which is widely used for special purposes where fusion and pressure welding are impossible or impractical.

The simplest arrangement has one rectangular metal plate lying on the ground, and the other similar plate is held a few plate thicknesses above it by spacers at the corners. Powdered explosive is sprinkled uniformly on the surface of the upper plate. The explosive is initiated simultaneously along a line at one edge, so it drives the two plates into contact. Once the configuration becomes steady, the contact line moves at the same speed as the detonation front, and the angle and the velocity of the plate are constants. One condition for a good weld is that the plate velocity divided by the sine of the angle must be less than the sound velocity in at least one of the plates. A jet is probably formed at the intersection and all surface contaminants are likely to be removed in the jet. Jets are sometimes, but not always, observed under the conditions for a good weld, so the real facts are more complex than that. For a good weld, the impact pressure should be great enough to cause the metals to flow plastically, but not great enough to cause melting, which weakens the bond. The conditions for a good weld are also the conditions for waves to be formed at the interface, and microscopic studies of the welds show various kinds of waves, from small ones that appear almost sinusoidal to breaking waves that interlock.

No complete theory of explosive welding yet explains all the details of jet formation, wave formation, and weld strength. Much is known about the practical details of how to make good welds. An introduction, with extensive references to the papers on the subject is Meyers (1994).

1.8. Wave Shaping and Lenses

The detonation wave emanating from a detonator is approximately spherical and it continues to expand as it moves. Sometimes explosives are more effective if the shape of the wave can be changed. An important application is in jet penetrators, where the shape of the wave as it contacts the liner makes appreciable changes in the performance of the device. A crucial breakthrough in the development of the nuclear weapon was von Neumann's design of an explosive lens that redirected spherical shock waves expanding from multiple points of initiation into a single, uniformly spherical shock wave converging on a nuclear core [Serber (1992)].

The simplest system to understand is a line generator made from an equilateral triangle cut from sheet explosive, with round holes cut in it following the pattern in which bowling pins are arranged, but with more rows. Initiation by a detonator takes place at any vertex of the triangle, and spreads as a circle in the sheet, but not across the cut-out areas. At each intersection, it spreads down the new track. Every path to the base of the triangle has the same length, and so the detonation arrives at each bottom intersection at the same time. Point initiation is transformed into an imperfect line of detonation, with scallops in it as it arrives earlier at each intersection than at the points between intersections. Obviously, the quality of the line is improved if the holes are smaller and closer together. Commercial fabricated sheets are sold.

An example of the use of a line initiator might be for closing a pipe with explosive. Strips of flexible sheet explosive are wrapped like tape around the pipe where it is to be pinched closed. A line initiator with the base of the triangle equal to the circumference of the pipe is used to initiate the wrapped explosive simultaneously. The line generator may be decoupled from the pipe by wrapping the pipe with foam rubber and taping the line generator on over the foam. This arrangement yields a smooth implosion of the pipe. Carefully choosing the thickness of the wrapped sheet explosive will ensure that the pipe is closed and yet avoid a bounce at the center that opens it up again.

Another way to fabricate a line-wave generator is by means of explosive fibers, or narrow explosive ribbons, in an inert material. For example, a piece of plastic has grooves machined or molded in it with one long groove from the detonator to the base corresponding to one edge of the triangle in the one made from sheet explosive. At intervals along that line, a set of parallel grooves go off at an angle and extend to the base. The grooves are all filled with explosive, one in a form like putty so it can be packed into each groove. The detonator initiates the long groove, and the detonation branches off to the base at each new groove. The path length, and therefore the time, is the same on each path. There are many ways to make wave shapers using fibers or strips.

A more precise line-wave generator with no scallops in the output wave can be made using fast and slow explosive. Once again, the basic shape is an isosceles triangle, but the two equal sides are outlined by strips of fast explosive with detonation velocity D_f. The main body of the triangle is a slab of slow explosive with detonation velocity D_s. If the sine of the base angle equals D_s/D_f, the detonation wave, initiated at the apex of the triangle, will arrive at the end of the fast strips at the same time that it arrives at the center through the slow explosive. The fast explosive initiates the slow explosive as it propagates along the edge, so the wave at any point on the base arrives there at the same instant. If a high-density HMX explosive is used for the strips, with $D_f = 8700$ m/s, and a slow baratol explosive with $D_s = 4900$ m/s is used for the main body, the angle will be about 34°. Half of this generator, using a right triangle, can be made even more simply, because the detonator will not need to be located so accurately.

Wave shapers are not limited to generating a straight line wave. The same techniques can be used to generate other shapes.

One step beyond the line-wave generator is the plane-wave generator. The common form is a cone of slow explosive covered with a layer of fast explosive, initiated at the apex. This system is the embodiment of the line-wave generator described above, rotated around its symmetry axis. The most common plane-wave lenses are baratol with an outer layer of Composition B. They are often referred to as P-16, P-22, P-40, P-80, and P-120 lenses, where the number denotes the diameter in tenths of an inch (a P-16 is 1.6 in. in diameter, a P-40 is 4 in. in diameter, etc.). With extreme care, these lenses are fabricated so that the detonation wave emerges simultaneously within about 0.1 μs over the whole surface. Baratol has been used for many years; it is a mixture of TNT and barium nitrate. For environmental reasons, its use is being phased out, and it is replaced by calcitol, a mixture of TNT and calcium carbonate.

Another plane-wave generator is called the mousetrap. Its thin metal plate is supported at an angle to the piece of explosive that is to be initiated with the plane wave. Outside the metal plate is a layer of explosive that is initiated along the end farthest from the main explosive charge with a line-wave generator. The detonation drives the plate against the explosive and hits it hard enough to cause immediate detonation. The metal farthest away starts first, and, if the angle is chosen correctly, reaches the explosive at the same time as the metal nearest it. The angle is usually about 11°, and the tolerances for making this kind of plane-wave generator are very tight.

Some interesting examples of wave shaping are given by Busco (1970) and Benedick (1972).

1.9. Conclusions

Explosives are a safe, convenient source of extremely high-power density. They have been applied to many varied problems, and have been used for nearly a century and a half. Engineering design principles for explosive driven devices have been developed and tested. The rest of this book is devoted to explaining in detail the properties and the mechanics of explosives.

1.10. Problems (Hints and Solutions)

PROBLEM 1 (Fragments).
Input data:

$$C_D := 2, \qquad \rho_a := 1.3, \qquad \rho_m := 7800, \qquad h := 0.025,$$

$$L := \tfrac{1}{2} \cdot C_D \cdot \frac{\rho_m}{\rho_a} \cdot h, \qquad L = 150, \qquad 5 \cdot L = 750.$$

Perimeter:

$$2 \cdot \pi \cdot (5 \cdot L) = 4.712 \cdot 10^3 \text{ m}.$$

Area:

$$\pi \cdot (5 \cdot L)^2 = 1.767 \cdot 10^6 \text{ m}^2.$$

Reducing h by a factor of 4 will decrease the perimenter by a factor of 4, and the area by a factor of 16.

PROBLEM 2 (Jet Penetrator).
Input data:

$$v_{tip} := 10{,}000, \qquad v_{tail} := 2000, \qquad L_{jet}(D) := 8 \cdot D, \qquad d_{jet}(D) := \frac{D}{25},$$

$$\rho_x := 1800, \qquad L_{expl}(D) := 2 \cdot D, \qquad E_{spec} := 5 \cdot 10^6, \qquad \rho_m := 9000.$$

To reach a length of $8D$:

$$t_{str}(D) := \frac{8 \cdot D}{(v_{tip} - v_{tail})}.$$

Tip position:

$$x_{tip}(D) := v_{tip} \cdot t_{str}(D).$$

Tail position:

$$x_{tail}(D) := v_{tail} \cdot t_{str}(D).$$

Mass of jet per unit length:

$$M_{\mathrm{jet}}(D) := \frac{\pi}{4} \cdot d_{\mathrm{jet}}(D)^2 \cdot \rho_{\mathrm{m}}.$$

Velocity of jet:

$$V(x, D) := v_{\mathrm{tail}} + \frac{x}{8 \cdot D} \cdot (v_{\mathrm{tip}} - v_{\mathrm{tail}}).$$

Energy of jet:

$$E_{\mathrm{jet}}(D) := \int_0^{8 \cdot D} \tfrac{1}{2} \cdot M_{\mathrm{jet}}(D) \cdot (V(x, D)^2)\, dx.$$

Energy of explosive:

$$E_{\mathrm{expl}}(D) := \frac{\pi}{4} \cdot D^2 \cdot L_{\mathrm{expl}}(D) \cdot \rho_x \cdot E_{\mathrm{spec}}.$$

Ratio:

$$R(D) := \frac{E_{\mathrm{jet}}(D)}{E_{\mathrm{expl}}(D)}.$$

$$D := 0.1, \qquad R(D) = 0.132, \qquad E_{\mathrm{jet}}(D) = 1.87 \cdot 10^6,$$

$$D := 0.2, \qquad R(D) = 0.132, \qquad E_{\mathrm{jet}}(D) = 1.496 \cdot 10^7.$$

The ratio is not dependent on D.

References

Ayerst, R.P. and McLaren, M. (1980). The role of chemical engineering in providing propellants and explosives for the U.K. Armed Forces. In *Advances in Chemistry Series*, No. 190, *History of Chemical Engineering* (W.F. Furter, ed.). American Chemical Society, Washington, DC.

Assheton, R. (1930). *History of Explosions*. Institute of Makers of Explosives, Washington, DC.

Bailey, A. and Murray, S.G. (1989). *Explosives, Propellants, and Pyrotechnics*. Brassey, London.

Baker, W.E., Cox, P.A., Westine, P.S., Kulesz, J.J., and Strehlow, R.A. (1983). *Explosion Hazards and Evaluation*. Elsevier, Amsterdam.

Benedick, W.B. (1972). Detonation wave shaping. In *Behavior and Utilization of Explosives in Engineering Design* (L. Davison, J.E. Kennedy, and F. Coffey, eds.). New Mexico Section of the American Society of Mechanical Engineers.

Berthelot, M. (1871). *Sur la force de la poudre et des matières explosives*. Gauthier-Villars, Paris.

Berthelot, M. (1883). *Sur la force des matières explosives d'apres la thermochemie*. Gauthier-Villars, Paris.

Berthelot, M. (1892). *Explosives and Their Power*. Translated from the French by C.N. Hake and W. MacNab. Murray, London.

Birkoff, G., MacDougall, D.P., Pugh, E.M., and Taylor, G.I. (1948). Explosives with lined cavities. *J. Appl. Phys.*, **19**, 563.

Busco, M. (1970). Optical properties of detonation waves (Optics of explosives). *Proceedings of the Fifth Symposium (International) on Detonation*. Office of Naval Research ACR-184, pp. 513–522.

Carleone, J. (ed.) (1993). *Tactical Missile Warheads*. AIAA, Waldorf, MD.

Chapman, D.L. (1899). On the rate of explosion in gases. *Philos. Mag.*, **47**, 90–104.

Davis, T.L. (Part 1, 1941; Part 2, 1943). *The Chemistry of Powder and Explosives*. Wiley, New York. Reprinted in one volume by Angriff Press, Hollywood, CA 90078.

Dobratz, B.M. and Crawford, P.C. (1985). *LLNL Explosives Handbook*. UCRL-52997 Change 2. Lawrence Livermore National Laboratory, Livermore, CA.

Fedoroff, B.T. et al. (1960). *Encyclopedia of Explosives and Related Items*. U.S. Army Armament Research and Development Command (Picatinny Arsenal). Dover, New Jersey. PATR-2700, in ten volumes.

Fickett, W. and Davis, W.C. (1979). *Detonation*. University of California Press, Berkeley, CA.

Hino, K. (1959). *Theory and Practice of Blasting*. Nippon Kayaku, Tokyo.

Kennedy, D.R. (1983). History of the shaped charge effect; The first hundred years. DTIC AD-A-220-095.

Langefors, U. and Kihlstrom, B. (1963). *The Modern Technique of Rock Blasting*. Wiley, New York.

Mayseless, M., Erlich, Y., Falcovitz, Y., and Rosenberg, G. (1984). Interaction of shaped charge jets with reactive armor (1984). In *Proceedings of the International Symposium on Ballistics*, Orlando, FL, 23–25 October 1985.

McShane, E.J., Kelly, J.L., and Reno, F.V. (1953). *Exterior Ballistics*. University of Denver Press, Denver, CO.

Meyers, M.A. (1994). *Dynamic Behavior of Materials*. Wiley, New York.

Manson, N. (1986). Contribution de Paul Vielle a la connaissance des detonations et des ondes de choc. *Sciences et Techniques de l'Armament*, Tome 60, 2ieme fascicule.

Manson, N. (1987). Historique de la decouverte de l'onde de detonation. *J. Phys.*, Colloque C4, Supplement au No. 9, Tome 48.

Medard, L.A. (1989). *Accidental Explosions*, vols. 1 and 2, English edition. Ellis Horwood, Chichester.

Nelson, H.R. Jr. (1985). Explosive accidents involving naval munitions. NWSY TR 85-5, Naval Weapons Station Yorktown, Yorktown, VA.

Norman, M.L. and Winkler, K.-H.A. (1985). Supersonic jets. *Los Alamos Sci.*, **12**, 38–71.

Persson, P.-A., Holmberg, R., and Lee, J. (1993). *Rock Blasting and Explosives Engineering*. CRC Press, Boca Raton, FL.

Robinson, C.S. (1944). *Explosions and Their Anatomy and Destructiveness*. McGraw-Hill, New York.

Serber, R. (1992). *The Los Alamos Primer*. University of California Press, Berkeley, CA.

Smarr, L.L., Norman, M.L., and Winkler, K.-H.A. (1984). Shocks, interfaces, and patterns in supersonic jets. In *Fronts, Interfaces, and Patterns* (A.R. Bishop, L.J. Campbell, and P.J. Channel, eds.). North-Holland, Amsterdam.

Sobrero, A. (1847). Sur plusiers composés détonants produit avec l'acide nitrique et le sucre, la dextrine, la lactine, la marnite, et la glycérine. *C.R. Acad. Sci. Paris*, **25**, 247–248.

Walters, W.P. and Zukas, J.A. (1989). *Fundamentals of Shaped Charges*. Wiley, New York.

CHAPTER 2

Explosives Development and Fundamentals of Explosives Technology

Peter R. Lee

2.1. Introduction

This introductory chapter is intended to give a brief outline of the history of the development of some common explosives and point the reader toward more weighty texts where full details of each of the points made and many more may be found. It also seeks to provide an introduction to some of the fundamentals of explosives technology, acting as a basis for later chapters on theories of initiation and hazard and performance testing of explosives. The following bibliography is not exhaustive, but is a useful starting point from which to study in greater detail the concepts introduced here briefly.

- Urbanski, T.: *Chemistry and Technology of Explosives*, Pergamon Press, Oxford (1964). This is a comprehensive tome, now becoming somewhat dated, but giving very detailed chemistry of the older explosives types.
- Cook, M.A.: *Science of High Explosives*, Reinhold, New York (1959). Initiation concepts, especially by shock, are developed clearly and in some detail. Gives an insight into the revolutionary changes in understanding of the behavior of explosives which began as a result of the stringent requirements of nuclear weapons design.
- Bowden, F.P. and Yoffe, A.D.: *Fast Reactions in Solids*, Butterworth Scientific, London (1958). Provides an introduction to the work on thermal initiation by the Cambridge school, currently represented by Professor J.E. Field at the Cavendish Laboratory.
- AMC Pamphlet AMCP 706–177: *Engineering Design Handbook; Properties of Explosives of Military Interest*, US Army Materiel Command (1971). Historical record of the development of explosives, with useful materials and explosives properties detail.
- AMC Pamphlet AMCP 706–180: *Engineering Design Handbook; Principles of Explosive Behavior*, US Army Materiel Command (1972). An excellent all-round explosives theory source-book, this should be the first document in a required reading list for new entrants to the world of explosives.
- Marinkas, P.L. (ed.): *Organic Energetic Compounds*, Nova Science, Commack, NY (1996). This has chapters on recent advances in explo-

sives synthesis and useful articles on detection of explosives, toxicity, biohazard, and bioremediation.

During the Gulf War in 1991, the world was shown daily the remarkable effects of the application of technology to weapons of war. The accuracy of laser-guided bombs and the ability of cruise missiles to fly around buildings are, rightly, considerable technological achievements. However, little was said in commentary about the extraordinary behavior of the energetic materials used, for example, in rocket motors that operate reliably at minus, or plus, 40 °C, or in bombs which breach thick concrete without exploding until commanded by a fuze train. The world has become accustomed to expect and require that explosives, rocket and gun propellants, and pyrotechnics, remain dormant, without reacting until they are "woken up" by an initiation train. This is an enormous demand, when we consider that an explosive may have available more than 6 MJ of energy per kilogram for useful work in a weapon system. The energy may be extracted as a result of detonation of the charge over a period of a few microseconds, at a power of 10^{11} or 10^{12} W, which is equivalent to the output of between 100 and 1000 power stations, each generating at 1000 MW for the same few microseconds. Alternatively the explosive energy may be used rather more slowly, as in a gun propellant, or a fast-burning rocket propellant, say, where decomposition occurs in the order of one millisecond. This still implies a power output of more than 10^9 W for the millisecond of operation, or, using the power station comparison, the output of a one-thousand megawatt plant for this period of time. Rocket motors operating in the sustain mode may burn for tens of seconds, but still generate power at the rate of tenths of megawatts.

It is important in any appreciation of the behavior, performance, and safety of explosives to be aware of this power locked away in materials, which are relatively safe to handle and which can be stored for many years with great safety. It is a tribute to generations of chemists and chemical engineers that these materials are as reliable and safe as they are, since they are stored in, and travel through, our communities daily. It is the more remarkable that commercial blasting explosives share the reputation for high reliability of military high explosives, because, while the latter are manufactured in relatively small quantities at relatively high cost, commercial explosives are manufactured in at least two orders of magnitude greater quantities, and probably more, to very small cost margins.

2.2. Nomenclature

2.2.1. Units of Measurement

Despite the gradual adoption of the "Système Internationale," or SI system of units throughout the world, there is a significant database published in the United Kingdom and the United States, which uses the cgs (centimeter/gram/second) system. For example, it is possible to encounter kcal/mole, cal/g, and kJ/kg as measurements of energy or detonation pressure expressed in gigapascals (GPa), kilobars (kb), or pounds force per square inch (psi). We have attempted to quote data in the units most frequently encountered in the literature. Purists will not like this approach, but we consider that usage in the major source-books of data should be the guiding principle and, until the United States in particular, changes to the SI, most workers will find this approach more convenient than continually having to convert from one set of units to another.

Table 2.1 gives a list of common conversion factors as an aid to the interpretation of all systems of units used in this book.

Table 2.1. Some conversion factors between different systems of units.

To convert from Mass (m)—Dimension M	To	Multiply by
grain (gr)	kilogram (kg)	6.479891E − 05
pound (lb Avoirdupois)	kilogram (kg)	4.535924E − 01
pound (lb Avoirdupois)	ton (t)	2.240000E + 03
ton (t)	tonne (t)	9.821429E − 01
kilogram force-second squared per meter (kgf^2s/m)	kilogram (kg)	9.806650
Pressure–stress (p)—Dimensions $ML^{-1}T^{-2}$		
Standard atmosphere (atm)	Pascal (Pa)	1.013250E + 05
bar (b)	Pascal (Pa)	1.000000E + 05
kilobar (kb)	Pascal (Pa)	1.000000E + 08
millimeter of mercury at 0 °C (mmHg)	Pascal (Pa)	1.333220E + 02
dyne per square centimeter (dyn/cm^2)	Pascal (Pa)	1.000000E − 01
Newton per square meter (N/m^2)	Pascal (Pa)	1.000000
pound per square inch (lb/in^2-psi)	Pascal (Pa)	6.894757E + 03
poundal per square foot (pdl/ft^2)	Pascal (Pa)	1.488164
Length (l)—Dimension L		
inch (in)	meter (m)	2.540000E − 02
foot (ft)	centimeter (cm)	3.048000E + 01
yard (yd)	meter (m)	9.144000E − 01
Volume (v)—Dimensions L^3		
cubic inch (in^3)	cubic meter (m^3)	1.638706E − 05
cubic inch (in^3)	cubic centimeter (cm^3)	1.638706E + 01
Energy (E or Q)—Dimensions (MLT^2)		
calorie (cal)	joule (J)	4.186800
Acceleration (a)—Dimensions LT^2		
gravity—standard free fall (g)	ms^{-2}	9.806650

2.2.2. Definitions of Physical, Chemical, and Materials Properties of Explosives

A (s^{-1})	Pre-exponential frequency factor in rate expression $k = k_0 \cdot A \cdot \exp(-E/RT)$ for the decomposition reaction.
C_v (cal/g · °C)	Heat capacity at constant volume (also J/kg·K).
D (m/s)	Velocity of detonation (N.B.: km/s is the same as mm/μs).
P (GPa)	Detonation pressure (may also be in kilobars).
ρ (g/cm^3)	Density (kg/m^3 is obtained from g/cm^3 by multiplying by 10^3).
MW	Molecular weight, from which is calculated the mass of a mole.
E (kcal/mole)	Activation energy for the decomposition reaction (also kJ/kg).
Q (kcal/mole)	Heat of decomposition for the decomposition reaction (also kJ/kg).
λ (cal/cm · °C · s)	Thermal conductivity (also J/m·K·s).

2.3. The Nature of Explosions

2.3.1. Physical Explosions

A liquid, such as water, sealed into a steel vessel and heated, will vaporize and pressurize the container. If heated for long enough to a sufficiently high temperature, such a vapor pressure will be generated in the container that it will fail suddenly and fragment. The release of the internal pressure provides sufficient driving force for the fragments to be propelled at high velocities with damaging consequences from them and the intense high-pressure pulse, the shock wave, also generated as the container fails. Highly compressed gases also constitute significant blast and fragment hazards if the containers they are stored in burst. For example, about 4 kg of helium pressurized to 600 bar in a 450 mm diameter sphere, of 48 l volume, will produce the same blast effects as the detonation of about 0.55 kg of TNT, if the sphere fails suddenly. If the container has thick metallic walls, a fragment hazard is added to the blast hazard. The effects are significantly greater if the compressed gas, or liquid, is capable of reacting with oxygen in the air after the bursting of the container and a source of ignition is available. The fuel–air explosion which ensues is chemical. If the energy associated with the damaging effects of an explosion has to be supplied from outside the system which fails, this is regarded as a physical explosion: the water had to be heated and the helium gas had to be compressed. A chemical explosive has its energy locked up within its molecules and the process of initiation seeks to extract that energy reliably and on demand, rather than unexpectedly and in circumstances leading to danger to personnel and property.

2.3.2. Chemical Explosions

The rapid reaction of a chemical, which results in the generation of heat may lead to its ignition and burning. If the burning reaction rate increases further, it becomes more violent and may become associated with the generation of a pressure wave, either in, or emanating from, the reactant. If the decomposition is also accelerated by increasing pressure the reaction will accelerate further. In the limit, the burning may become so rapid that the pressure pulse ahead of the burning front, which becomes steeper and steeper as the rate of pressurization increases, suddenly assumes a step condition, generating a shock wave. Material under these conditions is said to have undergone a transition from deflagration to detonation. Some materials require confinement in stout metal cases in order to be able to react fast enough to cause an explosion. Even then the event may not be a detonation, but simply, as in the case of a physical explosion, through overpressurization of the container, be due to the rapid chemical decomposition of the reactant. Gunpowder behaves like this. It produces a very loud noise and fragments from the container, but the process is essentially a rapid burning rather than a detonation. If sufficient confinement is available, certain explosive substances may burn so rapidly that they undergo deflagration to detonation transition (DDT). Explosives can also be induced to detonate without confinement, provided that the charge is of a sufficient diameter, i.e., greater than the "failure diameter," and the initiation stimulus is the output from another detonating charge, such as a detonator.

2.3.3. Nuclear Explosions

These are physical explosions, where the "external" energy supply comes from conventional explosives. They drive the radioactive elements at the heart of the device together to provide the conditions for the physical chain processes associated with the transmutation of matter into different elements and energy. The energy appears as multispectral radiation, some of which is heat, which is little different in principle from the heat generated as a result of the detonation of a conventional explosive, although it is several orders of magnitude greater.

2.4. What Are Explosives?

Any number of unstable chemicals and mixtures exist which may be capable of exploding with a range of degrees of violence. The vast majority of them are so unstable as to be highly dangerous and are of no use whatsoever, except as laboratory curiosities. In the United Kingdom, prior to 1983, an explosive was defined in law (Explosives Act, 1875, and the mass of subordinate legislation associated with it) as "... a substance used or manufactured with a view to produce a practical effect by explosion." Materials such

Table 2.2. Explosive materials classification.

Explosive matter

- Explosives
  - High explosives
    - Primary (initiating) explosives
      - PbAzide
      - PbStyphnate
      - HgFulminate
      - Tetrazene
    - Secondary explosives
      - Military explosives
        - TNT
        - RDX
        - HMX
        - PETN
        - CL-20
        - Mixtures
        - Boosters
      - Industrial explosives
        - Gelignites/dynamites
        - Permitted explosives
        - ANFO
        - Slurries/emulsions
  - Propellants
    - Gun propellants
      - Single base
      - Double base
      - Multiple-based
      - Black powder
    - Rocket propellants
      - Double base
      - Composites
      - Plastic
      - Modified DB
      - Liquids
  - Pyrotechnics
    - Flashes
    - Flares
    - Noise generators
    - Smokes
    - Fireworks
- Industrial chemicals which can form explosive mixtures
  - Ammonium nitrate
  - Sodium chlorate
  - Gas generating materials used in foamed plastics
  - Organic peroxides used as catalysts in polymerization
  - NG and PETN used in pharmacy
  - Salts of nitrated organic acids
  - Swimming pool chemicals
  - Solid fuel for camp stoves

as nitroglycerine, gelignite, TNT, etc., fit the definition, whereas nitrogen triiodide, explosive as it is, is not an "explosive" under the act. Table 2.2 shows a breakdown of energetic materials into families designed for the specific purposes associated with different types of weapon applications and civilian uses of explosives.

For an explosive to be of practical value it must be adequately safe and reliable, i.e.,

- be stable under anticipated storage conditions, i.e., be unaffected by the environment or other materials stored with it;
- burn, explode, or detonate only when required in use;
- be sufficiently sensitive to be initiated as required and precisely when required; and
- the initiation stimulus should be small compared with the output of the explosive.

The material must also be "effective" under the United Kingdom act. It must be capable of doing useful work on the environment (the target). In the military sense, this requires the explosive to be capable of converting the heat energy and compressed gas output from its decomposition into kinetic energy of the air in a blast wave, a rocket, a bullet or shell, fragments, the soil, etc. A commercial explosive, too, must be capable of moving rocks, soil, buildings, etc., by the effects of its detonation.

2.5. A Short History of Explosives

2.5.1. Greek Fire

The earliest energetic material referred to by name is "Greek Fire." It was actually developed by the Byzantines of Constantinople around the seventh century A.D. and consisted of a petroleum distillate, thickened by dissolving resinous and other combustible materials in it. Its exact composition, if it ever existed, is unknown.

2.5.2. Gunpowder, or Black Powder

It is certain that the first explosive was gunpowder, or black powder. It was first described under the generic name of "huo yao" in a Chinese manuscript on the military arts called "Wu Ching Tsing Yao," dated by scholars as A.D. 1040, which also describes explosive grenades and bombs able to be projected by catapult.

Potassium nitrate, known as "Saltpeter," which is an important constituent of gunpowder, first appeared in Europe via Arab sources in the late twelfth and early thirteenth centuries. However, it and European sources of the chemical which were discovered and worked later, were all impure, the European material more so than the Chinese and Indian. The Chinese made advances in the development of gunpowder relatively early, because they had learned how to recrystallize the impure nitrate early in their researches. It was not until about 1250 that Roger Bacon described the purification of saltpeter by recrystallization and not until about 1280, that Hasan-al-Rammah, described the use of wood ashes in the process and relatively pure saltpeter became available outside China. Despite apparently possessing the capability to make gunpowder, there is no evidence that the Muslims used it at any time during the whole of the period of the Crusades, between 1097 and 1291.

The use of gunpowder as a propellant had obviously advanced significantly in Europe by 1326, because a decree of the Council of Florence provided for the appointment of men to make iron bullets and bronze cannon for the defense of the castles and villages of the republic. Froissart, in France, repeatedly mentioned the use of guns from 1340 onward and in 1399 the English Crown possessed a number of guns. In addition, there is a record of a small stock of gunpowder and of saltpeter from the same time.

Gunpowder recipes were gradually optimized for different purposes. Furthermore, in order to achieve control over ballistic performance of cannon, even when they were using gunpowder optimized for propellant effects, studies took place to modify the physical form of the powder. These studies led, some time in the fifteenth century, to the process of "corning," which enabled the gunpowder to be produced in a granular form, so that the burning surface of the charge could be made more reproducible. Further advances were made in the sixteenth century, including "moderation" of the surfaces of the grains by glazing them with graphite. This disturbs the local proportions of saltpeter, sulfur, and carbon at the time of ignition, so that it is less violent and less liable to lead to disruption of the grains. Grain disruption causes massive increase in surface area of the charge and the burning rate to increase uncontrollably, often with catastrophic results. Many early guns blew up because of overpressurization caused by this uncontrolled burning. As the science of physical moderation of grain surfaces advanced, it became possible to increase the controlled burning rate of charges by increasing the proportion of saltpeter, which was also being made purer and more readily available. At the end of the eighteenth century and the beginning of the nineteenth century the composition became more or less standardized as KNO_3/C (charcoal)/S (75/15/10 w/w). Improvements in quality of powder in the late eighteenth century enabled the charge weight to be reduced from about one-half to one-third of the weight of the shot in British naval guns.

2.5.3. Brief Chronology of the Discovery and Development of the Commoner High Explosives and Propellants

The opportunities to discover new high explosive molecules were given to synthetic chemists by the development of techniques to make nitric and sulphuric acids relatively cheaply and in commercial quantities. Much of the interest in the early part of the nineteenth century was in the effects of nitration of relatively common materials, such as silk, wool, resins, wood, cotton, and a range of other materials. Toward the end of that century, a new industry arose from the development of new synthetic dyes from coal-tar products.

Table 2.3 provides a reference guide to the chemical formulas and structures of some of the more common explosives in use now, and Table 2.4 lists some of their more important explosive properties.

2.5.3.1. Nitroglycerine

The Languedoc Tunnel was built in France over a 15-year period starting in the late 1860s using black powder as a blasting agent, which proved reasonably effective in breaking up relatively weak, brittle, or porous rock, but not sufficiently powerful to be used effectively against granites. A more powerful explosive was needed to blast the rock forming the barriers in the first tarnsalpine tunnels. An Italian, Sobrero, was the first to make nitroglycerine, in 1846, but despite its undoubted power, he did not pursue its use, possibly because of the hazards associated with it. The Nobels, father Immanuel and his son Alfred, also made nitroglycerine, but were unable to initiate it reliably from flame, as was the custom with black powder at the time. They were disconcerted by its unpredictability and the dangers of handling it in its normal liquid state. Alfred solved these problems by the invention of the

Table 2.3. Names and formulas of some of the explosives discussed.

I. Nitrate Esters (–O–NO_2 links)

Name	Structural formula	Molecular formula
Nitrocellulose	$[-HC(CH_2\cdot ONO_2)(HC-O)\ldots CH-O-;\ O_2NO\cdot HC-CH\cdot ONO_2]$	$[C_6H_7N_3O_{11}]_n$
Nitroglycerine	$CH_2\cdot ONO_2$–$CH\cdot ONO_2$–$CH_2\cdot ONO_2$	$C_3H_5N_3O_9$
Pentaerythritol Tetranitrate (PETN)	$O_2NO\cdot CH_2-C(CH_2\cdot ONO_2)_2-CH_2\cdot ONO_2$	$C_5H_8N_4O_{12}$

II. Nitro-Compounds (–NO_2 links)

Name	Structural formula	Molecular formula
Trinitrotoulene (TNT)	benzene ring with CH_3, O_2N, NO_2, NO_2	$C_7H_5N_3O_6$
Trinitrophenol (Picric acid)	benzene ring with OH, O_2N, NO_2, NO_2	$C_6H_3N_3O_7$
Triaminotrinitrobenzene (TATB)	benzene ring with NH_2, O_2N, NO_2, H_2N, NH_2, NO_2	$C_6H_6N_6O_6$
Hexanitrostilbene (HNS)	O_2N–ring(NO_2, NO_2)–CH = CH–ring(O_2N, O_2N)–NO_2	$C_{14}H_6N_6O_{12}$

(*continued*)

Table 2.3 (*continued*)

III. Nitramines ($-N-NO_2$ links)

Name	Structural formula	Molecular formula
Trinitrophenyl methylnitramine (Tetryl, CE)	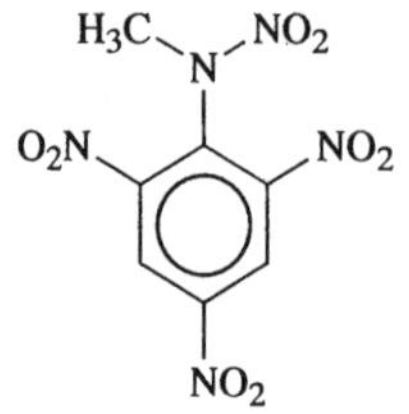 2,4,6-trinitrophenyl ring bearing $N(CH_3)NO_2$	$C_7H_5N_5O_8$
Cyclotrimethylene-trinitramine (RDX)	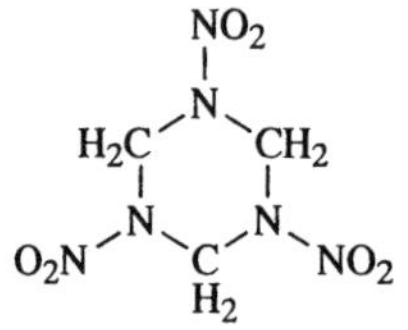 six-membered ring of alternating CH_2 and $N-NO_2$ groups	$C_3H_6N_6O_6$
Cyclotetramethylene-tetranitramine (HMX)	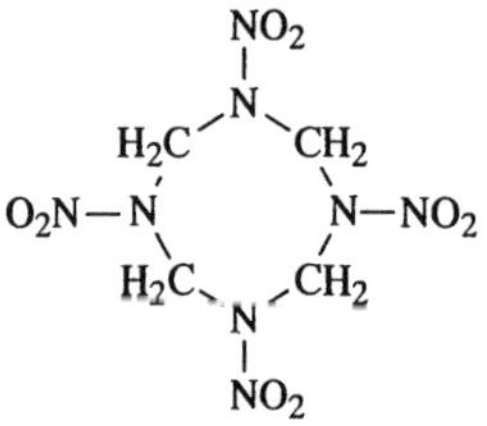 eight-membered ring of alternating CH_2 and $N-NO_2$ groups	$C_4H_8N_8O_8$
Nitroguanidine (Picrite)	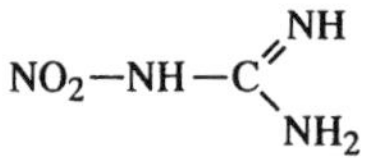 $NO_2-NH-C(=NH)-NH_2$	$CH_4N_4O_2$

(*continued*)

mercury fulminate detonator in 1864 and by absorbing the liquid explosive on to kieselguhr, a diatomaceous earth, in 1867. It is probably true to say that these two inventions changed completely the course of explosives science. Nitroglycerine absorbed on to kieselguhr is known as dynamite and it was found to be adequately powerful for all mining, quarrying, and civil engineering projects of the times. Alfred Nobel also discovered, in 1875, that nitroglycerine can be gelled with nitrocellulose, another powerful explosive, and that the pair of materials could be safely used as "blasting gelatine" and "gelatine dynamites."

In 1888 an important discovery from a military explosives standpoint was made when it was found that "dough" made from gelled nitroglycerine and nitrocellulose, could be mixed with other additives and then dried to form highly combustible, but stable propellants. These propellants were found to

Table 2.3 (*continued*)

IV. Metal Derivatives

Name	Structural formula	Molecular formula
Lead styphnate	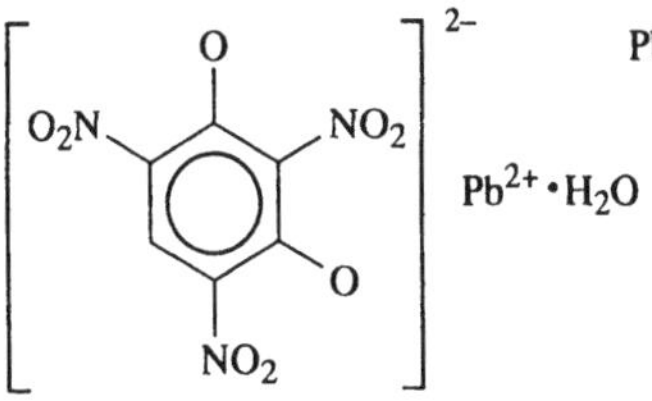 $[O_2N, NO_2, NO_2, O, O$ benzene ring$]^{2-}$ $Pb^{2+} \cdot H_2O$	$PbC_6H_3N_3O_9$
Lead 2,4-dinitroresorcinate	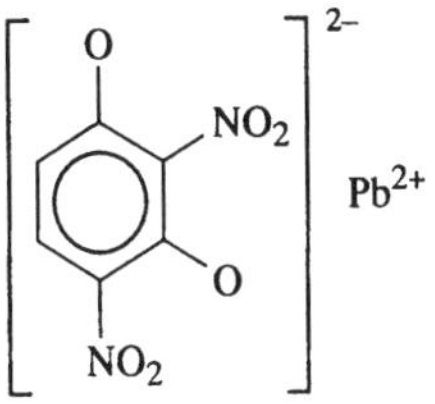 $[O, NO_2, O, NO_2$ benzene ring$]^{2-}$ Pb^{2+}	$PbC_6H_2N_2O_6$
Lead azide	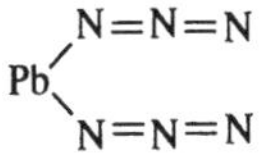 $Pb(N{=}N{=}N)_2$	PbN_6
Silver azide	$Ag{-}N{=}N{=}N$	AgN_3

V. Organic Primary Explosive

Name	Structural formula	Molecular formula
Tetrazene	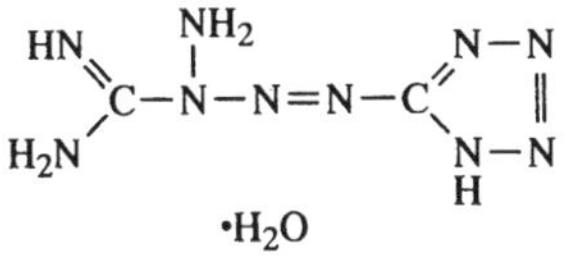 $(HN{=})(H_2N)C{-}N(NH_2){-}N{=}N{-}C$ (tetrazole ring N−N, N−NH) $\cdot H_2O$	$C_2H_8N_{10}O$

burn reliably and safely with great energy and soon found use as the basis for gun propellants. The first one was called "ballistite."

2.5.3.2. Nitrocellulose

Nitrocellulose can be made from the nitration of wood or cotton. Early work used paper or cotton as the raw material and, based on the work of Pelouze in 1838 and Schonbein in 1845, it was gradually developed as an explosive and as a propellant. Pelouze obtained a nitrocellulose with only about 5–6% of nitrogen, whereas, Schonbein used mixed nitric and sulfuric

Table 2.4. Physical, chemical, and explosive properties of some common explosive molecules.

Explosive	Charge density (ρ) g/cm^3	Velocity of detonation (D) m/s	Detonation pressure, GPa	Molecular weight	Activation energy (E) kcal/mole	Heat of decomposⁿ (Q) cal/g
NC	1.58	7300	21.0 (calc)	$(262.6)_n$		
NG	1.59	7650	25.2	227.1		
NQ	1.78	7147	28.2	104.1	57.1	500
Picric acid	1.60	7260	26.5 (calc)	229.1		
TNT	1.62	7045	18.9	227.13	34.4	300
Ammonium picrate	1.63	6850		246		
Tetryl	1.614	7479	22.64	287.15	38.4	500
PETN	1.67	7975	31.0	316.15	47.0	300
RDX	1.767	8639	33.79	222.13	47.1	500
HMX	1.89	9110	39.5	296.17	52.7	500
HNS I	1.60	6800	20.0 (calc)			
TATB	1.847	7660	25.9	258.18	59.9	600
PbAzide	4.38	5500		291		
PbStyphnate	3.02	5200		468		
AN	1.72	1500 at 0.7 density		80.05		
Black powder	1.92	1350 (deflagration)				
Octanitrocubane	2.1 (calc)	9898 (calc)	48.9 (calc)	464		

(a) *LASL Explosive Property Data*: Gibbs, T.R. and Popolato, A., University of California Press (1980).
(b) *LLNL Explosives Handbook*: Dobratz, B.M., UCRL-52997, Lawrence Livermore National Laboratory, Livermore, CA (1981).
(c) Mader, C.L.: Private communication.
N.B.: Units are as quoted in the sourcebooks.
* At constant pressure.

Table 2.4 (*continued*)

Pre-explosion freq. fact. (A) s^{-1}	Thermal conductivity (λ) cal/cm·°C·s	Heat capacity* (C_p) cal/g °C	(Source(s) of data): Notes, appearance, etc.
	5.5×10^{-4} (12% N)	0.247 at 25 °C	(b): 13.35% N
		0.356 35–200 °C	(b): liquid
8.75×10^{22}	0.00098 (25–50 °C)	$0.269 + 0.0007T$	(a): $37 < T$ °C < 167: High-bulk density, recrystalized from water
	2.4×10^{-4}	0.234 at 0 °C	(b) yellow: obsolete
2.51×10^{11}	$6.22 \times 10^{-4} < 45$ °C $5.89 \times 10^{-4} < 75$ °C	$0.254 + 0.00075T$ $0.309 + 0.00055T$	(a): $17 < T$ °C < 67 $97 < T$ °C < 150
			(b): WWII material still in service
2.51×10^{15}	7×10^{-4}	$0.211 + 0.00026T$	(a): $-100 < T$ °C < 100: yellow
6.3×10^{19}		$0.239 + 0.008T$	(a): $32 < T$ °C < 127
2.02×10^{18}	2.53×10^{-4}	$0.232 + 7.5 \times 10^{-7}T$	(a): $32 < T$ °C < 167
5×10^{19}	0.0012 (25 °C)	$0.231 + 0.00055T$	(a): $37 < T$ °C < 167
			(b): Types II and IV also
3.18×10^{19}	1.3×10^{-3}	$0.215 + 0.01324T$	(a): yellow
	4.2×10^{-4}	0.09	(b):
			(b):
	$2.9\text{–}3.9 \times 10^{-4}$	0.4 at 0 °C	(b): hygroscopic
			(b): black–grey
			(c): Not yet made

acids and produced highly nitrated cotton. However, it also gradually fell out of favor after factory accidents in England, France, and Austria, over the period 1846–1865. The problem arose because of impurities, which remain in the capillary fibers of the nitrated cellulose and which are liable to spontaneous ignition in air. Abel was able to eliminate these hazardous by-products by boiling the product for long periods in hard water, followed by "pulping." This involved finely chopping the product in water in a paper mill beater, to remove the lower boiling point products of nitration and then washing to remove water-soluble hazardous materials. Highly nitrated guncotton compressed into blocks was found to be detonable from fulminate detonators and this enabled it to be used as a commercial blasting explosive. It was also used by army engineers worldwide for nearly 80 years.

2.5.3.3. Gun Propellant Developments

It was not until 1864 that a Prussian, Schulze, developed a nitrocellulose propellant using wood as the base material. This was not suitable for rifles, but was used in shotguns, where its smokelessness was appreciated by users. It took a further 20 years before a nitrocellulose propellant was developed for use in rifled barrel guns. This was first made by Paul Vieille, in France and he called it "Poudre B." He made a paste from nitrocellulose gelled with a mixture of ether and alcohol, rolled it into sheets and cut it into small squares before drying. This material burned more slowly than untreated nitrocellulose and was used as the first single-base composition for small arms. It is referred to as *single base* to distinguish it from *double-base* materials where nitrocellulose is used to gel nitroglycerine and *triple base* propellants where nitroguanidine is also added. Military and commercial propellant development advanced rapidly in the mid-to-late nineteenth century and gun design and performance kept pace.

Single base propellants were developed extensively for use in the commercial propellant industry, but also in the United States armed services, where nitroglycerine was regarded unfavorably for some time, especially in the US Navy. The United States procedure for the preparation of nitrocellulose propellants was very similar to that of Vieille, using alcohol and ether as solvents, but incorporating diethylamine as a stabilizer. Thin cords were extruded, but immediately chopped into short lengths before being dried. Although the powders were smokeless they were hygroscopic and also tended to lose residual solvent, which was not completely removed from them in the first drying stage, so they tended to crack and shrink. They also suffered from muzzle flash, which has been one of the major problems with these and other gun propellants. Much of the work to improve gun propellants of all types has been connected with attempts to reduce muzzle flash and burning temperature, while, at the same time, maintaining energy content and vivacity of charges.

The British Government appointed a committee to recommend a replacement to service gunpowder as a propellant in the late 1880s. A *double base* system was adopted by Abel, Kellner, and Dewar at Woolwich Arsenal using a method in which nitrocellulose was incorporated with nitroglycerine in the presence of a solvent (acetone plus a little mineral jelly) to assist gelatinization. They used nitrocellulose of 13.1% nitrogen content and produced a dough which they were able to extrude into long cords, and which, after drying formed sticks of "cordite," was approved for service in 1896. The mineral jelly proved to be a lucky choice. Its original purpose had been to act as lubricant in the gun barrel, which turned out to be illusory, since it was consumed by the combustion process. However, fortuitously, it acted as a stabilizer for the propellant in storage. It neutralized the acidic products of decomposition of the nitrocellulose and nitroglycerine and was used until 1935, when it was replaced by "carbamite." The first cordite had a high-flame temperature and caused erosion of gun barrels and significant muzzle flash. Modified cordite was produced next, with a lower-flame temperature, but a still-unacceptable muzzle flash. This latter problem was reduced significantly by the addition of finely ground nitroguanidine, known as picrite, to produce so-called flashless cordite. The picrite lowered the flame temperature further and generated combustion products rich in incombustible gases which did not burn outside the barrel.

Solvents used in cordite manufacture tend to cause larger diameter (ca. 5 mm) sticks to shrink and distort as they dry. German explosives technologists, before and during World War I, used a technique involving water wet mixing, followed by gelatinization of the nitrocellulose/nitroglycerine mixture on hot rollers. They used nitrocellulose of lower nitrogen content than the British, and were able to extrude a "solventless" cordite from a heated cylinder loaded with rolled, gelatinized sheets, now referred to as "carpet rolls" from the hot rolls. This process was later developed in the United Kingdom and was used to produce a wide range of rocket grains for unguided rockets in World War II.

An extremely important discovery, said to be fortuitous, was made in the United States shortly after World War II. It was that the addition of 1–2% w/w of certain lead compounds to propellant compositions of low-heat content, i.e., ca. 1.8–3.6 MJ/kg, caused the rate of burning to be modified in an advantageous manner. This phenomenon was called "platonization" in the United States because the propellant burning rate was found to remain reasonably constant over a range of burning pressures, and a plot of rate as ordinate against pressure as abscissa was found to be more or less horizontal. These lead compounds became to be known as ballistic modifiers and are still used today, along with more modern materials.

The rapid improvement in gun propellant technology was accompanied by much research into improving the performance of explosives suitable for inclusion in shell to be fired using the new, more powerful propellants and guns. However, the problem proved difficult to resolve, because few new explosive molecules being developed at the time seemed to be capable of being made cheaply on a large scale, or adequately safe or capable of being moderated for use in high-explosive shell, especially under the severe mechanical stressing generated on firing.

2.5.3.4. Picric Acid (2,4,6-Trinitrophenol)

Picric acid was first prepared in 1771 by Woullf, who found that the action of nitric acid on natural indigo dye yielded another dye, this time yellow. Laurent carried out a systematic study on the nitration of phenol and reported it in 1841. He prepared dinitrophenol and picric acid by nitrating phenol with nitric acid. Schmidt and Glaz (1869) were the first to show that the nitration process could be made to proceed more smoothly if the phenol were first dissolved in sulfuric acid. The phenol is converted to phenol-4-sulphonic acid and is the first example of the use of mixed nitric and sulfuric acids as the nitrating medium, which is used today in the manufacture of TNT and RDX. It was found that picric acid could be made to detonate and Abel used it as a filling for artillery shells, employing a picric acid/potassium nitrate (40/60) mixture as a booster. Picric acid shell filling was known as "Lyddite," after the range area in England where tests were carried out. A further development was called "Shellite," which consisted of a mixture of picric acid and dinitrophenol, which, because of the low melting point of the latter could be poured as a hot slurry into shells.

There were several significant problems associated with the use of picric acid, which derive from its acidity. It tends to form sensitive compounds with such metals as copper, which was used in parts of fuzes and detonators. However, its major drawback in use was that picric acid did not seem always to detonate completely, possibly due to inadequate boostering. Whatever the cause, there was a distinct need, especially in the United Kingdom, where picric acid was widely used either as Lyddite or Shellite, to improve the performance of shell with a material which detonated more effectively and reliably, and was not likely to form sensitive compounds with parts of the fuze. The United Kingdom did not develop an alternative material based on picric acid to eliminate this problem. The United States did, and ammonium picrate is still in service in the US Navy as "Explosive D" which is used as the filling of the 15 in. shells in the recommissioned World War II battleships. Ammonium picrate is extremely safe and reliable as an explosive and has a very long shelf-life, but it is not particularly powerful.

2.5.3.5. Tetryl (2,4,6-Trinitrophenylmethylnitramine)

Tetryl was first described by Michler and Meyer in 1879 and van Romberg and Martens studied its properties and correctly deduced its structure at about the same time. It was first made by the nitration of dimethylaniline. This is a simple process and there was plenty of dimethylaniline available from the new artificial dyestuffs industry. Tetryl was found to be a very powerful explosive, but doubts were expressed and proven in tests that it was too sensitive to be used as an artillery shell filling alone. At one time it was thought that it could be used as a booster explosive to ensure that picric acid could be made to explode more reliably, but this was not possible, because the picric acid reacts with tetryl over a long period and produces unwanted nonexplosive products. It almost escaped being used until, in World War I, it was discovered that it could be used with TNT as a powerful and reasonably insensitive shell filling which could be cast as a hot slurry of tetryl in liquid TNT, known as Tetrytol. Part of the tetryl dissolves in the TNT to form an eutectic mixture tetryl/TNT (55/45), which freezes at 67.5 °C.

The main use of tetryl has been as a booster explosive, both as pressed charges, below the shutter in a safe and arm unit, and as a lower density stemming in a fuze above the shutter. It has a high-shock sensitivity and once deflagration has been initiated in it, it builds rapidly and reliably to detonation. It was used for many years in many countries until the mid-1970s when it was found to be mutagenic and the manufacture and processing of it were discontinued in most Western nations.

2.5.3.6. TNT (Trinitrotoluene)

TNT was first manufactured in 1863 from toluene by a laborious three-stage nitration process, using steadily stronger nitric acid. It was not until 1891 that Hauserman manufactured it in industrial quantities, but, by then, its characteristics were known and required for high explosive shell fillings. German scientists had been able to produce TNT by nitrating toluene obtained from their coal-tar industry. British and French military technologists emulated the German work, the French using indigenous coal-tar by-products as the source of toluene and the British by extracting it from petroleum from Borneo. The nitration process, which is relatively simple in all-glass apparatus, was difficult to industrialize because of the requirement to use nitric acid at high temperatures in the presence of oleum, which is hydrogen sulfate (anhydrous sulfuric acid) with excess sulfur trioxide dissolved in it. Initially, TNT was made by a batch process, but soon the exigencies of World War I drove chemists to develop continuous TNT manufacturing processes in most of the major belligerent countries. TNT was sometimes mixed with ammonium nitrate to form "Amatols" to eke it out and with tetryl, to increase the performance of some shell. It was used by the German army as a filling for its artillery shell as early as 1902 and is very convenient for this purpose, since it can be melted by steam at 80–90 °C and can be poured into shell cases at about 80–85 °C. However, it undercools significantly below its melting point before crystallization and contracts by about 10% of its volume on solidification. Thus, whenever a munition is filled with TNT, care must be taken to ensure that there is a continuous supply of molten explosive available at the filling mouth of the store while the majority of the filling is solidifying, in order to be able to supply the extra volume of material required as the explosive shrinks in the casing. This requires heated headers and complex control over the cooling process of the filled store. Filling quality is enhanced and undercooling virtually eliminated if the TNT is poured as a partly solidified slurry of fine TNT crystals (often added as ground powder), to the liquid at about 82 °C or hexanitrostilbene is used as a nucleating agent. The result is a finely crystalline charge, in contrast to the relatively large crystals formed when it is allowed to cool naturally from the clear melt without nucleating aids.

2.5.3.7. RDX (Cyclotrimethylenetrinitramine)

RDX was synthesized first by Henning in 1899, by dissolving hexamine (hexamethylenetetramine) in concentrated nitric acid and pouring the solution into cold water. The RDX appeared as small white crystals. Several nations were interested in replacement of TNT after World War I and RDX, along with PETN, was considered as a candidate. RDX was preferred because:

- it is more powerful than PETN;
- it is more stable, i.e., it has an higher melting point;
- it is less sensitive to impact and friction;
- the raw materials for its manufacture were more readily available; and
- it was cheaper to make.

It was made successfully in the United Kingdom, Germany, and the United States during World War II by methods which were all similar to the United Kingdom process known as the "Woolwich Method." This involves continuous nitration and dilution, followed by continuous filtration. This material received its name because, like all other British explosive materials developed at the Research Department at the Woolwich Arsenal in East London, it was to be assigned an "RD" number. It was not possible to assign the number when required and so its research appellation "X" was attached, hence "RD-X."

During World War II the demand for RDX was so enormous in the United States that a search for a cheaper, possibly continuous method was made. The "Bachmann Process" was the result. This process does not include the continuous filtration technique used in the Woolwich Process and the RDX so produced contains up to 12% HMX. United States "RDX" from that time contains a variable amount of HMX, which confers on "Bachmann" RDX an higher density and detonation velocity than material made by the Woolwich Process. United States Composition B, which is composed of 60% (w/w*) of Bachmann RDX and 40% (w/w) of TNT is, likewise, more powerful than United Kingdom RDX/TNT (60/40). It is erroneous to call the United Kingdom material "Composition B" and the United States explosive "RDX/TNT (60/40)." They are also different in characteristics. Composition B is less viscous at around the temperature at which charges are poured, ca. 85 °C. This confers slightly better charge characteristics and higher percentages of the theoretical maximum density can be achieved with this material compared with United Kingdom RDX/TNT (60/40). The presence of HMX appears to catalyze the growth of Composition B charges on thermal cycling, which may be advantageous in some circumstances and a problem in others.

* w/w refers to composition by weight: v/v refers to composition by volume, usually reserved for liquids.

2.5.3.8. HMX (Cyclotetramethylenetetranitramine)

The continuous filter used in the Woolwich process for RDX was found to trap white crystals of a material, which had the same elemental composition as RDX, but an higher molecular weight. Only a very small proportion of the RDX was found to exist in the form exhibited by these crystals (ca. 0.1%), but the material was of great interest, since it was very similar to, but not the same, as RDX. Tests showed that it had a similar explosive output to RDX, when corrected for molecular weight. Four crystalline forms could be recrystallized from suitable solvents, α, β, γ, and δ, of which the β form was found to be the most stable. It was dubbed High Molecular Weight Explosive (HMX) by the British to be consistent with the nomenclature for RDX, although, because its melting point is also higher than that of RDX, it is sometimes, erroneously, referred to as high melting point explosive, or His or Her Majesty's explosive in some quarters.

The detonation performance of this new explosive was found to be superior to that of RDX, based on its higher density and its value in the nuclear weapon field was clearly indicated. Hence, the United States authorities identified means by which the Bachmann Process for the manufacture of RDX could be modified to optimize the production of HMX. This process was also required to remove the RDX and to ensure that the HMX appeared in the stable β form and uncontaminated by the other forms. As a result of these modifications HMX is several times more expensive than RDX, but it is still used where the very highest explosive or propulsion performance is required, as in shaped charge warhead explosives and high specific impulse motor grains. A premium is exacted, since it is somewhat more sensitive to shock and mechanical stimuli compared with RDX.

2.5.3.9. PETN (Pentaerithrytoltetranitrate)

Pentaerithrytol was first made by Tollens and Wigand in 1891 by alkaline condensation of acetaldehyde and formaldehyde in aqueous medium, and in 1894 a German chemical company Rheinisch–Westfälische Sprengstoff AG first made PETN by direct nitration of pentaerithrytol. PETN is about as powerful as RDX, but it is more impact and friction sensitive than both it and HMX, a problem it shares with other nitric esters, such as nitroglycerine. It also has a lower density and melting point and is less stable to heat than RDX, but it is a valuable material for use in explosives devices not expected to undergo long storage, such as in blasting detonators, or in weapons stored under strictly controlled conditions. It may be desensitized by mixing with TNT and TNT/PETN (50/50) ("Pentolite" 50/50) is a well-known composition, which was used extensively in World War II, especially underwater.

2.5.3.10. HNS (Hexanitrostilbene)

Reich, Wetter, and Widmer first produced this explosive in 1912, by treating boiling trinitrobenzyl bromide with alcoholic potassium hydroxide. It is relatively thermally stable, having a melting point, with dissociation, at 318 °C, although its performance is very similar to that of TNT. If care is taken in preparation of a shock-sensitive crystal variant (HNS-IV), it can be made to detonate in relatively thin channels. It finds use in high-temperature environments, where thermal survivability is at a premium. For example, as a filling for shaped charges used to bring oil wells on stream, where the temperature at the site of the cap on the well may be elevated due to depth underground. It is also used as the filling for metallic sheathed line cutting cords, for missile break-up units, and aircraft cockpit escape systems.

The molecule of HNS has a slight twist, i.e., the two benzene rings are not coplanar. The angle they make to each other is close to the angle between TNT benzene rings in crystals of that explosive. HNS is, for this reason, an excellent nucleating agent for TNT and is used to reduce TNT crystal sizes in melt-cast TNT and RDX/HMX–TNT explosives.

2.5.3.11. TATB (Triaminotrinitrobenzene)

TATB is made by reacting 1,3,5-trichlorobenzene with SO_3, in the form of 30% oleum and sodium nitrate, to give 1,3,5-trichloro-2,4,6-trinitrobenzene. This reaction mixture is then quenched with a large volume of ice, and the 1,3,5-trichloro-2,4,6-trinitrobenzene is recovered by filtration and reacted with ammonia gas in the presence of toluene to give TATB. One problem is the amount of chlorine which remains occluded in the crystals and which interferes with the long-term stability of the explosive in storage.

At one time, in the mid-1970s this explosive was held to be the answer to the problems which had been experienced by thc US Navy in the Gulf of Tonkin during the Vietnam War, when three aircraft carriers were badly damaged by premature weapon functioning on board aircraft awaiting launch on flight decks. It is a very insensitive material, but it is expensive to procure and process into weapons, because the preferred manner of fabricating charges is by isostatic pressing and machining. However, it has set a standard for low vulnerability at a performance level similar to that of Composition B. It melts at 448 °C.

2.5.3.12. HNB (Hexanitrobenzene)

Hexanitrobenzene can be made by partial reduction of trinitrobenzene, followed by further nitration, using first nitric acid, then nitric acid with chromium trioxide as oxidation medium. It is a very powerful explosive, which suffers from the disadvantage of being highly hygroscopic. If it were possible to cure this problem, this could be a highly effective filling for a range of munitions.

2.5.3.13. Commercial Explosives

The cost of commercial explosives based on nitroglycerine, the dynamites, gelignites, blasting gelatins, etc., was steadily reduced before World War II following the introduction of the Schmidt continuous nitration process in

1929 and after it by the injector nitration method and continuous centrifugal separation of product. Nevertheless, commercial blasting explosives based on nitroglycerine were still somewhat more expensive than many quarries and other potential users were capable of paying. Cook produced his important book in 1959 which revolutionized the nonmilitary explosives field, by describing ammonium nitrate/fuel oil (ANFO) and slurry explosives. The former consists of prilled ammonium nitrate to which has been added a stoichiometric quantity of fuel, in the form of a petroleum derivative, e.g., diesel oil (7% w/w), etc. Slurries are multiphase mixtures in which ammonium nitrate, in the form of a supersaturated aqueous solution is gelled by a cross-linking agent holding the excess ammonium nitrate and various fine particulate fuels in suspension. Fuels include carbonaceous material, coal dust is a good example, and aluminum. Granular TNT, methylamine nitrate, or very fine aluminum powder were required in the early forms of slurry compositions as shock initiation sensitisers, but later, cheaper alternatives, such a glass microballoons were found to be effective. The nitroglycerine explosives still have a vital role to play where there is a need for a relatively high shock output from the explosive, whereas ANFOs and slurries tend to be reserved for use in large-scale quarrying, or soil heaving roles.

There has been some crossing of borders recently between military and commercial explosives and the uses of weapon technologies, especially in the offshore oil business. Here, it is often necessary to cut steel underwater, or to pierce thick steel targets, sometimes at high temperatures associated with borehole work at great depths, e.g., to bring oil from capped wells on stream. Military linear and conical-shaped charge technology has been harnessed for these purposes and, where high temperatures have been encountered, charges have been filled with hexanitrostilbene (HNS). Explosive bolt technology, from missile and other weapon delivery systems has frequently been used in civil space, diving, and other activities in hostile environments, in which it is necessary to carry out remote separation of equipments from each other. Lately, increased interest has arisen in explosive welding and forming, since it has been found possible to make very low density explosives by using finely granulated military explosives in, say, a polyurethane matrix, to which has been added a precisely calculated quantity of water. Normally water is anathema to a polyurethane matrix during curing, because it encourages the formation of gases. However, if the water content is controlled, the gasification can also be controlled. The result is the ability to reduce the density of the explosive to a small fraction of that of water and the detonation velocity to less than 1500 m/s. This is ideal for the explosive welding of unlikely and unpromising pairs of metals, enabling important new materials properties to be accessed by the composites.

2.5.3.14. Future Explosives Developments

A large measure of the military explosives research and development resources of many nations is devoted to the search for means of reducing the

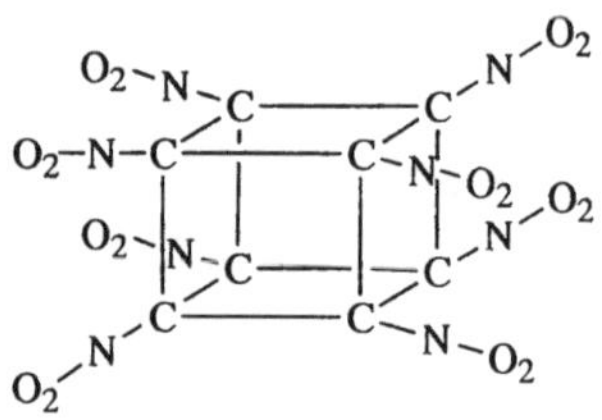

Figure 2.1. Octanitrocubane, $C_8N_8O_{16}$.

vulnerability of present and future weapon inventories to initiation as a result of accidents, or the effects of enemy strikes on weapon stocks. New explosives molecules are being sought, especially at the US Naval Weapon Center, China Lake, CA, USA, and at Xiang, PRC, with some success. Current aims are to synthesize more powerful and lower vulnerability explosives. It is likely that more powerful explosives will emerge first as a result of new routes aimed at synthesizing three-dimensional cage molecules. One of the aims is to produce the so-called "ultimate explosive" Octanitrocubane, shown in Figure 2.1. Some cubanes have been prepared already and other three-dimensional cage molecules may have been made using novel methods.

Another school of thought argues that HMX, hexanitrobenzene and, possibly, some of the new very powerful explosives emerging from China Lake, may be all that is required, provided a better understanding of the processes of initiation is achieved. It is argued that, provided the energetic molecules can be protected adequately within some type of polymeric matrix, which prevents the explosives being initiated by other than a deliberate stimulus emanating from the fuze, there is ample scope for increase in performance. The method proposed is via the use of polymeric binders of different types, which start off merely as the means for the protection of the energetic material crystals, but which might become involved with the evolution of new explosives along the lines shown in Figure 2.2.

Polymeric binders already exist which contribute energy to the matrix, e.g., polyNIMMO, developed by the Defence Research Agency in the United Kingdom. Energetic plasticizers, such as FEFO, BDNPA-F, polyglycidyl nitrate, etc., have been developed in the United States, mainly for use in the propellant field. There is still scope to increase the energy output from an HMX-based PBX by exploiting these technologies with the added possibility that either the plasticizer, or some cement used to bind the energetic materials crystals into the polymeric matrix should decompose endothermically under the influence of thermal stimuli. Such a material would not only be significantly more powerful than currently available explosives, but would also be able to withstand some degree of accidental thermomechanical stimulus without decomposing rapidly. Such a material is likely to be several years away, but it may be less of a forlorn hope than to pursue the Octanitrocubane route.

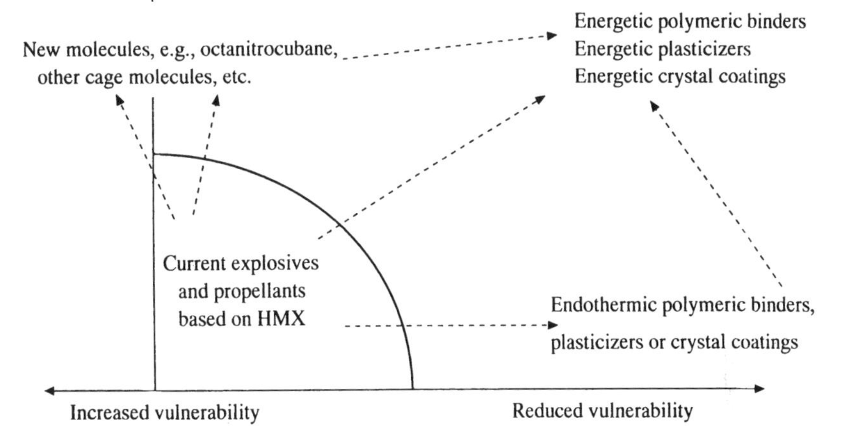

Figure 2.2. Evolution of new explosives.

A review by Spear and Dagley of the synthesis of organic energetic compounds, published as a chapter in the book edited by Marinkas referred to at the start of this introduction, is required reading. It is fully up to date on all developments in explosives synthesis and includes sections on all the principal families of new explosives, e.g., caged nitramine molecules, of which Octanitrocubane is an extreme example. References are made to predicted performance and hazard characteristics of the new molecules and possible areas of use, e.g., propellants, "insensitive" explosives are suggested. The references and bibliography at the end of the review are an important resource for further reading.

CHAPTER 3

Shock Waves; Rarefaction Waves; Equations of State

William C. Davis

3.1. Introduction

In a fluid, gradients in pressure, velocity, and/or temperature will induce motions in the fluid. Energy is transferred within the fluid, and may also be transferred to the materials that surround the fluid. Energy transfer processes include actual motion of the fluid, viscous flow processes, heat conduction, surface tension, diffusion of matter, radiation, etc. One of these processes often dominates the energy transfer, and the others can be neglected. In high-speed flow of compressible fluids, such as that produced by detonating explosives or by high-speed impact, the transfer of energy is almost all by motion of the fluid, with very little influence of viscosity and other mechanisms. In what follows all energy transfer by means other than fluid motion is neglected. A further simplification is to treat solids like aluminum or iron as fluids, with the strength completely neglected. This drastic approximation is also used throughout this chapter, and usually agrees well with experiment.

The treatment presented in this chapter focuses on the simplest model of compressible fluid flow, with all energy transfer accomplished by fluid motion. Other processes are neglected. In many applications of the model, it works very well. Still, we must not forget that it makes severe approximations, and must always ask whether the model is adequate for an application outside the usual range, or one requiring special accuracy.

The motions considered in this chapter are all one dimensional; that is, only slab symmetry is considered, and only one space variable is used. The real world is three dimensional, and there are some effects that appear in three dimensions that do not in one dimension. However, most important behaviors can be found in one dimension. Quantitative treatment of the interesting wave motions is very mathematical, and the algebra is difficult enough in one dimension.

The reader who is a beginner in the study of wave motions and equations of state will do well to repeat the steps in the development of the equations, carefully considering the underlying assumptions, and will also benefit by doing the problems. He or she will not be comfortable with the subject until it is made his or her own by spending time putting pencil to paper.

The physical laws that apply are the usual ones of Newtonian mechanics: the conservation of mass, the conservation of momentum, and the conservation of energy. The fluid is a continuum; that is, the infinitesimal volume in the equations is still very large compared with the atomic and molecular structure of matter, or, usually, large even compared with the particle size of granular materials. This restriction gives good results except in extraordinarily rarefied gases, or very coarse granular materials. The equations for flow with slab symmetry (plane one-dimensional flow) neglecting all energy transfer except by actual fluid motion are

$$\dot{v} - vu_x = 0, \tag{3.1}$$

$$\dot{u} + vp_x = 0, \tag{3.2}$$

$$\dot{\mathscr{E}} + v(up)_x = 0, \tag{3.3}$$

where the dot denotes the time derivative along the paths of particles, so that, for example,

$$\dot{v} = \frac{\partial v}{\partial t} + u\,\frac{\partial v}{\partial x},$$

and the subscript denotes the partial derivative, so $u_x = \partial u/\partial x$, and $\mathscr{E} = E + \frac{1}{2}u^2$, the total specific energy, and v = specific volume, p = pressure, u = velocity of material, E = specific internal energy, x = distance, and t = time. The density $\rho = 1/v$ is often used in this work, making it seem as if there were an additional variable; this extra variable is an unfortunate accident, but any worker in the field must expect to find it used.

Equations (3.1)–(3.3) are simply the conservation equations for mass, momentum, and energy in a continuous medium, with the neglect of all energy transfer except by fluid motion. They are often called the Euler equations. If the effects of heat conduction and viscosity are added to these equations, the equations are called the Navier–Stokes equations.

3.2. Notation and Units

The notation in the field of shock wave physics and detonations is an absolute mess; no standard has been generally accepted by workers in their papers. This chapter uses the notation of Davis (1985) and Menikoff and Plohr (1989), but other authors have used different choices.

Perhaps the most confusion arises with the derivatives called here the adiabatic gamma γ and the Gruneisen gamma Γ. Some have used the same Greek letters but interchanged them. Occasionally the adiabatic gamma is γ_S and the Gruneisen gamma is γ or γ_G. Sometimes k or n is used for one or the other. Some authors have used Γ for the fundamental derivative, here denoted by $\mathscr{G}$. The use of u_s and u_p for shock and particle velocity does not seem to cause much trouble.

The reader must exercise extreme care in deciphering books and papers about shock waves and detonations.

The units in this chapter are the standard SI units, using the meter, the kilogram, and the second as the units, and they form a consistent set. Another common set often found in computer codes uses the gram, the centimeter, and the microsecond as the fundamental set of units, which leads to the megabar (which is 10^{11} Pa) as the unit of pressure, and the megabar-cc/g (where cc means cm^3 and the unit is 10^8 J/kg) as the specific energy unit. The kilobar, which is 0.1 GPa, is very frequently used, as is the atmosphere, which is 101325 Pa.

The reader must be careful when using tables of parameters for explosives and inerts to get values for calculations. Perhaps the most common error is to forget to change g/cm^3 to kg/m^3. Another common error is to use tables that give energy as energy per unit volume when we expect specific energy as energy per unit mass.

3.3. List of Symbols

a	Piston acceleration.
a	Constant in the van der Waals equation of state.
a	Pressure constant in the Tait equation of state.
b	Covolume (excluded volume) in equations of state.
c	Sound speed.
$\hat{c}$	Constant term in $U = \hat{c} + su + qu^2 + \cdots$.
e	Energy per unit volume for the JWL equation of state.
j	Subscript indicating a C–J state.
k	Thermal conductivity.
m	Mass fraction.
n	Exponent in the Abel equation of state.
n	Exponent in inverse power potential.
p	Pressure.
q	Coefficient in $U = \hat{c} + su + qu^2 + \cdots$.
r	Radial distance between molecules.
r	Energy ratio in the mixture equation of state.
s	Coefficient in $U = \hat{c} + su + qu^2 + \cdots$.
t	Time coordinate.
u	Particle velocity.
v	Specific volume.
x	Space coordinate.
A	Area
A	Constant in the JWL equation of state.

A Viscosity term in the hydrodynamic equation.
B Constant in the JWL equation of state.
B Viscosity and heat conduction term in the hydrodynamic equation.
B Coefficient in the virial expansion.
B_T Isothermal bulk modulus in the Hayes equation of state.
C Constant in the JWL equation of state.
C Coefficient in the virial expansion.
C_p Specific heat at constant pressure.
C_v Specific heat at constant volume.
E Specific internal energy.
F Specific Helmholtz free energy.
H Subscript denoting values on the Hugoniot curve.
N Exponent in the Hayes equation of state.
Q Stored specific chemical or configuration energy.
R Specific gas constant.
R_1, R_2 Constants in the JWL equation of state.
S Specific entropy.
T Temperature.
U Shock velocity.
V Volume ratio for the JWL equation of state.
X Piston position.

$\mathscr{E}$ Total specific energy.
$\mathscr{G}$ Fundamental derivative of gas dynamics.

α Exponent in the BKW equation of state.
α Constant in the exponential-6 equation of state.
β Volume expansion coefficient.
γ Adiabatic gamma.
Γ Grüneisen gamma.
ε Energy constant in the equations of state.
λ Lamé constant.
θ Temperature constant in the BKW equation of state.
μ Viscosity.
μ Lamé constant.
ρ Density.
τ Piston time.
φ Intermolecular potential in the equations of state.
Φ Arbitrary function.
ω Exponent in the JWL equation of state.

3.4. Shock Waves

A shock wave is the dividing surface between moving material and stationary material. When a force is applied to a material surface, the material adjacent to that surface begins to move, while the material farther from that surface is still at rest. Necessarily, the material in motion is compressed, and occupies less space than it did initially. This statement seems self-evident, but when it is applied to real cases it often leads to configurations far removed from usual experience. In almost all usual experience, a force applied to a metal object can be thought of as moving the object as a rigid body; the far end of the object moves as soon as the force is applied. This common sense intuitive model is an approximation valid most of the time. The sound speed, or shock-wave speed, through the object is very fast, and the delay between application of the force and motion of the far end can easily go undetected. Only when the delay is important, or when the compression of the material is considerable, or when the deformation of the material is large, do we need to consider shock waves. Common experience that rigid-body mechanics, or mechanics with small deflections, describes the real world is deeply engrained in all of us, and study of shock-wave mechanics requires a suspension of disbelief. As an example, consider a large plane explosive charge in contact with a plane steel plate. The early motion of the steel after detonation of the explosive is not affected by the attachment of the plate, whether the plate is free, or bolted to a massive structure, or an integral part of the structure of a battleship. The later motion will, of course, be very different in these three cases.

A plane steady shock wave and the motion of the material can be diagrammed in x–t space. (Plane means that the whole system has slab symmetry in one space dimension, and steady means that the speeds are independent of time.) The back boundary can be chosen arbitrarily as some surface of identifiable particles; it is often called the piston interface, even though no actual piston is involved. Figure 3.1 is such a diagram. The x–t plane is useful for tracking waves and particle paths in flows with slab symmetry; it is commonly drawn with x as the abscissa and t as the ordinate, so slower waves have steeper curves, as is evidenced by the relative slopes of the fast shock wave and the slower piston path. A particle path in Figure 3.1 is represented, for a particle originally at some value of x, as a vertical line, corresponding to no motion, until it intersects the shock path, after which time it is bent to be parallel to the piston path, representing motion with the same speed as the piston. The speed, as well as the pressure and the density, jump suddenly as the particle crosses the shock path.

Determining the jump in speed, pressure, and density across the shock wave gives a set of equations known as the Rankine–Hugoniot equations. They can of course be obtained directly from (3.1)–(3.3), but the physical ideas are more easily understood by applying the conservation conditions in the simple conceptual apparatus shown in Figure 3.2. It can be thought of as a section of constant area taken out of a system with slab symmetry, so that the tube walls are frictionless and rigid, and the pistons consist of layers of marked particles. The material in the tube is initially at rest, with pressure p_0, density ρ_0, and specific internal energy E_0; the initial particle speed

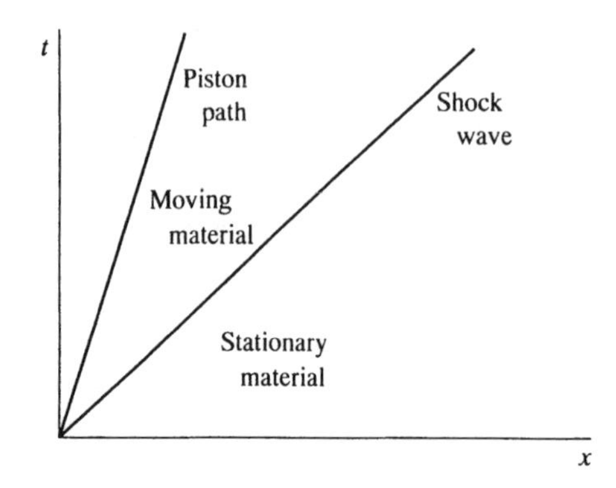

Figure 3.1. Diagram in the x–t plane showing the shock path, the piston path, and the regions containing moving and stationary material. Material behind the piston path is outside the universe of this diagram. At a given instant of time, say half-way up the diagram, material that was originally between the point on the shock path and the vertical axis has been compressed to fit between the piston and the shock, and all that material is in the same state and moving at the piston speed.

$u_0 = 0$. Distance along the tube is denoted by x, and time by t, with $x = 0$ at the inner face of the left piston. At $t = 0$ the left piston suddenly begins to move at speed u, driving a shock wave down the tube at speed U. At time t the material that was initially between the piston face at $x = 0$ and the position $x = Ut$ of the shock wave at time t is now between the piston face at $x = ut$ and $x = Ut$. Its density ρ must have increased for the same amount of matter to fit in the smaller space. If the area of the tube is A, then

$$\rho_0 AUt = \rho A(U - u)t \qquad \text{or} \qquad \frac{\rho_0}{\rho} = \frac{U - u}{U}. \tag{3.4}$$

The momentum of the material has changed also. The mass of material is $\rho_0 AUt$, and its speed has changed from zero to u, so the change in momentum is $\rho_0 AUtu$. The impulse, which is equal to the change in momentum, is

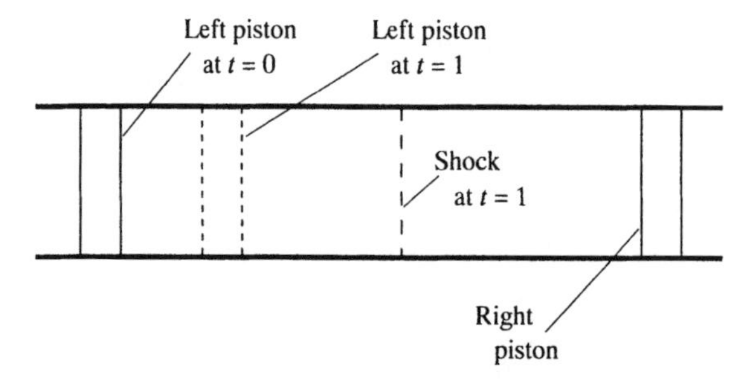

Figure 3.2. Diagram of the conceptual apparatus used to develop the Rankine–Hugoniot equations.

the net force times the time. The net force is the difference between pA on the left piston and p_0A on the right piston, so on equating the impulse to the change in momentum we find

$$(p - p_0)At = \rho_0 AUtu \qquad \text{or} \qquad p - p_0 = \rho_0 Uu. \tag{3.5}$$

The change in internal energy is equal to the change in specific internal energy times the mass $(E - E_0)\rho_0 AUt$, and the change in kinetic energy is the mass times $\frac{1}{2}u^2$ which is $\frac{1}{2}\rho_0 AUtu^2$. The work done by the left piston is the force times the distance it moved $pAut$. Equating the work done to the change in energy gives

$$pAut = (E - E_0)\rho_0 AUt + \tfrac{1}{2}\rho_0 AUtu^2 \qquad \text{or} \qquad E - E_0 = \frac{pu}{\rho_0 U} - \tfrac{1}{2}u^2. \tag{3.6}$$

From (3.5) we see that $\rho_0 U = (p - p_0)/u$ and can substitute into (3.6) tc get

$$E - E_0 = \frac{pu^2}{p - p_0} - \tfrac{1}{2}u^2 = \tfrac{1}{2}u^2\left[\frac{2p}{p - p_0} - 1\right]$$

$$= \tfrac{1}{2}u^2\left[\frac{p + p_0}{p - p_0}\right]. \tag{3.7}$$

These three independent expressions of the conservation laws are conditions on the jump in values across any shock wave. They are three equations for five variables, p, ρ, E, u, and U. Any three variables can be related in an equation obtained by eliminating one variable or another from combinations of (3.4), (3.5), and (3.6). Sometimes the algebra is not easy. There are ten equations, which may be written as:

$$\frac{v}{v_0} + \frac{u}{U} = 1, \tag{3.8}$$

$$p - p_0 = \rho_0 Uu, \tag{3.9}$$

$$E - E_0 = \tfrac{1}{2}(p + p_0)(v_0 - v), \tag{3.10}$$

$$p - p_0 = \rho_0^2 U^2 (v_0 - v), \tag{3.11}$$

$$E - E_0 = \tfrac{1}{2}u^2\left(\frac{p + p_0}{p - p_0}\right), \tag{3.12}$$

$$u^2 = (p - p_0)(v_0 - v), \tag{3.13}$$

$$E - E_0 = \frac{p^2 - p_0^2}{2\rho_0^2 U^2}, \tag{3.14}$$

$$E - E_0 = \tfrac{1}{2}u^2 + \frac{p_0 u}{\rho_0 U}, \tag{3.15}$$

$$E - E_0 = \tfrac{1}{2}u^2 + p_0(v_0 - v), \tag{3.16}$$

$$E - E_0 = \tfrac{1}{2}U^2\left(1 - \frac{v}{v_0}\right)^2 + p_0(v_0 - v). \tag{3.17}$$

Only three of these equations are independent; the other seven can be obtained from any selected three.

There are five dependent variables, so two more equations are needed to provide a solution. The fourth equation is some description of material properties, and the fifth is a value of one of the variables for a particular shock wave.

The material description can be as simple as a fit to some experimental data, such as a relationship between u and U. For example, it might be

$$U = \hat{c} + su + qu^2 + \cdots . \tag{3.18}$$

Combining (3.18) and (3.9) defines a curve in the p–u plane

$$p - p_0 = \rho_0(\hat{c}u + su^2 + qu^3 + \cdots), \tag{3.19}$$

that is, the locus of all states attainable in single shock waves of different strengths; this curve is called the Hugoniot curve. A choice of some variable, say U, used in (3.9), defines a line in the p–u plane

$$p - p_0 = \rho_0 U u, \tag{3.20}$$

and the intersection of this line with the Hugoniot curve gives a unique solution for the shock state. This line is called the Rayleigh line. Values of v and E can be obtained from (3.8) and (3.10). The Hugoniot curve in the p–v plane can be plotted by finding values for p and v corresponding to those for p and u on the Hugoniot curve in the p–u plane. The Rayleigh line in the p–v plane is given by (3.11).

The material description alternatively might be given by an expression such as

$$E - E_0 = \frac{pv - p_0 v_0}{\gamma - 1}, \tag{3.21}$$

where γ is a constant. Using (3.10) to eliminate $E - E_0$ gives an expression involving p and v which can be solved (either algebraically or by iteration with a computer) for a sequence of values for the Hugoniot curve in the p–v plane. (Notice that when $p \gg p_0$ the p's cancel from both sides of the equation; this indicates that there is an asymptote for v at $v/v_0 = (\gamma - 1)/(\gamma + 1)$. There is a limit to the compression that can be attained in a single shock.) The values for u and U can be found using the other equations to draw the Hugoniot curve in the p–u plane, or the U–u plane.

The p–v plane is useful for thermodynamics, the p–u plane for shock matching at interfaces, and the U–u plane for plotting experimental data. The x–t plane, shown in Figure 3.1, can also be plotted, with the values for all the variables known in the constant state regions.

PROBLEM 1. The material properties for a single shock in 2024 Dural, with initial density 2785 kg/m^3 and initial pressure 101325 Pa (1 atm), are satisfactorily described by the relationship $U = 5328 + 1.338u$ (m/s). Make a table of values of u, U, p, v, ρ, and $E - E_0$ for several values of u up to 4000 m/s. Plot the Hugoniot curve in the p–u plane, the p–v plane, and the U–u plane. Plot the Rayleigh line in each plane for $U = 8000$ m/s. Plot the x–t plane, showing the shock trajectory and the piston path, for $U = 8000$ m/s. How long does it take the shock to move through a Dural plate 25 mm thick? What is the thickness of the compressed plate when the shock reaches the surface? The power per unit area transmitted by the piston into the Dural is the pressure times the particle velocity; calculate the power input for $U = 8000$ m/s. Compare this value with the total electric generating capacity of the United States, which is about 10^{11} W, and calculate the area of the piston that could be driven by that electric power.

PROBLEM 2. The material properties of air for moderate shocks can be described by

$$E - E_0 = \frac{pv - p_0 v_0}{\gamma - 1},$$

with $\gamma = 1.4$, and this expression can be used with (3.10) to find, after a lot of algebra, the Hugoniot curve in the p–v plane as

$$\left(\frac{p}{p_0} + \frac{\gamma - 1}{\gamma + 1}\right)\left(\frac{v}{v_0} - \frac{\gamma - 1}{\gamma + 1}\right) = 1 - \left(\frac{\gamma - 1}{\gamma + 1}\right)^2.$$

Assume $p_0 = 10^5$ Pa and $\rho_0 = 1.29$ kg/m^3. Calculate values for a table of u, U, p, v, ρ, and $E - E_0$ for several values of p up to 10^7 Pa. Plot Hugoniot curves and one or two selected Rayleigh lines in the p–v plane, p–u plane, and U–u plane. Plot trajectories in the x–t plane for the highest pressure. Compute input power for the highest pressure and compute the area of the piston that could be driven by the 10^{11} W of electric power generated by the United States.

A table (page 105) with representative values for ρ_0, $\hat{c}$, s, and Γ_0 is an appendix to this chapter. The values there are for illustration, and should not be used for serious work without checking the original published papers to be sure that the numbers used really apply in the region of interest. Using wrong numbers gives wrong answers.

3.5. Rarefaction Waves

A rarefaction wave, like a shock wave, separates moving and stationary material, but there is an important difference. The shock wave was treated in the preceding section as a discontinuity, and, even more importantly, as a steady part of the flow. Even with all the physical effects that are neglected in that treatment, a real physical shock wave is nearly a discontinuity and is very closely steady. A rarefaction wave, on the other hand, is spread out in

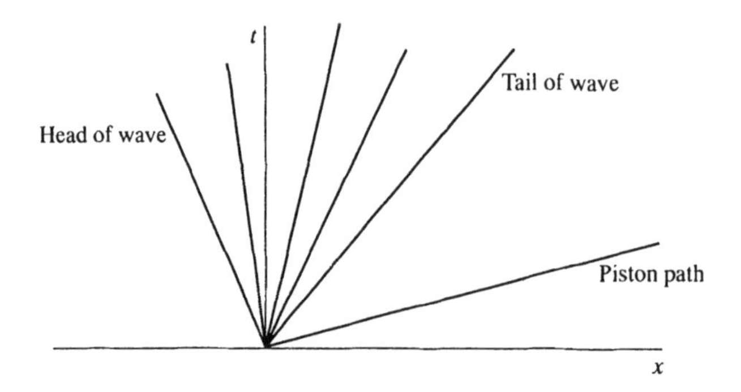

Figure 3.3. Diagram in the x–t plane showing the path of the wave head as it moves to the left into the stationary material, the piston path as it moves to the right, the expansion region between the wave head and the wave tail where the material expands, and the constant state region between the tail of the wave and the piston path where material in a constant state moves at piston speed.

space, and continues to spread with time. The expansion of the high-density material to a lower density does not take place instantaneously.

A diagram in the x–t plane is shown in Figure 3.3. Initially all the matter was to the left of $x = 0$, and at $t = 0$ the right boundary started to move at constant speed. The head of the rarefaction wave, moving at sound velocity in the high-density material, goes to the left. The expansion wave is shown as a fan of lines, and along each line the values of the state variables are constant. The final line is the locus where the velocity of the material equals that of the moving boundary, and between it and the boundary the state is constant. In the conceptual apparatus of Figure 3.2, this diagram corresponds to the right piston suddenly moving to the right at constant speed, with the left piston held fixed. During the time of interest, before any wave reaches the left piston, the material at the left is at rest in the initial condition, and the material at the right is moving at the piston speed, and both these regions are in a constant state. The conditions within the rarefaction wave, and the relationships between the piston speed and the state variables in the region near the piston, must be obtained by solving the differential equations for the conservation of mass, momentum, and energy, given in (3.1)–(3.3). In the section on shock waves, it was possible to find algebraic relations connecting the initial and final states on the Hugoniot curve. It is impossible to find similar relations for the rarefaction wave because the expansion region is not steady, but is increasing in width, and the mass of material in the region increases with time. That material has mass, momentum, and energy, and its distribution of states must be calculated in detail. The differential equations must be solved.

Equation (3.3), the energy equation, leads immediately to the realization that, if there are no shock waves or other discontinuities in the variables, the flow is all isentropic. If the x derivative (the parenthesis in that equation) is written out and the terms are collected in a different order, we obtain

$$\dot{E} + p(vu_x) + u(\dot{u} + vp_x) = 0. \tag{3.22}$$

Substituting into the first parentheses from the first equation, and the second parentheses from the second equation, gives

$$\dot{E} + p\dot{v} = 0, \tag{3.23}$$

which is equivalent, as can be seen by comparison with the thermodynamic relationship $dE = T\,dS - p\,dv$, to the definition of an isentropic relationship with $dS = 0$. Then

$$\left(\frac{\partial E}{\partial v}\right)_S = -p \qquad \text{becomes} \qquad \frac{dE}{dv} = -p. \tag{3.24}$$

This means that the potential $E(S, v)$ with $S =$ constant depends only on v. As a result, all the variables depend only on v, and all possible states, in the constant regions and in the rarefaction wave, lie on a single isentropic curve, where

$$p = p(v). \tag{3.25}$$

Using this result, we can define a new variable c

$$c^2 = -v^2 \frac{dp}{dv} \qquad \text{or} \qquad dp = -\left(\frac{c^2}{v^2}\right) dv, \tag{3.26}$$

and c is later identified with the sound speed. From this definition, it follows that

$$p_x = -\left(\frac{c^2}{v^2}\right) v_x \tag{3.27}$$

and

$$p_t = -\left(\frac{c^2}{v^2}\right) v_t. \tag{3.28}$$

The equation for the conservation of mass, (3.1), can be expanded as

$$v_t + uv_x - vu_x = 0 \tag{3.29}$$

and the equation for the conservation of momentum, (3.2), can be expanded as

$$u_t + uu_x + vp_x = 0. \tag{3.30}$$

Using (3.27) and (3.28) in (3.29) to eliminate derivatives in v, we find

$$p_t + up_x + \left(\frac{c^2}{v}\right) u_x = 0. \tag{3.31}$$

Now, if (3.30) is multiplied by $+(c/v)$ and added to (3.31), the result is

$$p_t + (u+c)p_x + \left(\frac{c}{v}\right)[u_t + (u+c)u_x] = 0. \tag{3.32}$$

Along a line $dx/dt = u + c$, (3.32) is the expression for total derivatives of p and u, so

$$\frac{dp}{dt} + \frac{c}{v}\frac{du}{dt} = 0 \qquad \text{along the line} \qquad \frac{dx}{dt} = u + c. \tag{3.33}$$

If (3.30) is multiplied by $-c/v$ and the same procedure followed, then

$$\frac{dp}{dt} - \frac{c}{v}\frac{du}{dt} = 0 \qquad \text{along the line} \qquad \frac{dx}{dt} = u - c. \tag{3.34}$$

From Figure 3.3, the diagram of the rarefaction wave in the x–t plane, we can see that the lines drawn there are those of (3.34), with $dx/dt = u - c$. Also, the state variables are constant along those lines, so $dp/dt = 0$, and $du/dt = 0$, satisfying the other part of (3.34). Equation (3.33) then states that, along lines crossing $dx/dt = u - c$, p and u are related by

$$\frac{dp}{du} = -\frac{c}{v}. \tag{3.35}$$

Using (3.26), (3.35) can be rewritten as

$$du = -\frac{v\,dp}{c} = \frac{c\,dv}{v} = \frac{dc}{\dfrac{v}{c}\dfrac{dc}{dv}}. \tag{3.36}$$

This set of equations relates the mechanical variable u and the thermodynamic variables p, v, and c throughout the expansion wave. Notice that dc/dv on the isentrope is negative. That is, as the particle velocity increases and the material expands, the material cools and the sound speed decreases. These equations (3.36), along with (3.24), connect the five variables E, p, v, c, u in the rarefaction wave in a way similar to the jump conditions for the shock. However, they must be solved for the whole flow.

As an example of how they are used, assume the simplest isentropic relationship, the polytropic gas, with

$$p = p_0\left(\frac{v_0}{v}\right)^{\gamma}, \tag{3.37}$$

where the adiabatic exponent γ is a constant. Then the definition $c^2 = -v^2\,dp/dv$ yields $c^2 = \gamma p v$, and differentiating again and rearranging terms gives

$$\frac{v}{c}\frac{dc}{dv} = -\frac{\gamma - 1}{2} \tag{3.38}$$

and (3.36) can be written as

$$du + \frac{2\,dc}{\gamma - 1} = 0, \tag{3.39}$$

which can be integrated to give

$$u = \frac{2}{\gamma - 1}(c_0 - c). \tag{3.40}$$

The constant of integration has been chosen to match the initial conditions, making $u = 0$ where $c = c_0$. The lines $dx/dt = u - c$ are lines where all the variables are constant. Any chosen value of c can be substituted into (3.40) to find the corresponding value of u, and into the relationship $c^2 = \gamma p v$ and the isentropic relation (3.37) to find values for p and v. Integrating $dx/dt = u - c$ with u and c constant yields $x = (u - c)t$, which is then used to relate the other variables to a position in the x–t plane.

PROBLEM 3. Show that the definition of c in (3.26) agrees with the usual concept of sound velocity. Write $p = p_0 + \tilde{p}$, $v = v_0 + \tilde{v}$, and $u = 0 + \tilde{u}$, where the tilde superscript denotes a small perturbation of the constant state variable. Substitute these values into (3.30) and (3.31) and neglect terms with a product of two small perturbations, to find

$$\tilde{u}_t + v_0 \tilde{p}_x = 0 \qquad \text{and} \qquad \tilde{p}_t + \left(\frac{c^2}{v_0}\right)\tilde{u}_x = 0.$$

Differentiate the first with respect to x, and the second with respect to t. Multiply the first by c^2/v_0 and subtract the result from the second, to find

$$\tilde{p}_{tt} - c^2 \tilde{p}_{xx} = 0.$$

Show that $\tilde{p} = f(x - ct) + g(x + ct)$ is a general solution of this equation, and that the two terms correspond to waves of arbitrary shape traveling in the positive and negative directions with wave speed c.

PROBLEM 4. The expansion isentrope for air at moderate pressure is well represented by (3.37). Assume $p_0 = 10^5$ Pa, $\rho_0 = 1.28$ kg/m^3, $c_0 = 331$ m/s, $\gamma = 1.4$, and the piston is withdrawn at a speed $u_p = 100$ m/s. Write the expressions for the boundaries between the rarefaction wave and the two constant state regions. Find the state variables, c, p, and v in the low-pressure constant state. Write expressions for these variables as functions of x and t in the rarefaction wave. For $t = 10^{-3}$ s, plot c, p, ρ, v, and u as functions of x; be sure to include the constant state regions.

PROBLEM 5. Show that in material described by (3.37) when $c = 0$, then necessarily $p = 0$ and $\rho = 0$. For the material described in Problem 4, find the piston speed that reduces the density to 0. What happens when the piston is withdrawn at a speed faster than that? Plot the state variables and u for a piston speed of 2000 m/s at $t = 10^{-3}$ s. The limiting speed of the gas is called the escape speed.

PROBLEM 6. The equation $dx/dt = u$ describes the motion of a material particle. Using the $u(x)$ found as part of the solution of Problem 4, find the path of a particle initially at $x = -0.1$ m, in the original constant state, as it passes through the rarefaction wave, and in the final constant state.

PROBLEM 7. A useful description of an imperfect gas along its expansion isentrope is

$$p = p_0 \left(\frac{v_0 - b}{v - b} \right)^n, \tag{3.41}$$

where b and n are constants. The covolume b is the volume of the material when it has been compressed as far as it can be. This simple expression of nonideality of a gas is not correct for all v, but it works well when v in the interesting region is not too close to b, and it is simple enough to be manipulated to get expressions in closed form. Show that

$$\frac{v}{c}\frac{dc}{dv} = -\left(\frac{n+1}{2} \cdot \frac{v}{v-b} - 1 \right), \tag{3.42}$$

and

$$dc = -\left(\frac{n+1}{2} \cdot \frac{v}{v-b} - 1 \right) \left[np_0 (v_0 - b)^n \frac{v^2}{(v-b)^{n+1}} \right]^{1/2} \frac{dv}{v}. \tag{3.43}$$

Use these expressions in (3.36) and integrate to show that

$$u = \frac{2}{n-1} \cdot \frac{v_0 - b}{v_0} \cdot \sqrt{\frac{np_0 v_0^2}{v_0 - b}} \cdot \left[1 - \left(\frac{v_0 - b}{v - b} \right)^{(n-1)/2} \right], \tag{3.44}$$

where the square root term is c_0, the sound speed in the undisturbed medium. Show also that

$$u = \frac{2}{n-1} \cdot \frac{v_0 - b}{v_0} \cdot c_0 \left[1 - \left[\frac{p}{p_0} \right]^{(n-1)/2n} \right], \tag{3.45}$$

$$u = \frac{2}{n-1} \left(\frac{v_0 - b}{v_0} \cdot c_0 - \frac{v - b}{v} \cdot c \right), \tag{3.46}$$

$$u_{esc} = \frac{2}{n-1} \cdot \frac{v_0 - b}{v_0} \cdot c_0, \tag{3.47}$$

where u_{esc} is the escape speed.

These wave motions are the mechanisms for transfering energy in high-speed flow of fluids. Thermodynamic derivatives have been defined to help relate the properties of the material to the flow equations. The adiabatic gamma, γ, which is sometimes called the dimensionless sound speed, is defined as

$$\gamma = -\frac{v}{p}\left(\frac{\partial p}{\partial v}\right)_S = \frac{c^2}{pv}. \qquad (3.48)$$

Adiabatic gamma is a function of whatever two thermodynamic variables are being used, as $\gamma = \gamma(p, v)$ in (3.48). Confusion often results when adiabatic gamma is used. In traditional thermodynamics, the same symbol is used for the ratio of the specific heats. Adiabatic gamma is not the ratio of specific heats, but to add to the confusion adiabatic gamma and the ratio of specific heats are equal for the special case of an ideal gas with constant specific heats. Adiabatic gamma is not a constant, but it is often a convenient simplification to use a constant value for illustrative calculations, and it is constant for the special case of an ideal gas with constant specific heats. As with all thermodynamic derivatives, there are many ways to express adiabatic gamma. For example, in wave physics we often have a description of the material properties that is incomplete because the thermal properties temperature and entropy are not included. The incomplete equation of state may be $p = p(E, v)$, and then adiabatic gamma may be written as

$$\gamma = -\frac{v}{p}\left[\left(\frac{\partial p}{\partial v}\right)_E - p\left(\frac{\partial p}{\partial E}\right)_v\right]. \qquad (3.49)$$

The fundamental derivative of gas dynamics is defined as

$$\mathcal{G} = 1 - \frac{v}{c}\left(\frac{\partial c}{\partial v}\right)_S, \qquad (3.50)$$

or

$$\mathcal{G} = \frac{1}{2}\left[\gamma + 1 - \frac{v}{\gamma}\left(\frac{\partial \gamma}{\partial v}\right)_S\right], \qquad (3.51)$$

or

$$\mathcal{G} = \frac{1}{2}\left[\gamma + 1 - \frac{v}{\gamma}\left(\frac{\partial \gamma}{\partial v}\right)_E + \frac{pv}{\gamma}\left(\frac{\partial \gamma}{\partial E}\right)_v\right]. \qquad (3.52)$$

With these definitions of γ and $\mathcal{G}$, (3.36) can be rewritten as

$$du = -\sqrt{\frac{v}{\gamma p}}\,dp = \sqrt{\frac{\gamma p}{v}}\,dv = -\frac{dc}{\mathcal{G} - 1}. \qquad (3.53)$$

The physical significance of $\mathcal{G}$ is discussed in Section 3.7, where it is shown that compression waves will develop into shock waves if and only if $\mathcal{G} > 0$.

3.6. Reference Frames

In the preceding sections the velocities were always positive, and the initial velocity was taken to be zero. In the real world, things are not always that simple. We must be able to treat a problem such as a projectile striking a target, or a rarefaction wave arising when a shock wave arrives at a free surface. Sometimes a series of events occur, each giving the material some residual velocity. It is often not possible to pick a reference frame just to simplify the equations.

The simplest case is the one where the coordinate system is such that the initial medium is at rest, but the motions are in the negative direction. This would have happened if the right piston in Figure 3.2 had been used for the shock wave, and the left piston for the rarefaction wave. A quick look at (3.8)–(3.17) shows that U and u enter as squares, products, or ratios, so as long as both are taken either positive or negative there is no problem. Where trouble comes is with material descriptions like (3.18) and (3.19), especially when they are put into some computer program without thinking carefully about the signs. If the velocities are negative, a plot in the p–u plane will have the curves to the left of the origin, and will appear quite different.

If the medium is in motion at a velocity u_0, all the velocities measured in the laboratory frame must have that u_0 subtracted from them, and that value substituted into the equations, with the proper regard for signs of the measured velocities. For example, suppose a projectile with initial velocity u_0 hits a rigid wall. The wall acts as the piston to send a shock wave into the material of the projectile. Measured in the laboratory frame, where the rigid wall is at rest, $u = 0$ in the shocked material. The shock wave moves in the direction opposite to u_0 with measured velocity U. In the momentum equation, (3.9), they will appear as

$$p - p_0 = \rho_0(U - u_0)(u - u_0) \tag{3.54}$$

and similarly in the other equations. If instead of hitting a rigid wall the projectile had been hit from behind by another faster projectile, U and u would both be positive. Another possibility would be to choose a "laboratory reference frame" moving with the projectile at velocity u_0 and then the measured speeds would be the ones to enter into the equations. Confusion arises when the reference frame is not actively chosen, but just tacitly assumed, so that it is easy to change it in the middle of the calculation. It always helps to make sketch plots of the arrangement in the x–t plane, the p–u plane, and the p–v plane. A sketch of the U–u plane, with an extra set of axes with origin at (u_0, u_0) will also help; the U–u plane seen with the added axes should look like that plane for the medium initially at rest. Note that the U–u plot may be in either the first or the third quadrant, depending on the signs of the velocities.

The description of the initial state, in addition to its velocity relative to the laboratory measurement frame, includes p_0, ρ_0, and E_0. The Hugoniot curve is said to be centered at that initial state. An Hugoniot curve centered at one state must not be used for another problem that should be centered elsewhere. The density and pressure seldom cause difficulty, but the initial energy does. There is no absolute scale for energy in most shock wave problems, and the zero can be chosen at will. Usually it is chosen to make $E_0 = 0$ in the initial state, but sometimes, particularly to describe detonation product gases, it is chosen zero for full expansion of the gases to zero pressure, or for expansion to atmospheric pressure. Authors sometimes just assume that the reader knows what they have assumed; we can waste a lot of time trying to see what went wrong.

For rarefaction waves, the effects of having waves with the velocity negative is more subtle. After deriving (3.33) and (3.34) it was necessary to decide which of the two characteristic lines would have $dp/dt = 0$ and $du/dt = 0$, and which would have $dp/du = \pm\rho c$. The choice was made there for $u \geq 0$; the opposite choice is proper for u in the negative direction. The result is equivalent to changing the signs of x and t wherever they appear. A wave with the velocities negative is sometimes called a forward-facing or right-facing wave, and one with velocities positive a backward-facing or left-facing wave; this nomenclature usually causes more confusion than clarity.

For compression waves, which may occur when the piston is not impulsively started but smoothly accelerated, u and c have the same sign if u is positive, opposite from the rarefaction wave. We must take great care to get all the signs right.

If the medium is in motion, then the assumption made in deriving the equations must be modified. The initial velocity of the medium must be added to u, always being careful of the signs. Drawing all the diagrams will help.

PROBLEM 8. Adapt the results of Problem 1 to a problem where the medium has an initial velocity of +500 m/s, and the piston is moving with velocities from +500 m/s to −3500 m/s. Make all the plots required in Problem 1.

PROBLEM 9. Adapt the results of Problem 4 to a problem where the medium is moving at +200 m/s and the left piston is withdrawn from it at a speed of +100 m/s. Draw the diagrams asked for in Problem 4. Draw a particle path corresponding to the one asked for in Problem 6.

3.7. Sharp Shocks and Diffuse Rarefactions

In the discussion so far it has been stated that shock waves are almost instantaneous jumps in all the variables, from the initial state to the final state. That fact made it possible to treat the waves very simply, and to find algebraic relationships among the variables. Similarly, in the section on rarefaction waves it was stated that rarefaction waves are spread out, and grow wider with time. In this section, the statements are made more precise. There can be exceptions to the rule, but they occur only under special circumstances. One case of departure from the usual rule is in the immediate neighborhood of the triple point in a material with a phase change.

To see how the system works, consider again the conceptual apparatus in

Figure 3.2. This time imagine that the left piston is started from rest with constant acceleration, rather than given an impulsive start at finite velocity. At the instant the acceleration is applied, when the piston velocity is zero, a sound signal propagates into the medium with speed c. As the piston accelerates to velocity u, the signal propagates with speed $u + c$, with c the new value at the piston. This second signal will then, after some time, overtake the first, if u is positive, and will fall behind if u is negative. If the acceleration is in the positive direction, there will be a signal path with speed $u + c$ for each value of u as the piston moves. One of these will be the first to overtake another. Along each of these paths the state variables are constant, with the value they have at the piston. When overtake occurs, the state is double valued, and thus physically impossible. What happens then is that a shock wave is formed.

If the constant acceleration is a, then at a time τ the position of the piston is $X = \frac{1}{2}a\tau^2$ and the piston speed is $u_p = a\tau$. Integrating (3.53) with $\mathscr{G}$ constant gives $c_p = c_0 + (\mathscr{G} - 1)u_p$ and the equation for the characteristic curve along which the variables are constant is $x = X + (u_p + c_p)(t - \tau)$. Notice that this equation has both t and τ as variables. The line in the x–t space starts from a point on the piston path at a particular value of τ and the values at the piston are constant along the line at the specified values of x and t. Because the value of $u_p + c_p$ is increasing (if the acceleration is positive) with τ the successive lines will intersect. For a full treatment of the shock formation problem, see Courant and Friedrichs (1948).

If $u + c$ increases with pressure, a shock will be formed in a compression wave, but a rarefaction wave will be diffuse. From (3.53), after changing the reference frame, it is easy to show that $d(u + c) = \mathscr{G}\sqrt{v/\gamma p}\, dp$. From this expression, it is easy to see that $u + c$ will increase with p as long as $\mathscr{G} > 0$. For almost all materials over most of the space of interest this condition is satisfied and compression waves will grow to be shock waves, and rarefaction waves will be diffuse and occupy space.

In passing, it is worth noting that any acceleration of a piston into matter will produce a shock. For example, consider the arrangement used to derive the equations for the rarefaction, where the right piston in Figure 3.2 was moved out to the right. If that piston, after the flow was established, was slowed down, a shock would form in the matter behind it.

3.8. Transmission and Reflection of Waves at Interfaces

Consider a projectile striking a target. On impact, a shock wave will travel into the target, and a shock wave will travel back into the projectile. The projectile will be slowed by the impact, and the target accelerated; however, only the material in the projectile that has been traversed by the shock traveling back into it has been slowed, and only the material in the target that has been traversed by the shock traveling into it has been accelerated. At the

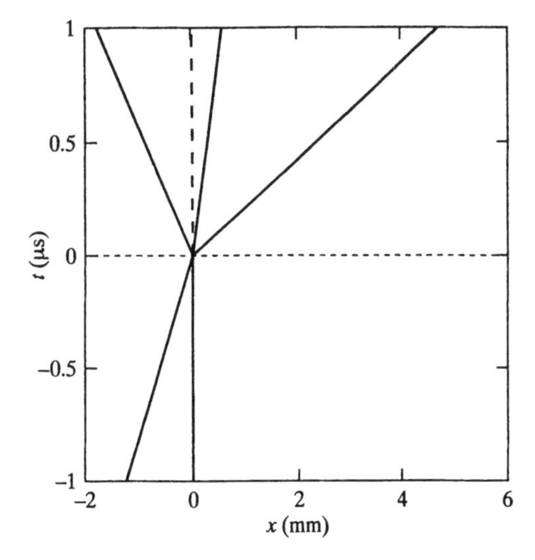

Figure 3.4. Diagram of the x–t plane for a lead projectile striking an iron target. The collision occurs at $x = 0$, $t = 0$. At times less than 0, below the dotted x-axis, the face of the projectile, the line with positive slope, is approaching the target at 1250 m/s (1.25 mm/μs). The target is stationary at (0, $t < 0$), represented by the solid line at $x = 0$. After the collision, a shock wave, the left-hand line in the region $t > 0$, is sent back into the moving projectile at -1751 m/s. The interface, the middle line, moves in the positive direction, at 599 m/s, and a shock wave, the right-hand line, moves into the target at 4720 m/s.

interface, the velocity of projectile and target must be equal, and the pressure in both must be equal. The x–t diagram is shown in Figure 3.4.

Ordinary experience tells us that a projectile will strike a target and bounce back, and that the projectile and target will separate. Certainly they will, but only after waves from the rear of the projectile and front of the target have traveled in to reach the interface. The problem considered here is the very early stage of the collision, while the projectile and target are affected only by the intial shock wave. Only the front surface of the projectile and the back surface of the target are shown in Figure 3.4. Eventually the shock waves will reach the other surfaces, and rarefaction waves will propagate into the materials. For now, the later stages of the collision are not considered.

It is important to choose a reference frame and stick with it. The frame chosen for this projectile–target problem has the target initially stationary in the reference frame, and the projectile is moving in the positive direction. Both materials are shocked, so their states will be represented by points on their respective Hugoniot curves. The condition at the interface is that pressure and particle velocity are equal across the interface. The proper plane for matching across the interface is thus the p–u plane. The projectile is in

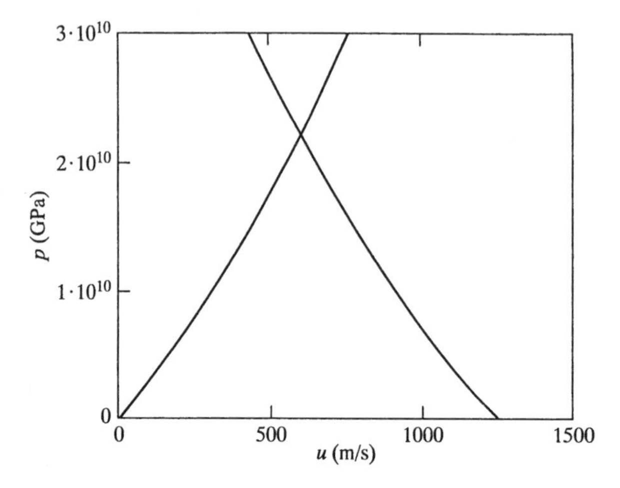

Figure 3.5. Hugoniot curves in the p–u plane for a lead projectile striking an iron plate. The lead Hugoniot curve is on the right with an initial speed of 1250 m/s, and the iron target curve is on the left, initially at rest. After the impact, the speed and pressure are the values at the intersection of the two Hugoniot curves.

motion, with an initial velocity u_0, so its Hugoniot curve must have its zero pressure point at u_0. The projectile is slowed by the collision, so the Hugoniot curve must be drawn so that the velocity decreases as the pressure increases. The right-hand Hugoniot curve in Figure 3.5 is the one for the projectile. The target is initially at rest at zero pressure, and it is given a velocity in the collision, so its Hugoniot curve must be drawn so the velocity increases as the pressure increases. The target Hugoniot curve is the left-hand one in Figure 3.5.

The condition at the interface after the impact is that pressure and particle velocity must be equal in both materials. The condition is satisified at the intersection of the two Hugoniot curves in Figure 3.5. The equations are written as described in the preceding section as (3.54).

Problem 10. A lead projectile traveling 1250 m/s impacts on a stationary iron target. Assume that the Hugoniot curves for both materials are adequately described by the relationship $U = \hat{c} + su$. Find the pressure and velocity at the interface after the collision. For lead, the density is 11,350 kg/m^3, $\hat{c} = 2050$ m/s, and $s = 1.46$; for iron, use 7850 kg/m^3, 3570 m/s, and 1.92.

The special case of a symmetric collision, with the projectile and target made of the same material, is of particular interest, because it allows us to determine experimentally the material description, finding values of shock velocity corrsponding to values of particle velocity. If we look at Figure 3.5

and imagine that the two Hugoniot curves are for the same material, it is obvious that the diagram is perfectly symmetric; the two curves are mirror images of each other. Therefore, the particle velocity in each material is $\frac{1}{2}u_0$; a measurement of the projectile speed allows us to determine the particle velocity. A measurement of the transit time for the shock through the target, along with knowledge of the thickness of the target, allows us to obtain the shock velocity as the thickness divided by the time. The conservation of momentum condition, (3.7), $p - p_0 = \rho_0 Uu$, then gives the pressure on the Hugoniot curve. Repeated measurements map out the whole curve in the region of interest. Usually the measured data are plotted in the U–u plane, and fitted with a low-order polynomial.

PROBLEM 11. One experiment is performed to get a first approximation for the Hugoniot curve of a new material; a projectile made of that material with a velocity of 1850 m/s strikes a target of the same material, and the shock in the target travels 10 mm in 1.98 μs. Sound measurements give the bulk sound velocity to be 3726 m/s. Use the bulk velocity as the value for $\hat{c}$, and find s for use in $U = \hat{c} + su$. The material density is 8450 kg/m^3; find the interface pressure in the experiment.

Once the symmetric collision experiments have been done for a few standard materials, say aluminum and copper to give a spread in the shock impedance, projectiles of those materials can be used with relatively small pieces as targets to determine the Hugoniot curves of other materials, as shown in Figures 3.4 and 3.5.

A somewhat more complicated system consists of two different materials in contact, with a shock wave traveling through one of them and crossing the interface. In this case, the initial state for the first material is the shock state. At the interface, a shock wave enters the second material, and a wave, which may be either a shock or a rarefaction, is reflected back into the first material. A case where a shock is reflected is considered first. The x–t plane is shown in Figure 3.6, and the p–u plane in Figure 3.7. The shock state in the first material is shown as a dot on its Hugoniot curve, which is below the Hugoniot curve for the second material. The shock state for the reflected shock in the first material will lie on an Hugoniot curve for that material, but one centered at the shock state, not at the usual low-pressure ambient state. The reflected shock Hugoniot curve may be calculated if the material properties are known or assumed, or it may be measured in a series of experiments like the one being considered. The state of the first material, behind the reflected shock, is given by the intersection of the reflected shock Hugoniot curve with the single shock Hugoniot curve for the second material. Note that the first material has two distinct Hugoniot curves to describe its state, because it has two shocks, while the second material has only one Hugoniot curve necessary to describe its state with just one shock.

The case where a rarefaction wave is reflected back into the first material

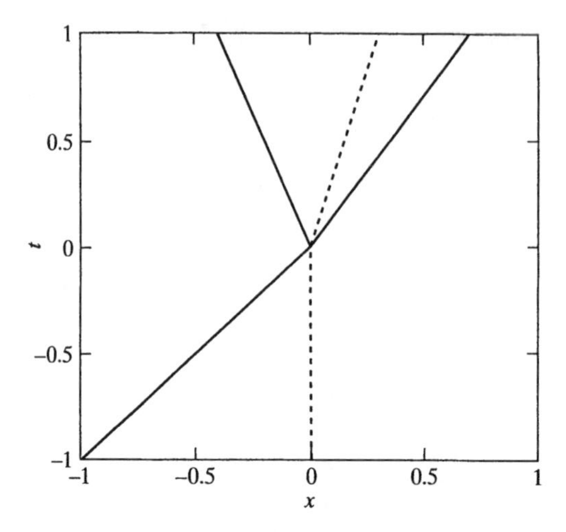

Figure 3.6. Diagram of the x–t plane for a shock wave passing from one material to another of higher shock impedance, so that a shock wave is reflected back into the first material, compressing it further. The interface is shown as a dotted line. The incoming shock has normalized speed 1, and the transmitted shock slows to speed 0.7. The interface speed is 0.3, and the speed of the reflected shock measured in the laboratory frame is −0.4.

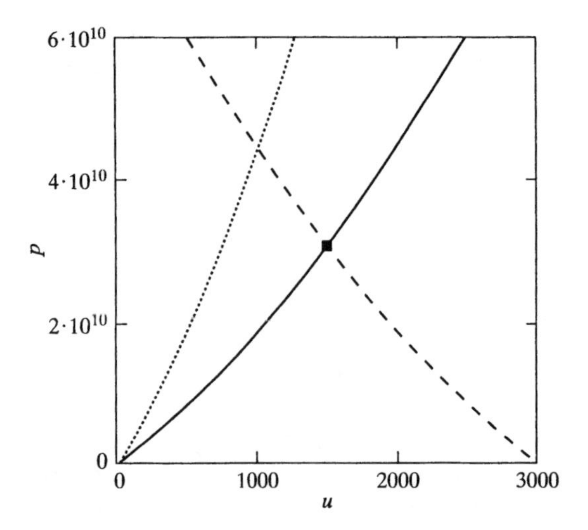

Figure 3.7. Diagram of the p–u plane for a shock wave transmitted from Dural to brass. The solid line is the Hugoniot curve for Dural, the dotted line for brass, and the dashed line is the Hugoniot curve for Dural centered at the shock state.

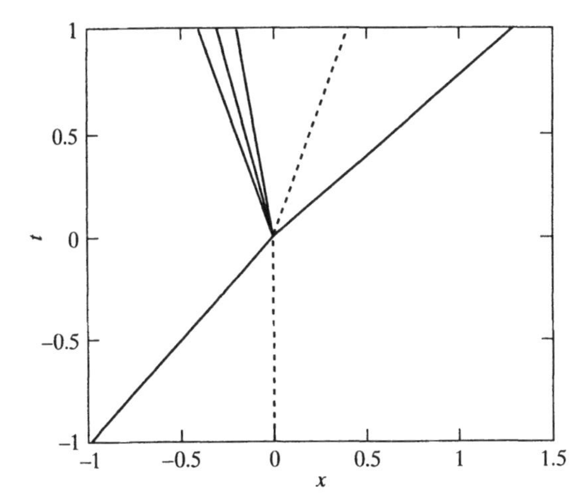

Figure 3.8. Diagram of the x–t plane for a shock wave in one material passing into another material of lower shock impedance, so that a rarefaction wave is reflected into the first material.

has its x–t plane diagram in Figure 3.8, and its p–u plane diagram in Figure 3.9. The x–t diagram looks much like the previous case with a reflected shock; the only qualitative difference is that the reflected wave is diffuse because it is a rarefaction wave. In the p–u plane the curve from the shock state in the first material goes to lower pressure. It is an isentrope, not an Hugoniot curve. It can be calculated if the material properties are known, or measured in experiments like the one being described. Again, there are two curves for the first material and one for the second. The isentrope centered at the shock state, and the Hugoniot curve centered at the same state, are tangent at second order, and make a smooth curve through the centering point.

In many cases, particularly with metals that are not too different in shock impedance, for the Hugoniot curve for reflected shocks, and for the expansion isentrope for reflected rarefactions, the single-shock Hugoniot curve reflected about the shock state as diagrammed in Figure 3.10 gives a very good approximation to the curves for the reflected states. In even more cases, while it is not as good an approximation, it is still better than any available information. It is often used for planning experiments even though better data will be available from the experiments after they are completed.

We must remember that the reflected curve is being used as an approximation to a reflected-shock Hugoniot curve in its upward branch, and to an expansion curve in its lower branch. The specific volume (or density) and specific internal energy must be obtained using the relevant equations. For example, the reflected shock Hugoniot curve centered at p_1 and u_1 is given

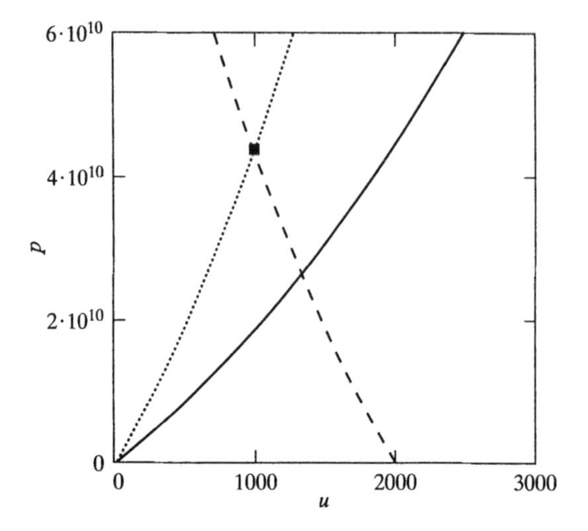

Figure 3.9. Diagram of the p–u plane for a shock wave passing from brass to Dural, reflecting a rarefaction wave back into the brass. The dashed curve is the expansion isentrope for brass, passing through the shock state marked with a dot.

by

$$p = \rho_0(2u_1 - u)[\hat{c} + s(2u_1 - u)], \tag{3.55}$$

which is used with the Hugoniot jump relationship (3.13) rewritten as

$$(u_1 - u)^2 = (p - p_1)(v_1 - v) \tag{3.56}$$

to find

$$v_1 - v = \frac{(u_1 - u)^2}{p - p_1} = \frac{(u_1 - u)^2}{\rho_0\{(2u_1 - u)[\hat{c} + s(2u_1 - u)] - u_1[\hat{c} + su_1]\}}. \tag{3.57}$$

In principle, (3.55) and (3.57) could be solved to give $v(p)$, but they are perhaps just as useful left as they stand.

The expansion curve for the rarefaction wave is also given by (3.53). Again, it is not an Hugoniot curve; it is an expansion isentrope, and it must be used as such. Equation (3.35) is

$$\frac{dp}{du} = -\frac{c}{v} \tag{3.58}$$

and used with (3.53) gives

$$\frac{c}{v} = \rho_0[\hat{c} + 2s(2u_1 - u)]. \tag{3.59}$$

Equation (3.26) is just the definition of c

$$\frac{dp}{dv} = -\left(\frac{c}{v}\right)^2 \tag{3.60}$$

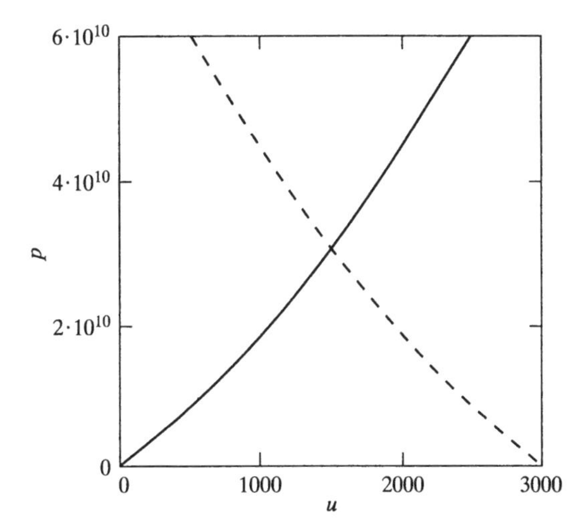

Figure 3.10. Hugoniot curve for Dural, with the same curve reflected about the vertical axis so that they intersect at 1500 m/s.

and can be used with (3.58) to find

$$\frac{dv}{du} = \frac{v}{c} \tag{3.61}$$

and by substituting (3.59) and integrating finally getting $v(u)$, and with (3.59), a relationship between c and u to use to find the characteristics $dx/dt = u + c$.

PROBLEM 12. A 30 GPa shock in lead reaches an interface with iron. Use the constants from Problem 9 to find the Hugoniot curves for each material, and reflect the Hugoniot curve for lead as an approximate reflected shock Hugoniot curve centered at the shock state. Find the shock pressure, shock velocity, and particle velocity in the iron. Find the apparent shock velocity (in the laboratory reference frame) in the lead. Sketch the x–t diagram. In the p–v plane, in a small region containing the initial shock state and the reflected shock state in the lead, sketch the single shock and reflected shock Hugoniot curves. The reflected shock Hugoniot curve lies to the left of the single shock Hugoniot curve because less entropy is produced by compression to the same pressure in two shocks than in one.

PROBLEM 13. A 30 GPa shock in lead reaches a free surface, an interface with vacuum. Use the constants from Problem 10 to find the approximate rarefaction isentrope. Sketch the x–t diagram, showing the escape velocity, the lead characteristic back into the shocked lead, and the characteristic where the pressure is 15 GPa.

3.9. The Shock Tube

The shock tube is a long metal tube, divided initially into two sections by a diaphragm. The left section is filled with gas at high pressure, and the right section is filled with gas at low pressure. At $t = 0$ the diaphragm is broken or removed. Gas flows from the high-pressure section, driving a shock wave in the low-pressure gas. A rarefaction wave propagates back into the high-pressure gas. There is a contact surface between gas originally at high pressure and gas originally at low pressure. Velocity and pressure must be equal across that surface, but density and temperature have a discontinuity there. In the treatment given here, it is assumed that the flow is strictly uniform across a diameter of the tube so it is a one-dimensional plane flow problem.

To make the problem as simple as possible, we assume that the gas is an ideal gas with constant specific heats, so the expansion isentrope for the high-pressure section is given by (3.35) as

$$p = p_0 \left(\frac{v_0}{v}\right)^\gamma, \tag{3.62}$$

where γ is a constant, and the relationship between sound speed and particle velocity is given by (3.40) as

$$u = \frac{2}{\gamma - 1}(c_0 - c). \tag{3.63}$$

The relationship between c and p is obtained from $c^2 = \gamma p v$ and (3.62) as

$$\frac{c}{c_0} = \left(\frac{p}{p_0}\right)^{(\gamma-1)/2\gamma} \tag{3.64}$$

and from (3.63) the isentrope in the p–u plane is obtained as

$$u = \frac{2c_0}{\gamma - 1}\left[1 - \left(\frac{p}{p_0}\right)^{(\gamma-1)/2\gamma}\right]. \tag{3.65}$$

These are the needed equations for the high-pressure gas. Notice that the subscript zero refers to the initial state in the high-pressure gas.

The description of the low-pressure gas also uses the ideal gas with constant specific heats, and the equation of state is

$$E - E_0 = \frac{pv - p_0 v_0}{\gamma - 1}. \tag{3.66}$$

Now notice that the subscript zero refers to the initial state in the low-pressure gas. We must be careful to keep track of which is which.

A relationship between U and u is convenient for the shock wave, as are the three independent Hugoniot jump conditions we choose (3.8), (3.9), and (3.15), which are

$$v = v_0\left(1 - \frac{u}{U}\right), \tag{3.67}$$

$$p = p_0 + \frac{Uu}{v_0}, \tag{3.68}$$

$$E - E_0 = \tfrac{1}{2}u^2 + p_0 v_0 \frac{u}{U}. \tag{3.69}$$

Substituting these expressions for v, p, $E - E_0$ into (3.66) and collecting terms gives the quadratic equation

$$U^2 - \tfrac{1}{2}(\gamma + 1)uU - \gamma p_0 v_0 = 0 \tag{3.70}$$

and its solution is

$$U = \frac{\gamma + 1}{4}u + \sqrt{\left(\frac{\gamma + 1}{4}u\right)^2 + \gamma\, p_0 v_0}. \tag{3.71}$$

The Hugoniot curve in the p–u plane is then

$$p = p_0 + \rho_0 Uu \tag{3.72}$$

with the expression for U substituted into it. The intersection of this shock Hugoniot curve from (3.72) and (3.71) with the expansion isentrope for the high-pressure gas, (3.65), gives the values for u and p in the shocked region of the low-pressure gas, and for the constant state following the expansion in the rarefaction wave in the high-pressure gas. The shock front in the low-pressure gas moves at the speed given by (3.71), and the contact discontinuity surface moves at speed u. The rarefaction wave head moves into the high-pressure gas with speed c_0, and the rarefaction wave is described in detail by the characteristics $dx/dt = u - c$, and u and c are related by (3.63), and are related to p by (3.64) and (3.65), and v is given by (3.62).

PROBLEM 14. A shock tube, with the high-pressure reservoir on the left, is filled with air at room temperature, 293 K. The high pressure is 5 MPa (about 50 atm) and the low-pressure is 50 kPa (about 0.5 atm). The density of the high-pressure air is 60 kg/m^3, and the density of the low-pressure air is 0.6 kg/m^3. Assume a constant adiabatic $\gamma = 1.4$. Sketch the x–t plane using accurate values for the speeds of the head and tail characteristics of the rarefaction, the contact surface, and the shock in the low-pressure gas. Using $pv = RT$, compute the temperatures on either side of the contact surface. In this problem, be careful to keep track of the two distinct subscripts zero.

3.10. Detonation

In a detonation chemical energy released in or near the shock front supports the shock, and the detonation is a self-supporting flow, the front of which propagates at constant speed. The chemical energy is stored as internal energy of the unreacted explosive. If the explosive is an ideal gas with constant specific heats, the equation of state of the unreacted explosive is

$$E_0 = \frac{p_0 v_0}{\gamma - 1} + Q, \tag{3.73}$$

where Q is the stored specific chemical energy. If the explosive products are also a gas with the same value for adiabatic γ, the equation of state for the products is

$$E - E_0 = \frac{pv - p_0 v_0}{\gamma - 1} - Q. \tag{3.74}$$

Proceeding exactly as in the shock tube problem, we find the quadratic equation corresponding to (3.70) is

$$U^2 - \left[\frac{\gamma + 1}{2} u + \frac{(\gamma - 1)Q}{u}\right] U - \gamma p_0 v_0 = 0. \tag{3.75}$$

The detonation problem is much simpler if we assume that $\gamma p_0 v_0$ is negligible compared with U^2. See Fickett and Davis (1979) for the formulas without this approximation. In this approximation the constant term disappears and we find

$$U = \frac{\gamma + 1}{2} u + \frac{(\gamma - 1)Q}{u}. \tag{3.76}$$

This curve is sketched in Figure 3.11; U has a minimum value, and no solutions exist below that. This is a necessary feature for a self-supporting detonation. The Chapman–Jouguet hypothesis, which can be justified, see Fickett and Davis (1979), is that a detonation will propagate at that minimum speed. Values for the parameters for the Chapman–Jouguet detonation will be denoted by a subscript j.

It is easy to show that the minimum occurs at

$$u_j^2 = \frac{2(\gamma - 1)Q}{\gamma + 1}. \tag{3.77}$$

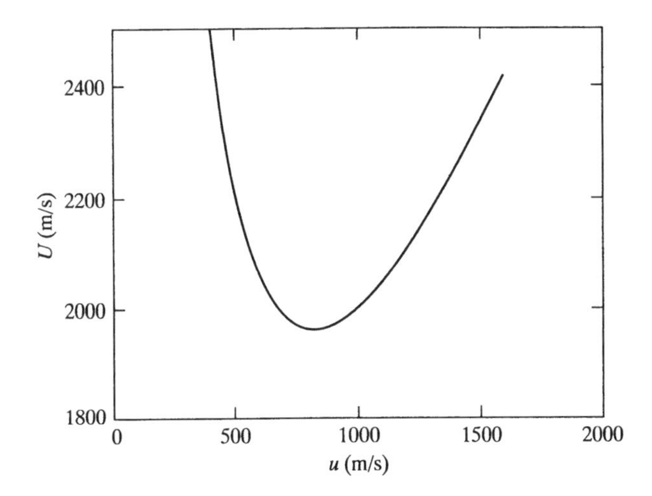

Figure 3.11. Plot of the Hugoniot curve for the product gases in the U–u plane. This curve is for the Hugoniot states of the products, but is centered at the initial state. Unlike Hugoniot curves for inerts, which are nearly straight lines with positive slope in this plane, the Hugoniot curve for detonation products has a minimum, meaning that there is a minimum possible detonation speed. The curve is drawn for a gas with $\gamma = 1.4$, initial density 1 kg/m^3, and $Q = 2$ MJ/kg.

Other useful expressions are

$$U_j^2 = 2(\gamma^2 - 1)Q, \tag{3.78}$$

$$u_j = \frac{U_j}{\gamma + 1}, \tag{3.79}$$

$$p_j = 2(\gamma - 1)\rho_0 Q = \frac{\rho_0 U_j^2}{\gamma + 1}, \tag{3.80}$$

$$\frac{v_j}{v_0} = \frac{\gamma}{\gamma + 1}, \tag{3.81}$$

$$c_j = \sqrt{\gamma p_j v_j} = \frac{\gamma U_j}{\gamma + 1}, \tag{3.82}$$

$$u_j + c_j = U_j. \tag{3.83}$$

The physical meaning of (3.83) is that a rarefaction wave can follow a detonation wave; the combined sound speed and particle speed are the wave speed, both for the detonation and for the head of the rarefaction. The detonation wave structure, then, is a leading self-supporting shock wave, followed by a rarefaction wave that matches into the rear boundary condition. The particle velocity (for a detonation wave propagating to the right) is positive right behind the shock front, but decreases in the rarefaction wave,

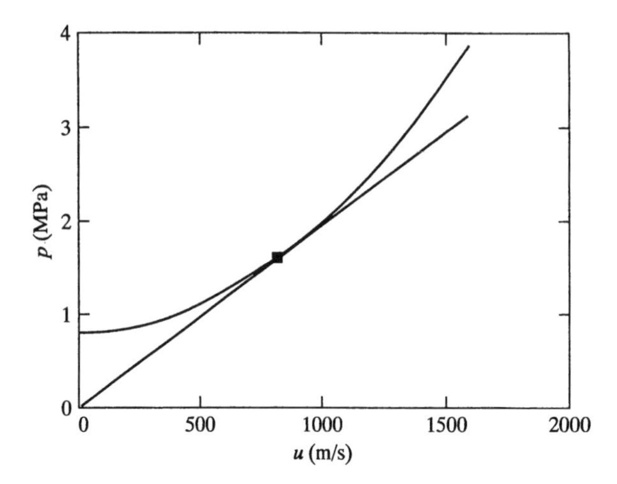

Figure 3.12. Diagram, in the p–u plane, of the Hugoniot curve and the tangent (minimum velocity) Rayleigh line, for the same explosive as in Figure 3.11. The C–J point is marked with a dot. The C–J pressure is 1.6 MPa, the detonation velocity is 1960 m/s, and the C–J particle velocity is 816 m/s, typical values for gas detonations.

and may become negative if the rear boundary is pushed back. A diagram of the p–u plane is shown in Figure 3.12.

The isentrope in the p–u plane is given by

$$u_j - u = \frac{2c_j}{\gamma - 1}\left[1 - \left(\frac{p}{p_j}\right)^{(\gamma-1)/2\gamma}\right]. \tag{3.84}$$

The expansion of the gas behind the detonation front is described by this expression. Note that u is positive for a distance behind the front, but then becomes negative far enough back. The total momentum of the gas must be zero, since the momentum was zero before the detonation started, and the forward momentum must be balanced by momentum in the opposite direction.

When the detonation has run through all the explosive, the product gases are ejected at high speed from the free surface. The isentrope for these gases, neglecting the effects of the rarefaction that is following the detonation front, is given by

$$u - u_j = \frac{2c_j}{\gamma - 1}\left[1 - \left(\frac{p}{p_j}\right)^{(\gamma-1)/2\gamma}\right]. \tag{3.85}$$

For gas expanding into vacuum, $p = 0$, in the same direction as the detonation was going, and using (3.79) and (3.82) to express u and c in terms of U, we can write

$$u_{\text{esc}} = u_j + \frac{2c_j}{\gamma - 1} = \frac{U_j}{\gamma + 1} + \frac{2}{\gamma - 1}\cdot\frac{\gamma U_j}{\gamma + 1} = \frac{3\gamma - 1}{\gamma^2 - 1}\, U_j. \tag{3.86}$$

For solid high explosives an appropriate value for γ is 3. For this value, the escape speed is equal to the detonation velocity.

3.11. Phase Changes

For many materials, shock-compression measurements show the Hugoniot curve in the U–u plane to have discontinuities in its slope. It often appears to be made up of two or three straight-line segments. As an example, experimental data for benzene [Dick (1970)] are shown plotted in Figure 3.13, and are fitted by three line segments. For some materials a jog in the curve like this is known to be the result of a phase change, either melting, freezing, or a change in crystal form. For benzene the change is believed to be the result of polymerization of benzene molecules. Usually the Hugoniot curves are fitted as in Figure 3.13 and used that way in computations.

Another approach is to write Hugoniot curves for the two phases. Let the Hugoniot curve for the initial phase be

$$p - p_0 = \rho_0(\hat{c}_0 u + s_0 u^2), \qquad U = \hat{c}_0 + s_0 u, \tag{3.87}$$

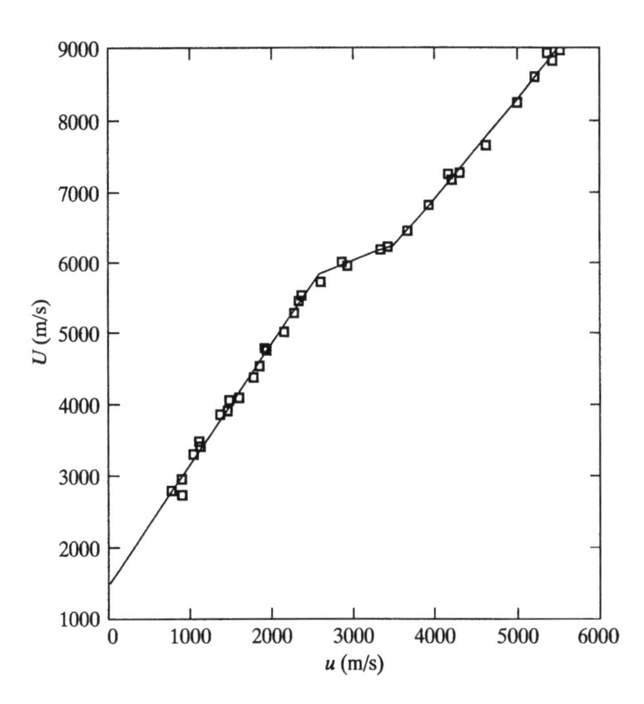

Figure 3.13. Graph of Hugoniot measurements for benzene, fitted with three line segments.

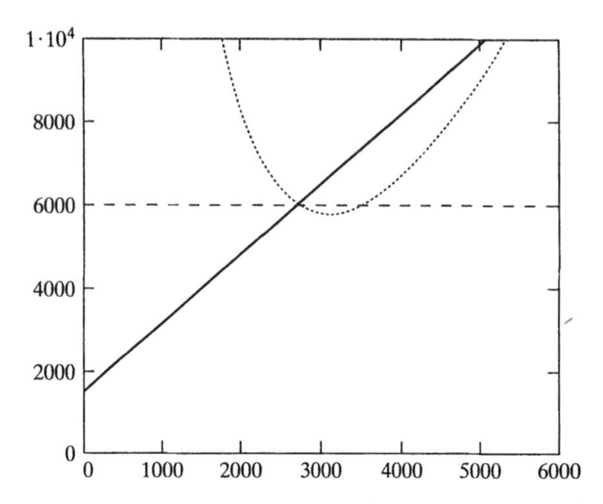

Figure 3.14. Diagram of the U–u plane, showing Hugoniot curves for the initial phase (solid straight line) and the high-pressure phase (dotted curved line), and also the Rayleigh line through the phase-change point.

and for the high-pressure phase

$$p - p_0 = \rho_0(Q_1 + \hat{c}_1 u + s_1 u^2), \qquad U = \frac{Q_1}{u} + \hat{c}_1 + s_1 u. \tag{3.88}$$

Both Hugoniot curves are centered at the same initial state, p_0, v_0, $u_0 = 0$. The points on the curves for the high-pressure phase are those that can be reached in a single shock starting from the initial point, with the change taking place. The constant Q_1 allows energy to be liberated or absorbed in the change.

A plot of the Hugoniot curves in the U–u plane is shown in Figure 3.14. For small values of u the states lie on the straight-line Hugoniot curve for the initial low-pressure material. At a large enough value of u the curves intersect; we assume that the change to the new phase takes place instantaneously at that state, and the states lie on the high-pressure curve. For a single shock to exist, the shock velocity is less than that at the intersection for small increases in u, until the curve intersects the Rayleigh line at constant U. Physically, the shock velocity does not decrease. Instead the shock wave splits into a wave traveling at the velocity for the first intersection, followed by a slower wave in the shocked material that matches the velocity of the piston. It is described by an Hugoniot curve centered at the first intersection, where the phase transition takes place. Between the two intersections of the high-pressure Hugoniot curve and the Rayleigh line, there is not a single shock, and the single-shock Hugoniot equations do not apply.

Fitting the parameters to data is straightforward. The values for $\hat{c}_0$ and s_0 are obtained as usual by fitting the data for $u < u_A$, where u_A is the value at the first intersection. For the other parameters, define point A as the first

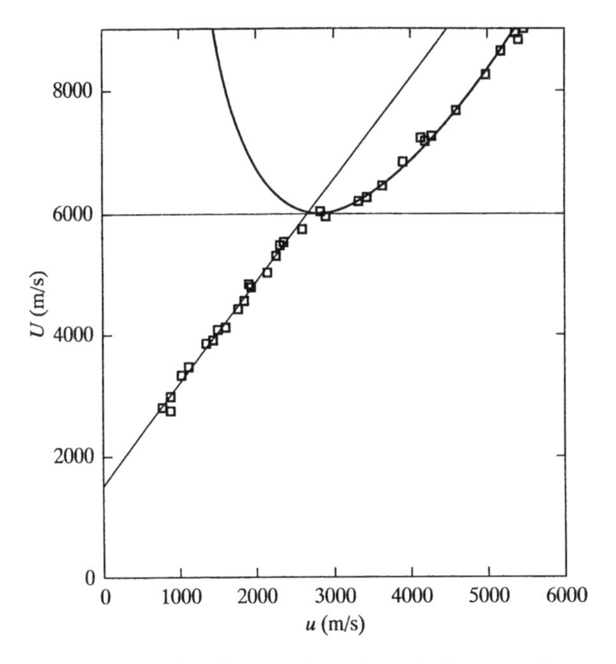

Figure 3.15. Plot of a two-phase fit to the benzene data.

intersection of the Hugoniot curves and the Rayleigh line, point B as the second intersection of the high-pressure Hugoniot curve and the Rayleigh line, and point C as a selected point in the data above point B. Then some tedious algebra shows that

$$Q_1 = \frac{(U_C - U_A)u_A u_B u_C}{(u_C - u_B)(u_C - u_A)}, \tag{3.89}$$

$$\hat{c}_1 = U_A - \frac{u_C(U_C - U_A)(u_A + u_B)}{(u_C - u_B)(u_C - u_A)}, \tag{3.90}$$

$$s_1 = \frac{(U_C - U_A)u_C}{(u_C - u_B)(u_C - u_A)}. \tag{3.91}$$

A fit to the benzene data is shown in Figure 3.15. The chosen points are A (6000 m/s, 2695 m/s), B (6000 m/s, 2900 m/s), and C (8320 m/s, 5000 m/s). The values for the parameters are $\hat{c}_0 = 1500$ m/s, $s_0 = 1.67$, $Q_1 =$ 18.73 MJ/kg, $\hat{c}_1 = -7408$ m/s, $s_1 = 2.396$. These values are just fitting constants; they have no real significance because the fitting form is simply fit to the data, and has no constraints outside the data region. The fit could be improved by using a more complicated form for the Hugoniot curve of the high-pressure material.

It is easy to see that the phase-change problem is similar in many respects to the detonation problem of the previous section. The possibility of a self-sustaining wave in a material that has a phase change was treated by Rabie, Fowles, and Fickett (1979).

In the discussion above the assumption that the change to the high-pressure phase is instantaneous allowed the choice of the point A to be made at the apparent end of the data points on the low-pressure phase Hugoniot curve. It is likely in the real world that there is a delay caused by the time needed for the phase change, and that the apparent end of the data on the Hugoniot curve is not the true equilibrium point. If so, the intersection of the two Hugoniot curves would lie at A′, a point below A.

3.12. Hydrodynamics and Thermodynamics

The hydrodynamic equations given as (1.1)–(1.3) are the Euler equations, and contain no terms for viscous effects or for heat conduction. The Navier–Stokes equations can be written as

$$\dot{v} - vu_x = 0, \tag{3.92}$$

$$\dot{u} + vp_x + A = 0, \tag{3.93}$$

$$\dot{E} + u\dot{u} + v(pu)_x = B, \tag{3.94}$$

where

$$A = -v\frac{\partial}{\partial x}(\tfrac{4}{3}\mu u_x), \qquad B = u\frac{v_0}{u_0}\frac{\partial}{\partial x}(\tfrac{4}{3}\mu u u_x + kT_x),$$

with μ the viscosity, k the thermal conductivity, and T the temperature. In computer codes, these terms are used to make the solution stable, and may have different forms. The equilibrium equation of state for the material can be written as

$$E = E(p, v), \tag{3.95}$$

so that

$$\dot{E} = E_p\dot{p} + E_v\dot{v}, \tag{3.96}$$

where

$$E_p = \left(\frac{\partial E}{\partial P}\right)_v, \qquad E_v = \left(\frac{\partial E}{\partial v}\right)_p. \tag{3.97}$$

With these definitions, (3.94) can be rewritten with substitutions from (3.92) and (3.93) as

$$\dot{p} + \frac{v(E_v + p)}{pE_p}\cdot\frac{p}{v}\cdot\dot{v} = \frac{v}{E_p}\cdot\frac{uA + B}{v}. \tag{3.98}$$

The coefficients that describe the material properties are so important that they have their own special names. They are defined as

$$\gamma = \frac{v(E_v + p)}{pE_p}, \tag{3.99}$$

which is called the adiabatic gamma, the same one that has been used throughout this chapter, and

$$\Gamma = \frac{v}{E_{\mathrm{p}}}, \tag{3.100}$$

which is called the Grüneisen gamma. These two functions are the parts of the material description that enter from the equation of state. The adiabatic gamma is the dimensionless sound speed

$$\gamma = \frac{c^2}{pv} \tag{3.101}$$

and governs the transfer of energy by wave motion. In (3.50), (3.51), and (3.52) the fundamental derivative of gas dynamics was defined, and it depends on the adiabatic gamma and its derivatives. The Grüneisen gamma describes the effect of processes, such as shock waves, where viscosity and heat conduction produce entropy.

A satisfactory approximate equation of state must be reasonably close to the real values of the adiabatic gamma, the Grüneisen gamma, and the fundamental derivative in the regions of state space where they are being used to model a real material, and where they are important. The details of the behavior of these functions determine just what sort of waves and flows will occur in a given material. Almost always they are well behaved, and pressure increases develop into shock waves while rarefaction waves are diffuse. Sometimes, near critical points and phase changes, the usual rules are violated. This topic has been explored in detail by Menikoff and Plohr (1989).

Thermodynamics uses four potential functions: specific internal energy $E(S, v)$, specific enthalpy $H(S, p)$, Helmholtz free energy $A(T, v)$, and Gibbs free energy $G(p, T)$. Much physics and chemistry uses the Gibbs free energy because pressure and temperature are the variables easily controlled in the laboratory. For shock-wave physics, however, the internal energy is more useful, because in many processes the entropy is constant or nearly so, and the volume is the measured variable. Each potential has two partial first derivatives. For the specific internal energy they are

$$\left(\frac{\partial E}{\partial S}\right)_v = T, \qquad \left(\frac{\partial E}{\partial v}\right)_S = -p. \tag{3.102}$$

There are three second derivatives

$$\frac{\partial^2 E}{\partial S^2} = \frac{\partial T}{\partial S} = \frac{T}{C_v} = \frac{gT^2}{pv}, \tag{3.103}$$

$$\frac{\partial^2 E}{\partial S\,\partial v} = \frac{\partial^2 E}{\partial v\,\partial S} = \frac{\partial T}{\partial v} = -\frac{\partial p}{\partial S} = -\frac{\Gamma T}{v}, \tag{3.104}$$

$$\frac{\partial^2 E}{\partial v^2} = -\frac{\partial p}{\partial v} = \frac{\gamma p}{v}. \tag{3.105}$$

The symbols for these three second derivatives, the reciprocal dimensionless specific heat g, the Grüneisen gamma Γ, and the adiabatic gamma γ, are defined by these equations. The mixed second derivative in (3.104) with the requirement that the result does not depend on the order of differentiation also shows the Maxwell equation

$$\left(\frac{\partial T}{\partial v}\right)_S = -\left(\frac{\partial p}{\partial S}\right)_v . \tag{3.106}$$

There are four third derivatives. The requirement that the result of differentiation does not depend on its order gives relationships similar to the Maxwell equation that are differential equations for the second derivatives. An equation of state proposed to describe material properties can be tested for consistency with these equations.

The three other thermodynamic potentials have their own sets of first and second derivatives. The manipulation of partial derivatives relates all these derivatives. The sets from each of the potentials can be expressed, sometimes after some difficult manipulation, in terms of those for any of the potentials. There are only three independent second derivatives; the choice used here is for the potential $E(S, v)$, as defined above.

Most experimental measurements on shock and rarefaction waves in solids and liquids made so far have been mechanical measurements. These can have nothing to say about temperature and entropy. The material description derived from these measurements is necessarily incomplete, and the specific internal energy is not known as $E(S, v)$ as it must be to be a complete potential, but as $E(p, v)$. Equations (3.99) and (3.100) show how Γ and γ can be obtained. The other second derivative g cannot be obtained from the incomplete equation of state.

3.13. Equations of State

In thermodynamics texts, equation of state means a relationship between p, v, and T, such as $pv = RT$, or a more complicated relationship such as the van der Waals equation of state. In addition, we must have experimental or theoretical information about the specific heat at constant volume. Then the energy equation, a general result of thermodynamics from manipulation of partial derivatives,

$$dE = C_V\, dT + \left[T\left(\frac{\partial p}{\partial T}\right)_V - p\right] dv \tag{3.107}$$

gives the specific internal energy as a function of T and v. The "first $T\,dS$ equation"

$$T\, dS = C_V\, dT + T\left(\frac{\partial p}{\partial T}\right)_V dv \tag{3.108}$$

gives the entropy as a function of T and v, and finally the complete potential $E(S, v)$ can be found by eliminating T.

In the field of shock-wave physics, "equation of state" seems to mean whatever information about the material properties will allow the user to proceed to a solution of the problem that is of interest at the moment. As a result, we must be extremely careful about choosing from equations of state found in books and papers to do a new problem. Some of the reasons authors may have used a particular form or calibration are:

(a) Illustration of how to proceed with some problem, to be done on the blackboard or in a textbook. It must not take more than five minutes or three or four lines of text; this chapter uses such forms.
(b) For a homework problem. The execution time limit goes up to about twenty minutes.
(c) To make a graph that illustrates some point being explained in a text. Realistic equations of state often have the curves so close together that they are indistinguishable, and the equation of state must be distorted to allow the reader to see the point.
(d) Smoothing data in a narrow region for finding interpolated values. The equation of state may be limited to the range of the data, and may work well only for the special problem that was done.
(e) For estimating the detonation state of an explosive from its molecular composition, as is done with BKW, JCZ3, etc. These calibrations work well for their intended purpose, but have not been calibrated for other applications such as calculating the performance of an explosive system as the explosive products expand.
(f) For a particular engineering problem. Often the calibration of the equation of state is adjusted to compensate for unrelated errors in the computer modeling.
(g) Because the form had been already programmed into the computer. It may be easier to use a poor choice of equation of state and adjust it in the region where it is important rather than put a new form into the program.
(h) In order to obtain an analytic result for some special problem by choosing an equation of state that simplifies the equations.

3.13.1. Ideal Gas and Polytropic Gas

It is important to understand the concepts of various forms of equations of state; without it a list of equations of state that have been used is at best confusing and may invite serious error. The simplest equation of state to start with is the ideal gas equation of state

$$pv = RT. \tag{3.109}$$

Application of (3.107), the energy equation, shows that the term in brackets on the right-hand side is zero. Another thermodynamic equation

$$\left(\frac{\partial C_{\mathrm{v}}}{\partial v}\right)_T = T\left(\frac{\partial^2 p}{\partial T^2}\right)_v \tag{3.110}$$

allows us to show that C_{v} is not a function of v. Experiment shows that for most gases at reasonable temperatures C_{v} increases with temperature, but not very strongly. The choice $C_{\mathrm{v}} = \text{constant}$ is thus thermodynamically consistent, and in reasonable agreement with experiment. With this choice, the equation of state is called the polytropic gas equation of state. It does not apply over the full range of possible temperatures, but for a limited range, so we can write:

$$E - E_0 = C_{\mathrm{v}}(T - T_0), \tag{3.111}$$

$$pv - p_0v_0 = \left(\frac{R}{C_{\mathrm{v}}}\right)(E - E_0), \tag{3.112}$$

$$E - \frac{pv}{(R/C_{\mathrm{v}})} = E_0 - \frac{p_0v_0}{(R/C_{\mathrm{v}})}. \tag{3.113}$$

This result is frequently written as

$$E = \frac{pv}{R/C_{\mathrm{v}}}, \tag{3.114}$$

but we must remember to restrict the range of application. Comparison with (3.100) shows that

$$\frac{R}{C_{\mathrm{v}}} = \Gamma \quad \text{and} \quad E = \frac{pv}{\Gamma}. \tag{3.115}$$

It follows that $E_{\mathrm{p}} = v/\Gamma$ and $E_{\mathrm{v}} = p/\Gamma$, and from (3.99)

$$\gamma = \frac{v}{p} \cdot \frac{E_{\mathrm{v}} + p}{E_{\mathrm{p}}} = \Gamma + 1 \tag{3.116}$$

so

$$E = \frac{pv}{\gamma - 1}, \tag{3.117}$$

which is the common way to write the polytropic gas equation of state.

Now consider the equation for the isentrope

$$\left(\frac{\partial E}{\partial v}\right)_S = -p \tag{3.118}$$

and use it in (3.117) to get, on the isentrope,

$$-p = \frac{v}{\gamma - 1} \cdot \frac{dp}{dv} + \frac{p}{\gamma - 1}, \tag{3.119}$$

where the total derivative notation is used to emphasize that this equation applies only on the isentrope. Integration of (3.119) gives

$$\frac{p}{p_0} = \left(\frac{v_0}{v}\right)^{\gamma} \tag{3.120}$$

the usual expression for the isentrope. Equation (3.108) can be solved to give the entropy as

$$S - S_0 = R \ln \frac{v}{v_0}\left(\frac{T}{T_0}\right)^{1/\Gamma}, \tag{3.121}$$

and we must remember that this expression is valid only where $C_v = \text{constant}$ is valid.

The ideal gas equation of state describes a gas of point particles; it will be a good approximation only for a gas whose molecules are far apart compared with their dimensions. In the atmosphere, the air molecules have an average spacing of about 3 nm, and the molecules are about 0.1 nm in some average dimension. The time they spend close enough together to interact is a very small fraction of the total time, and the ideal gas approximation works well. At the high gas densities of interest for detonation and shock problems, the approximation fails to describe the real gas. However, (3.117) is often used in limited regions with the value of γ adjusted to fit some data, but only to get a qualitative idea of how the system will behave using an easily manipulated form.

3.13.2. Abel

The Abel equation of state

$$p(v - b) = RT \tag{3.122}$$

is a simple modification of the ideal gas equation of state. The constant b, called the covolume, acts to limit compression of the gas. The particles are incompressible when $v = b$. The Abel equation of state is much like the van der Waals form

$$\left(p + \frac{a}{v^2}\right)(v - b) = RT. \tag{3.123}$$

The pressure term a/v^2 is small in most of the range of interest for detonations and shocks and is usually neglected. It is easily shown, using the procedure used for the polytropic gas equation of state, that it is consistent for the Abel form to assume $C_v = \text{constant}$. With the definition

$$\frac{R}{C_v} = n \tag{3.124}$$

and following the procedure for the polytropic gas we get

$$E = \frac{p(v-b)}{n-1}, \quad \gamma = n \cdot \frac{v}{v-b}, \quad \Gamma = (n-1) \cdot \frac{v}{v-b}, \quad \mathcal{G} = \frac{n+1}{2} \cdot \frac{v}{v-b}. \tag{3.125}$$

Notice that Γ and γ are not constants. They are like those for a polytropic gas at very large volume, but increase as the volume becomes comparable with b. The Abel equation of state is useful for illustrating how the covolume affects a flow. The equations can be solved analytically for most problems. In particular, the relationship between u and c in a rarefaction can be found, as illustrated in Problem 7.

3.13.3. Inverse Power Potential

It may seem contradictory that the introduction of the covolume, which acts to model the interaction of the molecules, still leaves the internal energy independent of volume. That is

$$\left(\frac{\partial E}{\partial v}\right)_T = T\left(\frac{\partial p}{\partial T}\right)_v - p = 0. \tag{3.126}$$

The reason is that the molecules in this model are incompressible; they occupy a fixed volume b and they cannot be compressed to store energy. If b is made a function of T, so that

$$p[v - \hat{b}(T)] = RT, \tag{3.127}$$

then

$$\left(\frac{\partial E}{\partial v}\right)_T = \frac{RT^2}{(v - \hat{b})^2} \cdot \frac{d\hat{b}}{dT}. \tag{3.128}$$

More insight into how molecular interaction affects the equation of state can be obtained by considering a gas in which spherical molecules repel each other with a force proportional to $r^{-(n+1)}$. The interaction energy is proportional to r^{-n} and the model is called a gas with an inverse power potential. It can be shown [Jeans (1954); Hirschfelder, Curtiss, and Bird (1954); Landau and Lifshitz (1958)] that the covolumes enter with $\hat{b} \propto T^{-3/n}$ so the equation of state may be written as

$$pv\left(1 - \frac{\tilde{b}}{vT^{3/n}}\right) = RT, \tag{3.129}$$

where $\tilde{b}$ is a constant. A typical value for n is 9. It can be shown that $\gamma \leq n/3 + 1$ and $\Gamma \leq n/3 - 1$. Many forms for equations of state have been proposed with covolume dependence on volume and temperature similar to that obtained for the inverse power potential.

3.13.4. Expansion of Equations of State in Powers of v

Any of these equations of state can be expanded as a series: the ideal gas is

$$\frac{pv}{RT} = 1, \tag{3.130}$$

the Abel equation of state is

$$\frac{pv}{RT} = 1 + \frac{b}{v} + \left(\frac{b}{v}\right)^2 + \cdots, \tag{3.131}$$

and the inverse power potential equation of state is

$$\frac{pv}{RT} = 1 + \frac{\tilde{b}}{vT^{3/n}} + \left(\frac{\tilde{b}}{vT^{3/n}}\right)^2 + \cdots. \tag{3.132}$$

More complicated equations of state can be expanded similarly, such as

$$\frac{pv}{RT} = 1 + \frac{B(T)}{v} + \frac{C(T)}{v^2} + \cdots, \tag{3.133}$$

where B and C are theoretical or calibrated functions of temperature. This form is called a virial equation of state, and is useful at moderate densities where the functions can be used to fit measurements. Equations of state intended for accurate representation of material properties are usually calibrated against tables of thermodynamic properties which are available for low pressure; this calibration replaces the overly simple choice of constant specific heat used for the polytropic gas equation of state.

3.13.5. Tait

The Tait equation of state

$$E - E_0 = \frac{(p + a)v}{\Gamma_0} - \frac{(p_0 + a)v_0}{\Gamma_0} \tag{3.134}$$

is useful for treating solids and liquids, because it allows the sound velocity to be chosen to fit the experimental value at zero pressure. The constant a is chosen as

$$a = \rho_0 c_0^2 - (\Gamma_0 + 1)p_0. \tag{3.135}$$

The important derivatives are found to be

$$\Gamma = \Gamma_0, \tag{3.136}$$

$$\gamma = \Gamma_0 + 1 + \frac{a}{p}, \tag{3.137}$$

$$\mathcal{G} = 1 + \tfrac{1}{2}\Gamma_0. \tag{3.138}$$

Note that while all three derivatives might have been functions of p and v, only γ varies, and it is a function only of p. The equation for the isentrope is

$$\left(p + \frac{a}{\Gamma_0 + 1}\right)v^{\Gamma_0 + 1} = \text{constant}. \tag{3.139}$$

Using (3.8), (3.9), and (3.15) and substituting them in the equation of state to find the Hugoniot curve in the U–u plane, after some algebra we find

$$U^2 - \mathscr{G}uU - c_0^2 = 0, \tag{3.140}$$

which is easily solved to give

$$U = c_0\left[\sqrt{1 + \left(\frac{\mathscr{G}u}{2c_0}\right)^2} + \frac{\mathscr{G}u}{2c_0}\right]. \tag{3.141}$$

This square root can be expanded to give

$$U = c_0\left[1 + \tfrac{1}{2}\mathscr{G}\frac{u}{c_0} + \tfrac{1}{2}\left(\frac{\mathscr{G}u}{2c_0}\right)^2 - \tfrac{1}{8}\left(\frac{\mathscr{G}u}{2c_0}\right)^4 + \cdots\right]. \tag{3.142}$$

The usual linear U–u relationship $U = \hat{c} + su$ can be fitted to this if $\hat{c} = c_0$ and $s = \frac{1}{2}\mathscr{G}$ or, equivalently, $\Gamma_0 = 4s - 2$.

3.13.6. BKW

The Becker–Kistiakowsky–Wilson (BKW) equation of state has been calibrated for estimating detonation properties. It has the form

$$\frac{pv}{RT} = 1 + \frac{\tilde{b}}{v(T+\theta)^\alpha}\exp\left(\frac{\beta\tilde{b}}{v(T+\theta)^\alpha}\right), \tag{3.143}$$

where β and $\tilde{b}$ are constants. Expanded in series it is

$$\frac{pv}{RT} = 1 + \frac{\tilde{b}}{v(T+\theta)^\alpha} + \beta\left(\frac{\tilde{b}}{v(T+\theta)^\alpha}\right)^2 + \tfrac{1}{2}\beta^2\left(\frac{\tilde{b}}{v(T+\theta)^\alpha}\right)^3 + \cdots. \tag{3.144}$$

For mixtures of gases the covolume $\tilde{b}$ is the weighted sum of the individual covolumes for the gases, weighted by the mole fraction. Mader (1979) gives a detailed description of the history, calibration, and application of the BKW equation of state to calculating detonation states. In addition to Mader's calibration, usually called BKW, others have recalibrated the constants for a better fit to other groups of calibration data, and their calibrations are called BKWR, BKWS, and so forth. A recent calibration for a large number of product species using a large database from many experiments is given by Hobbs and Baer (1993). The choice of the best fit for any particular purpose is a complicated one.

3.13.7. Intermolecular Potentials

The equations of state described above are developed from corrections to gas equations of state. At the very high densities reached in detonations the interactions are so strong that it might be better to treat the material with a solid-type equation of state. Jacobs, Cowperthwaite, and Zwisler [see Cowperthwaite and Zwisler (1976)] used approximate solid equations of state for the very high-pressure regions in equations of state called JCZ2 and JCZ3. They based their equations of state on a model where the molecules are held in a lattice, and where they interact according to an assumed potential function. JCZ2 uses what is called a 12–6 potential

$$\varphi(r) = 4\varepsilon^* \left[\left(\frac{r^*}{2^{1/6} r} \right)^{12} - \left(\frac{r^*}{2^{1/6} r} \right)^{6} \right], \tag{3.145}$$

where ε^* and r^* are constants. The first term is a repulsive force and the second term is an attractive force. At the equilibrium position $r = r^*$ the potential is $-\varepsilon^*$. JCZ3 uses the exponential-6 potential

$$\varphi(r) = \frac{\alpha\varepsilon^*}{\alpha - 6} \left[\frac{6}{\alpha} \cdot \exp\left[\alpha \left(1 - \frac{r}{r^*} \right) \right] - \left(\frac{r^*}{r} \right)^{6} \right], \tag{3.146}$$

where α is a constant. See Sentman, Strehlow, Haeffele, and Eckstein (1981).

Intermolecular potentials have been used with less restrictive approximations in statistical mechanics calculations to get improved equations of state, for example, by Ree and van Thiel (1985) in a code named CHEQ, and by Shaw and Johnson (1985). Calculations have also been done by Shaw (1993) using a calculation scheme that simulates the individual collisions among molecules rather than a statistical mechanics approximation. With the enormous increases in computer power we may expect new results from these methods and others. The validity of the equation of state depends, of course, on the accuracy of the assumed interaction potentials.

BKW, BKWR, BKWS, JCZ2, JCZ3, CHEQ, and many others are used in computer programs, such as TIGER, to find the equilibrium composition of the explosive products at the C–J state in a detonation. Equilibrium is found by minimizing the Gibbs free energy for a mixture of assumed products with respect to the proportion of each product present. When the calculations are done wisely the results are surprisingly good, usually within 5% of the experimental detonation velocity, and within 10% of the experimental pressure. Unfortunately, these good results are not good enough. It is very expensive to develop a new explosive, and managers who must choose among the new explosive candidates need to predict the gain in performance over old explosives, which is likely to be only a few percent, with higher accuracy. Work will go on to improve the predictions, but many workers think that the chemical reactions in detonations do not really go to complete equilibrium. If they are right, some conceptual change in the method will be required.

3.13.8. JWL

For design of devices using high explosives, simple equations of state that can be easily recalibrated are needed. The pressure–volume relationship on an isentrope is represented as a sum of functions, each of which is important over part of the pressure range, which can be written as

$$p_S = \sum \Phi_i(v). \tag{3.147}$$

An example of such an equation of state is the JWL form, Kury et al. (1965), which is

$$p_S = Ae^{-R_1 V} + Be^{-R_2 V} + CV^{-(1+\omega)}, \tag{3.148}$$

where A, B, C, R_1, R_2, and ω are all constants to be calibrated (note that A, B, and C have dimensions of pressure, while the other constants are dimensionless), and $V = v/v_0$. The last term on the right-hand side is the low-pressure (large-volume) term, to be compared directly with the polytropic gas isentrope $pv^\gamma =$ constant. The exponential terms are the high-pressure small-volume terms, and usually the user chooses $R_1/R_2 \approx 4$ to make the two terms important in different regions. The energy on the isentrope is obtained by integrating the thermodynamic equation

$$\left(\frac{\partial E}{\partial v}\right)_S = -p \quad \text{or} \quad \left(\frac{\partial e}{\partial V}\right)_S = -p \quad \text{with} \quad e = \rho_0 E,$$

to find

$$e_S(v) = \frac{A}{R_1} e^{-R_1 V} + \frac{B}{R_2} e^{-R_2 V} + \frac{C}{\omega} V^{-\omega},$$

where the subscript S indicates values on the isentrope. Now C can be eliminated using (3.148) to find

$$p = A\left(1 - \frac{\omega}{R_1 V}\right)e^{-R_1 V} + B\left(1 - \frac{\omega}{R_2 V}\right)e^{-R_2 V} + \frac{\omega e}{V}. \tag{3.149}$$

The subscripts S have been removed because we may think of (3.149) being a general incomplete equation of state $p(E, v)$, with C being the constant that selects a particular isentrope. Equivalently, we may obtain the same result by using the thermodynamic relationship

$$\left(\frac{\partial E}{\partial p}\right)_v = \frac{v}{\Gamma} \quad \text{or} \quad \left(\frac{\partial e}{\partial p}\right)_V = \frac{V}{\Gamma},$$

where Γ is the usual Gruneisen gamma, in a Taylor expansion

$$p = p_S(v) + \frac{\Gamma}{V} \cdot [e - e_S(v)] + \cdots,$$

where $p_S(v)$ and $E_S(v)$ are the values on the isentrope, with Γ held constant. This result identifies $\omega = \Gamma$. It is unfortunate that there is a proliferation of symbols, but the change in symbol emphasizes that the function is held constant. Tables of the constants for the JWL equation of state, calibrated for many explosives, are available in Dobratz and Crawford (1985), and the equation of state is easy to recalibrate to make it agree with experiments modeled with a computer program. The JWL equation of state has proven very useful for engineering calculations, and has been widely used.

3.13.9. Linear *U–u*

Solids, especially metals, can support many kinds of waves because they have strength. However, for many purposes, it is an adequate approximation to treat solids as fluids as long as the pressures are several times the yield strengths. We must never forget that solids are not fluids, and that the approximation is not always a good one. The most common description of solids uses the measured Hugoniot curve as a reference, and a Gruneisen relationship to get off the reference curve. For example, it is often true that the Hugoniot curve can be represented over a reasonably wide range of pressures as a simple linear curve in U–u,

$$U = \hat{c} + su, \tag{3.150}$$

which can be used with the Hugoniot relationships (3.8)–(3.17) to obtain, on the Hugoniot curve representing states accessible in a single shock from a given initial state, any of the five variables p, v, E, U, and u in terms of any other. Some useful relationships are:

$$p = p_0 + \rho_0 u(\hat{c} + su), \tag{3.151}$$

$$\frac{v}{v_0} = \frac{\hat{c} + (s-1)u}{\hat{c} + su}, \tag{3.152}$$

$$p - p_0 = \frac{\rho_0 \hat{c}^2 (1 - v/v_0)}{[1 - s(1 - v/v_0)]^2}, \tag{3.153}$$

$$E - E_0 = \frac{\frac{1}{2}\hat{c}^2 (1 - v/v_0)^2}{[1 - s(1 - v/v_0)]^2} + p_0(v_0 - v). \tag{3.154}$$

These equations apply only on the Hugoniot curve.

States off the Hugoniot curve are obtained by using the Hugoniot curve as a reference, and writing the Taylor expansion

$$p(E, v) = \tilde{p}(v) + \left(\frac{\partial p}{\partial E}\right)_v [E - \tilde{E}(v)] + \frac{1}{2}\left(\frac{\partial^2 p}{\partial E^2}\right)_v [E - \tilde{E}(v)]^2 + \cdots$$

for p at any E and v as a displacement in $E - \tilde{E}$ from the Hugoniot curve. Values on the Hugoniot curve are denoted with a superscript tilde to distinguish them. From the definition in (3.93) we have

$$\left(\frac{\partial p}{\partial E}\right)_v = \frac{\Gamma}{v}.$$

It is usually assumed that $\Gamma/v = \text{constant} = \rho_0 \Gamma_0$, so the higher-order terms in the expansion vanish. The equation of state then becomes

$$p(E, v) = \tilde{p}(v) + \rho_0\Gamma_0[E - \tilde{E}(v)]. \tag{3.155}$$

Consider the case of a plane shock wave in a material reaching a boundary of high impedance, so that a shock wave is reflected back into the material. The shocked states will lie on an Hugoniot curve centered at the initial single-shock state, and in the p–u plane the second-shock Hugoniot curve will be to the left of the original curve (as the material is slowed in the reflected shock) and at higher pressure than the first shock. The Hugoniot relation (3.10) is, with subscript 1 denoting the state behind the first shock, and 2 the state behind the reflected shock,

$$E_2 - E_1 = \tfrac{1}{2}(p_2 + p_1)(v_1 - v_2) \tag{3.156}$$

and the result, after substituting (3.156) into (3.155) is

$$p_2 = \frac{\tilde{p}(v_2) + \rho_0\Gamma_0[\frac{1}{2}p_1(v_1 - v_2) - [\tilde{E}(v_2) - E_1]]}{1 - \frac{1}{2}\rho_0\Gamma_0(v_1 - v_2)}. \tag{3.157}$$

The change in particle velocity from state 1 to state 2 is, from (3.13),

$$u_2 - u_1 = -\sqrt{(p_2 - p_1)(v_1 - v_2)}. \tag{3.158}$$

Knowing the state 1, and assuming a series of values for v_2, makes it possible to calculate the Hugoniot curve for the reflected shock. Other state variables can be obtained by using the Hugoniot jump conditions (3.8)–(3.17).

Similarly, if the shock reaches a low-impedance boundary, a rarefaction wave will be reflected. The pressure will decrease and the velocity will increase. To solve for the pressure on the isentrope, with $p(E, v)$ known, we must find $(\partial p/\partial v)_S$. The differential is $dp = (\partial p/\partial E)_v\, dE + (\partial p/\partial v)_E\, dv$ and also $(\partial p/\partial E)_v = \rho\Gamma = \rho_0\Gamma_0$ and $(\partial E/\partial v)_S = -p$, so

$$\left(\frac{\partial p}{\partial v}\right)_S = -\rho_0\Gamma_0 p + \left(\frac{\partial p}{\partial v}\right)_E.$$

Now using the equation of state, (3.155), to find the derivative on the right-hand side, we find $(\partial p/\partial v)_E = d(\tilde{p} - \rho_0\Gamma_0\tilde{E})/dv$ and finally obtain the ordinary differential equation on the isentrope

$$\frac{dp_S}{dv} + \rho_0\Gamma_0 p = \frac{d}{dv}(\tilde{p} - \rho_0\Gamma_0\tilde{E}). \tag{3.159}$$

The solution of this equation is

$$p_S(v_2) = p_S(v_1)e^{-\rho_0\Gamma_0(v_2 - v_1)} + e^{-\rho_0\Gamma_0 v_2}\int_{v_1}^{v_2} e^{\rho_0\Gamma_0 v}\frac{d}{dv}(\tilde{p} - \rho_0\Gamma_0\tilde{E})\, dv. \tag{3.160}$$

The expressions for $\tilde{p}$ and $\tilde{E}$ as functions of v are given in (3.153)–(3.154). The integration has to be carried out numerically. The particle velocity on the isentrope is obtained by using (3.36) $du = c\, dv/v$ and (3.26) $c^2/v^2 =$

$-dp/dv$ to find

$$u_2 - u_1 + \int_{v_1}^{v_2} \sqrt{-\frac{dp_S}{dv}}\, dv, \tag{3.161}$$

where the term in the square root is obtained from (3.159).

Some representative values for ρ_0, $\hat{c}$, s, and Γ_0 are given in an appendix (page 105) to this chapter. The values there are for illustration, and should not be used for serious problems. It is really essential to make sure that the numbers used apply in the region of interest, and that the data are reliable. We must go to the original work to be sure the data are applicable. This check takes time and is a nuisance, but getting the wrong answers is worse.

3.13.10. Walsh Mirror Image

It was noticed long ago that the reflected-shock Hugoniot curve calculated this way, and the expansion isentrope calculated similarly, usually lie very close, in the p–u plane, to the single-shock Hugoniot curve reflected about the first-shock state. That is, the calculated states for shocks or rarefactions centered at u_H are approximately given by

$$p(u;\, u_H) = p_H(2u_H - u), \tag{3.162}$$

where $p_H(u) = p_0 + \rho_0 u(\hat{c} + su)$, the Hugoniot curve for the first shock. There seems to be no general evaluation of the approximation, and values are obtained by computation. This approximation was extremely useful in the early days of shock-wave physics when computing the integrals in (3.160) and (3.161) was much more difficult than it is now. It is still very useful because it allows us to see quickly an approximate result, and it is usually within the accuracy with which the material description is known. A result of special interest is that the velocity for full expansion to zero pressure, called the free-surface velocity, is just twice the particle velocity at the shocked state. The free-surface velocity is accessible for measurement, and the approximation $u_{fs} = 2u_H$ has been widely used.

When using the mirror-image description, we must never forget that it is an approximation to the Gruneisen equation of state (3.139); there will be an inconsistency if it is compared with an accurate calculation.

It is also possible to accept the mirror image description as the correct one for the material, and use it to determine an equation of state. Enig (1963) studied the properties of such an equation of state, getting the thermal parts by assuming a constant thermal expansion coefficient, determined experimentally at ambient pressure. (See Enig's paper for references to the early work by Walsh and others.) He showed that the specific heat goes to infinity in the region of interest, so the equation of state is not a good description throughout the region. Nonetheless, it is useful as long as we know the deficiencies.

3.13.11. Hayes

Hayes (1976) proposed a complete equation of state, giving both thermal and mechanical properties, by assuming that

$$C_v = C_v^0, \qquad \rho\Gamma = \rho_0\Gamma_0, \qquad B_T v^N = B_t^0 v_0^N,$$

where the quantities with subscript or superscript 0's are constants, and $B_T = -v(\partial p/\partial v)_T$ is the isothermal bulk modulus. These assumed quantities give the complete set of three second derivatives for the Helmholtz free energy which are

$$\left(\frac{\partial^2 F}{\partial v^2}\right)_T = \frac{B_T}{v}, \qquad \left(\frac{\partial^2 F}{\partial v\, \partial T}\right) = -\frac{C_v\Gamma}{v}, \qquad \left(\frac{\partial^2 F}{\partial T^2}\right)_v = -\frac{C_v}{T}.$$

The thermodynamic equations can be integrated to get the expression for the free energy

$$F - F_0 = C_v^0\left[(T - T_0)\{1 + \rho_0\Gamma_0(v_0 - v)\} + T\ln\left(\frac{T_0}{T}\right)\right]$$
$$+ \frac{B_T^0 v_0}{N(N-1)}\left[\left(\frac{v_0}{v}\right)^{N-1} - (N-1)\left(1 - \frac{v}{v_0}\right) - 1\right]. \tag{3.163}$$

The initial-state quantities T_0 and v_0 enter as integration constants; p_0 is missing because Hayes chose $p_0 = 0$. Equation (3.163) is a complete and thermodynamically consistent equation of state. The constant N is adjustable to fit experimental shock-wave data; the other constants are chosen to fit thermal and mechanical properties measured at low pressure.

Problem 15. Find $S = -(\partial F/\partial T)_v$ and $p = -(\partial F/\partial v)_T$ for the Hayes equation of state given in (3.163). Use $E - E_0 = F + TS - F_0 - T_0 S_0$ to find $E(T, v)$, and then use the expression for $p(T, v)$ to eliminate T, finally arriving at an expression for $E(p, v)$.

$$E - E_0 = \frac{1}{\rho_0\Gamma_0}\left\{p - \frac{B_T^0}{N}\left[\left(\frac{v_0}{v}\right)^N - 1\right]\right\} - \Gamma_0 C_v^0 T_0\left(1 - \frac{v}{v_0}\right)$$
$$+ \frac{B_T^0 v_0}{N(N-1)}\left[\left(\frac{v_0}{v}\right)^{N-1} - (N-1)\left(1 - \frac{v}{v_0}\right) - 1\right].$$

Note that using the expression for S to eliminate T would give a complete equation of state $E(S, v)$. Use the Hugoniot relations (3.6), (3.7), and (3.13) to make the function $E(p, v)$ into an expression for the Hugoniot curve in terms of U and u, finally finding

$$U^2 = \frac{\left\{\frac{B_T^0}{\rho_0 \Gamma_0 N}[(1-y)^{-N} - 1] + \Gamma_0 C_v^0 T_0 y - \frac{B_T^0 v_0}{N(N-1)}[(1-y)^{-(N-1)} - (N-1)y - 1]\right\}}{\frac{y}{\Gamma_0} - \frac{1}{2}y^2},$$

where $y = u/U$. Now calibrate this expression by choosing N to match the experimental shock data $U_{\text{exp}} = 2690 + 1.72u$ m/s. The parameters used by Hayes are $\rho_0 = 1840$ kg/m^3, $B_T^0 = 12.6$ GPa, $C_v^0 = 1110$ J/kg K, $\Gamma_0 = 1.074$, $T_0 = 293$ K. Hayes chose $N = 5.6$; try to find a better value. Note that the best choice depends on the range where it is used.

3.13.12. Davis

Davis (1985, 1993) proposed an incomplete equation of state for detonation products

$$p = \frac{E}{v}\left[k - 1 + \left\{1 + b\left(1 - \frac{E}{E^S}\right)\right\}F\right], \tag{3.164}$$

where k and b are constants, $E^S = E^S(v)$ is the specific internal energy on the principal isentrope, and $F = F(v)$ is a function chosen to be small when v is large and to increase as v decreases, finally reaching a constant value at very small v. It is easy to see that for v large and $F \sim 0$ the equation of state becomes that for the polytropic gas given by (3.117) with k the large volume adiabatic gamma.

Davis (1993) calibrated this equation, using

$$F(v) = 2a\left(\frac{v}{v_c}\right)^{-n} \Big/ \left[\left(\frac{v}{v_c}\right)^n + \left(\frac{v}{v_c}\right)^{-n}\right],$$

with a, v_c, and n as calibration constants, for detonation products and for overdriven detonations of PBX-9404 and LX-17.

Notice that the expression for p in (3.164) is quadratic in E off the principal isentrope, unlike most other proposed equations of state. Interesting discussion of the thermodynamics and hydrodynamics is given in the papers.

3.13.13. Williamsburg

Byers Brown and Braithwaite (1991) at a meeting held in Williamsburg, VA, proposed an equation of state for detonation products, and have discussed it further in Byers Brown and Braithwaite (1993a, b). The equation of state is based on statistical mechanics. It has been calibrated for PETN and ANFO, and compared with the results of the calibrations of other equations of state. The forms are more complicated than those for other equations of state discussed here, and the reader should refer to the original papers for more information. The work is informative because the papers discuss the underlying physical basis is detail.

3.13.14. Summary

The equations of state discussed above are a representative sample of some that have been used, but the sample is by no means inclusive of all the proposed equations of state. It should be obvious that choosing an equation of state for some particular purpose is not an exercise that has a single definite answer. The choice always depends on what the chooser thinks he or she knows best. The first group of equations of state were modifications to the ideal gas equation of state, and the user must know what form of correction is necessary. The equations of state based on a statistical-mechanics treatment with known forces between the molecules require that the user can make a good approximation for those forces, as well as perform the statistical mechanics accurately, which usually requires making the real forces spherically symmetric, and often making corrections because only forces between just two molecules are included. The Hayes' equation of state requires that we know a good approximation for the second derivatives of the potential. The JWL equation of state is just some arbitrarily chosen terms that can be calibrated to a library of experimental results. Other equations of state are similar constructions. All apply only over a limited range of the thermodynamic variables, and the interesting range for any application is determined by the particular circumstances. Whatever equation of state we may choose for an application, certainly we must look, after the calculations have been done, at the regions of thermodynamic space used, and wonder whether the equation of state is really applicable there. While it is often not clear which equation of state will best serve some purpose, there is no mystery about the equations of state themselves although there may be lengthy algebra.

3.14. Equations of State for Mixtures

Often we need an equation of state for some common material for which no experimental equation of state is available, wet sand, for example. If data for water and quartz are known, it should be possible to find the equation of state, or at least a good approximation, for wet sand. This section discusses the ideas underlying the approximations that have been used. See Meyers (1994) and McQueen (1991) for some other approaches.

Most important, the mixture is treated as a continuum; the structure is not considered. For example, wet sand is considered to move in a shock wave as if the sand grains and the water are homogeneous. The grains are treated as far too small to cause important shock roughness, and their effects on the response of the material, for example, their local contact stresses, are not considered. The Hugoniot curve for wet sand, with 50 μm sand grains, is also the Hugoniot curve for boulders 1 m in diameter in water. A second consequence of the continuum assumption is that the material is all assumed to be moving together; that is, there is just one particle velocity, and it is the same in all the materials of the mixture. No flow of one material past another is taken into account, and the treatment is one dimensional, so no transverse motion is allowed. A third consequence is that there is just one pressure; all the materials are at the same pressure throughout their structure.

A second assumption is ideal mixing; that is, the volume of the mixture at any shock state is the sum of the volumes of the components at that state. There are no effects of attraction or repulsion at the interfaces between the different materials, even if one material dissolves in another. Chemical reaction is not allowed. Consistent with these assumptions, we can take $E_0 = 0$ for the components and the mixture. Also, since the assumptions are reasonable only at high pressures, where the forces between grains or particles are negligible, we also take $p_0 = 0$.

To illustrate the process, we limit the mixture to just two components. Then ideal mixing gives

$$v = m_1 v_1 + m_2 v_2,$$

where m_i $(i = 1, 2)$ is the mass fraction of material i and v_i is the volume at an appropriate pressure and energy, determined from the equation of state, and v is the total specific volume of the mixture. The Hugoniot jump conditions, given by (3.11) and (3.12) apply to the mixture, and are

$$p = \frac{u^2}{v_0 - v},$$

$$E = \tfrac{1}{2}u^2.$$

The energy of the mixture is the sum of the energies of the components, so

$$E = m_1 E_1 + m_2 E_2.$$

Then since the pressure is equal in the two components, the individual equations of state give

$$p = p_1(E_1, v_1)$$

and

$$p = p_2(E_2, v_2).$$

At this point there are six equations and eight variables: v, v_1, v_2, E, E_1, E_2, p, u. One more equation amounts to choosing a particular point to calculate, but still one equation is missing. Another physical assumption is needed. The problem is that energy is shared between the two components. If u is specified, the total specific internal energy of the mixture is known, but how it is shared is not known. One assumption, commonly made, is that the temperature is the same in the two components. This is reasonable if there is time for heat to flow from the hotter one to the cooler one within the shock thickness that might be resolved in the particular experiment. If a complete equation of state that gives $T(E_1, v_1)$ and $T(E_2, v_2)$ is known, which is unlikely, the needed equation is provided. Another assumption is that no energy is transferred by heat conduction from one component to the other; this is reasonable if the experiment is over within a very short time. Under this assumption, $E_1 = E_2 = \frac{1}{2}u^2$. Clearly, some approximation that allows us to choose either of these two, or something in between, is needed. If the specific internal energy were purely thermal, and did not depend on the specific volume, and if the specific heat were constant, then the temperature would be $T = E/C_v$ and setting the temperatures equal would mean that $E_1 = rE_2$ where r is the ratio of the specific heats $r = C_{v_1}/C_{v_2}$. For real experiments and real equations of state, r is nearer to one than to the ratio of specific heats. The required equation is

$$E_1 = rE_2, \qquad r \cong 1,$$

and, with this equation, and a good choice for r, the set of equations is complete. Selecting a value for some variable, usually p or u, allows us to calculate all the others on the Hugoniot curve for the mixture. Depending on the forms for the equations of state for the components, the calculation may require iteration, but programs like Mathematica, Maple, or MathCad do it easily.

The equations combine to give

$$p_1(E_1, v_1) = \frac{u^2}{v_0 - v}, \tag{3.165}$$

$$p_2(E_2, v_2) = \frac{u^2}{v_0 - v}, \tag{3.166}$$

with the auxiliary equations

$$v = m_1 v_1 + m_2 v_2,$$

$$E_1 = \tfrac{1}{2}u^2 \cdot \frac{r}{1 - m_1(1 - r)},$$

and

$$E_2 = \tfrac{1}{2}u^2 \cdot \frac{1}{1 - m_1(1 - r)}.$$

After eliminating v, E_1, and E_2 from (3.165) and (3.166), for any choice of u these are a pair of simultaneous, nonlinear, algebraic equations for v_1 and v_2.

The result of the calculation is a set of values on the single-shock Hugoniot curve for the mixture. The quantities u, p, E, and v are obtained from the equations above; U is obtained from (3.8). Values for states off the Hugoniot curve are obtained from

$$p(E, v) = p_H(v) + \rho_0\Gamma_0[E - E_H(v)],$$

as usual, with

$$\frac{1}{\rho_0\Gamma_0} = \frac{m_1 v_{01}}{\Gamma_{01}} + \frac{m_2 v_{02}}{\Gamma_{02}}.$$

Selection of a reasonable value for r is not as perplexing a problem as it might at first seem. The reason is that it has a relatively small effect on the result. Results for a mixture of materials that are quite similar, copper and zinc 67/33 as in brass, are shown in Figures 3.16 and 3.17. The solid curve is for $r = 1$, the upper dotted curve is for $r = 0.1$, and the lower dotted curve is for $r = 10$, in both figures. While the chosen value for r does make a difference, any value near $r = 1$ will give a useful approximation.

Even for a mixture of materials that are quite different, the value of r has only a modest effect. Figures 3.18 and 3.19 show the results for a 90/10 mixture of copper and polystyrene.

Problem 16. Plot mixture equations of state in the U–u, p–v, and p–u planes for wet sand. Assume the sand with density 2204 kg/m^3 is described by $U = 6456 + 1.491u$ m/s, with $\Gamma = 0.90$. For water, the density is 1000 kg/m^3 and $U = 1800 + 1.600u$ m/s, and $\Gamma = 0.50$. Use mixtures that are

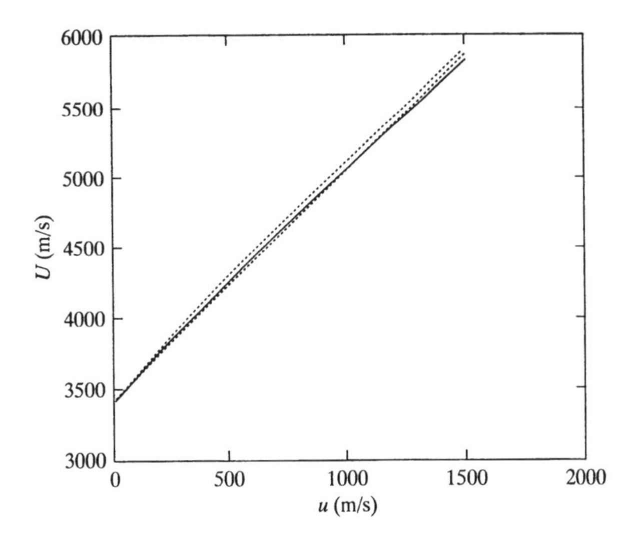

Figure 3.16. Plot of the U–u plane for a brass-like mixture of 67/33 copper and zinc. The solid curve is for $r = 1$, and the dotted curves are for $r = 0.1$ and 10, with the $r = 0.1$ curve the higher one. Copper is material 1.

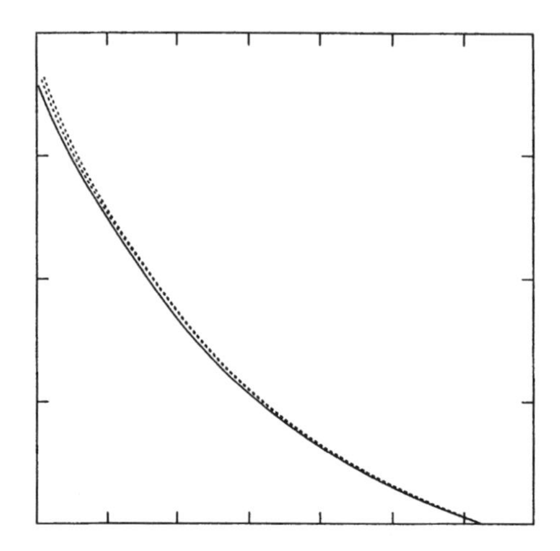

Figure 3.17. Plot of pressure versus volume (in m^3/kg times 10^4) for a brass-like mixture of copper and zinc. The solid curve is for $r = 1$, the upper dotted curve is for $r = 0.1$, and the lower dotted curve is for $r = 10$. Copper is material 1.

75%, 85%, and 95% quartz by weight. See (3.153) and (3.154) for help with the equation of state.

3.15. The Adiabatic Gamma, the Grüneisen Gamma, and the Fundamental Derivative

For the smooth parts of the flow, that is, the flow in regions where there are no shock waves, values are needed for the adiabatic gamma γ and the Grüneisen gamma Γ, as was shown in (3.98), which can rewritten as

$$\frac{\dot{p}}{p} + \gamma \frac{\dot{v}}{v} = \Gamma \cdot \frac{uA + B}{pv}, \tag{3.167}$$

where A represents the viscosity terms and B represents the viscosity and heat conduction terms. Accurate values for the two gammas are not easy to obtain.

The sound speed behind a shock wave can be measured by introducing a rarefaction wave behind the shock, and measuring the time it takes to overtake the shock. The technique is to accelerate a thin plate to the desired speed and then let it fly until the internal ringing from the acceleration has died away, and then let it impact on the material to be measured, producing a shock wave in the material, which is followed by the rarefaction produced when the reflected shock reaches the back surface of the flyer. The rarefaction wave travels at sound speed in the material, and γ can be obtained from the

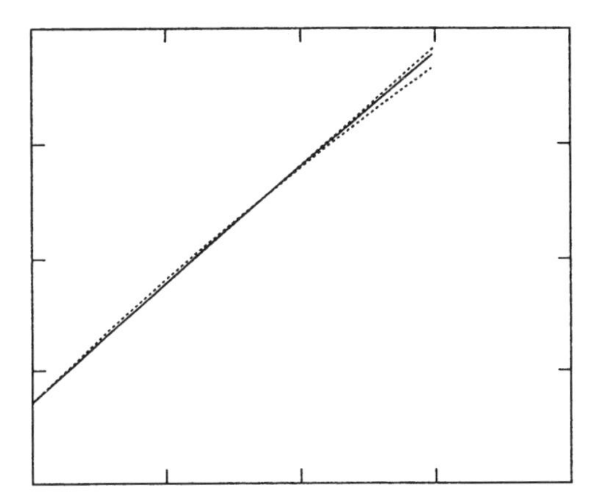

Figure 3.18. Plot of the U–u plane for a 90/10 mixture of copper and polystyrene. The solid curve is for $r = 1$, the lower dotted curve is for $r = 0.1$, and the upper dotted curve is for $r = 10$. Copper is material 1. Note how low U is for small u.

relationship

$$\gamma = \frac{c^2}{pv}. \tag{3.168}$$

The method is described in detail by McQueen (1983, 1991) and by Fritz et al. (1996a). Unfortunately, measurements have been made only for a few materials.

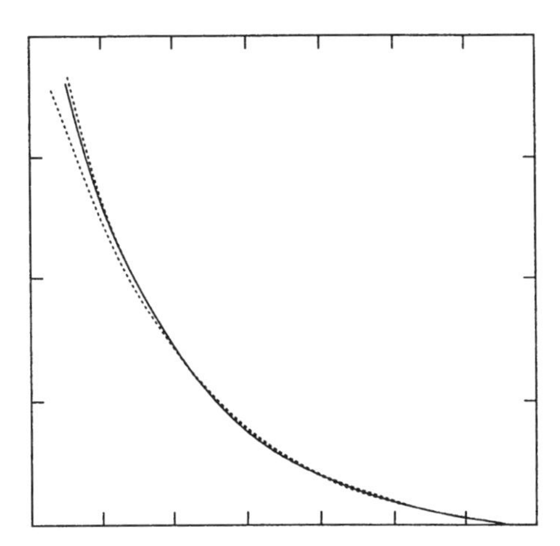

Figure 3.19. Plot of pressure (in GPa) versus volume (in m^3/kg times 10^4) for a 90/10 mixture of copper and polystyrene. The solid curve is for $r = 1$, the upper dotted curve is for $r = 10$, and the lower dotted curve is for $r = 0.1$. Copper is material 1.

The sound speed can also be obtained using a relationship involving the slope of the Hugoniot curve and the Grüneisen gamma. The definitions of the two gammas, when the specific internal energy is expressed as a function of p and v are

$$\Gamma = \frac{v}{E_p} \qquad \text{and} \qquad \gamma = \frac{v}{p} \cdot \frac{E_v + p}{E_p},$$

from (3.98). The total derivative

$$dE = E_v \, dv + E_p \, dp,$$

can be written

$$dE = \frac{(\gamma - \Gamma)p}{\Gamma} \, dv + \frac{v}{\Gamma} \, dp. \tag{3.169}$$

On the Hugoniot curve, then, we can write

$$\frac{dE_{\mathrm{H}}}{dv} = \frac{(\gamma - \Gamma)p_{\mathrm{H}}}{\Gamma} + \frac{v}{\Gamma} \cdot \frac{dp_{\mathrm{H}}}{dv},$$

where the subscript H denotes values on the Hugoniot curve. Then the Hugoniot relationship (3.10)

$$E_{\mathrm{H}} - E_0 = \tfrac{1}{2}(p_{\mathrm{H}} + p_0)(v_0 - v)$$

can be used to eliminate the derivative of E. Rearranging terms gives

$$\gamma p_{\mathrm{H}} = \left(-v \frac{dp_{\mathrm{H}}}{dv}\right)[1 - \tfrac{1}{2}\rho\Gamma(v_0 - v)] + \tfrac{1}{2}\rho\Gamma(p_{\mathrm{H}} - p_0)v. \tag{3.170}$$

This expression gives γ at a point p_{H}, v on the Hugoniot curve. It is convenient to write γp_{H} because it remains finite for a solid or a liquid when the pressure goes to 0 and the sound speed remains constant. Now, if Γ is known, and the Hugoniot curve is known, the adiabatic gamma γ can be obtained. If the Hugoniot curve is obtained from $U = \hat{c} + su$, the equation becomes

$$\gamma p_{\mathrm{H}} = \frac{\hat{c}^2 v[v_0 + s(v_0 - v)]}{[v_0 - s(v_0 - v)]^3} \cdot \left[1 - \frac{\rho\Gamma s(v_0 - v)^2}{v_0 + s(v_0 - v)}\right], \tag{3.171}$$

and the usual assumption is

$$\rho\Gamma = \rho_0\Gamma_0 = \text{constant}.$$

A plot of the sound speed c versus p is shown in Figure 3.20. The values of $\hat{c}$

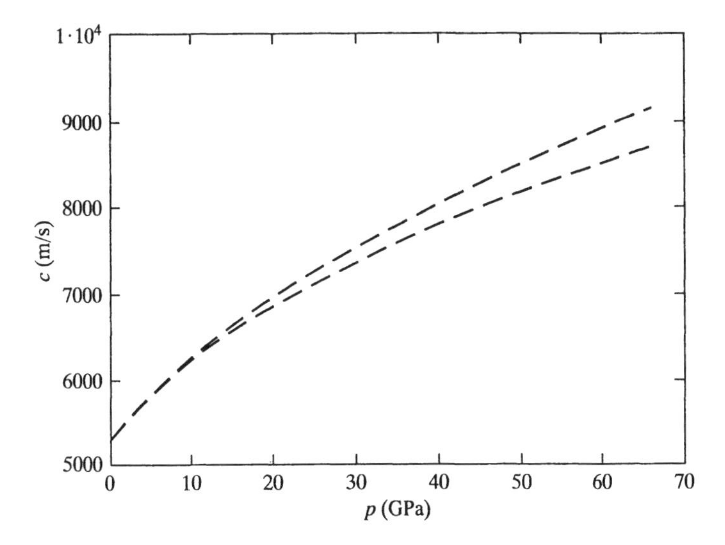

Figure 3.20. Plot of sound speed on the Hugoniot curve as a function of pressure for Dural. The upper curve is for $\Gamma = 1$, and the lower curve for $\Gamma = 2$.

and s are those for Dural, and the upper curve is for $\Gamma = 1$, and the lower curve is for $\Gamma = 2$. It can be seen that even though Γ may be poorly known, the sound speed will still be useful for the lower pressure range. The slight dependence of c on Γ at low pressures occurs because the weak shock waves make little entropy.

It is clear that, if accurate measurements of the sound speed (and therefore of γ) are available, (3.170) can be solved for the Grüneisen gamma Γ. It is equally clear, from Figure 3.20, that the sound speed must be known very accurately to get an accurate measure of Γ.

Usually Γ must be obtained from other measurements. The Grüneisen gamma may be written many different ways; for example,

$$\Gamma = \frac{v}{\left(\dfrac{\partial E}{\partial p}\right)_v} = \left(-\frac{v}{T} \cdot \frac{\partial T}{\partial v}\right)_S = \left(\frac{v}{T} \cdot \frac{\partial p}{\partial S}\right)_v, \tag{3.172}$$

as can all thermodynamic functions. Another possibility is

$$\Gamma = \frac{\beta c^2}{C_{\text{p}}}, \tag{3.173}$$

where β is the volume expansion coefficient $(1/v) \cdot (\partial v/\partial T)_{\text{p}}$, C_{p} is the specific heat at constant pressure, and c is the sound speed. These quantities can be obtained from laboratory measurements, and shock-wave experiments are not needed. Usually the measurements are made at ambient pressure, and the value obtained from them is Γ_0, and the additional assumption $\rho\Gamma = \rho_0\Gamma_0$ is used to find Γ. However, there is a complication; the sound speed is the bulk sound speed appropriate to simple hydrodynamic forces in a fluid. Solids present a problem because they can support more than one type of sound wave.

An isotropic solid, that is, a solid whose properties are the same in any direction, has two independent elastic constants; these are often written as λ and μ, and are called the Lamé constants. A fluid has only one elastic constant, equal to $\lambda + \frac{2}{3}\mu$. (A triclinic crystal, the one with the least symmetry, may have 21 distinct elastic constants, and a cubic crystal has 3; here only isotropic solids are considered.) The isotropic solid can propagate two types of sound waves. The speed of the longitudinal wave is c_{L} and the speed of the transverse wave is c_{T}. It can be shown [Davis (1988)] that

$$c_{\mathrm{L}}^2 = \frac{\lambda + 2\mu}{\rho},$$

$$c_{\mathrm{T}}^2 = \frac{\mu}{\rho},$$

$$c_{\mathrm{B}}^2 = \frac{\lambda + \frac{2}{3}\mu}{\rho} = c_{\mathrm{L}}^2 - \tfrac{4}{3}c_{\mathrm{T}}^2,$$

where c_{B} is the desired bulk speed. Both longitudinal and transverse speeds can be measured in the laboratory.

A disturbing feature of this method for determining Γ is that we might expect to find that $c_{\mathrm{B}} = \hat{c}$, the limiting speed extrapolated back from shock-wave measurements. In fact, $\hat{c}$ for solids is found to be several percent larger than c_{B}, and for liquids as much as 25%. For a discussion, see Pastine and Piascesi (1966). It may be that the high-frequency limit of the sound speed sampled by a shock wave is high in any dissipative material, as discussed by Lighthill (1978).

The fundamental derivative $\mathscr{G}$ is defined in (3.51) and (3.52) as

$$\mathscr{G} = \frac{1}{2}\left[\gamma + 1 - \frac{v}{\gamma}\left(\frac{\partial \gamma}{\partial v}\right)_S\right],$$

or

$$\mathscr{G} = \frac{1}{2}\left[\gamma + 1 - \frac{v}{\gamma}\left(\frac{\partial \gamma}{\partial v}\right)_E + \frac{pv}{\gamma}\left(\frac{\partial \gamma}{\partial E}\right)_v\right].$$

With accurate values for γ these equations will give accurate values for $\mathscr{G}$, but the values for γ may not be available. Fortunately, the derivative terms are usually quite small away from phase changes, and $\mathscr{G} = \frac{1}{2}(\gamma + 1)$ is often adequate.

Any worker in the field of shock-wave physics would do well to be suspicious of values given for γ, Γ, and $\mathcal{G}$ in tables, or those used by other workers.

Hugoniot Curve Data [Fritz (1996b)]

	ρ_0 (kg/m^3)	$\hat{c}$ (m/s)	s	Γ_0
Antimony	6700	1983	1.652	0.60
Barium	3704	760	1.550	0.95
Beryllium	1851	7998	1.124	1.16
Cadmium	8639	2434	1.684	2.27
Copper	8930	3940	1.489	1.99
Gold	19280	3071	1.536	2.97
Iron	7850	3955	1.580	2.01
Lead	11350	2050	1.460	2.77
Magnesium	1745	4536	1.237	1.42
Mercury	13510	1730	1.731	1.96
Nickel	8900	4563	1.445	1.93
Silver	10490	3229	1.595	2.38
Sulfur	2020	3223	0.959	—
Tantulum	16654	3402	1.220	1.82
Tin	7287	2608	1.486	2.11
Tungsten	19256	4022	1.260	1.65
Uranium	18960	2487	1.539	2.18
Zinc	7138	3005	1.581	1.96
Brass	8450	3737	1.428	2.28
24ST Dural	2785	5328	1.338	2.00
304 Stainless Steel	7926	4480	1.510	2.18
PMMA	1186	2650	1.494	0.97
Neoprene	1439	2785	1.419	1.39
Polyethylene	954	3303	1.410	1.64
Teflon	2204	2081	1.623	0.59
Sand	2204	6456	1.491	0.90
Sodium Chloride	2165	3528	1.343	1.60
Water	1000	1800	1.600	0.50

Warning: The data in this table are intended for the practice problems in the text. For other work, please look up the original Hugoniot curve measurements to be sure of having data for your region of interest.

3.16. Problems (Hints and Solutions)

PROBLEM 1 (Shock Wave in Dural).

$$\rho_0 := 2785, \qquad U(u) := 5328 + 1.338u, \qquad p(u) := \rho_0 \cdot U(u) \cdot u,$$

$$v(u) := \frac{1}{\rho_0} - \frac{u^2}{p(u)}, \qquad \rho(u) := \frac{1}{v(u)}, \qquad E(u) := \frac{u^2}{2},$$

$$i := 0 \ldots 7, \qquad u_i := 500(i+1), \qquad j := 0 \ldots 5.$$

$$M_{i,0} := u_i, \qquad M_{i,1} := U(u_i), \qquad M_{i,2} := p(u_i),$$

$$M_{i,3} := v(u_i), \qquad M_{i,4} := \rho(u_i), \qquad M_{i,5} := E(u_i).$$

$$M = \begin{bmatrix} 500 & 5.997 \cdot 10^3 & 8.351 \cdot 10^9 & 3.291 \cdot 10^{-4} & 3.038 \cdot 10^3 & 1.25 \cdot 10^5 \\ 1 \cdot 10^3 & 6.666 \cdot 10^3 & 1.856 \cdot 10^{10} & 3.052 \cdot 10^{-4} & 3.277 \cdot 10^3 & 5 \cdot 10^5 \\ 1.5 \cdot 10^3 & 7.335 \cdot 10^3 & 3.064 \cdot 10^{10} & 2.856 \cdot 10^{-4} & 3.501 \cdot 10^3 & 1.125 \cdot 10^6 \\ 2 \cdot 10^3 & 8.004 \cdot 10^3 & 4.458 \cdot 10^{10} & 2.693 \cdot 10^{-4} & 3.713 \cdot 10^3 & 2 \cdot 10^6 \\ 2.5 \cdot 10^3 & 8.673 \cdot 10^3 & 6.039 \cdot 10^{10} & 2.556 \cdot 10^{-4} & 3.913 \cdot 10^3 & 3.125 \cdot 10^6 \\ 3 \cdot 10^3 & 9.342 \cdot 10^3 & 7.805 \cdot 10^{10} & 2.438 \cdot 10^{-4} & 4.102 \cdot 10^3 & 4.5 \cdot 10^6 \\ 3.5 \cdot 10^3 & 1.001 \cdot 10^4 & 9.758 \cdot 10^{10} & 2.335 \cdot 10^{-4} & 4.282 \cdot 10^3 & 6.125 \cdot 10^6 \\ 4 \cdot 10^3 & 1.068 \cdot 10^4 & 1.19 \cdot 10^{11} & 2.246 \cdot 10^{-4} & 4.453 \cdot 10^3 & 8 \cdot 10^6 \end{bmatrix}.$$

PROBLEM 2 (Shock Wave in Air).

$$p_0 := 10^5, \qquad \rho_0 := 1.29, \qquad \gamma := 1.4, \qquad v_0 := \frac{1}{\rho_0},$$

$$i := 0 \ldots 9, \qquad p_i := 10^6 \cdot (i+1), \qquad j := 0 \ldots 5,$$

$$v(p) := v_0 \cdot \left[\frac{1 - \left(\dfrac{\gamma-1}{\gamma+1}\right)^2}{\dfrac{p}{p_0} + \dfrac{\gamma-1}{\gamma+1}} + \frac{\gamma-1}{\gamma+1} \right], \qquad u(p) := [(p - p_0) \cdot (v_0 - v(p))]^{1/2},$$

$$U(p) := \frac{p}{\rho_0 \cdot u(p)}, \qquad \rho(p) := \frac{1}{v(p)}, \qquad E(p) := \tfrac{1}{2} \cdot u(p)^2.$$

$$M_{i,0} := u(p_i), \qquad M_{i,1} := U(p_i), \qquad M_{i,2} := p_i,$$

$$M_{i,3} := v(p_i), \qquad M_{i,4} := \rho(p_i), \qquad M_{i,5} := E(p_i).$$

$$M = \begin{bmatrix} 717.411 & 1.081 \cdot 10^3 & 1 \cdot 10^6 & 0.203 & 4.918 & 2.573 \cdot 10^5 \\ 1.075 \cdot 10^3 & 1.442 \cdot 10^3 & 2 \cdot 10^6 & 0.167 & 6.003 & 5.782 \cdot 10^5 \\ 1.342 \cdot 10^3 & 1.733 \cdot 10^3 & 3 \cdot 10^6 & 0.154 & 6.486 & 9.005 \cdot 10^5 \\ 1.564 \cdot 10^3 & 1.983 \cdot 10^3 & 4 \cdot 10^6 & 0.148 & 6.758 & 1.223 \cdot 10^6 \\ 1.758 \cdot 10^3 & 2.204 \cdot 10^3 & 5 \cdot 10^6 & 0.144 & 6.934 & 1.546 \cdot 10^6 \\ 1.933 \cdot 10^3 & 2.406 \cdot 10^3 & 6 \cdot 10^6 & 0.142 & 7.056 & 1.869 \cdot 10^6 \\ 2.094 \cdot 10^3 & 2.592 \cdot 10^3 & 7 \cdot 10^6 & 0.14 & 7.146 & 2.192 \cdot 10^6 \\ 2.243 \cdot 10^3 & 2.765 \cdot 10^3 & 8 \cdot 10^6 & 0.139 & 7.215 & 2.515 \cdot 10^6 \\ 2.382 \cdot 10^3 & 2.929 \cdot 10^3 & 9 \cdot 10^6 & 0.138 & 7.27 & 2.837 \cdot 10^6 \\ 2.514 \cdot 10^3 & 3.083 \cdot 10^3 & 1 \cdot 10^7 & 0.137 & 7.314 & 3.16 \cdot 10^6 \end{bmatrix}.$$

Problem 4 (One-Dimensional Expansion of a Gas). Using (3.37) and the definition $c^2 = \gamma p v$, it is straightforward to show that $p/p_0 = (c/c_0)^m$ where the exponent $m = 2\gamma/(\gamma - 1)$:

$$p_0 := 10^5, \qquad c_0 := 331, \qquad \gamma := 1.4, \qquad \rho_0 := 1.28, \qquad v_0 := \frac{1}{\rho_0}.$$

Using (3.40)

$$c(u) := c_0 - \frac{\gamma - 1}{2} \cdot u, \qquad p(u) := p_0 \cdot \left(\frac{c(u)}{c_0}\right)^{(2\cdot\gamma/(\gamma-1))},$$

$$v(u) := v_0 \cdot \left(\frac{p_0}{p(u)}\right)^{1/\gamma}, \qquad \rho(u) := \frac{1}{v(u)}.$$

Within the rarefaction wave the state variables are constant along the lines $dx/dt = u - c$, and integration with boundary conditions $x(0) = 0$ gives $x = (u - c)t$. If we choose a value of u, then c is easily found, and then p, v, and ρ can be found all along the line given by $x = (u - c)t$.

When $u = 100$, $c(100) = 311$, from the equation above. The line $x = (100 - 311)t$ defines the end of the expansion. The piston path is given by $x = 100t$, and the region between these two lines is a constant state where the particle velocity is 100 m/s, and the sound speed is 311 m/s. The other variables have values given by the expressions above.

$$c(100) = 311, \qquad p(100) = 6.464 \cdot 10^4,$$

$$v(100) = 1.067, \qquad \rho(100) = 0.937.$$

In the initial state the particle velocity is 0, and the sound speed is 331 m/s. The rarefaction head moves into the undisturbed gas along the line $x = (0 - 331)t$, or $x = -331t$. Thus the reduction in pressure takes place between the lines $x = -331t$ and $x = -211t$. To the left of $x = -331t$, the state is the constant initial state. To the right of $x = -211t$, and to the left of the piston path $x = 100t$, the state is the final expanded state given by the values above for $u = 100$ m/s.

For any choice of u, values for the other variables can be found from the equations above, and they will be constant along the line $x = (u - c)t$. For example, for $u = 40$,

$$c(40) = 323, \qquad p(40) = 8.426 \cdot 10^4, \qquad v(40) = 0.883, \qquad \rho(40) = 1.133,$$

and these are all constant along the line $x = (40 - 323)t$. At $t = 0.001$ s, $x = -0.283$ m.

Problem 10 (Projectile Hitting a Target). The input data are:

$$u_0 := 1250, \qquad \rho_{\text{Pb}} := 11{,}350, \qquad c_{\text{Pb}} := 2050, \qquad s_{\text{Pb}} := 1.46,$$

$$\rho_{\text{Fe}} := 7850, \qquad c_{\text{Fe}} := 3570, \qquad s_{\text{Fe}} := 1.92.$$

The projectile is described by

$$p_{\mathrm{Pb}}(u) := \rho_{\mathrm{Pb}} \cdot [c_{\mathrm{Pb}} + s_{\mathrm{Pb}} \cdot (u_0 - u)] \cdot (u_0 - u).$$

The target is described by

$$p_{\mathrm{Fe}}(u) := \rho_{\mathrm{Fe}} \cdot (c_{\mathrm{Fe}} + s_{\mathrm{Fe}} \cdot u) \cdot u.$$

After the collision the pressures and particle velocities are equal in both projectile and target.

Guess value: $u := 625$,

$$u_{\mathrm{m}} := \mathrm{root}(p_{\mathrm{Pb}}(u) - p_{\mathrm{Fe}}(u), u), \quad u_{\mathrm{m}} = 598.73, \quad p_{\mathrm{Fe}}(598.73) = 2.218 \cdot 10^{10},$$

$$p_{\mathrm{Pb}}(598.73) = 2.218 \cdot 10^{10}.$$

See Figure 3.5 for a diagram of the p–u plane.

PROBLEM 11 (Symmetric Collision). Input data:

$$\rho_0 := 8450, \quad u_0 := 1850, \quad u := \frac{u_0}{2}, \quad U := \frac{0.010}{1.98 \cdot 10^{-6}}, \quad c := 3726.$$

The results are

$$s := \frac{U - c}{u}, \quad s = 1.432, \quad p := \rho_0 \cdot U \cdot u, \quad p = 3.948 \cdot 10^{10}.$$

PROBLEM 12 (Transmitted Shock). The shock in lead is described as

$$p_{\mathrm{Pb}}(u) := 11350(2050 + 1.46 \cdot u) \cdot u.$$

For $p = 30$ Gpa

Guess value: $u := 1000$,

$$u_{\mathrm{s}} := \mathrm{root}(p_{\mathrm{Pb}}(u) - 3 \cdot 10^{10}, u), \quad u_{\mathrm{s}} = 815.599,$$

$$U_{\mathrm{s}} := 2050 + 1.46 u_{\mathrm{s}}, \quad U_{\mathrm{s}} = 3.241 \cdot 10^3,$$

$$\rho_{\mathrm{s}} := \frac{11350 U_{\mathrm{s}}}{U_{\mathrm{s}} - u_{\mathrm{s}}}, \quad \rho_{\mathrm{s}} = 1.517 \cdot 10^4.$$

The reflected shock in lead is described as

$$pr_{\mathrm{Pb}}(u) := 11{,}350[2050 + 1.46 \cdot (2 \cdot u_{\mathrm{s}} - u)] \cdot (2 \cdot u_{\mathrm{s}} - u),$$

$$v_{\mathrm{s}} := \frac{1}{\rho_{\mathrm{s}}}, \quad p_{\mathrm{s}} := 30 \cdot 10^9, \quad vr_{\mathrm{Pb}}(u) := v_{\mathrm{s}} - \frac{(u_{\mathrm{s}} - u)^2}{pr_{\mathrm{Pb}}(u) - p_{\mathrm{s}}}.$$

The shock in iron is described by

$$p_{\mathrm{Fe}}(u) := 7850(3720 + 1.92 \cdot u) \cdot u.$$

The match between lead and iron is

$$u_{\mathrm{m}} := \mathrm{root}(pr_{\mathrm{Pb}}(u) - p_{\mathrm{Fe}}(u), u), \quad u_{\mathrm{m}} = 778.697, \quad p_{\mathrm{Fe}}(u_{\mathrm{m}}) = 3.188 \cdot 10^{10}.$$

When drawing the reflected shock Hugoniot curve in the p–v plane, remember that it is centered at $p = 30$ GPa and the shock density, 15,170 kg/m^3.

PROBLEM 13 (Reflected Hugoniot Curve as Expansion Isentrope).
Input data:

$$c_{\text{hat}} := 2050, \qquad s := 1.46, \qquad \rho_0 := 11350, \qquad v_0 := \frac{1}{\rho_0}.$$

The 30 Gpa shock state is

$$p_1 := 30 \cdot 10^9, \quad u_1 := 815.599, \quad \rho_1 := 15{,}170, \quad v_1 := \frac{1}{\rho_1}, \quad U_1 := 3241.$$

On the expansion isentrope

$$p(u) := \rho_0 \cdot [c_{\text{hat}} + s \cdot (2 \cdot u_1 - u)] \cdot (2 \cdot u_1 - u).$$

For the lead characteristic back in to the shocked lead the value for c at the shocked state is

$$c_1 := \rho_0 \cdot v_1 \cdot (c_{\text{hat}} + 2 \cdot s \cdot u_1), \qquad c_1 = 3.316 \cdot 10^3.$$

The lead characteristic in the x–t plane is given by

$$x(t) := (u_1 - c_1) \cdot t.$$

For the other information we must integrate (3.59) to find

$$v(u) := v_1 + \frac{v_0}{2 \cdot s} \ln\left[\frac{c_{\text{hat}} + 2 \cdot s \cdot u_1}{c_{\text{hat}} + 2 \cdot s \cdot (2 \cdot u_1 - u)}\right],$$

$$c(u) := \frac{v(u)}{v_0} \cdot [c_{\text{hat}} + 2 \cdot s \cdot (2 \cdot u_1 - u)],$$

$$\frac{1}{v(2 \cdot u_1)} = 1.121 \cdot 10^4, \qquad c(2 \cdot u_1) = 2.075 \cdot 10^3, \qquad p(2 \cdot u_1) = 0.$$

Note that the density in the material near the free surface is 11,210 kg/m^3, just a little less than the initial density.

For the 15 GPa state

Guess value: $u := 1000$

$$u_{15} := \text{root}(p(u) - 15 \cdot 10^9, u), \qquad u_{15} = 1.151 \cdot 10^3,$$

$$p(1151) = 1.499 \cdot 10^{10}, \qquad c(1151) = 2.878 \cdot 10^3,$$

$$v(1151) = 7.346 \cdot 10^{-5}, \qquad \frac{1}{v(1151)} = 1.361 \cdot 10^4,$$

$$x_{15}(t) := (1151 - 2878) \cdot t.$$

PROBLEM 14 (Shock Tube).

$$p_{\mathrm{H0}} := 5 \cdot 10^6, \qquad \rho_{\mathrm{H0}} := 60, \qquad p_{\mathrm{L0}} := 5 \cdot 10^4, \qquad \rho_{\mathrm{L0}} := 0.6,$$

$$\gamma := 1.4, \qquad c_{\mathrm{H0}} := \left(\frac{\gamma \cdot p_{\mathrm{H0}}}{\rho_{\mathrm{H0}}}\right)^{1/2}, \qquad c_{\mathrm{L0}} := \left(\frac{\gamma \cdot p_{\mathrm{L0}}}{\rho_{\mathrm{L0}}}\right)^{1/2},$$

$$c_{\mathrm{H0}} = 341.565, \qquad c_{\mathrm{L0}} = 341.565.$$

For the expansion of the high-pressure section

$$u_{\mathrm{H}}(p) := \frac{2 \cdot c_{\mathrm{H0}}}{\gamma - 1} \cdot \left[1 - \left(\frac{p}{p_{\mathrm{H0}}}\right)^{(\gamma-1)/(2\cdot\gamma)}\right],$$

or, equivalently,

$$p_{\mathrm{H}}(u) := \left[p_{\mathrm{H0}} \cdot \left(1 - \frac{\gamma - 1}{2} \cdot \frac{u}{c_{\mathrm{H0}}}\right)^{(2\cdot\gamma)/(\gamma-1)}\right].$$

The shock in the low-pressure section is given by

$$U_{\mathrm{L}}(u) := \frac{\gamma + 1}{4} \cdot u + \left[\left(\frac{\gamma + 1}{4} \cdot u\right)^2 + c_{\mathrm{L0}}^2\right]^{1/2},$$

$$p_{\mathrm{L}}(u) := p_{\mathrm{L0}} + \rho_{\mathrm{L0}} \cdot U_{\mathrm{L}}(u) \cdot u, \qquad p_{\mathrm{L}}(200) = 1.078 \cdot 10^5.$$

The expanded high-pressure gas has the same pressure and particle velocity as the shocked low-pressure gas at the interface

Guess value: $u := 500$,

$$u_I := \mathrm{root}(p_{\mathrm{H}}(u) - p_{\mathrm{L}}(u), u), \qquad u_I = 554.844, \qquad p_{\mathrm{H}}(u_{\mathrm{I}}) = 3.196 \cdot 10^5,$$

$$p_{\mathrm{L}}(u_{\mathrm{I}}) = 3.196 \cdot 10^5, \qquad c_{\mathrm{H}}(p) := c_{\mathrm{H0}} \cdot \left(\frac{p}{p_{\mathrm{H0}}}\right)^{(\gamma-1)/(2\cdot\gamma)},$$

$$c_{\mathrm{H}}(p_{\mathrm{H}}(u_{\mathrm{I}})) = 230.596, \qquad \rho_{\mathrm{L}}(u) := \rho_{\mathrm{L0}} \cdot \left(\frac{U_{\mathrm{L}}(u)}{U_{\mathrm{L}}(u) - u}\right),$$

$$c_{\mathrm{L}}(u) := \left(\frac{\gamma \cdot p_{\mathrm{L}}(u)}{\rho_{\mathrm{L}}(u)}\right)^{1/2}, \qquad c_{\mathrm{L}}(u_{\mathrm{I}}) = 484.6.$$

Notice that the sound speed in the gas that was originally at high pressure decreased as the gas expanded and cooled, while sound speed in the gas at originally low pressure increased as it was compressed and heated.

PROBLEM 15 (Hayes Equation of State).

Input data

$$\rho_0 := 1840, \qquad v_0 := \frac{1}{\rho_0}, \qquad B := 12.6 \cdot 10^9, \qquad C := 1110,$$

$$\Gamma_0 := \frac{1976}{\rho_0}, \qquad T_0 := 293.$$

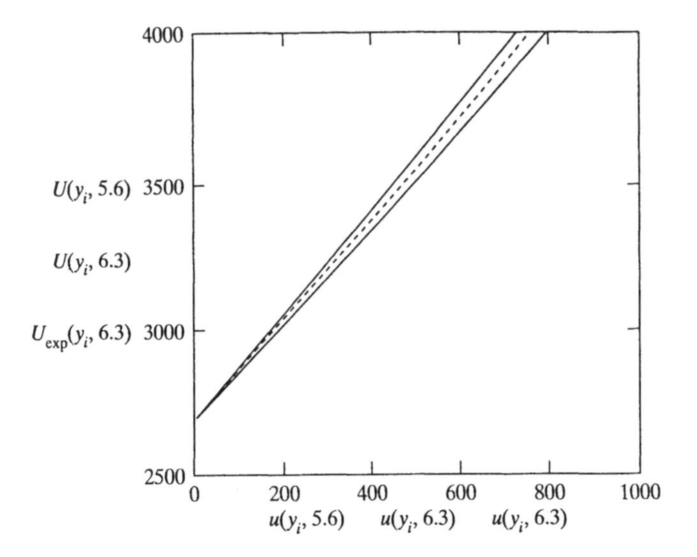

Figure 3.21. The dotted curve is the experimental fit. The low solid curve is for $N = 5.6$, and the upper solid curve is for $N = 6.3$. The lower curve is tangent at the origin. A curve for $N = 6$ is a better fit at larger u, but the differences are so small that the plot is confusing.

The experimental data are fit by

$$c := 2690, \qquad s := 1.72.$$

The Hugoniot curve for the Hayes equation of state can be broken into parts

$$A(y) := \frac{y}{\Gamma_0} \cdot \frac{y^2}{2},$$

$$B_1(y, N) := \frac{B}{\rho_0 \cdot \Gamma_0 \cdot N} \cdot [(1 - y)^{-N} - 1],$$

$$B_2(y) := \Gamma_0 \cdot C \cdot T_0 \cdot y,$$

$$B_3(y, N) := \frac{B \cdot v_0}{N \cdot (N - 1)} \cdot [(1 - y)^{-(N-1)} - (N - 1) \cdot y - 1],$$

$$U(y, N) := \left(\frac{B_1(y, N) + B_2(y) - B_3(y, N)}{A(y)} \right)^{1/2}.$$

To plot the results

$$u(y, N) := y \cdot U(y, N), \qquad i := 0 \ldots 100, \qquad y_i := 0.002 \cdot i,$$

$$U_{\text{exp}}(y, N) := c + s \cdot u(y, N).$$

References

Byers Brown, W. and Braithwaite, M. (1991). Analytical representation of the adiabatic equation for detonation products based on statistical mechanics and intermolecular forces. In *Shock Compression of Condensed Matter* (S.C. Schmidt, R.D. Dick, J.W. Forbes, and D.G. Tasker, eds.). Elsevier, Amsterdam, pp. 325–328.

Byers Brown, W. and Braithwaite, M. (1993a). Williamsburg equation of state for detonation product fluid. In *High Pressure Science and Technology* (S.C. Schmidt, J.W. Shaner, G.A. Samara, and M. Ross, eds.). AIP Press, New York, pp. 73–76.

Byers Brown, W. and Braithwaite, M. (1993b). Development of the Williamsburg equation of state to model non-ideal detonation. *Tenth Symposium (International) on Detonation.* Office of Naval Research, ONR 33395-12, pp. 377–385.

Cowperthwaite, M. and Zwisler, W.H. (1976). The JCZ equations of state for detonation products and their incorporation into the Tiger code. *Sixth Symposium (International) on Detonation.* Office of Naval Research, ACR-221, pp. 162–172.

Courant, R. and Friedrichs, K.O. (1948). *Supersonic Flow and Shock Waves.* Interscience, New York.

Davis, W.C. (1985). Equation of state for detonation products. *Eighth Symposium (International) on Detonation.* Naval Surface Weapons Center, NSWC MP 86-194, pp. 785–795.

Davis, W.C. (1993). Equation of state for detonation products. *Tenth Symposium (International) on Detonation.* Office of Naval Research, ONR 33395-12, pp. 369–376.

Davis, J.L. (1988). *Wave Propagation in Solids and Fluids.* Springer-Verlag, New York, Chap. 8.

Dick, R.D. (1970). Shock wave compression of benzene, carbon disulfide, carbon tetrachloride, and liquid nitrogen. *J. Chem. Phys.*, **52**, 6021–6032.

Dobratz, B.M. and Crawford, P.C. (1985). *LLNL Explosives Handbook.* UCRL-52997, Change 2, pp. 8-21 to 8-23.

Enig, J.W. (1963). A complete E, p, v, T, S thermodynamic description of metals based on the p, u mirror-image approximation. *J. Appl. Phys.*, **34**, 746–754.

Faber, T.E. (1995). *Fluid Dynamics for Physicists.* Cambridge University Press, Cambridge.

Fickett, W. and Davis, W.C. (1979). *Detonation.* University of California Press, Berkeley, CA.

Fritz, J.N., Hixson, R.S., Shaw, M.S., Morris, C.E., and McQueen, R.G. (1996a). Overdriven-detonation and sound speed measurements in PBX-9501 and the "thermodynamic" pressure. *J. Appl. Phys.*, **80**, 6129–6141.

Fritz, J.N. (1996b). Private communication.

Hayes, D.B. (1976). A $P^n t$ detonation criterion from thermal explosion theory. *Sixth Symposium (International) on Detonation.* Office of Naval Research ACR-221, pp. 76–81.

Hirschfelder, J.O., Curtiss, C.F., and Bird, R.B. (1954). *Molecular Theory of Gases and Liquids.* Wiley, New York, p. 154.

Hobbs, M.L. and Baer, M.R. (1993). Calibrating the BKW–EOS with a large product species data base and measured C–J properties. *Tenth Symposium (International) on Detonation.* Office of Naval Research ONR 33395-12, pp. 409–418.

Jeans, J.H. (1954). *Dynamical Theory of Gases*. Dover, New York, §172.

Kury, J.W., Hornig, H.C., Lee, E.L., McDonnel, J.L., Ornellas, D.L., Finger, M., Strange, F.M., and Wilkins, M.L. (1965). Metal acceleration by chemical explosives. *Fourth Symposium (International) on Detonation*. Office of Naval Research ACR-126, pp. 3–13.

Landau, L. and Lifshitz, E.M. (1958). *Statistical Physics*. Pergamon, London, pp. 91–92.

Lighthill, J. (1978). *Waves in Fluids*. Cambridge University Press, Cambridge, Chaps. 1.13 and 3.6.

McQueen, R.G. (1991). Shock waves in condensed media: Their properties and the equation of state of materials derived from them. *Proceedings of the International School of Physics "Enrico Fermi,"* Course 113 (S. Eliezer and R.A. Ricci, eds.). North-Holland, Amsterdam, pp. 101–215.

McQueen, R.G., Fritz, J.N., and Morris, C.E. (1983). The velocity of sound behind strong shock waves in 2024 Al. In *Shock Waves in Condensed Matter* (J.R. Asay, R.A. Graham, and G.K. Straub, eds.). North-Holland, Amsterdam, pp. 95–98.

Mader, C.L. (1979). *Numerical Modeling of Detonations*. University of California Press, Berkeley, CA.

Menikoff, R. and Plohr, B.J. (1989). The Riemann problem for fluid flow of real materials. *Rev. Mod. Phys.*, **61**, 75–130.

Meyers, M.A. (1994). *Dynamic Behavior of Materials*, Wiley, New York.

Pastine, D.J. and Piascesi, D. (1966). The existence and implications of curvature in the relation between shock and particle velocities for metals. *J. Phys. Chem. Solids*, **27**, 1783.

Rabie, R.L., Fowles, G.R., and Fickett, W. (1979). The polymorphic detonation. *Phys. Fluids*, **22**, 422–435.

Ree, F.H. and van Thiel, M. (1985). Detonation behavior of LX-14 and PBX-9404: Theoretical Aspect. *Seventh Symposium (International) on Detonation*. Naval Surface Weapons Center, NSWC MP 82-334, pp. 501–512.

Sentman, L.H., Strehlow, R.A., Haeffele, B., and Eckstein, A. (1981). Development of a single species equation of state for detonation products suitable for hydrocode calculations. *Seventh Symposium (International) on Detonation*. Naval Surface Weapons Center, NSWC MP 82-334, pp. 721–732.

Shaw, M.S. and Johnson, J.D. (1985). The theory of dense molecular fluid equations of state with applications to detonation products. *Eighth Symposium (International) on Detonation*. Naval Surface Weapons Center NSWC MP 86-194, pp. 531–539.

Shaw, M.S. (1993). Direct Monte Carlo simulation of the chemical equilibrium composition of detonation products. *Tenth Symposium (International) on Detonation*. Office of Naval Research ONR-33395-12, pp. 401–408.

CHAPTER 4

Introduction to Detonation Physics

Paul W. Cooper

In this chapter we will examine the propagation of a steady-state detonation wave and the parameters affecting some aspects of detonation. The non-steady-state aspects, those that pertain to initiating detonation, are described elsewhere.

When we look at the behavior of steady-state detonations we find four phenomena common to all explosives:

1. The propagation wave velocity is greater than the speed of sound of the unreacted material into which the wave is traveling.
2. The wave velocity is constant in any given specimen of explosive.
3. The wave velocity is proportional to the density of the explosive material.
4. The wave velocity is lower in smaller diameter specimens of the same explosive material, and propagation completely fails below some minimum critical diameter.

The first two of the above observations can be quantitatively described by a relatively simple model based on derivation from first principle physics. The third and fourth observations cannot be quantitatively modeled by first principle physics but can be correlated empirically to a body of experimentally derived data.

4.1. Nomenclature

The terms and symbols that are used in this chapter are defined as follows:

Term	Definition	Dimensions
A, A_L	constants in detonation velocity versus diameter relationships	(mm)
A, B, C	constant parameters in the JWL equation of state	(GPa)
d	diameter	(mm)
d_f	detonation failure diameter	(mm)

D	detonation velocity	(km/s)
D_i, D'	detonation velocity at TMD	(km/s)
D_∞	detonation velocity at infinite diameter	(km/s)
P	pressure	(GPa)
P_{CJ}	detonation pressure at the Chapman–Jouget state	(GPa)
R_1, R_2	constant parameters in the JWL equation of state	None
R, R_C	radii in D versus charge radius equation	(mm)
u	particle velocity	(km/s)
u_{CJ}	detonation particle velocity at the Chapman–Jouget state	(km/s)
U	shock-wave velocity	(km/s)
V	specific volume	(cm^3/g)
V_0	specific volume at initial or unreacted state	(cm^3/g)
V_{CJ}	specific volume at the Chapman–Jouget state	(cm^3/g)
ρ	density	(g/cm^3)
ρ_0	density at initial or unreacted state	(g/cm^3)
ρ_{CJ}	density at the Chapman–Jouget state	(g/cm^3)
γ	ratio of specific heats	none
ω	constant parameter in the JWL equation of state	none

4.2. The Simple Model

Let us look first at the physical model. To make the mathematics tractable, we will make some simplifying assumptions:

1. The model will deal with a uniaxial wave, that is, the wave front is planar and has no lateral boundaries (the wave front is an infinite plane), and is traveling in a direction normal to the wave front.
2. The wave front is discontinuous, it is handled exactly as the jump-discontinuity in simple nonreactive shock waves.
3. The reaction product gasses exiting behind the detonation front are in chemical and thermodynamic equilibrium and the chemical reaction is complete.
4. The chemical reaction is completed instantly, and therefore the reaction zone length is zero.
5. The detonation process is steady state; the wave velocity is constant and the products leaving the detonation front are at the same state regardless of the position of the wave in distance or time.

With these constraints, the detonation is seen as a shock wave moving through an explosive. The shock front compresses and heats the explosive, which initiates chemical reaction. The exothermic reaction is completed instantly. The energy liberated by the reaction feeds the shock front and drives it forward. At the same time, the gaseous products behind this shock wave are expanding. A rarefaction moves forward toward the shock front. The shock front, chemical reaction, and the leading edge of the rarefaction are all in equilibrium, so they are all moving at the same speed, which we call the *detonation velocity*, *D*. Therefore, the front of the shock does not change shape (pressure remains constant) with time, and the detonation velocity does not change with time.

Let us look at the detonation jump condition from the unreacted explosive to the shock compressed reaction product gases on the pressure–specific volume (P–V) plane, Figure 4.1.

As we can see in Figure 4.1, we are dealing with two materials in a detonation jump condition, the unreacted explosive and the completely reacted gaseous detonation products. Not only are we jumping from one physical state to another, but also to a new chemical state. In Figure 4.1 we see the initial state at point (A), the unreacted explosive; we see also the state at point (C) the jump condition to the fully shocked but as yet unreacted explosive; and on another Hugoniot, the state (B) of the detonation–reaction products.

The Chapman–Jouguet (C–J) state is at the point where the Rayleigh line (a straight line connecting the initial and final shock states on a P–V plane) is tangent to the Hugoniot of the detonation–reaction products. Chapman and

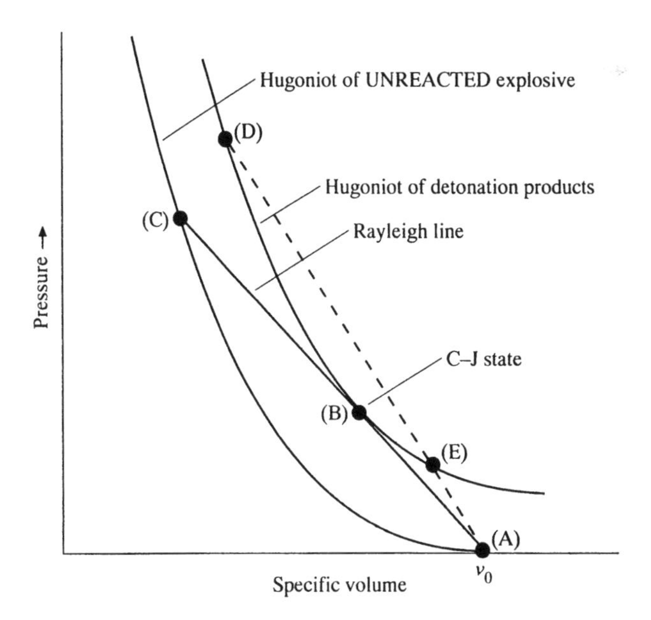

Figure 4.1. Detonation jump condition from the unreacted state to the C–J state on the P–V plane [copied with permission, Cooper (1996)].

Jouguet (at the beginning of the twentieth century) independently observed and reported on this phenomenon, noting that the wave velocity behind a detonation front must be at the local sonic velocity, that of the hot, highly compressed gaseous products. Also, they noted that in order to maintain equilibrium (steady state) this must be the minimum permissible velocity for these conditions. If the jump condition were such that the Rayleigh line intersected the products' Hugoniot at a slope greater than that of the tangent, then two states would be possible for the products, one at each of the two points where the Hugoniot is intersected by the dashed line in Figure 4.1.

The slope of the Hugoniot curve at any point is U^2/V^2; comparing slopes on this plane is therefore effectively comparing velocities. The slope of the Hugoniot curve of the detonation reaction products at state (D) in Figure 4.1 is greater than the slope of the jump Rayleigh line (the Rayleigh line slope for the detonation is D^2/V^2). Therefore, the reaction zone and rarefaction would be overtaking the shock front, thus violating our statement above that these are all at the same velocity. So this state is not in equilibrium and therefore not permissible under our original constraints.

At point (E) in Figure 4.1, the slope of the Hugoniot, and hence the rarefaction wave velocity, is lower than that of the Rayleigh line; therefore, the rarefaction would be slower than the detonation front, making the reaction zone continuously spread out in time. This state is not in equilibrium and therefore is not permissible according to our constraints. The only place on the Hugoniot of the products where the slope of the Hugoniot equals the slope of the Rayleigh line (and therefore the reaction zone, rarefaction front, and shock front are all at the same velocity), is at the tangent point, the C–J state.

We see that the C–J point is at the state of the products behind the detonation front. What about point (C) on Figure 4.1, the Rayleigh line intersection with the Hugoniot of the unreacted explosive? This point is called the Von Neumann spike and is the shock state that initiates reaction. For the purpose of this simple model, the Von Neumann spike is ignored and the reaction zone thickness is assumed to be zero.

The rarefaction wave which brings the product gases from the C–J state to the fully expanded state at ambient pressure is called the Taylor wave. This is shown in Figure 4.2.

If the explosive has heavy rear and/or side confinement, the gases cannot expand as freely as unconfined gases; thus, the Taylor wave is higher and longer than if the explosive were not confined. When the explosive is very thick (along the detonation axis) the Taylor wave is higher. When the explosive is very thin in the direction of the detonation, and there is little rear or side confinement, the Taylor wave is lower. The actual shape of the Taylor wave is determined by a combination of the isentrope for expansion of the detonation gases, the charge size, and the degree of confinement.

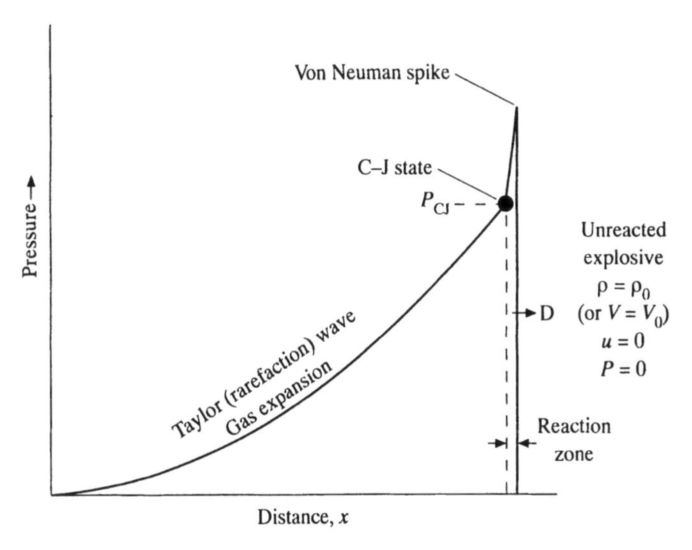

Figure 4.2. Profile of a detonation wave [copied with permission, Cooper (1996)].

4.3. The Jump Equations

As with nonreactive shock waves, the assumption of conservation of mass, momentum, and energy across the shock front leads to three simple equations, the Rankine–Hugoniot Jump equations:

the mass balance

$$\frac{\rho_{CJ}}{\rho_0} = \frac{V_0}{V_{CJ}} = \frac{D}{D - u_{CJ}}, \tag{4.1}$$

the momentum balance

$$P_{CJ} = \rho_0 u_{CJ} D = \frac{u_{CJ} D}{V_0}, \tag{4.2}$$

and the energy balance

$$E_{CJ} - E_0 = \tfrac{1}{2}(P_{CJ} + P_0)(V_0 - V_{CJ}), \tag{4.3}$$

where ρ_{CJ} and ρ_0 are density at the C–J and initial states, respectively;
V_{CJ} and V_0 are specific volume at the C–J and initial states;
D is the detonation velocity;
u_{CJ} is the particle velocity at the C–J state;
P_{CJ} is the detonation or C–J pressure; and
E_0 and E_{CJ} are the specific internal energy at the initial and C–J states.

From hydrodynamics we know that γ, the ratio of specific heats of a gas, is related to a change of specific volume by

$$\gamma = \frac{V}{\Delta V},$$

and so, in the case of the detonation, becomes

$$\gamma = \frac{V_{CJ}}{V_0 - V_{CJ}}. \tag{4.4}$$

Combining this with (4.1) and (4.2), allows us to express the momentum balance in terms of γ

$$P_{CJ} = \frac{\rho_0 D^2}{1+\gamma} = \frac{D^2}{V_0(1+\gamma)}. \tag{4.5}$$

Knowing the jump equations in combination with the P–V and P–u Hugoniots and Isentropes (the isentropic expansion from the C–J state) allows us, as with nonreactive shock waves, to calculate changes in energies (area under the Hugoniot on the P–V plane) and shock interactions (impedance matching on the P–u plane).

4.4. The Detonation Product *P–V* Isentrope

The quantitative evaluation or construction of the expansion isentrope from the C–J state cannot be done from the C–J state parameters alone. We need either an independent equation of state (EOS) or empirical data for the expansion of the gasses. This can be handled in several ways.

One method is to write all of the chemical equilibrium equations for all possible species in the reaction product gases and solve these with thermochemical analogues. The isentropic expansion can then be estimated having the equilibrium energy and gas quantities along with the Rankine–Hugoniot jump equations. This is what is done in thermochemical/hydrodynamic computer codes such as BKW (Mader, 1967) and RUBY (Levine and Sharples, 1962) and the latter's offspring TIGER (Cowperthwaite and Zwisler, 1973), CHEQ (Nichols and Ree, 1990), and CHEETAH (Fried, 1995).

There are several different forms of empirical equation of states which were fitted to experimental data for specific explosives at specific densities. Among these are the BKW (Becker–Kistiakowski–Wilson), JCZ (Jacobs–Cowperthwaite–Zwisler), and JWL (Jones–Wilkins–Lee). The last form is particularly adaptable to simple calculation of the isentrope if we can find the values of the constants in the equation of state:

$$P = Ae^{-R_1 V/V_0} + Be^{-R_2 V/V_0} + C\left(\frac{V}{V_0}\right)^{-(1+\varpi)}. \tag{4.6}$$

Table 4.1 shows values of the JWL constants which have been fitted to experimental data for several different explosives.

Table 4.1. The JWL equation of state (EOS) parameters [data from Dobratz (1981)].

	C–J parameters			JWL equation of state parameters					
Explosive	ρ_0 (g/cm^3)	P_{CJ} (GPa)	D (km/s)	A (GPa)	B (GPa)	C (GPa)	R_1	R_2	ω
Comp B	1.717	29.5	7.98	524.2	7.678	1.082	4.20	1.10	0.34
Comp C-4	1.601	28.0	8.193	609.8	12.95	1.043	4.50	1.40	0.25
Cyclotol 77/23	1.754	32.0	8.25	603.4	9.924	1.075	4.30	1.10	0.35
Detasheet C[a]	1.480	19.5	7.00	349.0	4.524	0.854	4.10	1.20	0.30
Explosive D[a]	1.42	16.0	6.5	300.7	3.94	1.00	4.3	1.2	0.35
H-6	1.76	24.0	7.47	758.1	8.513	1.143	4.9	1.1	0.20
HMX	1.891	42.0	9.11	778.3	7.071	0.643	4.20	1.00	0.30
HNS	1.00	7.5	5.10	162.7	10.82	0.658	5.4	1.8	0.25
HNS	1.40	14.5	6.34	366.5	6.750	1.163	4.8	1.40	0.32
HNS	1.65	21.5	7.03	463.1	8.873	1.349	4.55	1.35	0.35
NM	1.128	12.5	6.28	209.2	5.689	0.770	4.40	1.20	0.38
Octol 78/22	1.821	34.2	8.48	748.6	13.38	1.167	4.50	1.20	0.38
Pentolite 50/50	1.70	25.5	7.53	540.9	9.373	1.033	4.5	1.1	0.35
PETN	0.880	6.2	5.17	348.6	11.29	0.941	7.00	2.00	0.24
PETN	1.260	14.0	6.54	573.1	20.16	1.267	6.00	1.80	0.28
PETN	1.500	22.0	7.45	625.3	23.29	1.152	5.25	1.60	0.28
PETN	1.770	33.5	8.30	617.0	16.93	0.699	4.40	1.20	0.25
Tetryl	1.730	28.5	7.91	586.8	10.67	0.774	4.40	1.20	0.28
TNT	1.630	21.0	6.93	371.2	3.231	1.045	4.15	0.95	0.30

[a] No cylinder test data.

Another method of estimating the isentrope is to assume that γ is constant over the range from the C–J state down to the pressure to which the expansion ends, and then use the "gamma law"

$$P_1 V_1^\gamma = P_2 V_2^\gamma \qquad \text{or} \qquad P = P_{CJ}\left(\frac{V_{CJ}}{V}\right)^\gamma. \tag{4.7}$$

The advantage of the gamma law is that values of γ are easily estimated from the C–J state parameters. The disadvantage is that it does not track the real expansion accurately over large ranges of pressure, especially at the lower pressures, as seen in Figure 4.3.

However, for the purpose of energy calculations from one shock state to another, the constant gamma expansion provides a surprisingly close estimate.

4.5. Detonation Velocity and Density

At the beginning of this chapter we noted that one of the phenomena common to all explosives is "The wave velocity is proportional to the density of

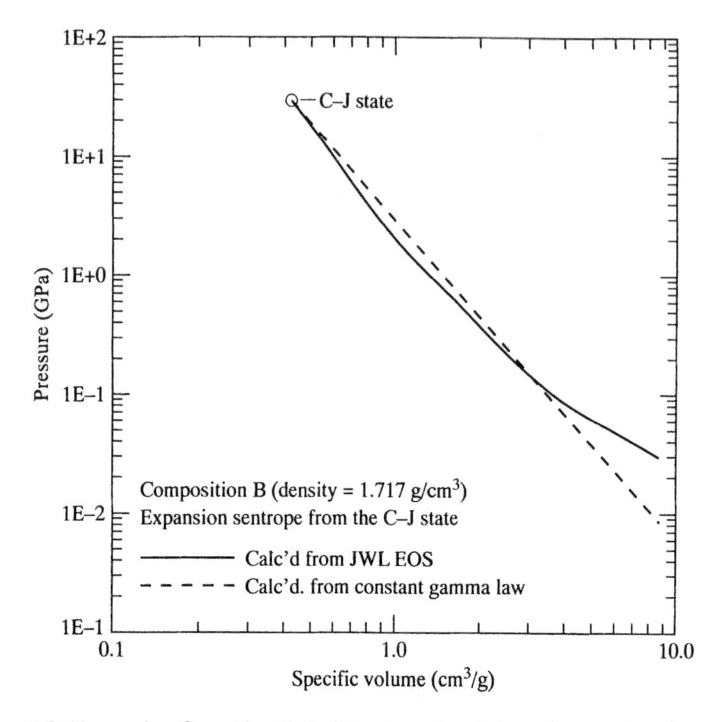

Figure 4.3. Expansion from the C–J state along the detonation product isentrope.

the explosive material." The dependence of detonation velocity upon explosive initial density is close to linear for most explosives over a fairly broad range of density, that is,

$$D = j + k\rho_0, \tag{4.8}$$

where j and k are constants particular to each explosive. For most explosives, k or $\Delta D/\Delta\rho$ is in the neighborhood of around 3 (km/s)/(g/cm^3). A typical example of this is seen in Figure 4.4 which shows data from over 180 experiments with 3 in diameter specimens of TNT.

The values of j and k cannot be predicted from first principles or from the C–J state parameters, but there is a reasonable body of data to which empirical correlations can be made. Table 4.2 provides values for several explosives.

There are other methods of estimating D at ρ. One method that relies only on the chemical structure (Rothstein and Petersen, 1979, 1981) yields values for D' (D at the theoretical maximum density, TMD, of the explosive)

$$D' = \frac{F - 0.26}{0.55},$$

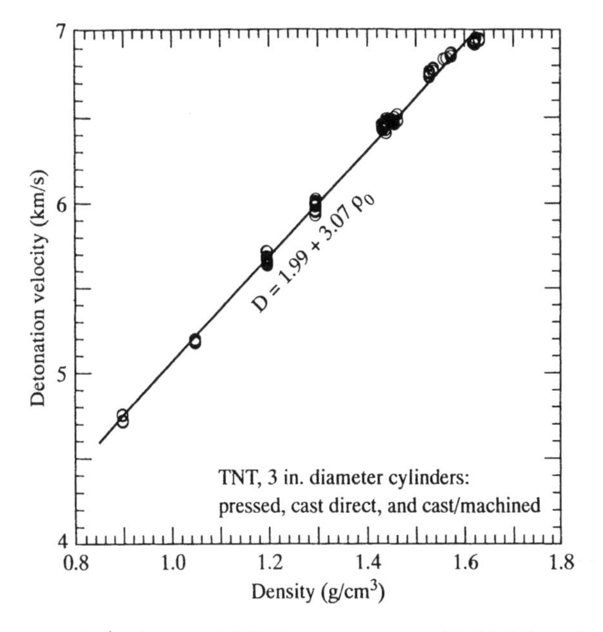

Figure 4.4. Detonation velocity of TNT over a range of initial densities [data from Mader (1982)].

and

$$F = \frac{100}{\text{MW}}\left[n(\text{O}) + n(\text{N}) + n(\text{F}) - \frac{n(\text{H}) - n(\text{HF})}{2n(\text{O})} + \frac{A}{3} - \frac{n(B/F)}{1.75} - \frac{n(C)}{2.5} - \frac{n(D)}{4} - \frac{n(E)}{5}\right] - n(G),$$

where MW = HE molecular weight;
G = 0.4 liquid explosives and 0 for solid explosives;
A = 1 for aromatic compounds, otherwise it is 0;

(the following are for 1 mol of the explosive composition):

$n(\text{O})$ = number of oxygen atoms;
$n(\text{N})$ = number of nitrogen atoms;
$n(\text{H})$ = number of hydrogen atoms;
$n(\text{F})$ = number of fluorine atoms;
$n(\text{HF})$ = number of hydrogen fluoride molecules that can possibly form from available hydrogen;
$n(B/F)$ = number of oxygen atoms in excess of those available to form CO_2 and H_2O and/or the number of fluorine atoms in excess of those available to form HF;
$n(C)$ = number of oxygen atoms doubly bonded directly to carbon (as in a ketone or ester);

Table 4.2. Effect of initial density on detonation velocity [data from Dobratz (1981)].

Explosive	Const. j	Const. k	Range of applicability
Ammonium	1.146	2.276	$0.55 < \rho < 1.0$
perchlorate	−0.45	4.19	$1.0 < \rho < 1.26$
BTF	4.265	2.27	
DATB	2.495	2.834	
HBX-1	0.063	4.305	
LX-04-1	1.733	3.62	
Nitroquanidine	1.44	4.015	$0.4 < \rho < 1.63$
PBX-9010	2.843	3.1	
PBX-9404	2.176	3.6	
PETN	2.14	2.84	$\rho < 0.37$
PETN	1.82	3.7	$0.37 < \rho < 1.65$
PETN	2.89	3.05	$1.65 < \rho$
Picric acid	2.21	3.045	
RDX	2.56	3.47	
TATB	0.343	3.94	$1.0 < \rho$
TNT	1.67	3.342	$1.2 < \rho$

$D = j + k\rho_0$; D in km/s; ρ_0 in g/cm^3.

$n(D)$ = number of oxygen atoms singly bonded directly to carbon (as in >C—O—R and where R can be —H, $—NH_4$, —C, etc.); and

$n(E)$ = number of nitrate groups existing either as a nitrate ester or as a nitric acid salt such as hydrazine mononitrate.

$$[n(\mathrm{H}) - n(\mathrm{HF})]/2n(\mathrm{O}) \quad \text{is } 0 \quad \text{if } n(\mathrm{O}) = 0 \quad \text{or if} \quad n(\mathrm{HF}) \geq n(\mathrm{H}).$$

Yet another method of estimating D at ρ (Kamlet and Jacobs, 1968) invokes the thermochemical properties of an idealized detonation reaction and uses these for estimation of both D and P_{CJ}. The idealized reaction assumes that all of the oxygen available in the explosive molecule is used in the formation of water and carbon *di*oxide, and that none of the oxygen is used in the formation of carbon *mon*oxide.

$$\mathrm{C}_c\mathrm{H}_h\mathrm{N}_n\mathrm{O}_o \rightarrow (n/2)\mathrm{N}_2 + (h/2)\mathrm{H}_2\mathrm{O} + (o/2 - h/4)\mathrm{CO}_2 + (c - o/2 + h/4)\mathrm{C}.$$

Their empirical relationships are

$$\Phi = NM^{1/2}Q^{1/2},$$

$$D = 1.01\Phi^{1/2}(1 + 1.3\rho_0),$$

and

$$P_{\mathrm{CJ}} = 1.56\rho_0^2\Phi,$$

where N = moles of gaseous detonation products per g of HE (mol gas/g HE);
M = average molecular weight of detonation product gas (g gas/mol gas); and
Q = chemical energy of detonation reaction (cal/g).

The term Q is the heat of reaction of the idealized reaction shown above and is found from

$$Q = \frac{\Delta H_f^0(\text{reaction products}) - \Delta H_f^0(\text{HE})}{\text{HE formula weight}},$$

where ΔH_f^0 is the standard thermochemical heat of formation.

A relatively accurate method of estimating detonation velocities for CHNO explosives only (Stine, 1990) is based upon using the atomic composition of either a pure or mixed explosive, along with the explosive's density and heat of formation. In this method the explosive composition is defined as $C_aH_bN_cO_d$, where a, b, c, and d are atomic fractions (i.e., a is the number of carbon atoms in the molecular formula divided by the total number of all atoms in the molecular formula, etc.)

$$D = 3.69 + (-13.85a + 3.95b + 37.74c + 68.11d + 0.6917\Delta H_f)\left(\frac{\rho}{M}\right),$$

where ρ is initial HE density (g/cm^3), ΔH_f is the heat of formation of the explosive (kcal/mol), and M is the molecular or gram formula weight.

The detonation velocity of mixtures of explosives with other materials including other explosives can be estimated by a simple summing equation (Urizar, 1981). This relationship is

$$D = \sum_i V_i D_i, \tag{4.9}$$

where V_i is the volume fraction of constituent i in the mixture, and D_i is its "Characteristic Velocity." The characteristic velocity of the explosive component is its detonation velocity at TMD. Table 4.3 lists characteristic velocities of several pure explosive compounds along with those of a number of inert additives.

This mixing scheme can also yield an estimate of detonation velocity dependence on density by solving equation (4.9) for the mixture of a pure explosive with a void (whose density is assigned, in Table 4.3, a value of 0 and whose velocity is given as 1.5 km/s). The relation obtained in this manner is

$$D = 1.5 + \left(\frac{D' - 1.5}{\rho_{\text{TMD}}}\right)\rho.$$

Knowledge of the detonation velocity is important because it is the easiest of the C–J state parameters to measure accurately; and, in combination with

Table 4.3. Densities and characteristic velocities D_i for use in the Urizer equation [data from Dobratz (1981)].

Material	Density, ρ (g/cm^3)	Characteristic velocity, D_i (km/s)
Polymers and Plasticizers		
Adiprene L	1.15	5.69
Beeswax	0.92	6.50
Estane 5740-X2	1.2	5.52
Exon-400 XR61	1.7	5.47
Exon-454 (85/15 wt% PVC/PVA)	1.35	4.90
FEFO (as constituent to ~35%)	1.60	7.20
Fluoronitroso rubber	1.92	6.09
Halowax 1014	1.78	4.22
Kel-F elastomer	1.85	5.38
Neoprene CNA	1.23	5.02
Paracril BJ (Buna-N nitrile rubber)	0.97	5.39
Polyethylene	0.93	5.55
Polystyrene	1.05	5.28
Sylgard 182	1.05	5.10
Teflon	2.15	5.33
Viton A	1.82	5.39
Inorganic Additives		
Air or void	Nil	1.5
Al	2.70	6.85
$Ba(NO_3)_2$	3.24	3.80
$KC10_4$	2.52	5.47
$LiC10_4$	2.43	6.32
LiF	2.64	6.07
Mg	1.74	7.2
Mg/Al alloy (61.5/38.5 wt%)	2.02	6.9
$NH4C10_4$	1.95	6.25
SiO_2 (Cab-O-Sil)	2.21	4.0
Pure Explosives at TMD		
DATB	1.84	7.52
FEFO (invalid when <35% present)	1.61	7.50
HMX	1.90	9.15
NC	1.58	6.70
NQ	1.81	8.74
PETN	1.78	8.59
RDX	1.81	8.80
TATB	1.94	8.00
TNT	1.654	6.97

initial unreacted HE density, can be used to determine all of the other C–J state parameters.

4.6. The C–J State

When we examine the Rankine–Hugoniot jump equations for a detonation, for the shock going from initially unstressed and unreacted explosive to the highly compressed hot gases at the C–J state, we see that there are seven variables involved: P_0, P_{CJ}, ρ_0, ρ_{CJ}, D, u_{CJ}, and ΔE. Two of these can be easily specified as boundary conditions, the initial pressure P_0 and the initial density ρ_0, this leaves five variables and three equations. In the case of non-reactive shocks we also have the empirical relationship of the U–u Hugoniot as a fourth equation, thereby leaving only one independent variable. There is a similar empirical relationship for the detonation jump condition and it relates the initial density to the C–J density.

If in the same experiment we independently measure both the detonation velocity and either the detonation particle velocity or detonation pressure, then from these data, combined with the mass and momentum equations, we can find the C–J density. This type of experiment has been done many times, and considerable data are available in open literature. In Figure 4.5 we see the results of some 230 experiments involving 83 different explosives and explosive mixtures (Cooper, 1992) where the C–J density is plotted versus initial density.

Fitting this data to a power relationship yields

$$\rho_{\mathrm{CJ}} = 1.386\rho_0^{0.96}. \tag{4.10}$$

This relationship can be combined with equation (4.4) to provide a means to estimate the value of γ at the C–J state, knowing only the initial density of an explosive,

$$\gamma = \frac{1}{(1.386\rho_0^{-0.04} - 1)}. \tag{4.11}$$

Combination of (4.10) with the mass and momentum balance equations (4.1) and (4.2) yields an accurate estimate of the C–J state pressure

$$P_{\mathrm{CJ}} = \rho_0 D^2\left(1 - \frac{\rho_0}{\rho_{\mathrm{CJ}}}\right) = \rho_0 D^2(1 - 0.7215\rho_0^{0.04}). \tag{4.12}$$

We see then that we can determine all of the C–J state parameters either by the use of computer codes, or estimates based on the chemical composition and/or structure, and thermochemical properties, or by simple experiments.

There is often a need to know the properties of detonation reaction products at the shock states created when an explosive detonates in contact with another material. Knowledge of the C–J state parameters is necessary

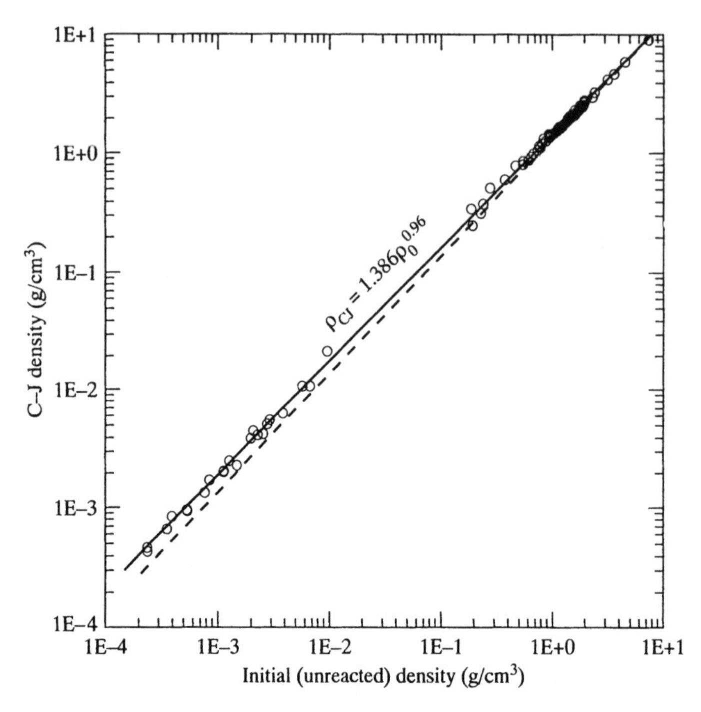

Figure 4.5. Density at the C–J state as a function of initial unreacted density (Cooper, 1992).

for this situation but is not, by itself, sufficient for such a calculation. For that, we must know the *P–u* Hugoniot of the detonation products.

4.7. The Detonation Product *P–u* Hugoniot

When an explosive detonates in contact with another material, it produces a shock wave into that material. The peak pressure of that shock wave depends upon both the properties of the detonation reaction products as well as properties of the adjacent target material. If the adjacent material has a shock impedance greater than that of the detonation reaction products at the C–J state, then the resulting pressure at the interface will be greater than the C–J pressure. That pressure lies along the shock adiabat of the detonation reaction products. Conversely, if the adjacent material has a shock impedance lower than that of the detonation reaction products at the C–J state, then the resulting shock pressure at the interface will be lower than the C–J pressure. That pressure lies along the expansion isentrope of the detonation reaction products. Although only the adiabat is usually referred to as the Hugoniot, for our purposes here, the Hugoniot is considered to be the combination of the adiabat and the isentrope, joined at the C–J point.

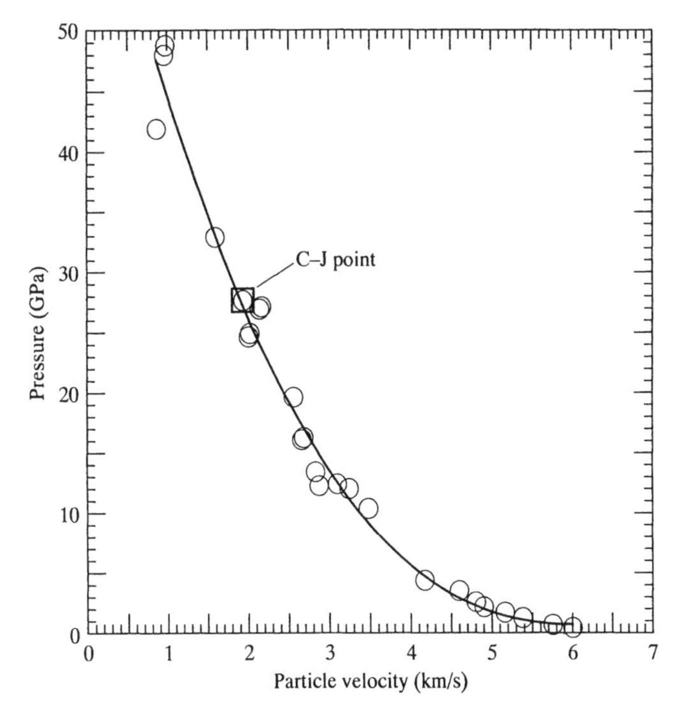

Figure 4.6. P–u Hugoniot of the detonation reaction products of TATB/T2 constructed from experimental data [data from Pinegree et al. (1985)].

As we discussed earlier with the expansion isentrope on the P–V plane, the detonation reaction product Hugoniot on the P–u plane can also be estimated by the various thermochemical/hydrodynamic computer codes. In addition, this Hugoniot can also be estimated by correlation to experimental data. The particular datasets of interest are from experiments where a particular explosive is detonated in contact with a number of targets each of different shock impedance, and the interface state measured. When such experiments are conducted for a given explosive with a variety of targets spanning the range from low- to high-target shock impedance, the Hugoniot of the detonation products of that particular explosive can be constructed. One such Hugoniot is shown in Figure 4.6, for an explosive consisting of TATB and a binder (Pinegree et al., 1985). The target materials used in these experiments were copper, aluminum, magnesium, transacryl, water, and argon gas at initial pressures ranging from 5 bar to 705 bar.

Data from similar sets of experiments, but for a broad variety of explosive types and densities, can be combined by normalizing the data using the C–J conditions as the reference state (Cooper, 1989). This is done in Figure 4.7, where we see the ratio of interface pressure to C–J pressure plotted versus the ratio of interface particle velocity to C–J particle velocity.

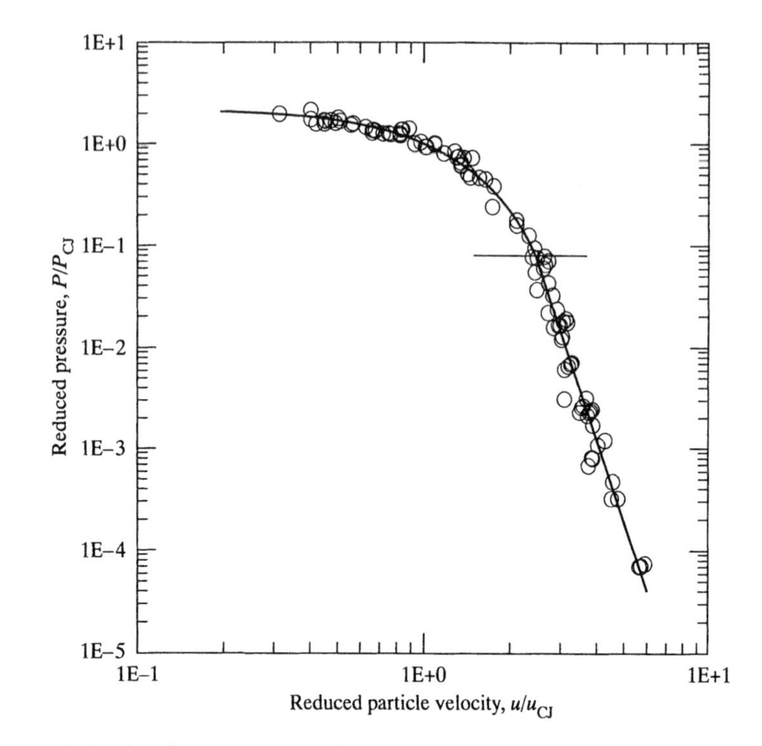

Figure 4.7. Reduced pressure plotted versus reduced particle velocity with the C–J condition as the reference state (Cooper, 1989).

Two fitting curves are used in correlating the data in Figure 4.7:

for $P/P_{\text{CJ}} > 0.08$;

$$P = (2.412P_{\text{CJ}}) - \left(1.7315\frac{P_{\text{CJ}}}{u_{\text{CJ}}}\right)u + \left(0.3195\frac{P_{\text{CJ}}}{u_{\text{CJ}}^2}\right)u^2, \tag{4.13}$$

and for $P/P_{\text{CJ}} < 0.08$;

$$P = \left(235\frac{P_{\text{CJ}}}{u_{\text{CJ}}^{-8.7126}}\right)u^{-8.7126}. \tag{4.14}$$

Now we can estimate the P–u Hugoniot for the detonation reaction products of any explosive for which we know the C–J state parameters. This Hugoniot is used in shock interaction calculations in exactly the same manner as is the reverse (left-going) Hugoniot of a nonreactive material.

4.8. Detonation Velocity and Charge Diameter

In the beginning of this chapter we stated that the fourth observation which we make of steady-state detonation is that "the detonation wave velocity is lower in smaller diameter specimens of the same explosive material and propagation completely fails below some minimum critical diameter." Unlike the other observations discussed above where we could either derive equations a priori, describing a particular behavior or predict that behavior from correlations with other known parameters, the effect of charge diameter on detonation velocity and failure to detonate below some critical diameter cannot be predicted from classical detonation theory, nor has it been reasonably correlated to other detonation phenomena.

This effect of detonation velocity dependence on charge diameter is believed to be related to the thickness of the detonation reaction zone. Explosives which are believed to have thicker reaction zones exhibit greater dependence on diameter. The effect of diameter is measured in rate-stick experiments. In these experiments velocity is measured in long charges of various diameters. An example is seen in Figure 4.8, where velocity data (Gibbs and Popolato, 1980) for Composition B explosive is plotted versus charge diameter.

Notice in Figure 4.8 that the detonation velocity appears to approach asymptotically a maximum value as the diameter increases. When this type of data is plotted as a function of reciprocal diameter, we see that the approach to maximum velocity becomes linear with $1/d$. This is seen in Figure 4.9, where the same Composition B data has been replotted in this manner.

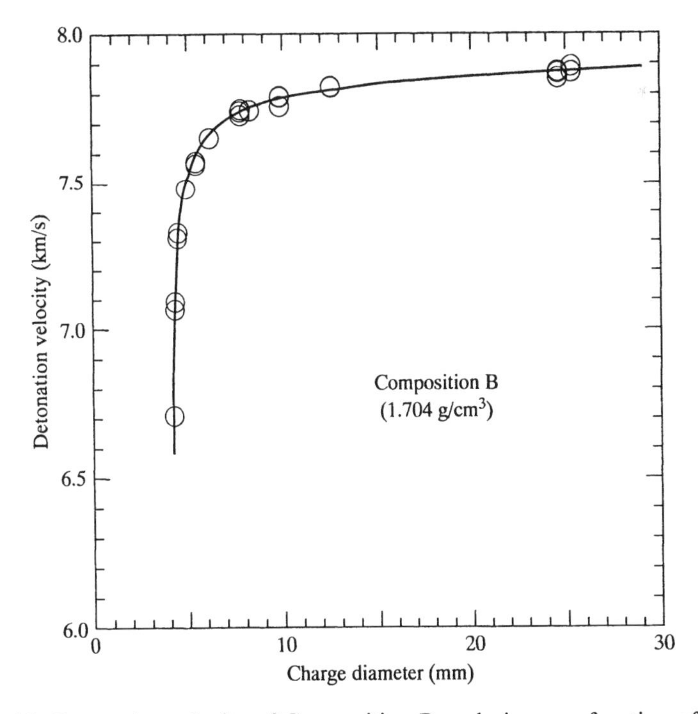

Figure 4.8. Detonation velocity of Composition B explosive as a function of charge diameter [data from Gibbs and Popolato (1980)].

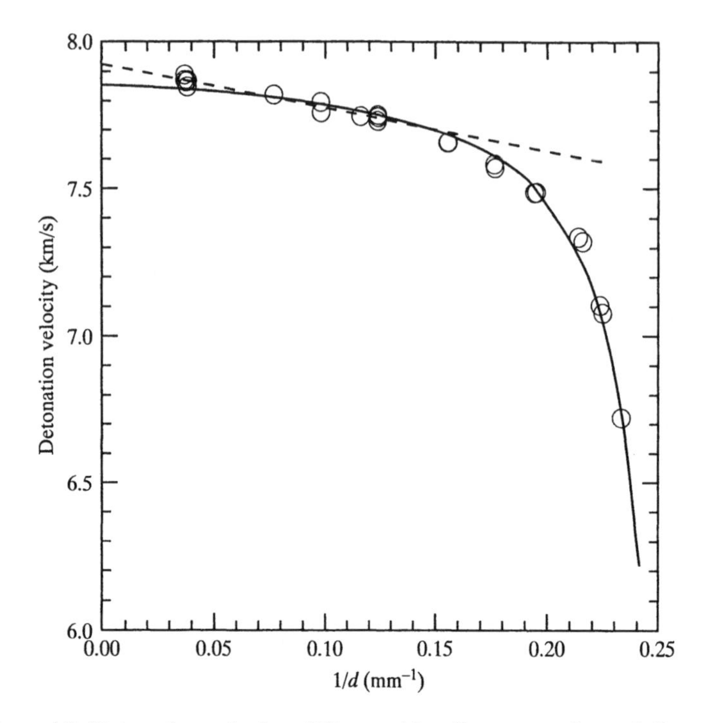

Figure 4.9. Detonation velocity of Composition B versus reciprocal diameter.

There are two methods of using this data; one is to fit a linear equation to the linear portion of the data where $1/d$ approaches zero (dashed line in Figure 4.9), and the other (Campbell and Engelke, 1976) to fit a bit more complex equation to the entire group of data (solid line in Figure 4.9). The two fitting equations are:

the linear portion fit

$$D = D_\infty\left(1 - A_L\frac{1}{d}\right), \tag{4.15}$$

where D_∞ is the detonation velocity at infinite diameter, A_L is a constant particular to each different explosive, and d is the charge diameter; and

the Campbell and Engelke fit

$$D = D_\infty\left(1 - A\frac{1}{R - R_C}\right), \tag{4.16}$$

where A and R_C are constants particular to each different explosive, and R is the charge radius.

Table 4.4 lists values for the parameters of equation (4.16) along with the corresponding failure radii for several different explosives.

Table 4.4. Parameters of the diameter-effect curve [data from Gibbs and Popolato (1980)].

Explosive	Data points/Dia.[b]	Density/TMD[a] (g/cm^3)	D_∞ (km/s)	R_C (mm)	A (mm)	Experiment failure radius (mm)[c]
Nitromethane (liquid)	9/5	1.128/1.128	6.213	−0.4	0.26	1.42
Amatex 80/20	4/4	1.613/1.710	7.030	4.4	5.9	8.5
Baratol 76/24	3/3	2.619/2.63	4.874	4.36	10.2	21.6
Comp. A	5/5	1.687/1.704	8.274	1.2	0.139	<1.1
Comp. B	26/12	1.700/1.742	7.859	1.94	0.284	2.14
Cyclotol 77/23	8/8	1.740/1.755	8.210	2.44	0.489	3.0
Dextex	7/4	1.696/1.722	6.816	0.0	5.94	14.3
Octol	8/6	1.814/1.843	8.481	1.34	0.69	<3.2
PBX 9404	15/13	1.846/1.865	8.773	0.553	0.089	0.59
PHX 501	7/5	1.832/1.855	8.802	0.48	0.19	<0.76
X-0219	8/6	1.915/1.946	7.627	0.0	2.69	7.5
X-0290	5/5	1.895/1.942	7.706	0.0	1.94	4.5
XTX 8003	162/4	1.53/1.556	7.264	0.113	0.0018	0.18

[a] TMD = Theoretical Maximum Density.

[b] Number of shots that propagated a steady wave/number of distinct diameters at which observations were made.

[c] R is the average of the radii from two go/no-go shots (all shots fired in air except NM which was in brass tubes with 3.18 mm thick walls).

The advantage of equation (4.16) is that the entire range of detonation–diameter behavior, including the drop to failure, can be described. The advantage of the linear fit is that fewer experiments need be conducted. The linear fit can also be used to roughly predict failure diameter. The quantity (A_L/D_∞) can be roughly correlated to the failure diameter (Cooper, 1993) as seen in Figure 4.10.

4.9. Conclusion

We have seen that the detonation wave can be modeled as a simple shock using the same jump equations which are used for shocks in inert materials. The quantitative aspects of the C–J state and effects of density are seen to be easily obtained by well-fitted correlations. Energy changes and shock interactions can be calculated from the Hugoniots, and the Hugoniot equations can be quantified by existing databases and estimating correlations. All of the above fit well with the assumption in simple detonation theory that the reaction zone thickness is zero. We also saw that some detonation phenomena, particularly those affected by charge diameter, cannot ignore the effects of a finite reaction zone, and these cannot be predicted by simple theory or by correlation to other independent parameters.

The following bibliography is offered for further reading on these topics.

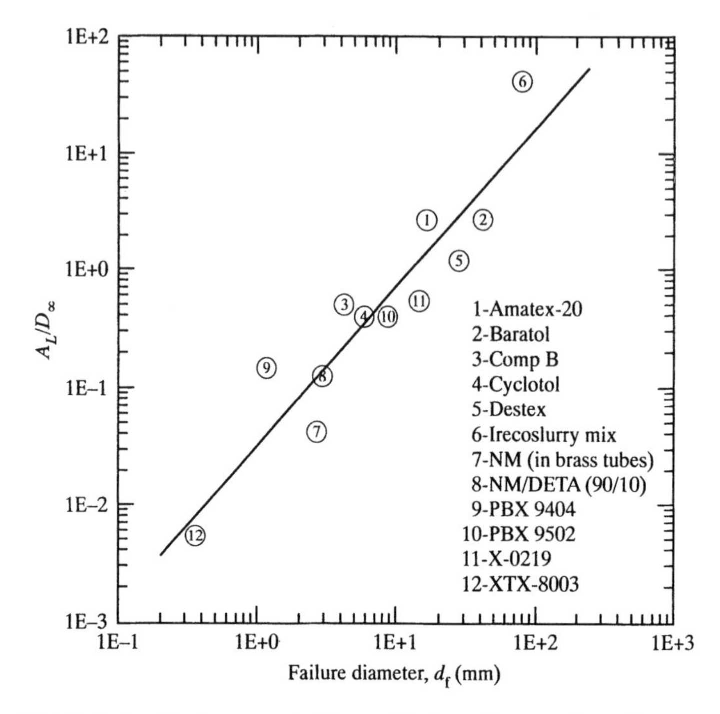

Figure 4.10. Relationship between A_L/D_∞ and failure diameter for a diverse group of explosive materials (Cooper, 1993).

Bibliography

Cheret, R. (1993). *Detonation of Condensed Explosives.* Springer-Verlag, New York.

Cook, M.A. (1958). *The Science of High Explosives.* Reinhold, New York.

Cooper, P.W. (1996). *Explosives Engineering.* VCH, New York.

Fickett, W. and Davis, W.C. (1979). *Detonation.* University of California Press, Berkeley, CA.

Fickett, W. (1985). *Introduction to Detonation Theory.* University of California Press, Berkeley, CA.

Johansson, C.H. and Persson, P.A. (1970). *Detonics of High Explosives.* Academic Press, London.

Mader, C.L. (1979). *Numerical Modeling of Detonations.* University of California Press, Berkeley, CA.

Zel'dovich Ya.B. and Raizer, Yu.P. (1967). *Physics of Shock Waves and High-Temperature Hydrodynamic Phenomena.* Academic Press, New York.

References

Campbell, A.W. and Engelke, R. (1976). The diameter effect in high-density heterogeneous explosives. *Proc. 6th Symposium (International) on Detonation*, Coronado, CA.

Cooper, P.W. (1989). Shock behavior of explosives about the C–J point. *Proc. 9th Symposium (International) on Detonation*, Portland, OR.

Cooper, P.W. (1992). Extending estimation of C–J pressure of explosives to the very low density region. *18th International Pyrotechnics Symposium*, Breckenridge, CO.

Cooper, P.W. (1993). A new look at the run distance correlation and its relationship to other non-steady-state phenomena. *Proc. 10th Symposium (International) on Detonation*, Boston, MA.

Cooper, P.W. (1996). *Explosives Engineering*. VCH, New York.

Cowperthwaite, M. and Zwisler, W.H. (1973). *TIGER Computer Program Documentation*. SRI Publication No. Z106.

Dobratz, B.M. (1985). *LLNL Handbook of Explosives*. UCRL-52997, Lawrence Livermore National Laboratory, CA (March 1981).

Fried, L.E. (1995). *CHEETAH 1.22 User's Manual*. UCRL-MA-117541, Lawrence Livermore National Laboratory, CA.

Gibbs, T.R. and Popolato, A. (1980). *LASL Explosive Property Data*. University of California Press, Berkeley, CA.

Kamlet, M.J. and Jacobs, S.J. (1968). Chemistry of detonations, I. A simple method of calculating detonation properties of CHNO explosives. *J. Chem. Phys.*, **48**, 23–35.

Levine, H.B. and Sharples, R.E. (1962). *Operator's Manual for RUBY*. UCRL-6815, Lawrence Livermore Laboratory, CA.

Mader, C.L. (1967). *FORTRAN BKW: A Code for Computing the Detonation Properties of Explosives*. LA-3704, Los Alamos Scientific Laboratory.

Mader, C.L., Johnson, J.N., and Stone, S.L. (1982). *Los Alamos Explosives Performance Data*. University of California Press, Berkeley, CA.

Nichols (III), A.L. and Ree, F.H. (1990). *CHEQ 2.0 User's Manual*. UCRL-MA-106754, Lawrence Livermore National Laboratory, CA.

Pinegree, M. et al. (1985). Expansion isentropes of TATB compositions released into argon. *Proc. 8th Symposium (International) on Detonation*, Albuquerque, NM.

Rothstein, L.R. and Petersen, R. (1979). Predicting high explosives detonation velocities from their composition and structure. *Propellants and Explosives*, **4**, 56–60.

Rothstein, L.R. and Petersen, R. (1981). Predicting high explosives detonation velocities from their composition and structure. *Propellants and Explosives*, **6**, 91–93.

Stine, J.R. (1990). On predicting properties of explosives—Detonation velocity. *J. Energetic Mater.*, **8**, 41–73.

Urizer, M.L. (1981). *LLNL Handbook of Explosives*. UCRL-52997, pp. 8.10–8.12, Lawrence Livermore National Laboratory (updated January 1985).

CHAPTER 5

The Chemistry of Explosives

Jimmie C. Oxley

5.1. Background

Explosive devices may be mechanical, chemical, or atomic. Mechanical explosions occur when a closed system is heated—a violent pressure rupture can occur. However, this doesn't make a heated can of soup an explosive. An explosive substance is one which reacts chemically to produce heat and gas with rapid expansion of matter. A detonation is a very special type of explosion. It is a rapid chemical reaction, initiated by the heat accompanying a shock compression, which liberates sufficient energy, before any expansion occurs, to sustain the shock wave. A shock wave propagates into the unreacted material at supersonic speed, between 1500 m/s and 9000 m/s.

Typical military explosives are organic chemicals, often containing only four types of atoms: carbon (C), hydrogen (H), oxygen, (O), and nitrogen (N) [Urbanski (1964, 1965, 1967, 1984); Gibbs and Popolato (1980); Dobratz (1981); Kaye (1960–1978); and Davis (1943)]. To maximize volume change, gas formation, and heat release, explosives are designed to be dense and have high oxygen content and positive heats of formation. In monomolecular organic explosives, which includes most military explosives, oxygen is available in NO_2 groups. Upon detonation, exothermic (heat releasing) reactions transform nitrogen atoms into nitrogen (N_2) gas, while the oxygen atoms combine with hydrogen and carbon atoms to form gaseous products (H_2O, CO, or CO_2). This is similar to a combustion process, but a detonation is different from burning in two ways. In combustion, there is an unlimited amount of oxidizer available, oxygen from the air. An explosive oxidizes so quickly that it usually must contain its own source of oxygen either in the same molecule, as with most military explosives (e.g., trinitrotoluene, TNT), or in a neighboring molecule, as in the intimate mixture of ammonium nitrate and fuel oil (ANFO).

$$\text{TNT} \quad 4C_7H_5N_3O_6 \rightarrow 7CO_2 + 6N_2 + 10H_2O + 21C,$$

$$\text{ANFO} \quad 37NH_4NO_3 + CH_3(CH_2)_{10}CH_3 \rightarrow 12CO_2 + 37N_2 + 87H_2O.$$

"Oxygen Balance" is a method of quantifying how well an explosive provides its own oxidant. There are various ways of defining oxygen balance (OB). We can balance the oxygen so that every carbon has one oxygen (balanced for CO) or so that every carbon has two oxygen (balanced for CO_2) [Kaye (1960–1978)]. [Note that carbons in the molecule already bonded to oxygen do not contribute to the energy balance ($-n_{COO}$).] We can also balance in terms of weight percent oxygen in the explosive (OB) or in terms of oxidant per 100 grams explosive (OB_{100}) [Kamlet (1976)]:

$$OB_{100} = [100*(2n_O - n_H - 2n_C - 2n_{COO})]/\text{mol. wt. compound}$$

(balanced to CO)

example nitroglycerin:

$$\begin{array}{l} CH_2ONO_2, \\ | \\ CHONO_2, \\ | \\ CH_2ONO_2, \end{array}$$

$$9 = n_O; \quad 5 = n_H; \quad 3 = n_C; \quad 0 = n_{COO},$$

$$\text{mol. wt.} = 9 * 16 + 5 * 1 + 3 * 12 + 3 * 14 = 227,$$

$$OB_{100} = [100 * (2 * 9 - 5 - 2 * 3 - 0)]/227$$

$$= (700/227) = 3.08,$$

$$OB(\text{balanced to } CO_2) = [1600 * (n_O - 2n_C - 0.5n_H)]/\text{mol. wt. compound},$$

$$\text{OB nitroglycerin} = [1600 * (9 - 6 - 5/2)]/227 = 800/227 = 3.52.$$

Another way in which a detonation differs from a fast burn (deflagration) is the manner in which the performance is evaluated. The performance of a fuel is directly related to the amount of heat released, while the performance of an explosive has a less direct relationship to heat released. Detonation is unique in the rapid rate at which energy is released. A detonating high explosive creates a tremendous power density compared to deflagration materials:

	W/cm^3
Burning acetylene	10^2
Deflagrating propellant	10^6
Detonating high explosive	10^{10}

The performance of an explosive cannot be expressed in a single characteristic. Performance is dependent on the detonation rate or velocity, the packing density, the gas liberated per unit weight, and the heat of explosion. Detonation velocity, itself, is dependent on packing density, charge diameter, degree of confinement, and particle size. Both the terms brisance and strength are used in describing performance of an explosive. When an explosive detonates there is a practically instantaneous pressure jump from the shock wave. The subsequent expansion of the detonation gases performs work, moves objects, but it is the pressure jump which shatters or fragments objects (see discussion of ideal and nonideal explosives, pp. 163–164). Brisance (from French for shatter) is a description of the destructive fragmentation effect of a charge upon its immediate vicinity. Since shattering effect is dependent upon the suddenness of the pressure rise, it is most dependent upon detonation velocity. Brisance is the term of importance in military applications. Brisance is often evaluated from detonation velocity, but there are "crusher" tests in which the compression of lead or copper blocks by the detonation of the test explosive is taken as a measure of brisance [Kaye (1960–1978)]. Strength is important in mining operations; it describes how much rock can be moved. The strength of an explosive is more related to the total gas yield and the heat of explosion. It is often quantified with the Trauzl lead block test, where 10 g of a test sample are placed in a 61 cm^3 hole in a lead block and initiated with a No. 8 blasting cap. Performance is evaluated from the size of the cavity created in the lead block [Meyer (1987)].

Explosives are often classified by the stimuli to which they respond and the degree of response. The term 'low explosives" is applied to propellants or deflagrating materials, such as black powder and smokeless powder (colloided nitrocellulose). Although they contain within themselves all oxygen needful for their combustion, their self-oxidization is slow and sensitive to degree of confinement. Detonating or "high" explosives are characterized by a very high rate of reaction and pressure release. TNT and nitroglycerin are examples of high explosives. The shock waves in high explosives travel at speeds in the range 5000–9000 m/s, compared to gun powder, a low explosive, at ~100 m/s. High explosives may be difficult to burn, but with sufficient impulse detonation can be initiated regardless of the degree of confinement.

Based on their ease of initiation, high explosives are subdivided into the designation primary or secondary. Primary explosives are detonated by simple ignition—spark, flame, or impact. Examples of primary explosives are lead azide, lead picrate, lead styphnate, mercury fulminate, diazodinitrophenol (DDNP), *m*-nitrophenyldiazonium perchlorate, tetracene, nitrogen sulfide (N_4S_4), copper acetylide, fulminating gold, nitrosoguanidine, potassium chlorate with red phosphorus (P_4), and the tartarate and oxalate salts of mercury and silver (of these, modern detonators tend to use lead azide or DDNP). Secondary explosives differ from primary explosives in not being initiated readily by impact or electrostatic discharge; they do not easily undergo a deflagration-to-detonation transition (DDT). They can be initiated by large shocks; usually they are initiated by the shock created by a primary explosive. A blasting cap and frequently a booster are required. (A booster is a sensitive secondary explosive which reinforces the detonation wave from the detonator into the main charge.)

In general, secondary explosives are more powerful (brisant) than primary

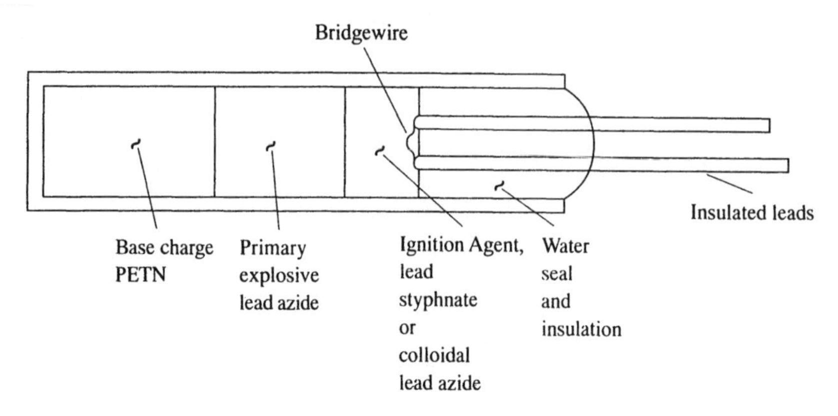

Low-energy detonator. Typical electric blasting cap.

explosives. Examples of secondary explosives include nitrocellulose, nitroglycerin, dynamite, TNT, picric acid, tetryl, RDX, HMX, nitroguanidine, ammonium nitrate, ammonium perchlorate, liquid oxygen mixed with wood pulp, fuming nitric acid mixed with nitrobenzene, compressed acetylene and cyanogen.

5.2. Conventional Explosives

Most military explosives are secondary explosives, and for the most part they fall in one of three categories all of which contain nitro (NO_2) groups. First, nitrate esters, such as nitroglycerin, nitrocellulose, PETN (active component in sheet explosives), contain $O—NO_2$ groups. These are the oldest type of explosives still used by the military. Nitration of alcohols was a popular research topic in the 1830s to 1840s. By the 1860s technological advances made nitroglycerin and nitrocellulose relatively safe and useful explosives.

Nitrate ester — Nitroarenes — Nitramine

ONO_2 ONO_2 ONO_2 ONO_2 O_2N CH_3 NO_2 NO_2 O_2N OH NO_2 NO_2 $NO_2—N$ $N—NO_2$ $NO_2—N$

PETN	TNT	Picric acid	RDX
1894	1863	1742	1899 discovery
1930	1900	1870	1940 used

Nitrate esters are the least stable military explosives; they lose NO_2 readily. Second, nitroarenes with a C—NO_2 linkage are typified by TNT (component of Composition B) or picric acid. Third, nitramines contain N—NO_2 groups; typical examples RDX and HMX are the active ingredients in plastic-bonded explosives such as Composition B, C-4, and Semtex [Urbanski (1964, 1965, 1967, 1984); Gibbs and Popolato (1980)].

The most general method of attaching a nitro (NO_2) group is treating the parent C—H, O—H, or N—H containing compound with mixed acid—equal volumes of concentrated nitric (HNO_3) and sulfuric (H_2SO_4) acids. Sulfuric acid protonates nitric acid forming nitronium cation (NO_2^+). Nitronium, a powerful electrophile, attacks the X—H bond (X = C, O, or N). Use of low temperature prevents undesired oxidation as well as violent reactions:

$$HNO_3 + H_2SO_4 \longrightarrow NO_2^+ + H_2O + HSO_4^-,$$

$$R_2N\text{—}H + NO_2^+ \longrightarrow R_2N\text{—}NO_2 + H^+.$$

Other reagents may be used for nitration depending on the species involved. Higher concentrations of nitric acid may be used (100% NHO_3 is produced by the distillation of KNO_3 with 98% H_2SO_4). Nitronium tetrafluoroborate (NO_2BF_4) is an expensive but gentle nitration reagent [Olah et al. (1989)]. Acetyl nitrate, formed in situ from an acetic acid solution of acetic anhydride and 98% NHO_3, is a hazardous but effective nitration reaction. The nitro group is usually introduced on to an arene by use of mixed acid. However, the presence of a nitro group or any other species on an arene ring influences and directs further substitution. Due to this consideration synthesis schemes may not be straightforward.

5.3. Nitrate Esters

Most alcohols can be converted into nitrate esters whose physical state is usually the same as the parent alcohol.

Methyl Nitrate is a powerful explosive, but of little practical use. Its vapors have a strongly aromatic odor and cause headaches if inhaled.

$$\underset{\text{methanol}}{CH_3OH} + HNO_3 \rightarrow \underset{\text{methyl nitrate}}{CH_3ONO_2}.$$

In World War II the Germans used land mines composed of sodium and methyl nitrate in separate compartments. Pressure facilitated mixing and then explosion.

Dinitroglycol or EGDN ethylene glycol dinitrate (bp, 197.5 °C; mp, −22.3 °C; p, 1.518 g/cc) is used as an ingredient in nonfreezing dynamites. It has many of the advantages of nitroglycerin and is somewhat safer to manufacture and handle. Its principal disadvantage is its greater volatility.

EGDN produces headaches similar to those produced by nitroglycerin, and in proportion to its greater volatility, the headaches are more violent but do not last as long.

$$\underset{\text{glycol}}{HOCH_2CH_2OH} \xrightarrow{\text{mixed acid}} \underset{\text{dinitroglycol}}{O_2NOCH_2CH_2ONO_2}.$$

With a drop height of 20 cm, dinitroglycol is less sensitive than nitroglycerin. It has a perfect oxygen balance:

$$O_2NOCH_2CH_2ONO_2 \rightarrow 2CO_2 + 2H_2O + N_2.$$

TEGDN Triethylene Glycol Dinitrate $O_2NOCH_2CH_2OCH_2CH_2OCH_2$-$CH_2ONO_2$ and *DEGDN Diethylene Glycol Dinitrate* $O_2NOCH_2CH_2OCH_2$-CH_2ONO_2 are nitrate esters often used as plasticizers in propellants.

Nitroglycerin was first prepared by Italian chemist Ascanio Sobrero in 1846. For several years after its discovery there was little interest in using it as an explosive. It was first used as a medicine. In the 1860s Alfred Nobel found a way to improve the safety of its manufacture and use in dynamites and, more importantly, developed a practical way to initiate it. This material would correctly be named as the trinitrate, but historical use prevails in the name "nitroglycerin."

$$\underset{\text{glycerin}}{\begin{matrix} CH_2OH \\ | \\ CHOH \\ | \\ CH_2OH \end{matrix}} \xrightarrow{\text{mixed acid}} \underset{\text{nitroglycerin}}{\begin{matrix} CH_2ONO_2 \\ | \\ CHONO_2 \\ | \\ CH_2ONO_2 \end{matrix}} \longrightarrow 3CO_2 + 2.5H_2O + 1.5N_2 + 0.25O_2.$$

Nitroglycerin is a yellow oil with boiling point (bp) estimated as 245 °C and melting point (mp) as 13.1 °C. Pure nitroglycerin is odorless at ambient temperature but has a faint odor above 50 °C. It has high vapor pressure and prolonged exposure results in severe and persistent headaches. It is extremely shock sensitive, and its sensitivity is greater if it is warmed. Drop height at ambient is 4 cm with a 2 kg weight.

Vapor pressure of nitroglycerin [Marshall and Peace (1916)]

mm	°C	mm	°C
0.00025	20	0.019	60
0.00083	30	0.043	70
0.0024	40	0.098	80
0.0072	50	0.29	93

Nitrated Sugars (*Mono- and Oligo Saccharides*). Sugars are polyhydric alcohols which contain an aldehyde or ketone group or a cyclic acetal or ketal arrangement. Their nitrate esters are easy to prepare but are less stable than those of simple polyhydric alcohols; this instability may be due to the difficulty in purifying them. *d*-Glucose pentanitrate is a viscous colorless syrup, insoluble in water. It is unstable above 50 °C.

$$
\begin{array}{c}
CH_2OH \\
—|—OH \\
HO—|— \\
—|—OH \\
—|—OH \\
CH_2OH
\end{array}
\xrightarrow{\text{mixed acid}}
\begin{array}{c}
CH_2ONO_2 \\
—|—ONO_2 \\
O_2NO—|— \\
—|—ONO_2 \\
—|—ONO_2 \\
CH_2ONO_2
\end{array}
$$

d-glucose *d*-glucose pentanitrate

Nitrated sugars can be used with nitroglycerin to make a nonfreezing dynamite. Nitrated sugars form many eutectics with nitroarenes.

Nitrated Polysacchrides: Guncotton, Nitrocellulose, Nitrostarch. Cellulose and starch are both polymers of glucose. They differ in the chemical linkage between glucose molecules. The chemical bonding in cellulose is such that 2000 to 3000 anhydroglucose units are linked in a straight chain which is essentially one dimensional. In starch, glucose monomers are joined by alpha-linkages which leads to a spiral structure. This results in a three-dimensional structure which, on the average, contains 25 to 30 anhydroglucose units. Starch is also susceptible to side chain branching.

In 1833, Braconnot found nitrated starch produced a varnish-like coating impervious to water. Schonbein and Bottger, independently, discovered the nitration of cotton in 1846. Between 1865 and 1868 Abel patented improved preparations, which required pulping. The pulping allowed impurities to be more easily washed out by "poaching" and resulted in improved stability. Nitrocellulose, like most nitrate esters, is intrinsically unstable, but decomposition at ambient temperature is slow if it has been thoroughly purified. Cotton fiber is practically pure cellulose, but equally useful cellulose can be produced from wood. Cellulose contains three hydroxyl groups per anhydroglucose unit and yields a trinitrate on complete nitration (14.4% N). However, complete nitration is difficult. Nitrocellulose is rated by percent N, and this is dependent on the method of nitration:

12.7% N—95% HNO_3,

13.5% N—mixed acid,

14.1% N—HNO_3, H_3PO_4, P_2O_5.

Nitrocellulose is used with nitroglycerin to form double-based propellants and smokeless powders. Nitrocellulose with TNT forms a eutectic (mp, 63 °C) mixture.

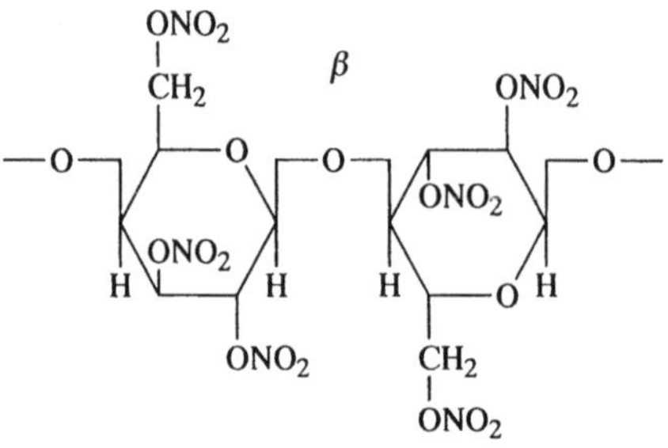

Nitro cellulose

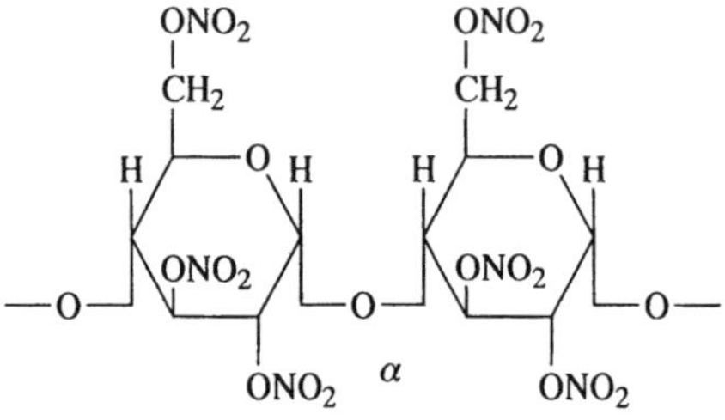

Nitro starch

5.3.1. Nitrate Ester Formulations

Smokeless Powder. The first successful smokeless powder appears to have been made by Captain Schultze of the Prussian Artillery in 1864. He used nitrated wood pulp (nitrocellulose). Nobel produced a smokeless powder in 1888 mixing nitroglycerin and soluble nitrocellulose. Over the years various proportions and ingredients have been mixed and various names applied, e.g., Ballistite, Cordite [Davis (1943)]. Most smokeless powders now contain about 1% diphenylamine (DPA) as a stabilizer. It removes acids formed during decomposition. The product of the reaction between DPA and the acid is diphenylnitrosamine, which itself decomposes at 110 °C. Excess diphenylamine (at the 10% level or above) produces unstable smokeless powders, and powders which contain 40% DPA combust spontaneously when heated in air at 110 °C. Ethyl centralite (diethyldiphenylurea) $[Ph(Et)N]_2CO$ is also used as stabilizer, a means to moderate the burn and reduce flash. Another way to produce a flashless powder is using cool explosives such as ammonium nitrate, guanidine nitrate, or nitroguanidine.

Dynamite. In 1863 Alfred Nobel invented the detonating blasting cap, making it possible to reproducibly initiate nitroglycerin. In 1866 he found that combining 75% NG with 25% kieselguhr (diatomaceous earth) made a relatively stable material, which he patented as "dynamite." After using the inactive base of kieselguhr, he developed dynamites with active bases, nitroglycerin absorbed by mixtures of materials such as nitrate salts and wood meal, charcoal, sugar, or starch. Early dynamites which contained only nitroglycerin froze at relatively high temperatures. When they were frozen, they were less sensitive to shock and initiation, but the intermediate state, half-frozen/half-thawed was quite sensitive. Furthermore, thawed explosives tended to exude nitroglycerin. A low-freezing dynamite was developed by nitrating a mixture of glycerin and ethylene glycol, but these were not available until the late 1920s when the production of ethylene glycol for antifreeze made the precursor chemical relatively inexpensive. Today no dynamite is produced using only NG as the active ingredient, rather mixtures of EGDN/NG, ranging 100/0 to 60/40; in addition, a non-NG dynamite is made using metriol trinitrate (MTN) and diethylene glycol dinitrate (DEGDN) as the active ingredients.

In 1875 Nobel incorporated collodion cotton (NC) into dynamites producing gelatinous formulations called blasting gelatin (6–8% NC) or less stiff formulations (< 2% NC), gelatin dynamites. These were more water resistant than straight dynamites. As of this writing, in the United States there is only one manufacturer of dynamite, Dyno Nobel in Carthage, MO; there are a variety of formulations, but the principle distinguishing ingredients are the amount of NG/EGDN, sodium nitrate (SN), ammonium nitrate (AN), salt, and nitrocellulose (NC) [Hopler (1993a)]. Blasting gelatin is basically NG/EGDN with NC; it is highly water-resistant. Straight dynamites are mixtures of NG/EGDN with SN and wood meal; these are quite sensitive and are sometimes termed "ditching" dynamites. Extra dynamite reduces the level of the NG/EGDN to 10–15% and uses high concentrations of AN as well as SN; and permissible dynamites incorporate salt (sodium chloride) to the extra dynamite formulation to lower the flame temperature for use in coal and other gaseous mines. A number of other ingredients called "dopes" may be used, usually at less than the 3% level: clear wheat, cob meal, balsa, starch, cork, guar gum, and calcium stearate. Chalk is used in all formulations as an anti-acid; and if hollow microspheres are added they are phenolic, rather than glass, which can cause friction problems. Most dynamites contain some nitrocotton, typically in the range of 0.1% to 0.2%; higher percentages of nitrocotton (0.4% to 2.5%) are used in dynamite products called "gelatins."

	Straight	Extra	Blasting gelatin	Nobel
NG/EGDN	40%	20%	90%	62.5%
NG only				
NC	0.1%	0.1%	7%	
AN (mixture course and fine)	15–30%	70%		
$NaNO_3$	20–30%			27%
Wood pulp	6–8%	6%	< 1%	8%

PETN or Pentaerythritol Tetranitrate. $C(CH_2ONO_2)_4$ is white crystalline solid. It is prepared by nitration of pentaerythritol with 95% HNO_3; PETN is insoluble in the acid and precipitates from the reaction mixture

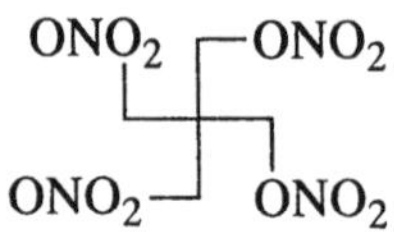

ONO_2 ONO_2

ONO_2 ONO_2

Among explosive nitrate esters, PETN is the most stable. Its high symmetry results in a high melting point (141 °C), which is often correlated with high thermal stability [Oxley et al. (1995)]. Its boiling point has been measured (200 °C), but it explodes at 205 °C. In contrast, its less symmetric relatives, 2,2-methyl-1,3-propane dinitrate $(CH_3)_2C(CH_2ONO_2)_2$ and metriol trinitrate (MTN) $CH_3C(CH_2ONO_2)_3$ (an ingredient in non-NG dynamites) melt at substantially lower temperatures, 24 °C and −60 °C, respectively.

PETN is extraordinarily sensitive to explosive initiation. It is detonated by 0.01 g lead azide, whereas tetryl requires 0.025 g. PETN is extensively used in detonators, detonating fuzes (Primacord), priming compositions, and DETA sheet (DuPont trademark). A one-to-one mixture of TNT and PETN, called Pentolite, is used in boosters.

5.4. Nitroarenes

Nitroarenes can be synthesized from direct nitration of arene rings, such as benzene (C_6H_6), toluene ($C_6H_5CH_3$), and phenol (C_6H_5OH). However, roundabout synthetic schemes are often necessary to account for various practical considerations such as vapor pressure of an intermediate or the fact that the presence of nitro groups on an arene ring discourages subsequent nitration.

Nitrobenzene is the simplest nitroarene ($C_6H_5NO_2$). Termed in old literature "oil of mirbane," it is a pale yellow liquid with boiling point 208 °C. It has an almond-like odor and can be absorbed through the skin or by breathing its vapors. Its toxicity is typical of nitroarenes although nitrotoluenes tend to be slightly less toxic than nitrobenzenes. (Further discussion of explosive toxicity can be found in Chapter 9.) Nitrobenzene is not generally considered an explosive, though it is an energetic material and is sometimes mixed with ammonium nitrate.

1,3,5-*Trinitrobenzene* is one of the simplest nitroarenes used as an explosive. It can be prepared directly from the treatment of benzene with anhydrous nitric and fuming sulfuric acids. However, the yield by this route is so low that it is usually prepared by oxidation of TNT. The methyl group in TNT activates the benzene ring toward substitution, allowing addition of three nitro groups to toluene, in contrast to benzene which under forcing conditions only adds two nitro groups.

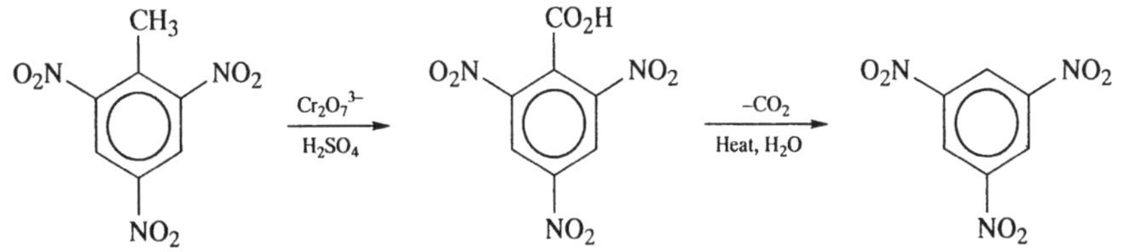

$$\text{TNT} \xrightarrow[H_2SO_4]{Cr_2O_7^{2-}} \text{2,4,6-trinitrobenzoic acid} \xrightarrow[\text{Heat, } H_2O]{-CO_2} \text{1,3,5-trinitrobenzene}$$

Symmetrical trinitrobenzene is a yellow solid with melting point 122.5 °C. Unsymmetrical isomers melt at 62 °C and 128 °C. These compounds are poisonous, as are all nitrobenzenes, and are readily oxidized. Trinitrobenzene is less sensitive to impact than TNT and more powerful and brisant, but its preparation is more expensive.

Hexanitrobenzene (HNB) at one time was sought as the ideal explosive, one with perfect oxygen balance. During World War II there were serious efforts to prepare it; the Germans succeeded using the route shown below. An easier route is the oxidation of pentanitroanaline by peroxydisulphuric acid ($H_2S_2O_8$) (fuming H_2SO_4 + 98% H_2O_2) to HNB in 90% yield.

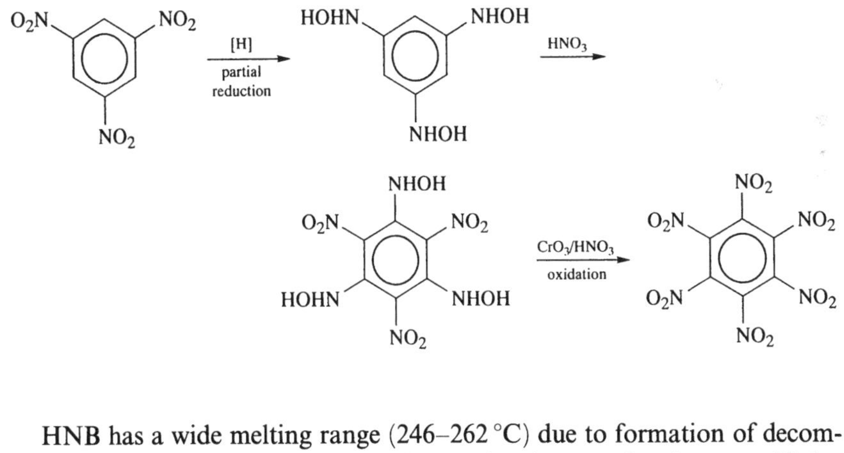

$$\text{1,3,5-trinitrobenzene} \xrightarrow[\text{partial reduction}]{[H]} \text{1,3,5-tris(hydroxylamino)benzene} \xrightarrow{HNO_3} \text{1,3,5-tris(hydroxylamino)-2,4,6-trinitrobenzene} \xrightarrow[\text{oxidation}]{CrO_3/HNO_3} \text{hexanitrobenzene}$$

HNB has a wide melting range (246–262 °C) due to formation of decomposition products near its melting point. In the absence of moisture and light, HNB shows good stability at room temperature, but irradiation produces isomerization of nitro groups to nitrites, which in turn are hydrolyzed to hydroxyl groups:

$$—NO_2 \rightarrow —ONO \rightarrow —OH.$$

Thus, HNB is easily hydrolyzed to pentanitrophenol $(NO_2)_5C_6(OH)$, tetranitroresorcinol, $(NO_2)_4C_6(OH)_2$, and eventually trinitrophloroglucinol $(NO_2)_3C_6(OH)_3$.

2,4,6-*Trinitrotoluene* (TNT) has been a popular explosive because it melts at a relatively low temperature (80.65 °C) without decomposition and, therefore, is readily cast. Compared to other military explosives (RDX, HMX, PETN), TNT has high chemical stability and low sensitivity to impact. It is reported that it may be distilled under vacuum (10–20 mm) at 210–212 °C without evidence of decomposition. Although TNT is relatively simple and safe to manufacture, it is no longer made in the United States or Canada, due to environmental problems with its manufacture, to the existing United States surplus, and to the availability of more powerful explosives. Due to its low melting point, TNT is used in admixture with many other high explosives. Various names are applied to these formulations. Tritonal is a 80/20 mixture of TNT and aluminum. Baratol is a mixture of $(BaNO_3)_2$ with TNT in various ratios from 20/80 to 76/24. Octols and cyclotols are mixtures of TNT with HMX and with RDX, respectively. Adding more methyl groups as in trinitroxylene (TNX) or trinitromesitylene reduces the strength of the explosive.

CH_3 $\xrightarrow[H_2SO_4]{HNO_3}$ CH_3, NO_2 $\xrightarrow{HNO_3\ /\ H_2SO_4}$ CH_3, O_2N, NO_2, NO_2

Nitrotoluenes are toxic, but their toxicity is lower than that of nitrobenzenes. It has been suggested that nitrotoluene is more easily oxidized by the body to nitrobenzoic acid, which is only slightly toxic. Slight TNT poisoning manifests itself by cyanosis, dermatitis, nose bleed, constipation, and giddiness. The severer form is characterized by toxic jaundice and aplastic anemia (see Chapter 9). Dinitrohydroxylaminotoluene may be detected in the urine.

The effect of the electron-withdrawing nitro groups on the arene ring is to strongly activate the methyl group. Consequently, most of TNT chemistry occurs on the methyl group. The methyl protons are fairly acidic for organic compounds (pK, 15.3). They may behave like the methyl hydrogens of acetone in certain condensation reactions

$$\text{TNT} + \text{PhCHO} \xrightarrow{0\,°\text{C}} \text{Pic-CH=CH-Ph} + H_2O$$

where Pic = picryl = trinitrobenzyl, and Ph = C_6H_5.

Hexanitrostilbene (HNS) formed by the oxidative coupling of two TNT molecules is an explosive in its own right. Its thermal stability makes it useful in many high temperature applications:

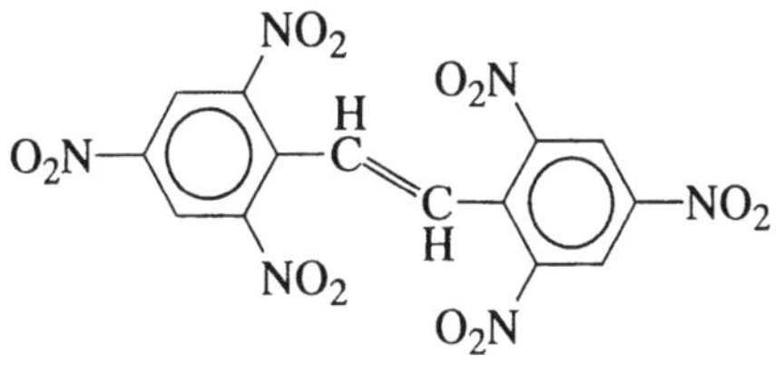

Picric Acid. Of all the nitroarenes, picric acid has been known the longest. It was first obtained in 1742 by Glauber with nitric acid on wool. In the second half of the nineteenth century, picric acid was widely used as a fast dye for silk and wool. At the same time the potassium salt was being used in Great Britain and the United States for filling shells. The general view was that picric acid was not an explosive, only its salts were. Eventually picric acid was accepted as a high explosive. Since it was high melting, it was mixed with various nitro compounds to depress its melting point. The acidic character of picric acid is corrosive to metal and results in formation of metal salts, which are very impact and friction sensitive. When picric acid and nitrophenols were used in ammunition, they were not allowed to come in contact with the metal. Thus, when TNT manufacture was simplified, picric acid use decreased; it has practically been eliminated from armaments.

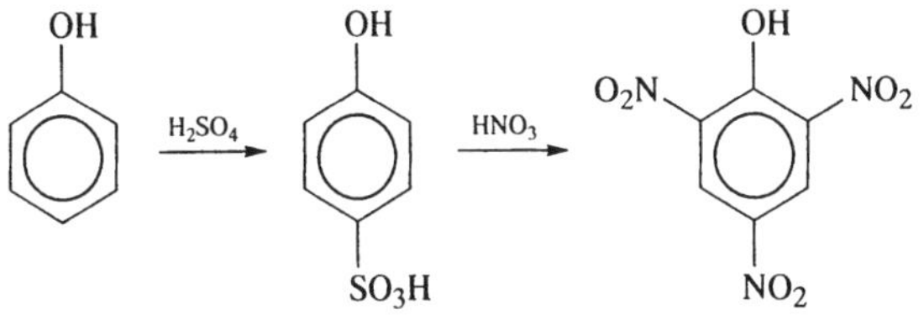

Picric acid could be made by direct nitration of phenol. However, due to the volatility of the intermediate nitrophenol, it is usually synthesized via the sulfate. Picric acid is a colorless or light yellow solid with melting point of 122.5 °C and a vapor pressure described by $\log P = 12.024 - (5729/T(\mathrm{K}))$ (58–103 °C). The trinitrophenol is called picric acid because the hydroxyl proton is very acidic (pK, 3); therefore, it is about 50% dissociated.

Salts of Picric Acid: Ammonium or Guanidine Picrate. Ammonium picrate is prepared by suspending picric acid in hot water and adding strong ammonium hydroxide until the acid completely dissolves. Ammonium picrate is less shock sensitive than picric acid and has a higher melting point (mp, 265–271 °C). It is sometimes referred to as Explosive D. Guanidine picrate is prepared from guanidine nitrate and ammonium picrate (mp, 319 °C).

Picramide or Trinitroaniline finds little use as a military explosive since there are other more powerful explosive which can be prepared from the same materials. It may be prepared in about 90% yield by nitrating aniline in glacial acetic acid or using mixed acid or via picric acid by the route shown below. The presence of HONO acid must be avoided since it attacks the amino group, replacing it by HO, thus, making picric acid. Picramide forms orange–red crystals with melting point 186 °C and vapor pressure at 150 °C of 0.0059 mm. Both picric acid an picramide are more powerful than TNT.

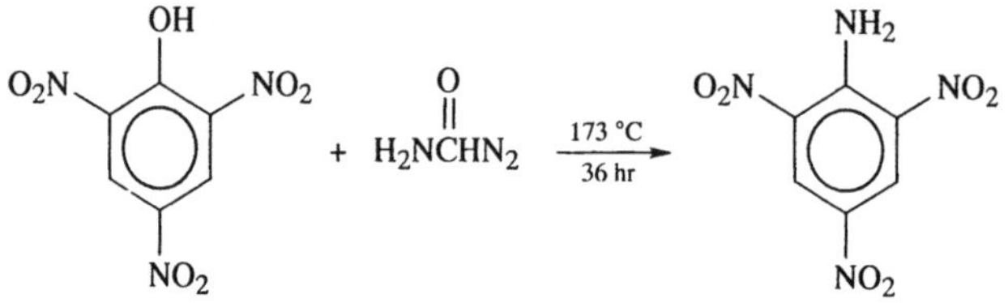

$$+ \; H_2NC(=O)NH_2 \xrightarrow[36\ \text{hr}]{173\ ^\circ\text{C}}$$

Styphnic Acid or Trinitroresorcinol is readily formed by nitration of resorcinol. This yellow solid melts at 175 °C. Styphnic acid is more expensive and less powerful than picric acid. However, its salts are notably more violent explosives than picrates. Lead styphnate has been used to facilitate the ignition of lead azide in detonators.

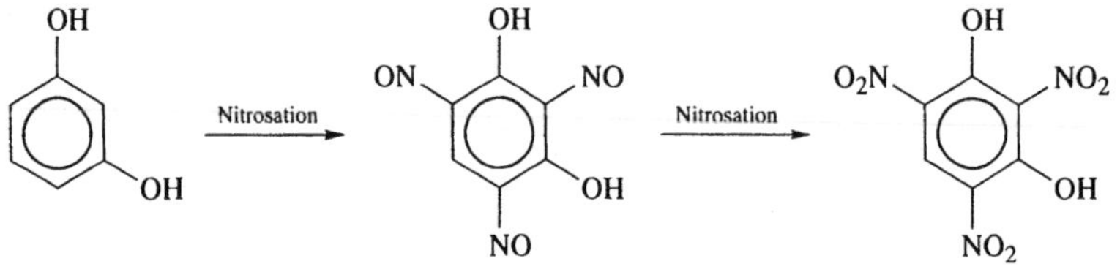

$$\xrightarrow{\text{Nitrosation}} \qquad \xrightarrow{\text{Nitrosation}}$$

Hexanitroazobenzene is an orange solid (mp, 215–216 °C) formed by the condensation of picryl (trinitrobenzyl) chloride and hydrazine. The process is carried out in water suspension in the presence of agents to bind HCl evolved

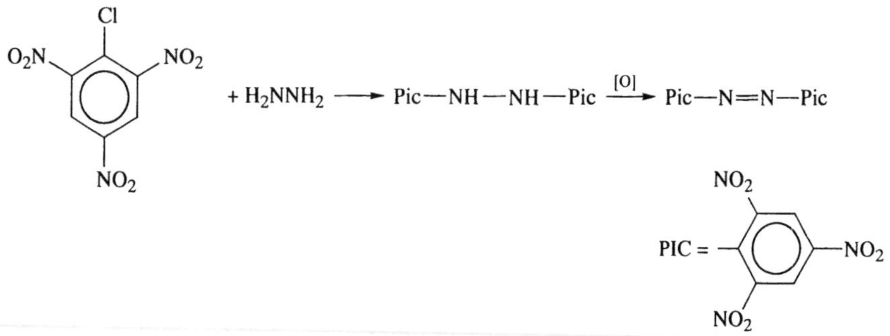

$$+ \; H_2NNH_2 \longrightarrow \text{Pic—NH—NH—Pic} \xrightarrow{[O]} \text{Pic—N=N—Pic}$$

PIC =

5.4.1. Thermally Stable Nitroarenes

There is a tendency for the melting point of a nitroarene to increase with increasing molecular weight. Therefore, in an effort to increase thermal stability, synthetic design has aimed for large molecules. HNS is in this category.

TATB or 2,4,6-*Triamino*-1,3,5-*trinitrobenzene* has excellent thermal stability and is resistant to impact and shock initiation. Its high molecular weight undoubtedly contributes to its elevated melting point, but a greater contributing factor is the graphite-like structure of TATB with alternating NH_2 and NO_2 around the ring. TATB is a yellow solid with melting point sometimes listed as 448–449 °C [Gibbs and Popolato (1980)] and sometimes as 350 °C [Meyer (1987)] with decomposition. Exposure to sunlight or UV turns TATB green, and prolonged exposure eventually turns it brown to black. It is presently the insensitive explosive of choice in nuclear weapons, which can only be as safe as the conventional explosive component. Its synthesis can be accomplished by direct amination because the NO_2 groups are electron withdrawing. The difficulty in its preparation is its insolubility in most organic solvents; it quickly precipitates with little control of particle size possible.

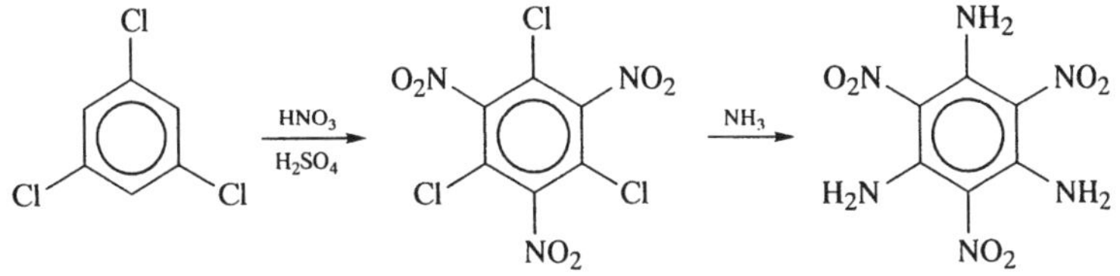

TATB can be oxidized by peroxydisulphuric acid ($H_2S_2O_8$) to hexanitrobenzene. This is a much easier route to hexanitrobenzene than the German synthesis shown earlier.

A number of other high molecular weight nitroarenes have been designed for high-temperature applications. They include DATB (2,4-diamino-1,3,5-trinitrobenzene), DIPAM (3,3′-diamino-2,2′,4,4′,6,6′-hexanitrodiphenyl; NONA, (2,2′,2″,4,4′,4″,6,6′,6″-nonanitro-terphenyl); and DuPont product, TACOT (tetranitro-2,3:5,6-dibenzo-1,3*a*,4,6*a*-tetrazapentalene). Some properties of these and other monomolecular explosives are shown in Table 5.1.

Table 5.1.

Explosive	Melting point, mp, °C	Density, g/cc	Detonation		Drop weight, H50 cm 12	Vapor pressure	
			Velocity, km/s	Pressure, kbar		log P (mmHg)	Temperature °C
AN	169	1.73			149		
AN							
ANFO	169						
AP (200 μm)					47		
AP (90 μm)					44		
ADN	94						
CL-20							
HNS	318	1.74	7			1.00E−09	100
HMX	285d	1.89	9.11	393	26	3.40E−05	150
NC							
NG	13	1.6	7.58	230			
NQ	245d	1.7	8.28	260	> 320		
NTO					86		
PETN	143	1.7	7.98	300		8.00E−05	100
RDX	204	1.77	8.64	347	22	0.013	150
TATB	448	1.85	7.66	275	> 320	1.00E−04	150
TNT	81	1.64	6.94	190	212		
TNAZ	101						
Tetryl	129	1.69	7.7		49	$13.71-6776/T$	
Styphnic acid					46		
Picric acid		1.76	7.68	265	79	$12.024-5279/T$	
Ammonium Picrate					152		
PYX	460	1.75			63		
DATB	286	1.78	7.6		> 320	$13.73-7314/T$	
TACOT	410	1.85	7.2				
NONA	440–450					2.50E−09	150
DIPAM	304	1.79					

Explosive	Isothermal kinetics			Moles gas/cmp	Differential Scanning Calorimetry (DSC) data			
	E, kcal/mol	A s^{-1}	Temp. range °C		Heat (DSC) cal/g	Exo(20°/min), Temp °C	E, kcal/mol	A s^{-1}
AN	26.8	1.6E+07	200–290	0.88	400	320	29.1	6.46E+08
AN	46.2	5.0E+14	290–380					
ANFO	35.2	2.40E+11	270–370		750	290, 370	47.3	3.2E+16
AP (200 μm)	21.3	4.81E+04	215–385	1.2	360	360	22.7	8.32E+05
AP (90 μm)					460	310		
ADN	38.3	9.57E+15		1.4	—	194	31.3	1.66E+17
CL-20	42.4	4.00E+17	146–220, acetone		—	—		
HNS								
HMX	52.9	2.46E+18	230–270	5.4	850	280		
NC				0.00	760	220		
NG								
NQ					250	230		
NTO	38.4	1.05E+13	240–280	2.3	700	260		
PETN				6.3	1000	210		
RDX	37.8	1.99E+14	200–250	3.9	960	257		
TATB				3.4	930	397		
TNT	32.8	1.19E+10	240–280	3.2	610	330	34	2.5E+11
TNAZ	46.6	3.55E+17	220–250	3.8	—	—		
Tetryl							38	2.5E+15
Styphnic acid					800	280		
Picric acid				4.10	1000	320		
Ammonium Picrate					1000	310		
PYX					1000	380		
DATB				3.40				
TACOT								
NONA								
DIPAM								

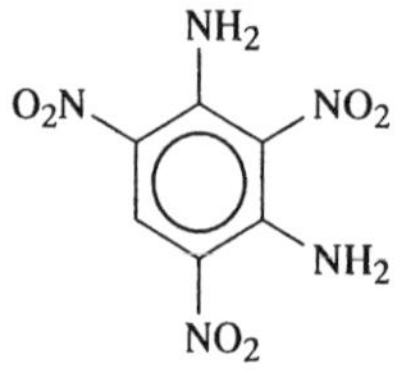

DATB

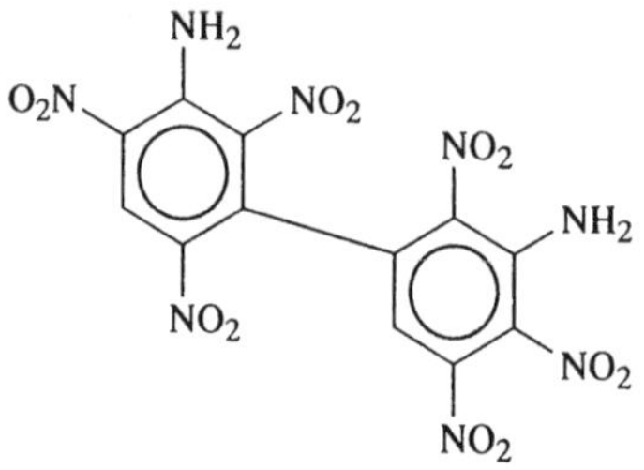

DIPAM

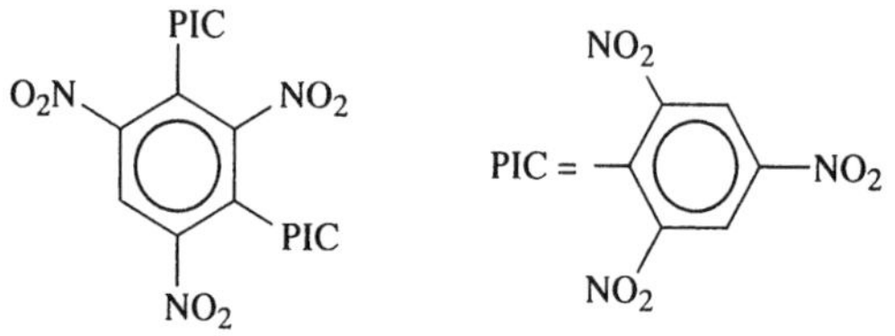

NONA

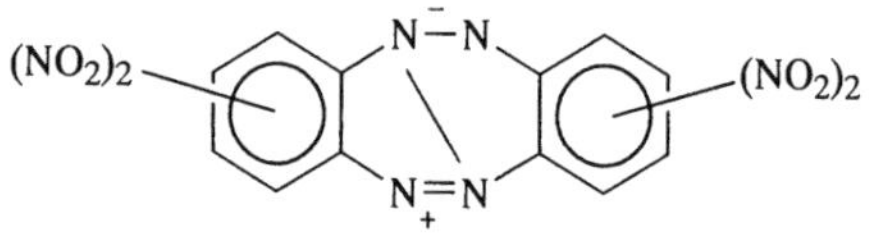

PIC =

TACOT
orange

5.5. Nitroalkanes

Nitromethane. CH_3NO_2 is one of the few nitroalkanes which finds occasional application as an explosive. It is a clear liquid with boiling point 101 °C, melting point −17 °C, density (p) 1.14 g/cc at 15 °C, and vapor pressure 37 mm (25 °C). Its industrial synthesis involves a very low yield (13%) vapor phase reaction:

$$CH_4 + HNO_3 \xrightarrow{400-450\,^{\circ}C} CH_3NO_2 + H_2O.$$

In the laboratory it is more readily made by the reaction of sodium nitrite on chloroacetic acid.

$$NaNO_2 + ClCH_2COOH \longrightarrow NaCl + CH_3NO_2 + CO_2.$$

Although the drop weight impact test (Chapters 10 and 11) shows that nitromethane reacts at fairly low drop heights (35 cm), nitromethane requires a strong initiator to detonate. Nitromethane can be made more sensitive to detonation by the addition of small amounts (5 wt%) of certain compounds—bases (aniline, ethylenediamine, methyl amine, diethylenetriamine) or acids (nitric or sulfuric acids). A nitromethane mixture, containing 95% nitromethane and 5% ethylenediamine [Kaye (1960–1978)] is called PLX (Picatinny Liquid Explosive). The methyl protons of nitromethane are acidic and in the presence of base the anion forms. The stabilization of the aci-ion by base has been suggested as the cause of their sensitization. Salts of nitromethane $NaCH_2NO_2$ are much more sensitive than CH_3NO_2.

$$CH_3NO_2 + NaOH \longrightarrow H_2O + CH_2{=}N\begin{matrix} O^- \\ \\ O^- \end{matrix}\ Na^+.$$

Tetranitromethane is a dense liquid (p, 1.64 g/cc) usually prepared by the method of Liang [Horning (1955)]. Although the boiling point of tetranitromethane is listed as 126 °C (mp, 14 °C) with an ambient vapor pressure of 13 mm, it is not distilled as it can decompose with explosive violence. Tetranitromethane is so rich in oxygen that it does not make a good explosive unless it is mixed with fuel such as nitrobenzene, toluene, or other arenes

$$\underset{\text{acetic anhydride}}{CH_3\overset{O}{\overset{\|}{C}}O\overset{O}{\overset{\|}{C}}CH_3} + \underset{\text{anhydrous nitric acid}}{HNO_3} \xrightarrow{10\,^{\circ}C} CH_3\overset{O}{\overset{\|}{C}}ONO_2 \xrightarrow{r.t.} (NO_2)_3C\overset{O}{\overset{\|}{C}}ONO_2$$

$$\xrightarrow{-CO_2} C(NO_2)_4.$$

Few other nitroalkanes are sufficiently thermally stable or readily preparable to be used as explosives.

1,2 *Dinitroethane* (bp, 135 °C; mp, 39 °C; p, 1.46 g/cc) is prepared by the action of NO_2 on ethylene. Dinitroethane $(NO_2)\ CH_2{-}CH_2(NO_2)$ is a powerful explosive, but it is so highly reactive that it has not been used as explosive.

Polynitroethylene—$[CH_2{-}CH(NO_2)]_n$—is made from the monomer nitroethylene, $CH_2{=}CHNO_2$. The monomer is a yellow–green lachrymatory liquid (bp, 35 °C at 70 mm Hg) which readily polymerizes to a white water-insoluble powder. The polymer burns without melting; it is a weak explosive and has no practical use due to its low thermal stability. This is one of the few examples given in this document of an energetic polymer, but there are many others.

Hexanitroethane (mp, 142 °C) has the ability to gelatinize nitrocellulose; therefore, an attempt was made to use it in smokeless powder. However, its high production costs and low thermal stability (decomposition at 75 °C) prevented it from being of practical use.

2,2 *Dinitropropane*, a white crystalline (mp, 51–52 °C), is generally considered too thermally unstable for use. Held at 75 °C for 48 hours in a closed container, it loses two-thirds of its weight.

Nitrocubanes. Octanitrocubane with no hydrogen atoms would be an ideal explosive from an oxygen balance point of view; ring strain would give it added detonation energy. Although the octanitrocubane has yet to be made, the synthesis of hexanitrocubane and heptanitromethylcubane have been achieved by the Eaton group at Chicago [Eaton (1992), Borman (1994a), Picatinny (1996, 1997); Nielsen (1995)].

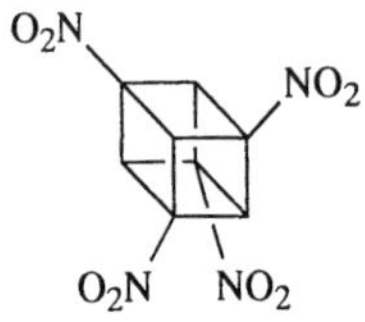

Tetranitrocubane

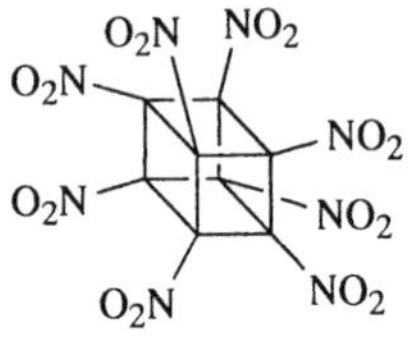

Octanitrocubane

5.6. Nitramines

Nitramine. NH_2NO_2 is the simplest nitramine, a white crystalline substance, melting at 72 °C. Unlike Me_2NNO_2, which has no hydrogens attached to the central N, nitramine is strongly acidic. Pure nitramine decomposes slowly on standing, forming nitrous oxide N_2O and water. It cannot be stored for more than a few days, and, therefore, is only of academic interest. The synthesis shown below is typical for all linear nitramines—dehydration of nitrate salt

$$NH_4NO_3 \xrightarrow{H_2SO_4} NH_2NO_2 + H_2O.$$

Urea Nitrate. Urea is manufactured by pumping ammonia and carbon dioxide into an autoclave where they are heated under pressure. Ammonium carbamate is formed first. This loses water to form urea. Urea is sometimes incorporated in blasting explosives for the purposes of lowering the temperature of explosion. It is also used as a stabilizer to remove acidity. Urea nitrate is formed by nitration of urea. It is stable, powerful, cool explosive; but it is corrosively acidic in the presence of moisture.

$$2NH_3 + CO_2 \longrightarrow \underset{\text{ammonium carbamate}}{NH_3{\cdot}HOC(O)NH_2} \xrightarrow{-H_2O} \underset{\text{urea}}{(NH_2)_2C{=}O}$$

$$\text{Urea} \xrightarrow{+HNO_3} \underset{\text{urea nitrate}}{(NH_2)_2C{=}O{\cdot}HNO_3} \xrightarrow{-H_2O} \underset{\text{nitrourea}}{NH_2\overset{O}{C}NHNO_2}\,.$$

Nitrourea is prepared by adding dry urea nitrate to concentrated sulfuric acid, while stirring and maintaining the temperature below 0 °C. Nitrourea decomposes in the presence of moisture.

Guanidine Nitrate (GN) is a intermediate in nitroguanidine manufacture and is itself a weak explosive sometimes used in mixture with other explosives, such as ammonium nitrate. Guanidine nitrate has a melting point of 215–216 °C and exists in two tautomeric forms:

$$\begin{array}{c} NH_3^+NO_3^- \\ | \\ C{=}NH \\ | \\ NH_2 \end{array} \longleftrightarrow \begin{array}{c} NH_2 \\ | \\ C{=}NH_2^+NO_3^- \\ | \\ NH_2. \end{array}$$

The synthesis shown below is the British aqueous fusion method which produces guanidine nitrate (GN) in a single reactor. The GN magma is pumped to the next reactor for dehydration to nitroguanidine

$$\underset{\text{calcium carbide}}{CaC_2 + N_2} \longrightarrow Ca(CN)_2 + CO_2 + H_2O \xrightarrow[118\,°C]{-CaCO_3} \underset{\text{cyanamide}}{NH_2C{\equiv}N}$$

$$\text{Cyanamide} \xrightarrow{+NH_4NO_3} \text{GN} \xrightarrow{-H_2O} \text{Nitroguanidine}$$

Nitroguanidine (NQ) is prepared by dehydration (with fuming H_2SO_4) of guanidine nitrate. Nitroguanidine is relatively stable to impact and heat, but it decomposes upon melting. Two melting points are reported 232 °C and 257 °C, and two tautomeric forms exist.

$$\begin{array}{c} NHNO_2 \\ | \\ C{=}N \\ | \\ NH_2 \end{array} \longleftrightarrow \begin{array}{c} NH_2 \\ | \\ C{=}NNO_2 \\ | \\ NH_2. \end{array}$$

The decomposition of nitroguanidine is reported to be accelerated by ammonia. Since its decomposition produces ammonia, along with melamine and possible cyanuric acid, the reaction is autocatalytic.

When nitroguanidine (NQ) is incorporated in nitrocellulose, the powder is flashless. However, flashless colloided powders containing NQ produce considerably more gray smoke than other flashless powders. NQ forms low melting eutectics with ammonium nitrate (AN) and guanidine nitrate (GN). It is an exception to the rule that most primary nitramine have no practical application as explosives.

20% NQ + 80% AN	132 °C,
41% NQ + 59% GN	167 °C,
17.5% NQ + 22.5% GN + 60% AN	113 °C.

RDX. Research Development Explosive, Cyclonite, Hexagen, cyclotrimethylenetrinitramine, and 1,3,5-trinitro-1,3,5,-tri-azacyclohexane are all names used to describe $[CH_2{-}N(NO_2)]_3$. RDX is more powerful than TNT and picric acid, replaces tetryl, and is less sensitive than PETN. In the original patent of 1899, the inventor suggested its might find medicinal use. RDX first became an important explosive in World War II. At the end of the war German production was 7000 tons/month and United States 15,000 tons/month.

There are two routes to RDX formation. Both require hexamethylene tetramine (also abbreviated hexamine), which is prepared as shown below

$$6CH_2O + 4NH_4NO_3 + 3(CH_3CO)_2O \longrightarrow \text{Hexamine} + 4HNO_3 + 6CH_3COOH$$

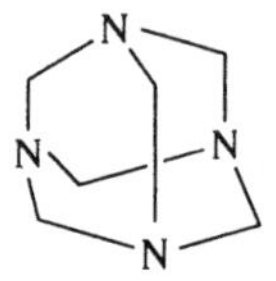

Hexamethylene tetramine

The British synthesis of RDX is direct nitration of hexamethylene tetramine, where two reactions proceed simultaneously:

$$(CH_2)_6N_4 + 4HNO_3 \rightarrow (CH_2NNO_2)_2 + 3CH_2O + NH_4NO_3,$$

$$(CH_2)_6N_4 + 6HNO_3 \rightarrow (CH_2NNO_2)_3 + 6H_2O + 3CO_2 + N_2.$$

The American synthesis, the Bachmann process, nitrates hexamethylene tetramine by a mixture of nitric acid, NH_4NO_3, acetic anhydride. Initially, hexamethylene tetramine dinitrate is formed by the low-temperature (below 15 °C) action of 50% nitric acid on hexamine. Further reaction forms RDX, HMX, and a variety other nitramines. By maintaining the temperature at 75 °C undesirable open-chain nitramines are destroyed and a yield of 75–80% RDX with some (10%) HMX contamination is obtained. At the present writing, RDX is synthesized in a continuous process at Holsten in Kingsport, TN,

$$(CH_2)_6N_4 \cdot 2HNO_3 + 2NH_4NO_3 \cdot HNO_3 + 6(CH_3CO)_2O$$

$$\rightarrow 12CH_3COOH + 2(CH_2NNO_2)_3.$$

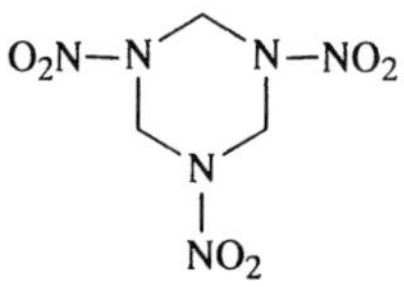

RDX 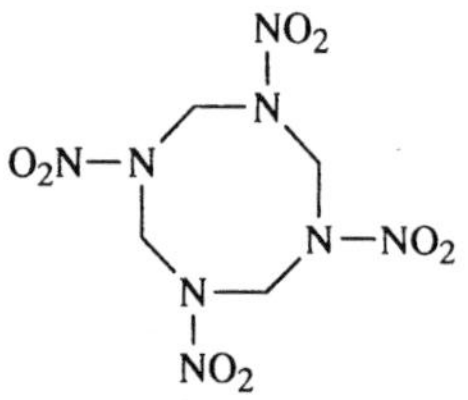 HMX

HMX. High Melting Explosive, cyclo-tetramethylenetetranitramine, octogen or 1,3,5,7-tetranitro-1,3,5,7-tetrazacyclo-octane $[CH_2—N(NO_2)]_4$ is a by-product (10–15%) of the RDX Bachmann synthesis. In fact, the synthesis of HMX uses the same reagents as that of RDX, but the temperature is held at 43 °C to favor HMX formation (9/1) over RDX. HMX is more powerful than RDX and more thermally stable. It has four polymorphs. Usually it is obtained in the beta form, which is the least impact sensitive. Neither RDX nor HMX are particularly water soluble (5 mg RDX dissolves in 100 ml water at 20 °C), [Gibbs and Popolato (1980)] and their chemistry is similar. However, HMX does not react with hydroxide, while RDX does; and HMX is more soluble than RDX in 55% nitric acid. Both of these reactions can be exploited as a means of separating them.

5.6.1. Nitramine Composites

Since the melting points of RDX and HMX are too high to permit melt casting, they are rendered shapeable by mixture with various additives. Such composites fall in three general categories. The examples given are not intended to be exhaustive of the possibilities.

A. Cyclonites Desensitized (Phlegmatized) with Wax
Molten (88 °C) wax 5–10% is stirred into the RDX. Dye and aluminum are often added.

B. Castable Mixture of Nitramine and TNT
TNT is melted and powdered or flaked nitramine is added to it. Such mixtures have various names and formulations:

COMP B	60% RDX/39% TNT/1% wax
COMP-B2	60% RDX/40% TNT
Cyclotol	75–70% RDX/25–30% TNT
Torpex-2	70% Comp B/12% TNT/18% Al
Octol	75–70% HMX/25–30% TNT

C. Plastic-Bonded Explosives
The nitramine is mixed with a plasticizer which allows it to be molded into the desired shape.

COMP C4 91% RDX, 9% plasticizer;
plasticizer = 5.3% di(-2-ethylhexyl)sebacate (recent, the sebacate has been replaced by the adipate);
2.1% polyisobutylene, 1.6% motor oil;

PBX 9407 94% RDX, 6% Exon 461;
Exon 461 copolymer = chlorotrifluoroethylene/tetrafluoroethylene/vinylidene fluoride.

[In April 1997, the US Congress ratified the International Commerical Aviation Organization (ICAO) Convention on the Marking of Plastic Explosives for the Purpose of Detection. When in effect, it will mandate the tagging of all new and stored C4 with a minimum of 0.1 wt% 2,3-dimethyl-2,3-dinitrobutane (DMNB).]

Tetryl 2,4,6-*trinitro-phenylmethylnitramine* or picrylmethylnitramine or 2,4,6,-trinitro-N-methylaniline was first described in 1879. N,N-dimethylaniline dissolved in concentrated sulfuric acid is slowly mixed with HNO_3. The resulting light yellow solid is recrystallized from benzene or acetone.

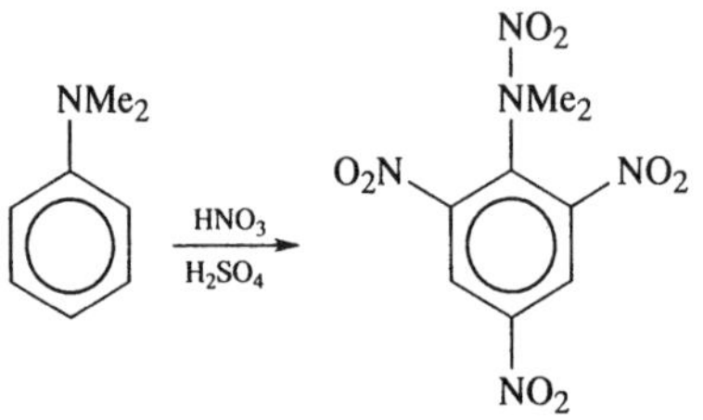

NMe_2 $\xrightarrow[H_2SO_4]{HNO_3}$ NO_2 / NMe_2 / O_2N / NO_2 / NO_2

Tetryl is more powerful and brisant than TNT and picric acid, but it is also more sensitive to shock. Tetryl is no longer commonly used as a United States military explosive due to the hazards of its preparation. Furthermore, if it is stored at 100 °C, it gives off nitrous fumes and formaldehyde; after 40 days a semiliquid mass is left.

CL-20 (China Lake compound 20) or *HNIW* are the usual abbreviations for hexanitrohexaazaisowurtzitane. This is a polycyclic nitramine which exists in several different polymorphs. One is thought to be the densest (p, 2.0g/cm^3) and most energetic explosive known [Borman (1994b); Brill et al. (1996)]. HNIW contains one 6- and two 5- and two 7-membered nitramine rings.

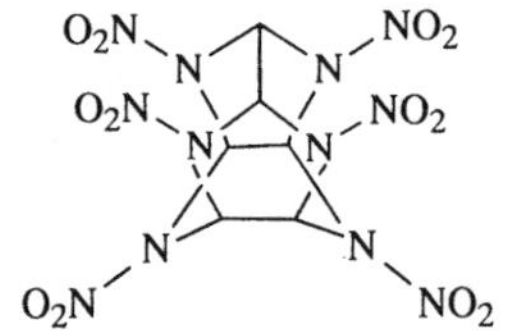

O_2N N N NO_2 O_2N N N NO_2 N N O_2N NO_2

CL-20

5.7. Heterocyclic Explosives

This is a not a true class of energetic material but is used here for explosives that don't fit well into the previous categories. Much effort is presently going into this area of explosive synthesis [Boyer (1986)]. These few compounds are meant to be representative of the type of compounds being prepared. Areas not discussed at all included proposed fluorinated energetic materials and energetic polymers.

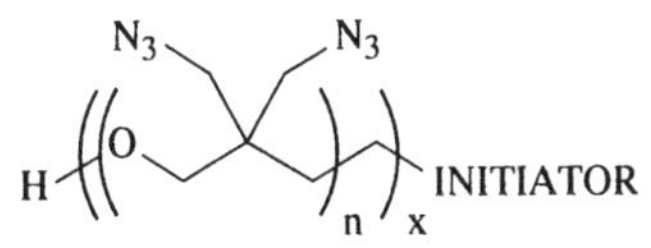

poly[bis(azidomethylene)oxetane]
(BAMO)

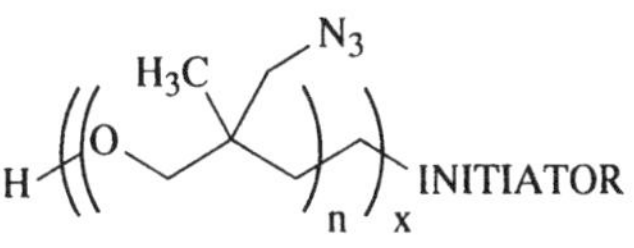

poly[(azidomethylenemethyl)oxetane]
(AMMO)

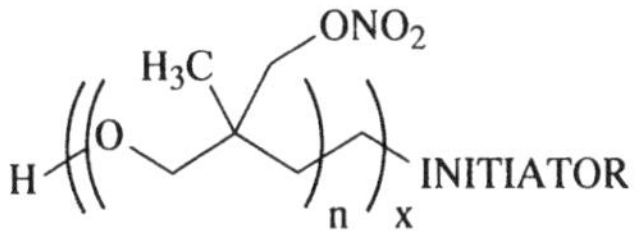

poly[(nitratomethylenemethyl)oxetane]
(NMMO)

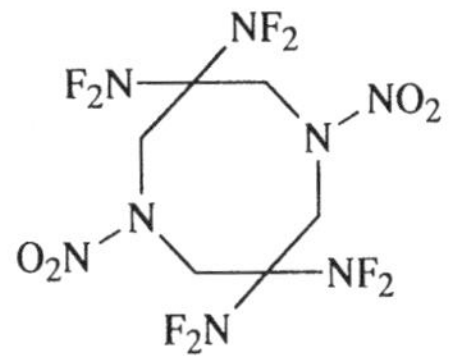

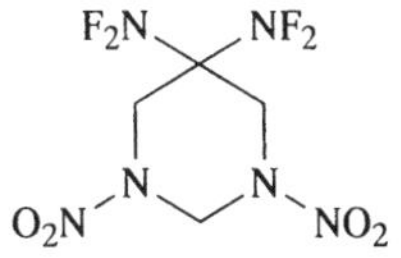

NTO. 3-Nitro-1,2,4-triazole-5-one is a relatively new explosive (1980s) with performance about 80% of TNT and sensitivity less than TNT. It is prepared by condensation of semicarbazide hydrochloride with formic acid, followed by nitration of the resulting tetrazole. NTO decomposes about 240 °C. Unfortunately, it is quite acidic, and this affects its chemical stability. Since it does not melt and has low solubility in most solvents, it or its derivatives are being considered in admixture with TNT or ammonium nitrate [Oxley et al. (1997a)].

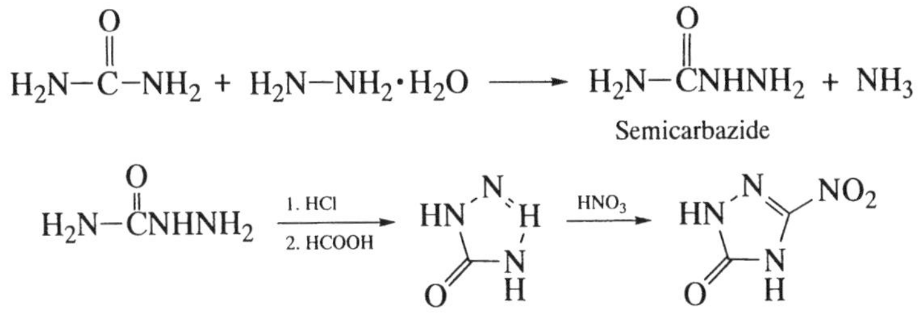

$$H_2N{-}C(=O){-}NH_2 + H_2N{-}NH_2 \cdot H_2O \longrightarrow H_2N{-}C(=O)NHNH_2 + NH_3$$

Semicarbazide

$$H_2N{-}C(=O)NHNH_2 \xrightarrow[\text{2. HCOOH}]{\text{1. HCl}} \text{TO} \xrightarrow{HNO_3} \text{NTO}$$

TO NTO

TNAZ. 1,3,3-Trinitroazetidine is a new (1984, Fluorochem) strained-ring explosive, which with a melting point of 101 °C is steam castable [Archibald et al. (1990); Lyer et al. (1991); Bottero (1996)]. Scale-up of its synthesis is in progress, improving a five-step route with low yield starting with epichlorohydrin and proceeding via the salt of *t*-butyl dintroazetidine. A discussion of one proposed synthetic route can be found in Coburn et al. (1997). Studies are underway to understand the thermal stability of TNAZ and to determine a mixture which would have a sufficiently low-melting point to retard sublimation [Oxley et al. (1997b); Picatinny (1997)].

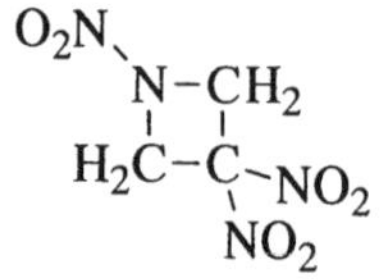

TNAZ

PYX. [2,6-Bis(picrylamino)-3,5-dinitropyridine] [Coburn et al. (1986)] was an early product of a Los Alamos National Laboratory effort to find useful peacetime uses for explosive. This particular material has high thermal stability and finds use in oil-well applications.

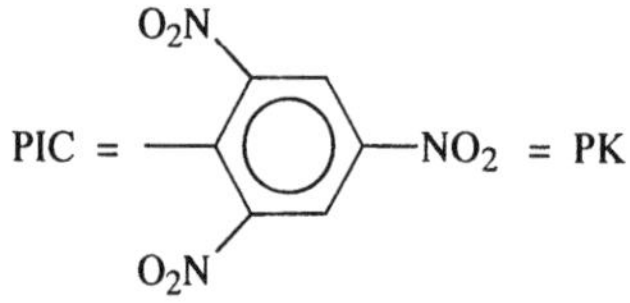

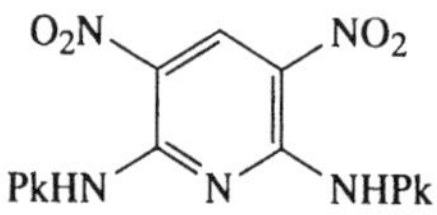

PIC = PK

PYX

5.8. Energetic Salts

Ammonium Nitrate (AN, NH_4NO_3) is prepared by reacting ammonia and nitric acid. Almost 18 billion pounds of AN are produced annually in the United States. Although its principal end use is fertilizer, about 20% is used in commercial explosives [Kirk-Othmer (1978)]. AN, ammonia, and nitric acid all rank in the top twenty chemicals produced [Chang and Tikkanen (1988)]. The synthesis is straightforward; ammonia vapor and nitric acid are mixed, and the exothermicity of the reaction causes the solution to boil and become more concentrated. It is further concentrated by vacuum evaporation. Variation in methods of removing the water produce different physical forms of AN. AN used in dynamites typically is concentrated in graining kettles, but in the late 1940s the development of a prilling process made the production of AN cheaper and faster. Today, AN, prepared for the fertilizer industry (FGAN) or the explosive industry (sometimes called industrial AN), typically, are prilled. A hot AN solution is pumped to the top of a prilling tower where it is sprayed into the cool air. As the AN spray falls, it forms small spherical pellets (1.5–2.5 mm) which are dried and dusted with talc to prevent sticking. Without coating, AN, being very hygroscopic, tends to form hard lumps. FGAN is prilled with less water (99% AN/1% water) than explosive-grade AN (95% AN/5% water). As a result, the explosive-grade AN requires a higher prilling tower (200 ft versus 60 ft for FGAN), and it is more porous than FGAN. The extra porosity helps the industrial grade AN soak up the fuel oil required to make it an explosive.

Some cite the Texas City disaster of April 1947, where two ships loaded with wax-coated AN detonated killing about 600 people, as the point at which the explosive power of AN was recognized. However, as early as 1867 two Swedish chemists patented an explosive which used AN alone or mixed with charcoal, sawdust, naphthalene, picric acid, nitroglycerin, or nitrobenzene. Nobel purchased the invention and used AN in dynamites. During World War I, Amatol, a mixture of AN and TNT in various proportions: 50/50, 60/40, or 80/20, was widely used. Amatols were not as brisant as TNT; detonation velocities decreased as the amount of AN increased.

	AN	TNT	Al flake	Stearic acid
German	54%	30%		16%
French	86%	8%	6%	

When combustible nonexplosives are added to AN, they react with the excess oxygen of AN to produce additional gas and heat and increase the power and temperature of the explosion. The combustible nonexplosive can be rosin, sulfur, charcoal, ground coal, flour, sugar, oil, or paraffin, but most often it is a fuel oil. AN was first used as the principle oxidizer in an explosive in 1953 by Akre of the Maumee Colliers in Terre Haute, Indiana. He mixed AN with coal dust. This process was used locally for several years before it was announced in 1955, but by 1958 it was widely accepted in the industry. In the Minnesota iron range, fuel oil replaced coal dust as the fuel. A mixture of AN with 5–6% fuel oil called ANFO is now a common blasting agent.

$$37NH_4NO_3 + CH_3(CH_2)_{10}CH_3 \rightarrow 12CO_2 + 13H_2O + 74H_2O + 37N_2.$$

AN may be mixed intimately with fuel in gels or emulsions. A typical aqueous emulsion contains 80% AN, 14% water, and 6% fuel mixed with an emulsifier such as sorbitan monooleate. AN becomes a more powerful explosive if mixed with an active fuel. An AN cartridge to which nitromethane is added just before use is sold commercially; AN with added hydrazine is a powerful liquid explosive, detonation velocity 6800 m/s. Because ANFO is easily prepared and was evidently used in the bombing of the Murrah building in Oklahoma City, April 19, 1995, there are a number of current research programs aimed at desensitizing commercially available AN.

AN has a melting point of 169 °C and begins to decompose as soon as it melts, the first step being dissociation into ammonia and nitric acid. At low temperature the decomposition mechanism is ionic, the slow step being the protonation of nitric acid. As a result, added acidic species, such as ammonium salts, accelerate its decomposition, while added basic species, such as the salts of weak bases, retard AN decomposition (see Chapter 9 for details).

AN/fuel formulations have almost completely replaced dynamites as a mining explosive [Hopler (1993b)]. As a result they are by far the explosives most widely used. AN formulations can be prepared which are quite powerful and cap-sensitive; nevertheless they perform as nonideal explosives. Ideal explosives such as TNT, RDX, HMX, and PETN have short (millimeter long) reaction zones; thus, upon initiation they release energy in a sharp pressure pulse supporting the detonation wave. This type of behavior has been modeled by a number of computer codes. Nonideal explosives have longer reaction zones than ideal explosives and, thus, release their energy more slowly. A smaller portion of their energy goes toward supporting the initial detonation wave and more of their energy is available for the following gas expansion which results in heaving power. Generally, heaving power is required from mining explosives although very hard rocks may require the penetrating power of an ideal explosive.

Ammonium Perchlorate (AP) is made by the electrochemical oxidation of sodium chloride NaCl to the chlorate $NaClO_3$ and on to the perchlorate $NaClO_4$. Metathesis with ammonium chloride produces ammonium perchlorae (AP) NH_4ClO_4. Energy-wise this is an expensive process, and both of the United States manufactures, Kerr–McGee and WECCO, are located near the Hoover Dam. Their combined yearly capacity is about 80 million pounds. Like ammonium nitrate, ammonium perchlorrate (greater than 45 μm particle size) has been classed as an oxidizer rather than as an explosive for purposes of shipping. The explosive capacity of AP was demonstrated in May 1988 when the one of the two United States AP manufacturing plants, PEPCON, blew up. Half the United States production capacity was temporarily lost, and two people were killed.

Both the French and Germans used AP explosives during World War I [Davis (1943)]. In World War II the United States used Galcit propellant that incorporated $KClO_4$ (75%) into molten asphalt (25%) [Kaye (1960–1978)]. It was the precursor of modern composite propellants in which AP is embedded in a polymer. The TNT equivalence of AP is about 0.31.

All the oxides of chlorine [hypochlorite (ClO^-), chlorite (ClO_2^-), chlorate (ClO_3^-), and perchlorate (ClO_4^-)] are energetic, but perchlorate salts are the most stable. Chlorates ClO_3^- are especially hazardous to handle. They decompose exothermically and are sensitive to heat, impact, and friction. Many chlorate mixtures, particularly those which contain sulfur, sulfide, or picric acid are extremely sensitive to blows and friction. The sensitivity can be reduced by phlegmatization in castor oil. Chlorate explosives with aromatic nitro compounds have higher detonation velocities and are more brisant than those in which the carbonaceous material is merely combustible. In 1885, 240,000 lb of a mixture of $KClO_3$ (79%) and nitrobenzene (21%) along with 42,000 lb dynamite were used to blast a portion of Hell Gate Channel in New York Harbor. Other similar mixtures are turpentine/phenol (90/10) absorbed on $KClO_3/MnO_2$ (80/20) or nitrobenzene/turpentine (80/20) absorbed on $KClO_3/KMnO_4$ (70/30). Flash powder is a mixture of potassium chlorate, sulfur, and aluminum. Mixtures of chlorate and fuel will spontaneously ignite with the addition of a drop of concentrated sulfuric acid (H_2SO_4). Spontaneous ignition or explosion can occur when alkali chlorates are combined with very reactive fuels (such as phosphorus, sulfur, powdered arsenic, or selenium) or with moist fuels. In fact, when powdered, dry, unoxidized $KClO_3$ and red phosphorus (Armstrong's powder) are pushed together, they ignite; this reaction has been tamed and utilized by use of separation and a binder in the common safety match. Armstrong's powder, wet with some volatile solvent such as methanol, has been used as an antipersonnel device. MnO_2 has been reported as a catalyst for the decomposition of chlorates and perchlorates.

Hypochlorites are generally highly reactive and unstable. The calcium salt (HTH), one of the most stable, ignites spontaneously with glycerin. Hypochlorite are used as liquid household bleach (an alkaline solution of NaOCl); as household dish washing detergents and scouring powders [$(Na_3PO_4 \cdot 11H_2O)_4 \cdot NaOCl$]; as a liquid bleach for pulp and paper bleaching [a mixture of $Ca(OCl)_2$ and $CaCl_2$]; and as a powdered swimming pool bleach [$Ca(OCl)_2/CaCl_2/Ca(OH)_2 \cdot 2H_2O$].

Hydroxylammonium Nitrate (HAN) NH_3OHNO_3 is a low-melting (42 °C) energetic salt which for some time has been considered as a possible propellant oxidizer. It is very difficult to maintain as a solid since upon exposure to air, it picks up to 95% of its weight in water and immediately dissolves. As a possible aqueous propellant, it is stabilized by an amine, but the salt is stable over a very narrow pH range, being sensitive to both acid and base. It has low thermal stability decomposing to water, N_2O, N_2, and HONO, an acid, which accelerates its decomposition.

Ammonium Dinitramide (ADN) $NH_4N(NO_2)_2$ is a low-melting (94 °C) energetic salt which was synthesized for the first time in the USSR (1971) and later in the United States (1991). Presently, it is considered the hot new candidate propellant, but much testing and formulating need to be done. This material is much less thermally stable than AN, probably due to its much lower melting point, and is light sensitive. One decomposition route produces ammonium nitrate and N_2O [Oxley et al. (1997c)].

Mercury Fulminate $Hg(ONC)_2$ is a primary explosive, sensitive to heat and friction. Its main use has been as a primer for initiating high explosives. Used in combination with $KClO_3$ it has a larger effect. It synthesis is mechanistically complex but in the laboratory simply involves dissolving mercury in concentrated nitric acid, adding ethanol, and thoroughly washing the white crystals of mercury fulminate when their formation is complete. Mercury fulminate is sensitive to light, undergoes marked decomposition above 50 °C, and is usually stored under water.

Azides are roughly divided into three class: stable ionic azides (alkali and alkaline earth azides); unstable covalent azides (haloazide) which frequently explode spontaneously, and heavy-metal azides [$Pb(N_3)_2$, AgN_3] that explode on shock. It is the alter group that are often are used as primers for initiating high explosives. The usually synthetic route is reaction of the metal nitrate with sodium azide [Fair and Walker (1977)].

5.9. Composite Explosives

Many potential explosives can be broadly classed as composite explosives. Rather than containing the oxidizer and fuel in a single molecule, as do the organic military explosives, composite explosives are formed by intimately mixing oxidizing compound(s) with fuel(s). These can be premixed or mixed just prior to use. In such mixtures there can be problems due to inhomogeneities; the finer the solid particle size and the more intimate the mix, the better the performance. Classic examples of composite explosives are black powder [a mixture of the oxidizer potassium nitrate (75%) and sulfur (10%) with the fuel charcoal (15%)] and ANFO. The tables below list oxidizers and fuels which can be combined to form composite explosives.

Oxidizers	
Oxygen and halogens	
Perchlorates	$KClO_4$, NH_4ClO_4, $NaClO_4$, and Ba and Ca salts
Chlorates	$KClO_3$, and Li, Na, Ba salts
Hypochlorite	$Ca(OCl)_2$
Nitrates	KNO_3 and NH_4, Na, Ba, Ag, Sr salts
Chromates	$PbCrO_4$ and Ba, Ca, K salts
Dichromates	$K_2Cr_2O_7$ and $NH_4Cr_2O_7$
Iodates	KIO_3 and Pb, Ag salts
Permanganate	$KMnO_4$
Metal oxides	BaO_2, Cu_2O, CuO, Fe_2O_3, Fe_3O_4, PbO_2, Pb_3O_4, PbO, MnO_2, ZnO
Peroxides	Na_2O_2, H_2O_2 (80%), dibenzoylperoxide

Fuels		
nitrobenzene	petroleum	halogenated-hydrocarbons
nitrotoluenes	turpentine	halogens
nitronaphthalene	naphtha	powdered metals
nitrocellulose	castor oil	carbon disulfide (CS_2)
picric acid	sugar	phosphorus (P_4)
	glycerin	sulfur (S_8)
	acetylene	
	wax, paraffin	
	sawdust	

5.10. Liquid Oxidizers and Explosives

The ultimate oxidizer is oxygen. Shortly after a machine was developed for the liquefaction of gases (1895) liquid oxygen explosives (LOX) came into use. LOX are formed by impregnating porous combustible materials, such as lampblack, with liquid oxygen. Two problems exist with liquid oxygen containing explosives: they lose their explosiveness as the liquid oxygen evaporates (bp, −183 °C); and they are easily inflamed. During World War I the Germans used LOX when other explosives ran low. In 1926, LOX were used for the first time in commercial rock blasting operations; their use was continued into the 1960s. The combination of liquid oxygen with the detonable fuel acetylene (acetylene 25%/O_2 75%) produces an explosive with detonation velocity comparable to military explosives (6000 m/s) [Kaye (1960–1978)].

Nitrogen tetroxide (N_2O_4) is another liquid oxidizer capable of forming powerful explosives. Liquid nitrogen tetroxide exists over the narrow temperature range between its freezing point (−11 °C) and 21 °C, where it dissociaties into nitrogen dioxide (NO_2), a toxic gas. Panclastites, explosives made with liquid N_2O_4 and combustible liquids (carbon disulfide, nitrobenzene, nitrotoluene, gasoline, halogenated hydrocarbons), were first considered when the Germans tested marine torpedoes containing sealed glass containers of N_2O_4 and CS_2 in the 1880s; set-back forces broke the glass containers generating the explosive mixture, and an impact fuse initiated detonation. Similar N_2O_4/fuel devices were used in both World War I and World War II. Nitrogen tetroxide oxidizers are still in use in space shuttle propellants. Panclastites are inexpensive and easy to prepare; some are more brisant and have better detonation velocities than TNT. However, though their performance is favorable, panclastites are too shock sensitive, too hard to handle, to find common military use. Their extreme sensitivity dictates that they be mixed just prior to use, and the corrosive nature of N_2O_4 requires special vessels. Mixtures of N_2O_4 with 64% nitromethane have a detonation velocity of 6900 m/s [Kaye (1960–1978)]. Nitrogen tetroxide explodes on contact with a number of fuels: acetic anhydride, liquid ammonia, methyl and ethyl nitrate, propene, hydrazine-type fuels.

Peroxides. Peroxides, with oxygen bonded to oxygen, can be violent oxidizers in the presence of fuel. For example, sodium peroxide (Na_2O_2) instantly ignites in the presence of moisture and a fuel (magnesium and sawdust or paper, or sulfur or aluminum). In addition to this feature, peroxides can also undergo a violent self-decomposition. Peroxide decomposition into water and oxygen can be catalyzed by small amounts of alkaline lead, silver, or manganese salts, or even saliva [Kaye (1960–1978); Oxley (1993)]. Pure hydrogen peroxide decomposes violently above 80 °C; therefore, it is sold as aqueous solutions. It is available at pharmacies as a 3% solution for use as a disinfectant or as a 40% solution for use as a hair lightener or as a gel to brighten teeth. Hydrogen peroxide pure and in concentrations as low as 86% is detonable. Solutions of 90.7% peroxide have reported detonation velocities of 5500 m/s to 6000 m/s. Mixtures of hydrogen peroxide vapor in air with as little as 35 mol% H_2O_2 are reported to detonate at 1 atmosphere with a velocity of 6700 m/s. Furthermore, hydrogen peroxide, pure or in water, is readily detonable when mixed with organic materials. H_2O_2/water/ethanol has a detonation velocity of 6700 m/s [Kaye (1960–1978)].

Austria made unsuccessful attempts to use H_2O_2 as an explosive in World War I. In World War II, the US Navy used it for propulsion in submarine torpedoes. Peroxide explosives have been successfully used in blasting operations. In addition to its monergolic application, hydrogen peroxide can be used with a number of fuels. Hydrogen peroxide mixed with fuels such as methanol, ethanol, or glycerol showed detonation rates as high as 6700 m/s. One of the propellant systems on the space shuttle uses the combination of hydrogen peroxide and unsymmetrical dimethylhydrazine. H_2O_2 (60%) with paraformaldehyde forms a crystalline compound of high brisance and sensitive (mp, 50 °C). Hydrogen peroxide (70%) with diesel fuel and gelling agent also makes a good explosive. H_2O_2 (83%) plus cellulose forms a gelatinous mass which is more powerful than TNT and insensitive to shock or friction. It has an ignition temperature of 200 °C; however, it cannot be stored over 48 hours without evolution of peroxide and loss of explosive power. Other patented peroxide explosives include H_2O_2 with water and glycerol, H_2O_2 (70%) with powdered boron (30%), and H_2O_2 used with hexamethylenetetramine and HCl [Kaye (1960–1978)].

Recently two peroxides have received wide publicity in the terrorist communities: triacetone triperoxide (TATP) and hexamethylene triperoxide diamine (HMTD). TATP can be synthesized from acetone, hydrogen peroxide, and sulfuric acid [Black Book (1977)]. The white crystalline solid which forms after standing 24 hours explodes violently upon heating, impact, or friction. It is highly brisant, very sensitive, and detonable under water. Its reported detonation velocity is 5290 m/s [Meyer (1987)]. It has been suggested for use in primers and detonators, but due to its volatility and sensitivity it has not found military application. HMTD is formed by a similar synthesis from hexamethylene, hydrogen peroxide, and acid.

$$CH_3\overset{\overset{O}{\|}}{C}CH_3 + H_2O_2 + H_2SO_4 \longrightarrow \begin{array}{c} O{-}C(CH_3)_2{-}O \\ | \qquad\qquad | \\ O \qquad\qquad O \\ | \qquad\qquad | \\ (CH_3)_2C{-}O{-}O{-}C(CH_3)_2 \end{array} \quad \text{TATP}$$

$$(CH_2)_6N_3 + H_2O_2 \longrightarrow \begin{array}{c} \ulcorner CH_2{-}O{-}O{-}CH_2 \urcorner \\ N{-}CH_2{-}O{-}O{-}CH_2{-}N \\ \llcorner CH_2{-}O{-}O{-}CH_2 \lrcorner \end{array} \quad \text{HMTD}$$

5.11. Unconventional Explosives

Metals. Some alkali metals spontaneously ignite on exposure to water or air. As the alkali metals increase in weight, their reaction to air becomes more violent. While potassium may oxidize so rapidly that it melts and ignites when pressure is applied (as in cutting), cesium burns in air as soon as it is removed from an inert oil covering. Moisture in the air serves to enhance further reactivity. Sodium and potassium form a eutectic (NaK) which is spontaneously ignitable. Sodium/potassium alloys are reported to react explosively upon contact with silver halides or to detonate upon contact with halogenated organic materials such as carbon tetrachloride. Potassium and heavier alkali metals burst into flame upon contact with water. Sodium too will inflame in water if it can be anchored in one spot long enough to allow the heat of reaction to ignite the hydrogen being produced:

$$Na + H_2O \rightarrow NaOH + \tfrac{1}{2}H_2.$$

Lithium is the least reactive alkali metal but will ignite if thrown on water as a dispersion. In World War II the Germans used land mines composed of sodium and methyl nitrate in separate compartments. Pressure brought the two together and into action.

Some finely-divided (powdered) nonalkali metals will also burst into flame in the presence of air. The best known are lead, iron, nickel, cobalt, and aluminum. These can be prepared by pyrolysis of their organic salts or by reduction of their oxides, or in some cases, by formation of a mercury amalgam. These metals may also explosively react with water, halogenated hydrocarbons, and halogens.

$$Al + CCl_4 \rightarrow AlCl_3.$$

Magnesium is used in a number of pyrotechnics. When a magnesium/silver nitrate mixture is moistened, it reacts explosively. Teflon $(C_2F_4)_n$ with powdered magnesium reacts explosively upon ignition. Devices of this composition are used as decoys for heat-seeking missiles.

$$(C_2F_4)_n + 2n\,Mg \rightarrow 2n\,C + 2n\,MgF_2.$$

Some methyl- and ethyl-substituted metals are spontaneously ignitable in air. The alkylated metals most frequently exhibiting this behavior are the alkali metals (Li, Na), aluminum, zinc, and arsenic or nonmetals such as boron and phosphorus. Many of these compounds also react explosively with water and with carbon tetrachloride (CCl_4). It is reported that triethylaluminum $[Al(C_2H_5)_3]$ in carbon tetrachloride reacts explosively when warmed to room temperature.

Thermite is generally the redox reaction between a metal oxide and a metal. However, the most important reaction and the one usually referred to by this name is that of aluminum and iron oxide:

$$8Al + 3Fe_3O_4 \rightarrow 4Al_2O_3 + 9Fe.$$

This reaction generates a tremendous amount of heat; molten iron is produced and its melting point is above 1530 °C. One peaceful application of this reaction is for welding in shipyards and railroads. With $KMnO_4$ in the metal mixture, reaction can be triggered with added glycerol. With sugar in the initial mix, reaction is triggered with a drop of concentrated H_2SO_4. Such initiating schemes are usually used in laboratory demonstrations. Thermite reactions using CuO or Mn_3O_4 are reported explosive. Mixtures of Pb, PbO_2, and PbO also undergo explosive thermite reactions.

Acetylides. A number of metal carbides exist which are explosive in their own right; most are termed acetylides rather than carbides. Copper acetylide and silver acetylide are most commonly prepared by teenagers. Being primary explosives, they explode violently upon heating, impact, or friction. Cuprous acetylide is the only acetylide which has been used in the explosives industry; it has been used in electric detonators. Acetylides can be formed by passing acetylene through a solution of the appropriate metal salt.

Self-Igniting Materials. Some chemicals are so reactive to the oxygen in air or to water that they spontaneously ignite. Three parameters affect the spontaneity of ignition in air: the dryness of the air, air pressure, and temperature. Most of these chemical systems cannot be classed as explosives, but if sufficient gas pressure and heat are evolved the effect could be catastrophic.

Hydrides. Phosphines, silanes, and boranes ignite on contact with air. Diphosphine (P_2H_4), a liquid at room temperature, can be made from the reaction of water with solid calcium phosphide (Ca_3P_2), which, in turn, can be formed from lime and red phosphorus. Adding water to calcium phosphide results in a mixture of phosphine and diphosphine, and a violent deflagration ensues. This reaction as been exploited in naval flares. Only mono- and di-silanes (SiH_4 and Si_2H_6) are stable to air at room temperature. The higher silanes decompose violently. Diborane (B_2H_6) is a gas available in cylinders or by the action of 85% phosphoric acid on $NaBH_4$. The gas is highly toxic, and, unless it is extremely pure, it reacts with oxygen at room temperature. Borane decomposition in oxygen is extremely exothermic; therefore, boranes, such as decaborane (14) ($B_{10}H_{14}$), have been seriously considered as a component in rocket fuel.

Phosphorus. White phosphorus tends to ignite with slight pressure or by contact with fuel. P_4 self-ignites in air above 34 °C; as a result, it is usually stored under water. The finely divided phosphorus left on the combustible material reacts exothermically with the oxygen in air:

$$P_4 + 5O_2 \rightarrow P_4O_{10}.$$

The heat of this reaction could initiate the reaction between carbon disulfide and air. Drying of the phosphorus can be delayed by addition of a high-boiling hydrocarbon such as gasoline or toluene. In contrast to white phosphorus, red phosphorus is nontoxic and less sensitive. Red phosphorus bursts into flames or explodes on mild friction or impact in mixture with chlorate, permanganates, lead dioxide (PbO_2), perchlorate, and other active oxidizers ($AgNO_3$).

Miscellaneous Energetics. Potassium permanganate and glycerin will ignite spontaneously after a small delay, due to the difficulty in wetting the $KMnO_4$ with viscose glycerin. Ethylene glycol, acetaldehyde, benzaldehyde, or DMSO could be used in place of glycerin. Potassium permanganate and concentrated sulfuric acid can readily inflame when in contact with fuels

$$14KMnO_4 + 4C_3H_5(OH)_3 \rightarrow 7K_2CO_3 + 7Mn_2O_3 + 5CO_2 + 16H_2O.$$

Fuels. In addition to the composite explosives resulting from direct mixing of fuels with the oxidizers, another class of explosives should be mentioned. Fuel/air explosives (FAE) involve ignition of fuel droplets dispersed in air. The first FAE bombs were used in VietNam to clear mine fields and open helicopter landings. They can produce very large static and dynamic impulses per weight fuel (since they do not have to carry their own oxygen) and can cause blase effects over large areas [Kaye (1960–1978)].

Ethylene oxide is a colorles gas (bp, 13.5 °C) with a wide detonable concentration range (3–100%). The reason it is detonable at any concentration is because oxygen is already incorporated in the ethylene oxide molecule. However, for complete combustion it requires additional oxygen from the air or from hydrogen peroxide

$$2\,\overset{\frown O \frown}{CH_2{-}CH_2} + 5O_2 \rightarrow 4CO_2 + 4H_2O.$$

Like ethylene oxide, acetylene is a colorless gas. Acetylene, with a carbon–carbon triple bond, releases a great deal of energy when these bonds are broken to combine with oxygen making CO_2. As a result, detonable limits of acetylene are nearly the same as its flammability limits [Burgess et al. (1968)]. Almost any mixture of the gas is flammable and detonable (from concentrations of 2.5% to 100% in air).

Cyanogen (C_2N_2) is a pungent gas that is easily prepared from mixing solutions of KCN and $CuSO_4$ or by heating $Hg(CN)_2$. It is used in warfare as a poisonous gas, but it should not be overlooked that cyanogen is also a high explosive. It explosive range is wide, from 7% to 73% in air.

References

Archibald, T.G., Gilardi, R., Baum, K., and George, C. (1990). *J. Org. Chem.*, **55**, 2920–2924.

Borman, S. (1994a). Military research on cubane explosive. *C&E News*, Nov. 28, p. 34.

Borman, S. (1994b). Advanced energetic materials emerge. *C&E News*, Jan. 17, p. 18.

Bottaro, J.C. (1996). Recent advances in explosives and solid propellants. *Chemistry and Industry*, April, pp. 249–252.

Boyer, J.H. (1986). *Nitroazoles.* Organic Nitro Chemistry Series, Vol. 1. VCH, Essen, Germany.

Brill, T., Russell, T., Tao, W., and Wardle, R., eds. (1996). *Decomposition, Combustion, and Detonation Chemistry of Energetic Materials.* MRS, Pittsburgh, PA.

Burgess, D.S., Murphy, J.N., Hanna, N.E., and Van Dolah, R.W. (1968). Large-scale studies of gas detonations. U.S. Bureau of Mines Report. Investigation No. 7196.

Chang, R. and Tikkanen, W. (1988). *The Top Fifty Industrial Chemicals.* Random House, New York.

Coburn, M.D., Harris, B.W., Lee, K.Y, Stinecipher, M.M., and Hayden, H.H. (1986). Explosives synthesis at Los Alamos. *I&EC Product Res. & Dev.*, **25**, 68.

Coburn, M.D., Hiskey, M.A., Oxley, J.C., Smith, J.L., Zheng, W., and Rogers, E. (1997). Synthesis and spectra of some ^2H, ^{13}C, and ^{15}N isomers of 1,3,3-trinitroazetidine and 3,3-dinitroazetidinium nitrate. *J. Energetic Materials* (to appear).

Davis, T.L. (1941, 1943, reprint of two volumes). *The Chemistry of Powder and Explosives.* Angriff Press, Hollywood, CA.

Dobratz, B.M. (1981). *LLNL Explosives Handbook: Properties of Chemical Explosives and Explosive Simulants.* Lawrence Livermore National Laboratory, UCRL-52997.

Eaton, P.E. (1992). Cubanes: Starting materials for the chemistry of the 1990s and new century. *Angew. Chem. Int. Ed. Engl.*, **31**, 1421–1436.

Fair, H.D. and Walker, R.F. (1977). *Energetic Materials*, vols. I and II. Plenum Press, New York.

Gibbs, T.R. and Popolato, A., eds. (1980). *LASL Explosive Property Data*. University of California Press, Berkeley, CA.

Hopler, R.B. (1993a). *The Story of Dynamite* (video). Ireco-Dyno.

Hopler, R.B. (1993b). Custom-designed explosives for surface and underground coal mining. *Mining Engineering*, October, pp. 1248–1252.

Horning, E. C., ed. (1955). *Organic Synthesis*, vol. 3. Wiley, New York, p. 803.

Kamlet, M.J. (1976). The relationship of impact sensitivity with structure of organic high explosives. I. Polynitroaliphatic explosives. *Sixth Symposium (International) on Detonation*. ONR ACR-211, pp. 312–322.

Kaye, S.M. (1960–1978). *Encyclopedia of Explosives and Related Items*. U.S. Army Armament Research and Development Command; PATR-2700. Dover, New York.

Kirk-Othmer (1978). *Encyclopedia of Chemical Technology*, 3rd ed., Wiley, New York.

Lyer, S., Eng, Y.S., Joyce, M., Perez, R., Alster, J., and Stec, D. (1991). *Proceedings of ADPA Manufacture Compatibility of Plastics and Other Materials with Explosives, Propellants, Pyrotechnics and Processing of Explosives, Propellants, and Ingredients*, San Diego, CA, p. 80.

Marshall and Peace (1916). Vapor pressure of glyceryl trinitrate. *J. Chem Soc. Ind.*, **109**, 298.

Meyer, R. (1987). *Explosives*, 3rd edn. VCH, Essen, Germany.

Nielsen, A.T. (1995). *Nitrocarbons.* Organic Nitro Chemistry Series, Vol. 5. VCH, Essen, Germany.

Olah, G.A., Malhotra, R., and Narang, S.C. (1989). *Nitration: Methods and Mechanisms.* Organic Nitro Chemistry Series, Vol. 3. VCH, Essen, Germany.

Oxley, J.C. (1993). Non-traditional explosives: Potential detection problems. *Terrorism and Political Violence*, **5** (2), 30–47.

Oxley, J.C., Smith, J.L., Ye, H., McKenney, R.L., and Bolduc, P.R. (1995). Thermal stability studies on homologous series of nitroarenes. *J. Phys. Chem.*, **99** (24), 9593–9602.

Oxley, J.C., Smith, J.L., Rogers, E., and Dong, X.X. (1997a). NTO decomposition products tracked with N-15 labels. *J. Phys. Chem.*, **101** (19), 3531–3536.

Oxley, J.C., Smith, J.L., Zheng, W., Rogers, E., and Coburn, M.D. (1997b). Thermal decomposition pathways of 1,3,3-trinitroazetidine (TNAZ), related 3,3-dinitrozetidinium salts, and ^{15}N, ^{13}C, and ^{2}H isotopomers. *J. Phys. Chem. A*, **101** (24), 4375–4483.

Oxley, J.C., Smith, J.L., Zheng, W., Rogers, E., and Coburn, M.D. (1997c). Thermal decomposition studies on ammonium dinitramide (ADN) and ^{15}N and ^{2}H isotopomers. *J. Phys. Chem.* (to appear).

Urbanski, T. (1964, 1965, 1967, 1984). *Chemistry and Technology of Explosives*, vols. 1–4. Pergamon Press, New York.

FEMEP (1977). *Field Expedient Methods for Explosive Preparation.* Desert Publications, Cornville, AZ.

IMBB (1981). *Improvised Munitions Black Book*, vols. 1–3. Desert Publications, Cornville, AZ.

Picatinny Arsenal (1996, 1997). "Fifteenth Annual Working Group Institute on Synthesis of High Energy Density Materials," June 12–13, 1996; and "Sixteenth Annual ... ", June 18–19, 1997.

CHAPTER 6

Theories and Techniques of Initiation

Peter R. Lee

6.1. Introduction

This chapter outlines the elementary theories of thermal and shock initiation of explosives and some generic initiation techniques for weapon systems.

6.2. Nomenclature

T (K)	Temperature, in degrees Kelvin (degrees Celsius, when referring to testing of weapons and weapon systems).
t (s)	Time in seconds or microseconds.
E (cal/mole)	Activation energy of the decomposition reaction (may also be in kJ/kg); this quantifies the energy required to bring unit mass of reactant to reaction. In detonation theory E is internal energy.
Q or q (cal/mole)	Heat output of a decomposition reaction (also kJ/kg); a mole is the molecular weight expressed in grams of the molecule decomposing.
V (cm^3)	Volume of the reacting material (may be in m^3).
ρ (g/cm^3)	Density of the reacting material (may be in kg/m^3).
λ (cal/cm · °C · s)	Thermal conductivity of the reacting explosive (may be in J/m · K · s).
A (s^{-1})	Frequency factor in the Arrhenius reaction rate expression $k = k_0 A \exp(-E/RT)$.
R (cal/°C · mole)	Universal Gas Constant (may also be J/K.kg).
χ (cal/cm^2 · °C · s)	Surface heat transfer coefficient (may also be J/m^2 · K · s) between the reactant and the surroundings.
S (cm^2)	Surface area of reactant (may also be m^2).
dq/dt (cal/s)	Rate of heat generation (may also be J/s).
e	Base of Naperian or natural logarithms (2.71828 . . .).
k (mole/s)	Rate of a chemical reaction (may also be kg/s).

c (g/cm^3)	Concentration of reactant (also density).
C_v (cal/°C · g)	Heat capacity at constant volume (also J/K · kg).
∇^2	Nabla squared is the Laplacian operator $\partial^2/\partial x^2 + \partial^2/\partial y^2 + \partial^2/\partial z^2$, where x, y, and z are the three principal axes.
κ	$\kappa = 0$, 1, or 2, dependent upon whether infinite planar, infinite cylindrical, or spherical reactant geometry is involved, respectively.
x (cm)	Linear coordinate in space.
r or r_0 (cm)	Characteristic linear dimension of the slab, cylinder or sphere, i.e., either half-width or radius.
z	Dimensionless linear coordinate given by x/r.
θ	Dimensionless temperature, given by $\theta = E(T - T_0)/RT_0^2$ where T is in the reactant and T_0 is the temperature of the surroundings.
δ	Frank-Kamenetskii Criticality Criterion (Dimensionless), given by $\delta = (Q/\lambda) \cdot (E/RT_0^2) \cdot r_0^2 \cdot Ac_0 \cdot \exp(-E/RT_0)$.
α	Biot Number, given by $\alpha = \chi r/\lambda$, relating internal to external resistances to heat flow.
υ (m/s)	Velocity of processes, including detonation.
v (cm^3/g)	Specific volume (this is the reciprocal of density).
u, W (m/s)	Particle velocity in a steady detonation wave.
τ (s)	Time during which a pressure, p, is exerted on a target by an impacting projectile (see Walker–Wasley Criterion).
d (m)	Diameter of shaped charge jet impacting against an explosive target (see Held Criterion).

6.3. Initiation Theories

Early in the history of the study of explosives and explosions, it became apparent that explosion is a special case of combustion. This conclusion arose in early experiments on gas phase explosions, particularly on the effect of oxygen on the combustion of a range of gaseous fuels. Some, such as hydrogen, exploded when ignited in the presence of oxygen, other mixtures burned at different velocities and with different effects on their surroundings, from highly damaging, shattering force, to quiet stable combustion. In all these cases the fuel and oxidant are in intimate contact, and an explosion or quiet combustion depends upon the chemical kinetics and energetics of the reactions between them. In the case of an explosion, the reaction takes place so rapidly in an expanding spherical shell about the initiation source that the heat generated as a consequence of the reaction cannot diffuse away from the reaction front. This heat serves as the thermal stimulus to heat the next spherical shell to ignition point and to expand the gaseous products of combustion. These effects may proceed so rapidly that the combustion processes compress the products progressively, with an increasingly steep pressure gradient, due to the inability of the products to escape rapidly enough from the reaction front. Eventually, during the reaction of certain mixtures, the burning front ceases to be associated with a smooth pressure increase. A step increase in pressure, or shock wave, is generated and a detonation wave is formed, which then moves through the rest of the unconsumed mixture at supersonic velocity. The velocity is supersonic compared with the speed of sound in the undisturbed medium. Therefore, all other factors being equal, we expect higher detonation velocities in denser materials, which have higher sound speeds.

The difference between quiet combustion and detonation was first observed around 1880 in France. The name "detonation wave" was given to the high-velocity flame front seen to propagate down a combustion tube in some instances and an analogy with shock waves was suggested. Around the turn of the twentieth century, Chapman (1899) and Jouguet (1906) independently used shock wave concepts to give a convincing explanation of the phenomenon. Their description has remained basic to all subsequent developments. The chemical reaction is assumed to take place in a narrow zone, referred to as the "reaction zone," which propagates at a constant velocity through the explosive, transforming it into detonation products. The detonation wave is, thus, a type of steady discontinuous wave like a shock wave, the physical behavior of which is governed solely by the properties of the unreacted and completely reacted energetic material on either side of the reaction zone. This enabled a theory to be developed without any need for detailed understanding of the chemical reactions taking place (see Section 6.5).

The processes taking place in nondetonative combustion, known as deflagration, are intimately connected with the detailed chemical kinetics of the reaction processes. These are somewhat unstable, because of the linkage of reaction velocity with thermal and pressure effects. Explosive deflagrations are useful, however, provided the reactions occurring can be stabilized through the use of additives and physical controls over burning rates. Much of the technology of gun and rocket propulsion centers around additives used to modify the output of deflagrating propellant compositions and to control their long-term slow decomposition in storage or during exposure to high temperatures prior to firing. Further control over reproducibility of burning characteristics is effected through size and shapes of individual propellant grains or sticks of gun propellants and rocket motor charges, also known, somewhat confusingly, as grains.

Deflagration is the usual instantaneous result of shock impact on an energetic material. Even in a so-called "prompt detonation" there is a period of time during which the reaction processes transition from being controlled by the decomposition kinetics of the energetic material to the physics of the shock wave. Growth to detonation may occur relatively slowly and unpredictably as a result of relatively low-intensity shock impact. In principle, the higher the intensity of the shock, the more rapidly the explosive reaches a stable detonation, provided it is capable of so doing in the dimensions of the charge under test. Similarly, it is possible, merely by heating a reactant, to generate a thermal explosion, which may, as a consequence of self-confinement from the bulk of the reactant, or due to the confinement afforded by a stout container, transition to detonation. However, in most cases detonation is initiated by shock and deflagration by thermal processes. The next sections of the chapter deal with thermal explosion theory and shock initiation theory, respectively.

6.4. Thermal Explosion Theory

There are two very useful guides to thermal explosion theory. Gray and Lee (1967) is exhaustive, although the review by Merzhanov and Abramov (1981) is more up to date. The reader is referred to both sources as essential reading on the subject.

Thermal explosion, or spontaneous ignition of an energetic material generally means the sudden inflammation of an uniformly heated mass of material. This may be a few millilitres of gas at low pressure, a few milligrams of initiatory explosive, an explosive charge in a warhead or rocket motor, bales of wool, a drum of polymer curing agent, or almost any quantity of any material which undergoes exothermic decomposition, either alone, or in combination with atmospheric oxygen. Thermal explosion theory is concerned with the competition between the heat generated in the exothermic reaction and the dissipative processes of conduction, convection, and radiation by which it is removed from the zone of reaction. The tendency of the temperature of the reacting medium to rise above the temperature of the surroundings during self-heating is opposed by the transfer of heat to its surface and subsequently to the surroundings and the reduction of the reaction rate as a result of reactant consumption. Thermal explosion theory can account for such diverse ignition phenomena as the spontaneous ignition of hay stacked when wet, of old gun propellant transferred to a warm environment and the effects of impact on a thin layer of explosive.

6.4.1. Semenov Theory

If, like Semenov (1928, 1930, 1932), we assume that the reaction rate of an exothermic reaction is an exponential function of temperature, according to Arrhenius, then, as the temperature of the surroundings is raised, the rate of production of heat increases exponentially, i.e., at a faster rate than the linear rate of heat loss from the reactant to the surroundings. Thus, as van't Hoff (1884) observed, there is an ambient temperature for any given mass

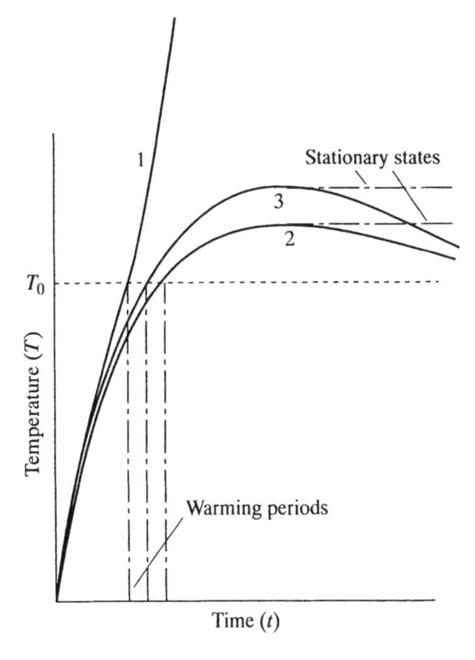

Figure 6.1. Curves of temperature as a function of time under different reaction conditions in a self-heating reactant in an hot environment.

of any exothermic material above which equilibrium, or quiet reaction, is not possible. This is the "Critical Temperature." Above this temperature heat evolution becomes the dominant process and rapid acceleration of the chemical decomposition reaction occurs, which culminates in ignition or explosion.

The critical conditions, in excess of which a self-heating system ignites or explodes are sharply defined and the first task of any theoretical or experimental investigation of the problem was to show how they could be determined. The accumulation of heat which leads to explosion in a system occurs in a finite time interval and, like the critical conditions, this induction period is an important characteristic of spontaneous ignition. Several alternative evolutions of temperature as a function of time in a decomposing exothermic reactant are shown in Figure 6.1. If heat production rate exceeds heat loss rate, explosion occurs after an induction period, curve 1. If heat loss exceeds heat production after some time, curve 2 results in which the temperature rises to a peak and then decays as reactant consumption reduces the reaction rate and heat production rate. There is a limiting curve, curve 3, which describes the last stable decomposition, i.e., the one producing the highest nonexplosive reaction temperature. This is the critical condition. A special case is the case of a zero-order reaction. Once the reactant reaches the temperature of the hot surroundings, the rate of heat production remains constant and a steady temperature excess over that of the surroundings is established. Two stationary states associated with a zero-order decomposition are shown in Figure 6.3 as the dot–dash lines extending parallel to the time axis. They do this until all reactant is consumed.

Semenov (1928) made the first mathematical representation of the critical condition, in a manner which is both elegant and simple. Critical conditions are related to the size and shape of the reacting material and its decomposition reaction kinetics. It gives an estimate of the critical reactant temperature rise, which is hardly altered by more complex theories and shows how it is possible for us to ignore reactant consumption in the case of self-heating of reactants such as stable explosives for which the activation energy of the decomposition is high. The theory can easily be adapted to deal with non-steady-state problems, where it enables the calculation of induction periods and with heterogeneous combustion problems.

The theory assumes that, due to a zero-order exothermic reaction with an Arrhenius-type rate constant, the reactant self-heats uniformly to a temperature $\overline{T}$, which is higher than its surroundings, at T_0. The bar over the T is a reminder that this temperature is uniform across the reactant. The physical interpretation of the uniform temperature is of a well-stirred fluid in a heated container, or of a small sample of a solid of relatively high thermal conductivity compared with that of the surrounding medium. If the reactant is a fluid, the surroundings are the walls of the containment vessel, if it is an explosive filling of a weapon, the weapon casing. In either case, the temperature of the surface of the reactant is assumed to be uniform. The exothermicity of the reaction is Q, its density ρ, and its volume V. The rate of production of heat is given by

$$\frac{dq_1}{dt} = QV\rho \cdot A \cdot \exp\left(-\frac{E}{R\overline{T}}\right), \tag{6.1}$$

where A is the pre-exponential factor, E is the activation energy, and R the universal gas constant. If the reactant loses heat to the surroundings according to a Newtonian Law of Cooling, i.e., there is a linear dependence of cooling rate on excess temperature,

$$\frac{dq_2}{dt} = \chi S(\overline{T} - T_0), \tag{6.2}$$

χ is the heat transfer coefficient, which is, strictly speaking, dependent upon temperature, but which, for the purposes of his analysis, Semenov regarded as independent of temperature, and S is the surface area. The use of this surface heat transfer coefficient, χ, implies, with the uniform temperature distribution in the reactant, all resistance to heat flow resides at the surface of the reactant. A steady thermal state is reached when the rates of heat production and heat loss are equal, i.e.,

$$QV\rho A \exp\left(-\frac{E}{R\overline{T}}\right) = \chi S(\overline{T} - T_0). \tag{6.3}$$

This is possible for a reaction of any order, but, for reactions of higher order than zero the steady state will be transitory. For a zero-order reaction, the steady state is a stationary state, which remains unaffected by reactant consumption, i.e., the temperature remains constant at $\overline{T}$ until the last molecule of reactant has been consumed (see Figure 6.1). A stationary state is unattainable experimentally, since the rates of all chemical reactions decrease as reactant is consumed. The term "steady state" can be applied for a limited time during non-zero-order decomposition of a reactant as the reactant temperature varies very slowly around its maximum prior to starting the decline to the temperature of the surroundings.

Figure 6.2 shows schematic plots of heat production curves, described by (6.1), and heat loss straight lines, described by (6.2), for an arbitrary system after Semenov. Curves 2, 3, and 4 show exponential heat production curves corresponding to different reaction concentrations in a given container, or different sized reactants of a given shape and density, over a relatively narrow temperature range. Over a wide temperature range the heat generation curves are standard exponential, S-shaped, but we are not interested in the higher temperatures at present, although they have a physical significance. Straight lines 1 and 5 show typical heat loss relationships as functions of temperature. The greater the reactant volume, i.e., the smaller the surface to volume ratio, the smaller is the gradient of the heat loss line.

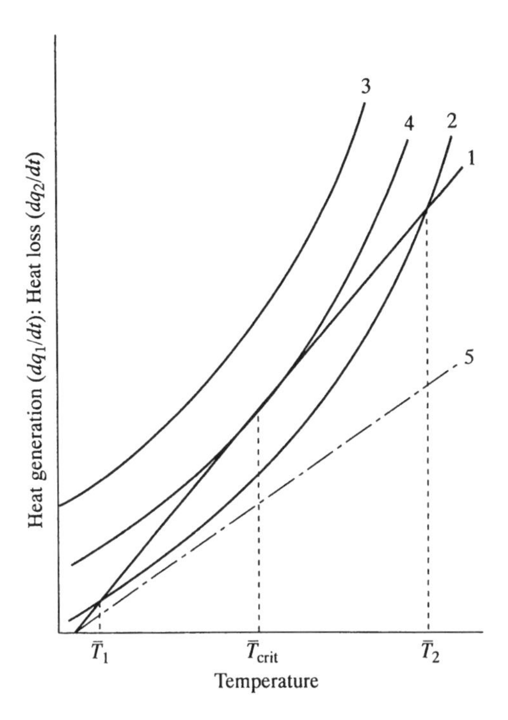

Figure 6.2. Semenov theory; heat production and heat loss as functions of temperature.

There are three possible cases to be considered:

- *Case* A. The heat loss line 1 cuts the heat production curve 2 in three points, two of which are shown here and define *Stationary States*. The third point of intersection occurs at very high temperatures and is outside our range of interest. The reactant will self-heat as long as the rate of heat production exceeds the rate of heat loss. At $\overline{T}_1$ a stable stationary temperature will be reached; any tendency for the temperature to exceed $\overline{T}_1$ is counteracted by an increase in the rate of heat loss above the rate of heat production and the system cools, reducing the rate of heat loss faster than the rate of heat production. $\overline{T}_2$ also defines a stationary state, but this one is unstable, since any tendency for the temperature to increase above this value results in the rate of heat production exceeding the rate of heat loss and explosion ensues. If the temperature should fall below $\overline{T}_2$ the rate of heat loss exceeds the rate of heat production and the temperature reverts to that of the stable stationary state, $\overline{T}_1$.
- *Case* B. Neither line 1 and curve 3, nor line 5 and curves 2, 3, or 4 have points in common. The rate of heat production exceeds the rate of heat loss at all temperatures and *only Explosive States can exist*. The relationship between lines 1 and 5, which represent, respectively, small and large reactants of a given density, or reactants of the same size, but small and large densities, or concentrations, indicates that, as the reactant size increases, the reactant density necessary to cause explosion decreases.
- *Case* C. The heat loss line 1 is tangential to the heat production curve 4 at temperature $\overline{T}_{\text{crit}}$, which defines the *Critical State, or the last possible steady or stationary state before explosion.*

6.4.1.1. Critical Conditions

The condition that the straight line 1 is tangential to curve 4 at $\overline{T}_{\text{crit}}$ is given by differentiating both sides of (6.3) and eliminating χS, using the equality of the right-hand sides of (6.1) and (6.2). The result is expressed in the form of the critical temperature rise given by

$$\Delta\overline{T}_{\text{crit}} = \overline{T}_{\text{crit}} - T_0 = \frac{R\overline{T}^2_{\text{crit}}}{E}, \tag{6.4}$$

or

$$\overline{T}_{\text{crit}} = \left(\frac{E}{2R}\right) \cdot \left[1 \pm \sqrt{\left\{1 - \left(\frac{4RT_0}{E}\right)\right\}}\right]. \tag{6.5}$$

Equation (6.5) shows that a self-heating reaction can only lead to ignition if $E \geq 4RT_0$. This is very easily achieved in the case of explosives, where E is of the order of 100 to 200 kJ/mole (25–50 kcal/mole) and T_0 is of the order of 500 K. The maximum value of $\overline{T}_{\text{crit}}$ occurs when $T_0 = E/4R$ and is given by $\overline{T}_{\text{crit}} = E/2R$, which is also the point of inflection of the curve of

$\exp(-E/R\overline{T})$ against $\overline{T}$. Thus, at very high temperatures, i.e., above $E/2R$ or 2000–10,000 K, no stationary states exist and explosion will always occur. Hence it is only necessary to consider the lower of the two values for $\overline{T}_{\text{crit}}$ implied in (6.5), i.e., the one with the minus sign before the square root. The larger of the two values of critical temperature also has a physical meaning. It corresponds to the extinction point in heterogeneous combustion in the high-temperature region where the reaction rate is diffusion-controlled, rather than kinetically controlled. $\overline{T}_{\text{crit}}$ can be determined by developing the square root in equation (6.5), as reported by Simchen (1964),

$$\overline{T}_{\text{crit}} = \left(\frac{E}{2R}\right)\sum_{i=1}^{\infty} b_i \left(\frac{RT_0}{E}\right)^i,$$
$$\overline{T}_{\text{crit}} \cong T_0 + \left(\frac{RT_0^2}{E}\right) + 2\left(\frac{R^2T_0^3}{E^2}\right) + \cdots, \tag{6.6}$$

where the next seven coefficients of the series are 5, 14, 42, 132, 429, 1430, and 4862. The series can be truncated after the second term to give

$$\overline{T}_{\text{crit}} - T_0 = \Delta\overline{T}_{\text{crit}} \cong \frac{RT_0^2}{E}, \tag{6.7}$$

with an error of about 5% in $\overline{T}_{\text{crit}}$ for a reaction with an activation energy of about 180 kJ/mole and $T_0 = 500$ K, where $\Delta\overline{T}_{\text{crit}}$ is of the order of 15–20 K. Physically, (6.7) implies that it is necessary only for an e, or 2.7183-fold increase in the reaction rate at T_0 to occur for the system to become unstable and explode. This is one of the most important relationships to have been developed in thermal explosion theory and has been confirmed by later studies and by experiment.

Semenov wrote $\exp(-E/R\overline{T}_{\text{crit}}) \cong \exp(-E/RT_0)$ in 1928, apparently as a result of making the approximation $\overline{T}_{\text{crit}} \cong T_0$. This is less than 4% in error, but the exponential is −300% in error, which is reflected in the rate of heat release predicted. The correct expression for the critical condition was first obtained by Groocock (1958), independently, and is given by

$$\frac{\chi S}{V} = eQ\rho A\left(\frac{E}{RT_0^2}\right)\exp\left(-\frac{E}{RT_0}\right). \tag{6.8}$$

For an mth order reaction, where reactant consumption is not taken into account

$$\ln\left(\frac{c_{\text{crit}}^m}{T_0^{m+2}}\right) = \frac{E}{RT_0} + k, \tag{6.9}$$

where k is given by

$$k = \ln\left(\frac{\chi S}{eEQ\rho VA}\right).$$

Equation (6.9) shows how thermal theory enables the activation energy of a reaction to be determined from the criticality criterion determined for a range of material sizes or concentrations. It is especially suited for studies on gaseous reactants.

Semenov's theory was the origin of the quantitative study of thermal explosions, from which a remarkably vivid interpretation of many aspects of thermal theory has been obtained. The physical picture of subcritical, critical, and supercritical states has scarcely been modified, the critical temperature rise of approximately RT_0^2/E has been confirmed by numerical solutions and more sophisticated theories and the condition for the possibility of thermal explosion has been shown to be $E/RT_0^2 \geq 4$. However, the theory applies only to systems with uniform temperature distribution such as well-stirred liquids. Merzhanov et al. (1961) studied well-stirred liquid DINA (dinitro-oxydiethylnitramine) and found that its critical temperature increment is closely in agreement with the Semenov theory. These physical conditions are not applicable to high explosives, or propellants in weapon cases in an heated environment, because the explosive material itself also provides a thermal barrier. The Semenov theory presumes that the only obstacle to the flow of heat resides at the surface of the reactant, which is either in contact with the container or casing, or is in direct contact with the surroundings.

6.4.2. Frank-Kamenetskii Theory

Frank-Kamenetskii's original work (1939) ascribed all resistance to heat flow between the reactant and the surroundings to the thermal conductivity of the reactant, which would be very small for an organic material, such as an high explosive or propellant. He asserted that thermal explosion theory as he understood it was a branch of heat conduction theory, with the extra factor of distributed heat sources, e.g., like the effects of an electric current being passed through a material of low electrical and thermal conductivity and generating heat by the application of Ohm's law. He also regarded Semenov's work as an approximation to his theory, rather than having its own perfectly valid physical reality. His assumption of conductive heat transfer as the sole means of heat dissipation within the reactant and between the reactant and its surroundings, means that it contrasts with the Semenov theory, where resistance is concentrated at the surface of the reactant. Frank-Kamenetskii's theory also requires that a distributed temperature exist in the reactant, whereas the Semenov theory requires an uniform temperature distribution. A distributed temperature in the reactant is realistic for low-conductivity materials such as high explosives, in containers, or bare in heated environments. Frank-Kamenetskii's theory is also more versatile than Semenov's theory and can deal with hot-spots as well as bulk heating. However, it has been shown since the development of the Frank-Kamenetskii theory that reality lies somewhere between it and the Semenov theory, since there is an element of resistance to heat flow both at the surface of a reactant and within it.

The Frank-Kamenetskii theory does, however, relate exactly to:

- unstirred, low conductivity, exothermically reacting fluids in containers with very thin, high thermal conductivity walls; and
- low conductivity solids, with or without high thermal conductivity, thin-walled containers;

both surrounded by media, usually fluids, of comparable thermal conductivity. Weapons stored at elevated temperatures are not described by the Frank-Kamenetskii theory precisely, because they have relatively thick containers of relatively low thermal conductivity, i.e., there is a temperature gradient in the wall as well as in the reactant.

The Fourier heat conduction equation in an inert isotropic material is

$$\operatorname{div} \lambda \cdot \operatorname{grad} T = C_{\mathrm{v}} \frac{dT}{dt},$$

where λ is the thermal conductivity of the material and C_{v} is its heat capacity, per unit volume. This equation is not soluble analytically. We can solve approximations to it which describe mathematically certain assumptions concerning the physical conditions under which the thermal processes can take place. These approximations will be dealt with individually.

It is reasonable to assume that λ is constant over the relatively small temperature range associated with a self-heating event, i.e., approximately 15 K. The equation transforms to

$$\lambda \nabla^2 T \cong C_{\mathrm{v}} \frac{dT}{dt}, \tag{6.10}$$

where ∇^2 is the Laplacian operator $\partial^2/\partial x^2 + \partial^2/\partial y^2 + \partial^2/\partial z^2$. If consideration is restricted to three idealized shapes, the infinite plane slab, the infinite circular cylinder, and the sphere, the Laplacian operator can be replaced by $\partial^2 T/\partial x^2 + (\kappa/x)(\partial T/\partial x)$, which involves only a single linear dimension instead of three. This dimension is the half-width of the infinite slab and the radius of the infinite cylinder and sphere and x is some distance from the center. If heat is generated via uniformly distributed heat sources in the material at each point, $P_{x,y,z}$, the rate is given by $S_{x,y,z}$ per unit volume per unit time in three dimensions, or P_x and S_x, respectively, in the case of the simplified shapes. If heat is generated by a chemical reaction with Arrhenius-type reaction kinetics, (6.10) becomes

$$\lambda \nabla^2 T - C_{\mathrm{v}} \frac{dT}{dt} = -Q \cdot f(c) \cdot A \exp\left(-\frac{E}{RT}\right), \tag{6.11}$$

where the Laplacian operator now represents ordinary differentiation and $f(c)$ is a function of the reactant concentration, or density, the exact form of which depends on the order of the reaction of the process causing the self-heating. These processes are usually first order.

Equation (6.11) is second order in distance, x, and requires two boundary conditions to be specified in order to generate a solution. These are that the temperature of the surface of the reactant is the same as that of the surroundings, or the hot bath into which the reactant has been placed, so that

$$T = T_0 \qquad \text{at} \quad x = r_0, \tag{6.12}$$

where r_0 is a characteristic dimension of the reactant, e.g., radius. The second boundary condition is an outcome of the geometry, since symmetrical heating implies a symmetrical temperature distribution within the reactant and this is expressed by

$$\frac{dT}{dx} = 0 \qquad \text{at} \quad x = 0. \tag{6.13}$$

The maximum temperature is at the center of the reactant, where there is no net heat flow.

Even the simplified equation (6.11) is insoluble analytically because of the time-dependence, and Frank-Kamenetskii did not have a computer available to him in 1939 to solve it numerically. He tried to simplify the problem by making physical and mathematical approximations so that the equations became more tractable. His simplifications were that he assumed:

- a stationary state, achievable only with zero-order reaction kinetics;
- that only three physical shapes could be considered, the infinite slab, infinite cylinder, and the sphere, each of which can be represented by the Laplacian operator, with $\kappa = 0$, 1, or 2, respectively; and
- that an approximation to the exponential term could be made.

The steady-state and zero-order kinetics have already been discussed in relation to the Semenov theory. The term "steady-state" can be applied only for an instant to a reaction of finite order, i.e., when $dT/dt = 0$, but this condition does not hold thereafter, except for zero-order reactions. As long as E and Q are large, the consumption of reactant up to ignition, i.e., where $\Delta \overline{T}_{\text{crit}} \cong RT_0^2/E$, can be regarded as negligible with little error and the assumption of zero-order kinetics is reasonable. Thus $f(c) = c_0$ or ρ_0 in (6.11). As soon as non-steady-state calculations are involved, however, some account needs to be taken of reactant consumption, or infinite induction periods are deduced. The restriction to the three idealized shapes is not severe, since analysis of heat flow in finite cylinders reveals that cylinders only two to three times their diameter in length have a significant part of their central section in which heat flow is one dimensional, dependent only upon the distance from the center.

The rates of chemical reactions are extremely sensitive to the temperature of the reactant, as represented in the Arrhenius kinetics expression which involves an exponential term in temperature. In thermal explosion theory we are often only interested in a temperature difference between that of the reactant and its surroundings of the order of 20 K, so relatively simple approximations are satisfactory. Frank-Kamenetskii made use of Semenov's result that the temperature difference due to self-heating of a reactant is approximately RT_0^2/E at criticality. He expanded the argument of the exponential $\exp(-E/RT)$ into a Taylor Series in terms of $(T - T_0)/T_0$, so that

$$\frac{E}{RT} \cong \left(\frac{E}{RT_0}\right)\left[1 - \left\{\frac{T - T_0}{T_0}\right\} + \left\{\frac{T - T_0}{T_0}\right\}^2 - \left\{\frac{T - T_0}{T_0}\right\}^3 + \cdots\right]. \tag{6.14}$$

He ignored powers of $(T - T_0)/T_0$ greater than one, justifying his choice by showing that, whenever $E \gg RT_0$, $(T - T_0)/T_0 \ll 1$. Hence,

$$\exp\left(-\frac{E}{RT}\right) \cong \exp\left(-\frac{E}{RT_0}\right) \cdot \exp\left(\frac{E(T - T_0)}{RT_0^2}\right).$$

According to Semenov $(T - T_0) = RT_0^2/E$ at criticality. Frank-Kamenetskii chose to use $E(T - T_0)/RT_0^2$ as a natural yardstick of temperature and wrote it as θ. Numerical comparison between the approximation and the unapproximated expressions show that, up to $\theta = 1$, agreement is good and satisfactory up to $\theta = 2$. Frank-Kamenetskii showed that we are only interested in $\theta < 2$ up to criticality.

The steady-state version of (6.11) is given by

$$\lambda\left(\frac{d^2T}{dx^2} + \frac{\kappa}{x}\frac{dT}{dx}\right) = -Qc_0A \exp\left(-\frac{E}{RT_0}\right) \exp\left(\frac{E(T - T_0)}{RT_0^2}\right), \tag{6.15}$$

using the approximations referred to above, where κ may be 0, 1, or 2 depending upon whether the shape of the reactant is an infinite slab, an infinite cylinder, or a sphere, respectively. The boundary conditions for the steady-state solution are as given in (6.12) and (6.13).

Equation (6.15) was first solved analytically by Frank-Kamenetskii, but, in order to do so and to simplify the analysis, he cast it in dimensionless terms by transforming the variables according to

$$\theta = \frac{E}{RT_0^2}(T - T_0) \qquad \text{and} \qquad z = \frac{x}{r_0},$$

θ is dimensionless temperature on a scale related to the Semenov critical temperature rise and z represents dimensionless distance as a proportion of the half-width of the infinite slab, or the radius of the infinite cylinder or the sphere. The equation to be solved becomes

$$\nabla_z^2\theta = -\delta \cdot e^{\theta}, \tag{6.16}$$

with boundary conditions

$$\theta = 0 \quad \text{at} \quad z = 1 \qquad \text{and} \qquad \frac{d\theta}{dz} = 0 \quad \text{at} \quad z = 0, \tag{6.17}$$

where $\delta = (Q/\lambda) \cdot (E/RT_0^2) \cdot r_0^2 \cdot Ac_0 \cdot \exp(-E/RT_0)$ is a parameter which

Table 6.1. Frank-Kamenetskii critical criteria, δ_{crit}, and maximum central temperatures, $\theta_{m,crit}$, for several common shapes.

Shape	δ_{crit}	$\theta_{m,crit}$	Shape	δ_{crit}	$\theta_{m,crit}$
∞ Slab	0.88	1.19	Cube	2.47	—
∞ Cylinder	2.00	1.39	Regular cylinder	2.79	—
Sphere	3.32	1.61	Brick	2.94	—

describes the explosive type, by summarizing its decomposition kinetics and its thermochemistry, and its shape and mass. There are solutions of (6.16) subject to boundary conditions (6.17) for a range of values of $\delta \leq \delta_{max}$, i.e., for a range of explosive materials, sizes, shapes, and concentrations. Each solution corresponds to a steady-state temperature distribution, or temperature profile across the reactant. For values of $\delta > \delta_{max}$ no such solution exists and δ_{max} is identified with the critical explosion criterion δ_{crit}. The value of δ_{crit} depends on the form of the Laplacian operator and has an unique value for each of the three special geometrical forms referred to above. All other characteristics of the reactant are swept up into the dimensionless form of δ. The critical condition is associated with the largest stable temperature possible at the center of the explosive sample. This is known as $\theta_{m,crit}$. Table 6.1 shows details of Frank-Kamenetskii critical conditions for the three shapes already discussed as well as three more, for which Gray and Lee (1967) calculated approximate critical criteria. The procedure depended upon the concept of the "equivalent sphere" to different reactant shapes, a concept first proposed by Wake and Walker in 1964 as part of their research into the spontaneous ignition of wool bales in warehouses. (Several other of Walker's group's papers are referred to for interest here because they were instrumental in enabling Gray and Lee to understand criticality in a range of different shaped reactants.)

Unfortunately data for the maximum central temperature rise for the cube, the regular cylinder, and the parallelepiped (brick) are not available.

Simple algebraic comparison of the critical criteria from the Semenov theory, as determined by Groocock (1958) and shown in (6.8), and the expression for δ_{crit} give

$$\delta_{crit} = \frac{1}{e} \cdot \frac{\chi r_0^2}{\lambda} \cdot \frac{S}{V} \qquad \text{or} \qquad \delta_{crit} = \frac{\chi r_0}{\lambda} \cdot \frac{(\kappa + 1)}{e},$$

where $\kappa = 0$, 1, or 2. If we substitute for both κ and δ_{crit} in the equation above, we obtain an expression for the effective surface heat transfer coefficient χ_{eff} given by

$$\chi_{eff} = \alpha_{\infty,k} \frac{\lambda}{r_0},$$

where $\alpha_{\infty,k}$ has values, 2.39, 2.72, or 3.01, when $\kappa = 0$, 1, or 2, respectively. Substitution of these values of χ_{eff} into the Semenov theory produces identical critical conditions to those obtained from the Frank-Kamenetskii theory. Thus, it is possible to treat any steady-state case involving distributed temperatures in terms of the simpler uniform temperature concept of the Semenov theory. However, this is a mathematical convenience and is not intended to imply any particular form of physical conditions in the reactant.

α_∞ is a Biot Number. Its significance is that it compares internal and external resistances to heat flow.

- If α is large, conduction in the reactant takes control, its surface temperature is the same as that of the heated surroundings. As $\alpha \to \infty$, the Frank-Kamenetskii condition of thermal resistance residing solely in the reactant is met.
- If α is small, heat loss from the surface of the reactant is inefficient, the surface temperature rises above that of the hot surroundings and the instantaneous temperature profile tends to become flat. As $\alpha \to 0$, the Semenov condition of uniform temperature distribution is met.

The significance of these features can best be appreciated by examination of the results of the numerical solution of the differential equations for a range of values of α and δ. The concept was first discussed by Thomas (1958) who showed that the Frank-Kamenetskii critical criterion and the center and surface temperatures at criticality are smooth functions of the Biot Number. This work, and that by Gray and Harper (1959, 1960) resolved the conflict between the apparently different physical realities of the Semenov and Frank-Kamenetskii models. The relationship between the models is best illustrated by means of a diagram from Gray and Harper's work, Figure 6.3.

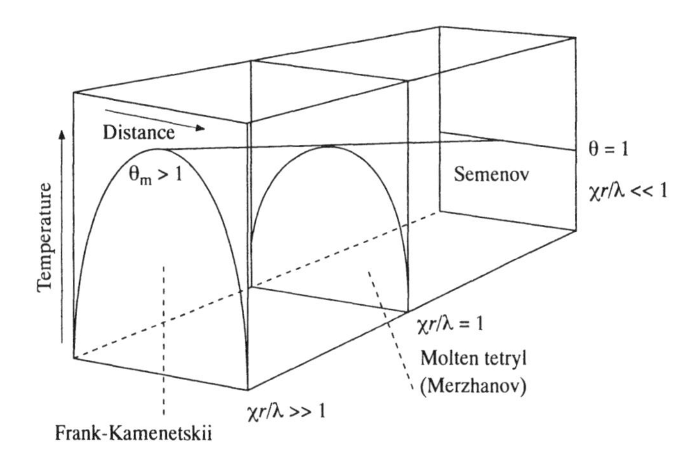

Figure 6.3. Relationship between the Semenov and Frank-Kamenetskii theories, after Gray and Harper (1959, 1960).

Merzhanov et al. (1959) studied the thermal explosion of molten tetryl in glass and stainless steel vessels. They measured critical temperatures of the reactant and the vessel walls at certain oven temperatures and found that they lay between the values they calculated assuming either the Semenov or the Frank-Kamenetskii theories. In addition they showed that not only was a temperature distribution set up in the reactant, but also there was a clear temperature difference between that of the reactant and the wall. The conditions they observed were similar to those relating to $\chi r/\lambda \cong 1$ in Figure 6.3.

It is not necessary to heat the whole of a reactant to a high temperature to cause a thermally initiated event. A hot-spot, or a localized region of very high temperature can be induced in an explosive by mechanical, rather than primarily thermal stressing. Hot spots may arise at density discontinuities in the explosive caused by entrapment of air bubbles, or grit, or they may occur at irregularities at the surface of a container, such as gritty inclusions, or defective surface machining. They may even be composed of the explosive itself, which is in some way heated intensely locally by some external stimulus, e.g., friction against a rough surface. There are two main types of hot-spot, reactive, where the heated material is the explosive itself, or inert, where it is the density discontinuity in the explosive which absorbs the heat.

Equation (6.11) is pertinent to thermal initiation by hot-spots as well as bulk initiation. However, the boundary conditions now become

$$T = T_i \quad \text{when} \quad r < r_0 \qquad \text{and} \qquad T = T_0 \quad \text{when} \quad r > r_0,$$

where r_0 refers to the characteristic dimension of the hot-spot. If only spherical hot-spots are considered, r_0 is the radius. T_i is the initial hot-spot temperature and could be several hundred degrees Kelvin above that of the bulk of the explosive, T_0. The approximation developed to assist in the analytical solution of the Frank-Kamenetskii theory cannot be used here, so analytical solutions are not possible. Time-dependent numerical solutions have been obtained for reactive hot-spots by Merzhanov et al. (1963) and for inert hot-spots by Lee (1992). In both cases, the authors used the initial reactant temperature, T_i, as the reference temperature in defining the dimensionless temperature, θ, instead of the temperature of the surroundings, T_0, chosen by Frank-Kamenetskii, but made no approximation to the exponential term. In the case of inert hot-spots, it is necessary to solve the differential equation for heat conduction in an inert sphere and introduce new boundary conditions at the inert/explosive surface. It is assumed that the amount of heat reaching the surface in a given time is equalled by the amount of heat leaving it, so as to prevent the build-up of temperature there. While the theory is interesting it has not, as yet, been able to provide more than qualitative insight into the initiation of explosives, such as in the work by Field et al. (1982) and by Heavens et al. (1974–1976).

Thermal explosion theory has proved to be of inestimable value in understanding the results of slow and fast cook-off weapons testing. This understanding has been made possible by pioneering numerical analysis carried out in the United States and the USSR, by Zinn and Mader (1960) and by Merzhanov et al. (1960, 1963). It is well documented by Boggs and Derr (1990) and referred to in more detail in Chapter 8 also by Peter Lee. The review of thermal theory in Boggs and Derr (1990) follows closely that by Merzhanov et al. (1981) and is an excellent in-depth analysis as well as providing a wealth of information on Insensitive Munitions Testing.

6.5. Elementary Detonation Theory

Gunpowder burns slowly, at a few millimeters per second in a compressed pellet, because the energy supply is controlled by conduction of heat from grain to grain in the pellet. When spread on the ground as a loose train, the same material may burn at several tens of meters per second. This is a result of the increase in heat transfer rate from grain to grain arising from forced convection of hot gases through the loose material. Despite the speed of the process, the effects of the burning of gunpowder on the ground are negligible, perhaps a slight scorching. However, when, in the second half of the nineteenth century, nitroglycerine or mercury fulminate was subjected to the same treatment, the burning process was found to be much faster and the effects on the surface on which the test was conducted were more severe; it was often shattered. The new high explosives, such as mercury fulminate and nitroglycerine all had this ability, but some were different in one major respect. The mercury fulminate would burn with shattering force in very small quantities, whereas nitroglycerine, in similar quantities might burn reasonably quietly. There seemed to be a critical mass for each explosive in excess of which it burned shatteringly, i.e., detonated. Less explosive merely deflagrated. Smaller quantities than the critical mass for detonation of a given explosive when unconfined could be detonated if the explosive were under confinement. Some explosives, like mercury fulminate, could transfer detonation to relatively small quantities of even unconfined nitroglycerine and cause them to detonate in quantities which would only burn if subjected to a flame under the same conditions.

Intensive studies on burning and detonation were carried out in the late nineteenth century with much effort devoted to the determination of the velocities of these processes. At the same time, mathematicians were studying shock-wave theory and had shown that compression waves of finite amplitude, if suitably maintained in a fluid, could be made to suffer progressive increase of pressure and velocity gradients at the wave front, which in the absence of thermal and viscous energy transfer, might become infinitely steep. A diagrammatic representation of the process is shown in Figure 6.4. A piston is pushed into a long column of fluid in a gradually accelerating rate. In fact, we can think of the acceleration as being in a set of incremental steps. As each step is carried out a compression wave moves ahead of the

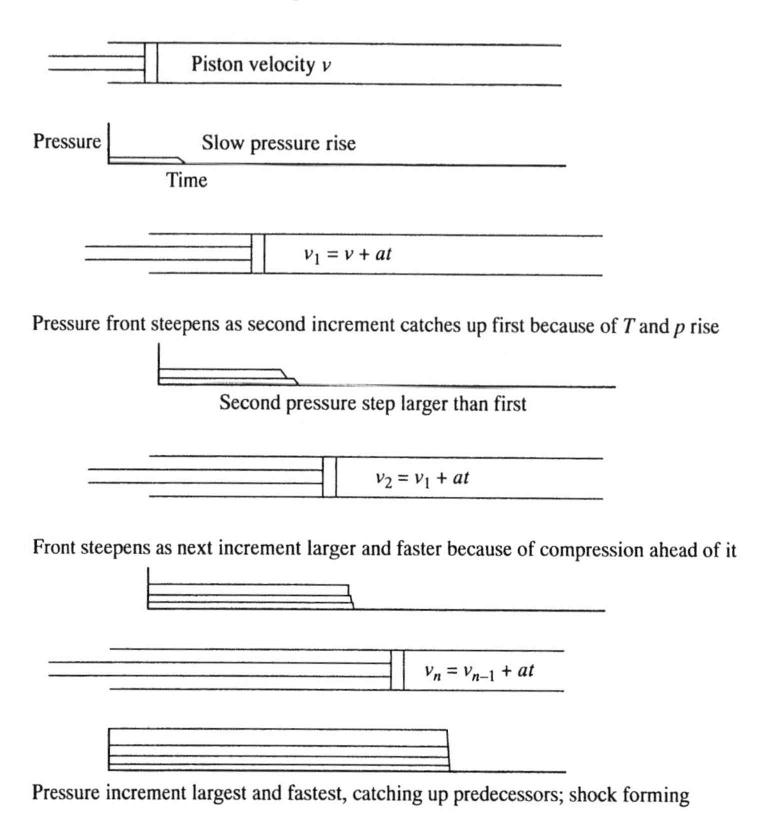

Figure 6.4. Generation of a shock wave by a piston. (Piston acceleration to the right in a series of steps.)

piston. This wave compresses the material ahead of the piston and heats it slightly. As the piston comes to compress previously heated material, its accelerated motion causes further heating and compression, so that the pressure wave forced ahead of the piston assumes a steeper front, leading to more intense heating and precompression, provided that the dissipative process associated with thermal diffusion and viscous transfer are still ignored. Eventually, the pressure front assumes an infinitely steep form and becomes a finite discontinuity between the fluid ahead of it and that behind it. This is a shock wave. It moves at supersonic velocity and is driven by the piston. If the piston continues to accelerate, so does the front. If the piston settles to a steady velocity, the front does so too. If the piston decelerates, a rarefaction wave is formed ahead of it, the front of which moves at the local velocity of sound in the fluid at each point and must, therefore, pass through the compression wave and overtake and weaken the shock front. The shock front loses speed and, if deceleration of the piston continues, the shock front will continue to be weakened until its velocity falls to sonic level.

The shock front in this example is driven by the piston behind it. It is unstable and relies on the acceleration of the piston for its energy. In a stable detonation, stability is achieved by the energy supply from the chemical conversion of the explosive into products. Rarefaction waves arise in the products of the detonation and shock-wave theory reveals that rarefactions which merge with shocks reduce their amplitude and slow them down. The question is how rarefactions remain detached from a stable detonation wave without overtaking it and weakening it. The answer is that in a detonation, the products stream in the same direction as the detonation wave. In a deflagration, or burning wave they stream away from it. (These points will be amplified in the next section.) Consequently in a detonation, the products assist in the maintenance of the pressure behind the front and prevent rarefactions reaching it. If the diameter of the column of detonating explosive is less than the critical diameter for stable detonation, the products of detonation cannot sustain this pressure behind the front because they expand sideways and permit rarefaction waves associated with them to coalesce behind the detonation front and decelerate it. If the diameter of the explosive exceeds the critical value, the rarefaction waves coalesce too far from the detonation front to affect it.

In order to be able to understand detonation theory the first requirement is to be clear what is meant by the "Equation of State" of reaction products. The pressures and temperatures at the reaction front in a detonating explosive are of the order of several hundred thousand atmospheres and several thousand degrees Kelvin. It is not possible to use the ideal gas equation of state, nor the van der Waals equation at these extreme pressures and temperatures, since the assumptions upon which they are predicated break down. More complex equations of state are required for the extreme conditions encountered in the detonation regime and they generally take the form

$$p = nRT \cdot \rho[f(\rho, T) + f_1(\rho, T)],$$

where $\rho = 1/v$, the reciprocal of the specific volume. The Becker–Kistiakowsky–Wilson (BKW) equation of state is sometimes called the Kistiakowsky–Wilson equation of state, but the latter (1941) attributed it to Becker (1921). Its final form is written

$$\frac{pv}{RT} = 1 + x \cdot e^{\beta x} \qquad \text{where} \quad x = \frac{K}{vT^{\alpha}} \qquad \text{and} \qquad K = \kappa \sum_{1}^{n} X_i k_i,$$

where α, β, and κ are parameters, K is the covolume made up of the individual covolumes of the components k_i. Values of $\alpha = 0.25$ and $\beta = 0.3$ were found to be suitable for the determination of the theoretical detonation velocities for a range of pure explosives and Cowan and Fickett (1956) used values of 0.5 and 0.09 to match the theoretical detonation velocity curve and the Chapman–Jouguet pressure with measured values for Composition B. κ was chosen as 11.85. There are several sources on equations of state. Mader's book (1979) is recommended, because an Appendix details the development of the BKW equation of state with comprehensive references and he is acknowledged to be the leading exponent of hydrodynamic modeling today.

The hydrodynamic theory of detonation, which depends upon the physics of mass, momentum, and energy balance across the reaction zone, also requires information on chemical energy changes. Although, as we shall see later, we do not need to know anything about the reaction mechanism or reaction kinetics, it is necessary to know the heat of explosion of an explosive, the quantities and types of the products, the temperature of the detonation, and the pressure reached in it. Standard thermochemistry techniques involving Hess' law and the use of calorimetric bomb tests or differential scanning calorimeter studies enable theoretical and experimental data on heats of explosion to be compared.

6.5.1. Shock Initiation

We usually think of shock initiation applying specifically to initiation of detonation, but it is perfectly feasible to initiate deflagration by shock, indeed, even so-called "prompt" detonation initiation consists of a very brief period of deflagration as the process builds up to detonation. It has already been shown that thermal explosions, or deflagrations, can be started from relatively weak input stimuli, e.g., a few hot-spots and that, given the right conditions, they may generate sufficient compression to lead to detonation via the process known as "deflagration–detonation transition" (DDT). The generation of "prompt" detonation requires that the receptor be shocked to a pressure of gigapascals (GPa). It is easier to study shock to detonation transition than DDT, mainly because the latter is unpredictable and demands the use of high-confinement systems, which are difficult to observe. On the other hand, shock inputs to a receptor can be controlled precisely and are easily observed. Furthermore, in the pure shock case, it is sufficient to consider shock amplitude and duration. Pioneering work by Hertzberg and Walker (1948) and Cachia and Whitbread (1958) reviewed by Jacobs (1960) showed how explosively generated shocks could be used to develop different types of gap sensitivity experiments (see Chapter 8 also by Peter Lee). In its current generic form the Gap Test consists of a donor explosive charge of standard size and well-known performance, which is detonated so as to send a known shock through a known inert barrier of given thickness into the receptor explosive. The shock that the receptor sees is varied by changing the thickness of the gap material from replicate experiment to experiment, and a statistical analysis is applied to the results to determine the thickness of barrier required to prevent the receptor charge detonating from the standard donor in 50% of tests. The critical gap is defined as the gap sensitivity (strictly speaking it is gap sensitiveness, see Chapter 8). There is a sharp cut-off between initiation and failure, exactly as was found in the case of thermal explosion. However, as was mentioned earlier, detonation starts some distance from the barrier after a run-up distance and time, which are dependent on the initiating explosive, the barrier, its thickness, and the receptor explosive. This effect was first shown by Cook et al. (1959) and their description is still considered to be definitive.

The whole edifice of shock initiation theory depends upon the interaction of hydrodynamics with chemical reaction rates. The shock entering the receptor explosive initiates a chemical reaction in it at the barrier surface. The rate of this reaction is dependent on the amplitude of the shock, i.e., the thickness of the barrier, and the chemical reactivity of the explosive. When the reaction has developed sufficiently to start to release substantial amounts of energy, the pressure will rise in the region near the barrier first and then further into the explosive charge, following on behind the shock. This pressurization causes compression waves to be generated behind the shock front in the already heated material. The reaction rate increases in this material because it is already hot and compressed from the passage of the shock and it increases further as the wave moves into the receptor, because it is also contributing its own energy. As the reaction rate increases, so does the speed of the pressure wave into the explosive and it may speed up sufficiently to overtake the shock and merge with it. If this happens, detonation takes over. If the deflagration does not catch up, the receptor does not detonate. If a wider gap is used, the initial shock transmitted into the receptor is weaker, so reaction in it is slower and the time and distance for detonation to become established increase. As the input shock level is decreased from test to test, the run-up distance to detonation increases until a threshold level of shock is reached for a given explosive in a given geometry, below which detonation cannot occur.

For weak shocks, the diameter of the receptor is small compared with the distance run to detonation. Lateral expansion of the receptor behind the primary shock, as well as rarefaction waves reflected from the sides of the receptor cause pressure decay, which competes with the pressurization due to the reaction-fueled compression wave and the transition process is slowed down or stopped. Large diameter receptor charges have shorter run distances to detonation than ones of smaller diameter, which are initiated from the same shock level, because the rarefactions do not arrive from the surfaces of the receptor stick as early. Thus, the stable detonation velocity of a receptor is a function of its diameter; receptor charges with diameters larger than a given value, detonate at an upper limit rate, which is a function of explosive type and density; receptor charges with diameters below a minimum value do not detonate at all.

These data pertain to non-steady-state shock initiation, the analytical solution of the equations of which has not yet been achieved. Indeed, even numerical solutions of the shock-initiation problem require knowledge of a large number of chemical and physical data (Mader, 1979), which are, as yet only partly available, with the consequence that a significant proportion of the predictive capability of current numerical work in this field requires empirical support.

6.5.2. The Hydrodynamic Theory of Detonation

6.5.2.1. Steady Detonation

Steady detonation theory is well developed, but, until about 1950, the only hydrodynamic data available on an explosive were:

- detonation velocity as a function of charge diameter and density: the detonation velocity, D, was found to be a linear function of density and a function of charge diameter;
- failure diameter; and
- gap test sensitivity (actually sensitiveness) for relatively small charges, principally because most explosives studied detonated reliably without confinement in sticks about $\frac{1}{2}$ in. (12.7 mm) in diameter.

Although a stable detonation wave is possible in any charge of diameter greater than the failure diameter, the smaller the diameter, the more curved the detonation front, as the lateral expansion of the products from the edges reduces the velocity there. Only in relatively large diameter sticks of explosive can we envisage a region with a plane, steady detonation front, as shown diagrammatically in Figure 6.5.

The reaction zone, where the explosive is transformed into gaseous detonation products with liberation of energy, propagates through the material with constant velocity, D. We assume that stable conditions have been achieved so that the thickness of the reaction zone and its velocity remain constant with time. This means that, if it were possible to travel alongside the detonation wave at exactly the same velocity, we should observe an unchanging picture of the transformation of explosive into products. Un-

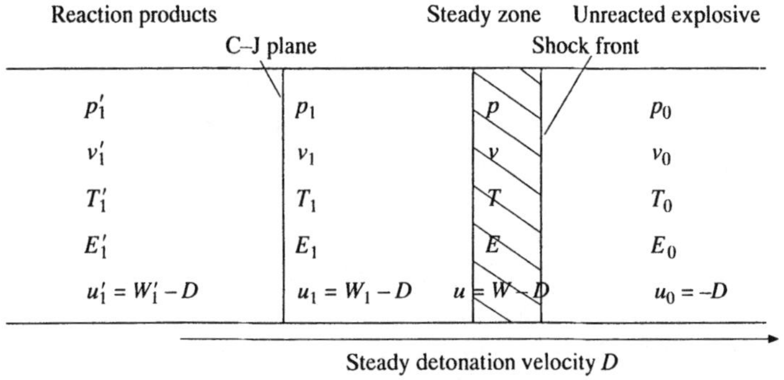

Figure 6.5. A stable plane detonation wave.

detonated explosive would seem to the observer to flow from the right at a velocity $u_0 = -D$. The pressure, p_0, temperature, T_0, specific volume ($v_0 = 1/\rho_0$), and internal energy per unit mass, E_0, at all points ahead of the shock front, change abruptly to p, T, v, and E at the shock and thereafter continuously as reaction proceeds. The stability of the front is maintained by energy liberated between the shock plane and some downstream plane, where conditions are marked by the subscript 1. Thus, conditions to the right of this plane on the diagram remain constant to the observer. This concept enabled Jones (1949) to use the Rankine–Hugoniot relations, which express the conservation of mass, momentum, and energy in the material stream flowing through the reaction zone, to relate the hydrodynamic variables across the reaction zone.

Conservation of mass, energy, and momentum across the reaction zone requires that

$$\frac{u}{v} = \frac{u_0}{v_0}, \tag{6.18}$$

$$\frac{u^2}{v} + p = \frac{u_0^2}{v_0} + p_0, \tag{6.19}$$

$$E + \tfrac{1}{2}u^2 + pv = E_0 + \tfrac{1}{2}u_0^2 + p_0 v_0. \tag{6.20}$$

Equations (6.19) and (6.20) assume that viscous and thermal transfer across the reaction zone can be neglected.

In the notation of Figure 6.5, based on the independent variables p and v, these equations become

$$u_0 = -v_0\sqrt{\frac{p - p_0}{v_0 - v}}, \tag{6.21}$$

$$u = -v\sqrt{\frac{p - p_0}{v_0 - v}}, \tag{6.22}$$

$$E - E_0 = \tfrac{1}{2}(p - p_0)(v_0 - v). \tag{6.23}$$

Equation (6.23) is one form of the Rankine–Hugoniot equation. It takes account of the fact that there is an entropy increase across the shock front (an isentropic change would be governed by $dE = -p\,dv$), and it implies that the increase in internal energy across the shock front is equal to the work done by the mean pressure during compression.

If we consider a coordinate system at rest in the unreacted explosive, equations (6.21) and (6.22) become

$$D = v_0\sqrt{\frac{p - p_0}{v_0 - v}}, \tag{6.24}$$

$$W = (v_0 - v) \cdot \sqrt{\frac{p - p_0}{v_0 - v}}, \tag{6.25}$$

where W is the flow, or streaming velocity in the new reference system. Equations (6.23)–(6.25) derive their origin from earlier ones for an inert shock wave. They differ from the latter only in the inclusion of the chemical energy term q in the energy conservation equation (6.23), which is defined as the difference in heats of formation of the explosive and the detonation products in the standard states to which E and E_0 refer.* Equations (6.23)–(6.25) also apply at any plane within the steady zone between the shock front and the C–J plane, with properties of the material there denoted by subscript 1. If we assume that the detonation products are in equilibrium, E may be expressed as a function of p and v only. In general q is also a function of p and v, since the equilibrium composition of the detonation products may also depend on their thermodynamic state. The quantity $E = e - q$ represents the effective internal energy of the detonation products, in the sense that this is the energy available for exploitation as external work and is the energy treated in thermodynamic arguments where chemical effects are not invoked.

Equations (6.23)–(6.25) define relationships which are best discussed with reference to Figure 6.6, in which are shown Rankine–Hugoniot curves. For a given initial density $\rho_0 = 1/v_0$ and detonation velocity, D, equation (6.24) describes a straight line in the p–v plane. This is known as the Rayleigh line (R), which defines the locus of all p, v states consistent with the prescribed velocity and conservation of mass and momentum. The Rayleigh line passes through the point representing the initial conditions, p_0, v_0.

Equation (6.23) is a purely thermodynamic relation defining a convex downward curve in the p–v plane. It is known as the Hugoniot curve (H) and is the locus of all p–v states attainable from p_0, v_0 by a discontinuous transition consistent with energy conservation. A family of "parallel" Hugoniot curves can be drawn based on one for an inert shock, which passes through p_0, v_0 (referred to as point A in Figure 6.6), which is also known as the Hugoniot for the unreacted explosive. As n, the fraction of reaction, increases from 0, for the inert shock, i.e., unreacted explosive, to 1,

* These standard states are the initial state in the unreacted explosive and the state when reaction is "complete." In one sense, reaction may not be complete after a detonation for a few tens of milliseconds, due to recombination reactions which may take place in the products as they expand and cool to atmospheric pressure. However, after a few tens of microseconds, unless there is aluminum present, most recombination reactions are all but complete. In this time the detonation wave will have moved up to half a meter from the position it occupied when the recombination reactions started. However, these latter parts of these reactions tend to contribute relatively little energy to the overall energy balance of the system, hence, conventionally, we assume that all useful energy has been extracted from the conversion of the explosive to products once the products reach a point downstream from the reaction zone which is referred to as the Chapman–Jouguet plane and is marked in Figure 6.4.

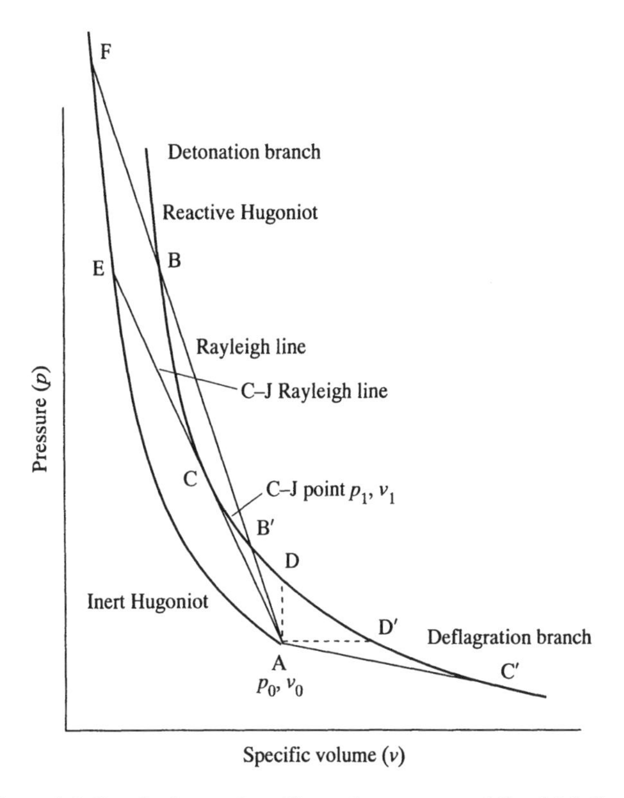

Figure 6.6. Steady detonation; Hugoniot curves and Rayleigh lines.

for the products of a fully reactive detonation wave, a series of curves can be drawn above the one representing the inert shock.

The precise form of the curves will depend upon the equation of state chosen to represent the p–v relationship between the products. Let us assume that the upper curve on the diagram represents completely reacted explosive where $n = 1$. The exothermic chemical energy term q defines the extent of the displacement of this curve above that for the unreacted explosive. Rayleigh lines may be drawn through p_0, v_0 (A) to meet the Hugoniot at B and B′. B represents the generalized position p_1', v_1' in the reaction products in Figure 6.5 and each Rayleigh line represents a different velocity of detonation, the steeper the line the greater the velocity.

From (6.24) and (6.25), elimination of v yields

$$DW = (p - p_0) \cdot v_0$$

or, eliminating p,

$$\frac{W}{D} = 1 - \frac{v}{v_0}.$$

If we assume that all the reaction is complete at the C–J plane (subscript 1 in Figure 6.5), and chemical equilibrium has been attained (6.23) defines the relationship between p_1 and v_1 that must be satisfied in order for the detonation wave to be stable. The velocity of the detonation wave is given by (6.24), rewritten as,

$$D\rho_0 = \sqrt{\tan\theta}.$$

It has been found experimentally that D, the detonation velocity, corresponds to the slope of the tangent from p_0, v_0 to the Hugoniot at C, where pressure and volume are given by p_1, v_1. This result, that the steady detonation velocity is the lowest velocity of a stable detonation wave starting in an explosive material at conditions p_0, v_0 and ending at p_1, v_1, is known as the Chapman–Jouguet Hypothesis, the point p_1, v_1 is at the Chapman–Jouguet condition and the plane where this occurs in Figure 6.5 is known as the Chapman–Jouguet (or C–J) plane. The hypothesis can be justified on stability grounds, since, in the detonation case, a detonation velocity associated with line AB in Figure 6.6 is unstable. Any point on the reaction products Hugoniot above the point of tangency, C, represents a pressure and density regime where the velocity of sound in the reacted material is greater than that in the detonation wave, so any rarefaction behind the detonation front would eventually catch it up and reduce the shock pressure. Such a rarefaction surely arises, because at some point behind the detonation front, the burnt gases lose pressure and density as they expand. Let us assume that the sound speed is c_1 relative to the fluid at point C, which itself has a velocity W_1, then, if $c_1 + W_1 > D$, the wave cannot be steady and must lose velocity. If $c_1 + W_1 < D$ this effect must persist some way into the region of the reaction zone and the chemical energy released in this short distance can have no effect on what happens ahead of it, so it cannot support the detonation front. In practice, this represents a reduction in the heat of reaction, which is associated with a drop in the wave velocity. Hence the condition of highest steady-state wave velocity can only be achieved when $c_1 + W_1 = D$, i.e., the detonation velocity is the sum of the sound speed and the particle velocity in the reactant and is valid only at the C–J point p_1, v_1.

The tangency condition defining the Chapman–Jouguet Hypothesis is written

$$-\left(\frac{dp_1}{dv_1}\right)_{\mathrm{H}} = \frac{(p_1 - p_0)}{(v_0 - v_1)},$$

where the suffix H denotes differentiation along the Hugoniot.

It is possible to extend the analysis further (see Chapter 3 by Davis), with some approximations, to derive a set of equations from which it is possible to calculate values for:

- the velocity of the detonation wave;
- the detonation pressure;
- the detonation temperature;

- the specific volume of the products; and
- the particle or streaming velocity of the products behind the detonation front.

However, while it is useful to be able to determine approximate values of p, v, and T from D this way, better, more refined, equations of state and parametric methods based on data derived from many hundreds of experiments, such as is the basis of the hydrocodes developed by Mader (1979) and others are likely to give results better than the $\pm 5\%$ accuracy expected from this technique.

The Hugoniot curve divides into an upper and a lower branch. In the upper branch, from point D to B, $p > p_0$, and $v < v_0$. This implies compression of the products, as expected in a detonation and as discussed earlier. The products of a detonation wave move in the same direction as the detonation wave and, at least up to the C–J plane, remain compressed and support the detonation front. In the lower branch of the Hugoniot curve, from point D′ to C′, it is also possible to draw a tangent to the curve with a positive slope, meeting it at point C′. In this case, $p' < p_0$ and $v' > v_0$. These conditions imply an expansion of the products and represent conditions consistent with a deflagration, where the products of the reaction move away from the burning front and do not provide compressive support for it. Consequently, the deflagration wave is slower and less damaging than a detonation wave, since it is not associated with intensely high pressures of products.

It is extremely important to understand what is meant by motion of products, either toward the detonation front, or away from it. Motion toward the front means motion in the direction of the front, but only while the reaction process started in the reaction zone goes to completion, i.e., between the shock and the C–J plane. Thereafter the products expand supersonically initially, when they do their major amount of damage by shock, then less rapidly as their energy dissipates, due to cooling of the fireball and decreasing pressure. In the case of deflagration, we speak of an increase in volume and a decrease in pressure at the equilibrium point. The increase in volume, relates to a decrease in density, which, if we consider a solid explosive, allows the products of deflagration still to be substantially compressed as gases, but insufficiently to encourage compressive waves to travel through the reaction products and overtake and support the burning front. Thus, a deflagration still produces compressed products which are capable of doing work on the surroundings, they simply do it somewhat slower than the products of a detonation.

If the Rayleigh line defining the C–J condition of tangency (AC) is produced to meet the inert Hugoniot, i.e., that for unreacted explosive, it does so at point E, where the pressure and specific volume are given by p_0', v_0'. There is clearly a discontinuity between the initial conditions, p_0, v_0 and p_0', v_0', where matter has been physically, but not chemically changed. This is a representation of a shock wave, not a detonation wave, because the material is unreacted. Just ahead of this shock the matter is at rest, but, as the detonation wave moves through the point, i.e., the shock wave impacts it, forces begin to operate on it to set it in motion with a velocity, W. This velocity is in the direction of the detonation wave, which is traveling at D. W will be less than D, since the shock front is engulfing new material at each moment, so it cannot be moving faster than the detonation. It can be shown that

$$W = \sqrt{(v_0 - v_0')(p_0' - p_0)}$$

from (6.25). In, say, TNT, W is about 1500 m/s, and D is 6900 m/s at a density of 1.57×10^3 kg/m^3. The explosive receives an enormous blow as the detonation front impacts each new portion of the charge. The blow is delivered by the discontinuous increase in the velocity of matter behind the front. This is less than the detonation velocity of the explosive, D, but of the same order of magnitude. So the blow is of the same order of magnitude as that delivered by a detonator or primer and, since the latter is sufficient to start detonation, it is possible that this effect, known as the "von Neumann Spike" (1952–1963), is sufficient to maintain it.

Zeldovich, Döring, and von Neumann (1940–1950) independently argued, according to von Neumann (1942), that, although the hydrodynamic theory of detonation was developed to define the thermodynamic state after the completion of energy release, it was also instantaneously applicable to any intermediate stage of the reaction. As has been remarked earlier, the two curves in Figure 6.6 refer to the inert, or nonreactive Hugoniot of the explosive and the fully reactive one. There are an infinity of intermediates between them corresponding to an infinity of extents of reaction between $n = 0$ and $n = 1$. Since the detonation process is assumed to be steady, the C–J condition for any fraction of reaction is still given by the tangency of the relevant Rayleigh line to the appropriate Hugoniot curve. Thus, the reaction zone for a detonation at the C–J point is represented by a continuity of states passing along the Rayleigh line from A to C or E to C. Of the two, the latter seems the more plausible, since it must be preceded by a pure shock transition from A to E, without chemical reaction, where the shock temperature alone would be sufficient to initiate chemical reaction. A continuous transition along the line AC implies that the chemical reaction is initiated spontaneously at ambient conditions, contrary to the known stability of practical explosives.

The Zeldovich–von Neumann–Döring (ZND) model assumes that in a detonation the chemical reaction is initiated by a pure shock wave (p and v follow the inert, unreacted explosive Hugoniot from A to E in Figure 6.6). Then, as the reaction proceeds, the pressure and density drop (from E to C), but the temperature continues to rise during this process. Hence, the reactive part of the wave is a deflagrative transition from the shocked explosive state (E) to the Chapman–Jouguet condition (C). The high shock pressure, in excess of the C–J pressure, at the front of the reaction zone is often referred to as the von Neumann spike. It was first observed by Kistiakowsky and Kydd (1954) in gaseous explosions and Duff and Houston (1955) in solid explosives. The ZND model also has a contribution to make on stability. If a detonation wave were to propagate at a higher velocity than the C–J velocity, say according to the Rayleigh line, AB′BF in Figure 6.6, it would represent a shock transition from A to F, followed by a deflagration transition to states B or B′. State B has already been excluded as unstable. A further transition to state B′ corresponds to a negative, or pressure-decreasing shock, which is hydrodynamically unstable, unless it is supported by further energy release. This cannot occur, because the line BB′ is outside the Hugoniot curve for complete reaction; there is simply no more energy available. Hence, state B′ is excluded as being hydrodynamically inaccessible in a steady shock plus deflagration process.

6.5.2.2. Transient Detonation Waves

The energy release in a C–J detonation maintains the detonation wave in a stable state which can persist indefinitely. Velocities of detonation can be increased by processes which feed extra energy into the explosive before it detonates, or into the detonation products. Experiments involving electrical discharges into the detonation products plasma during a detonation have been carried out in various laboratories, with some success. However, the greatest problem here is the packaging of the electrical energy, since an electrical power source, such as a capacitor, is significantly less compact than an explosive charge. On a more mundane level, it is possible to overdrive a detonation in a relatively slowly detonating explosive by initiating it by one with a significantly higher detonation velocity and detonation pressure. Subnormal detonation velocities can be observed if energy is extracted from the explosive, or, if for some other reason, the available energy is not fully released. However, in each case the mechanisms are transient. The most important transient detonation regime is in shock initiation, when a pure shock is generated in the explosive and this transitions to the detonation wave in a manner governed by the time dependence of the energy release from the chemical reaction, via the deflagration process.

Consider a relatively simple case first, where release of energy is instantaneous. The shock-generating mechanism imposes a certain pressure and particle velocity condition on the explosive surface. They may be in excess of, or less than, the corresponding values at the C–J condition. If they are in excess, the condition corresponds to some point on the branch CB of Figure 6.6, because this corresponds to a higher detonation velocity, temporarily overdriven to that state. If they are less, the condition lies on the adiabatic expansion path from C (the adiabat), which falls a little below CB′.

If the applied pressure is less than the C–J pressure, then a C–J detonation is immediately generated and the states at the detonation front and the explosive surface are connected by a rarefaction wave, which spreads out, but cannot, of course overtake the detonation front. This rarefaction was first analyzed by Taylor (1950) and is called the Taylor wave. If the applied pressure is higher than the C–J pressure, then the pressure in the detonation products remains at this value and an overdriven detonation wave is generated and maintained as long as the driving pressure is maintained. The work done on the explosive surface supplies additional energy capable of driving the detonation wave at a higher velocity, in contrast to the former case where the work done appears only in the energy distribution within the detonation products behind the detonation front. Overdriven detonation waves generated by high velocity plate impact have been studied by Skidmore and Hart (1965), who found that, within experimental error, their data conformed with numerical predictions made using a very simple equation of state, $pv = (\gamma - 1)e$.

Another example of overdriving is the convergent detonation wave, generated through engineering design of an explosive charge and its confinement. The inward radial flow of matter in the products of the detonation feeds energy through the detonation front, which causes continual acceleration of the wave.

It has been assumed that the chemical release rates in explosive decomposition are described by the Arrhenius equation, or rate law. The rate of the chemical reaction is an exponential function of the negative reciprocal temperature of the reactant. Energy release according to this law produces self-heating, as the thermal theory has been shown to predict. The reaction accelerates slowly at first, then suddenly progresses rapidly to explosion, or, in the present discussion, to completion via detonation. The process may be approximated by an initial temperature-dependent induction period, or an "adiabatic explosion time," during which the reactant remains quiescent, then seems, instantaneously, to go to complete reaction at the end. Observations on the shock initiation of homogeneous (liquid) explosives concur with this view of energy release, as shown by Campbell, Davis, and Travis (1961). A shock is generated and propagates into the explosive, raising its temperature. After the appropriate induction period at this temperature, the material near the surface, subject to the initiation stimulus, explodes and generates a detonation wave, which sets off to pursue the shock. This is a C–J detonation, but it is taking place in material in motion and at an higher pressure, density, and internal energy due to the precompressing shock and so has a higher detonation velocity than the undisturbed explosive. The detonation wave eventually overtakes the initiating shock and the interaction between the two generates an overdriven detonation, which persists until penetrating rarefactions reduce it to the steady C–J detonation state. This process was first reproduced numerically by Hubbard and Johnson (1959) for processes in a hypothetical explosive. Later calculations by Mader (1963) using data on nitromethane, gave good agreement with experiment.

The behavior of heterogeneous solid explosives was shown by Campbell, Davis, and Ramsay (1961) to be somewhat different. Here the initiating shock is found to accelerate slowly, until detonation occurs suddenly at or near the shock front and quickly settles down again to propagate in the forward direction as a steady C–J detonation. Sometimes a detonation wave is seen to propagate in the reverse direction, to consume shocked explosive. This is a "retonation" wave. This behavior as a whole does not conform to the Arrhenius reaction rate law. The slow acceleration of the shock in the early stages suggests that it is being supported by some type of energy release mechanism occurring very close to the shock front. It is likely to be a reactive shock, or a detonation with incomplete energy release. The mechanism of the process is thought to involve shock interactions at grain boundaries, density discontinuities, and other inhomogeneities, which generate hot-spots there. These hot-spots may easily initiate thermal decomposition, which at the intensely high pressures and temperatures involved in shock impact, induce very high reaction rates in the explosive. This local intensification of the energy release will accelerate the shock until detonation results. No satisfactory models exist for this part of the shock initiation process, although the "Forest Fire" (Mader, 1979) procedure is a step along the path to fuller understanding.

6.6. Relationships Between Thermal and Shock Initiation Theories

Many workers have attempted to quantify thermal or shock initiating stimuli in order to be able to rank the perceived hazard associated with the use of a given explosive under given accident conditions. There is a plethora of explosives hazard tests available, as reported in Chapter 8 but, it is not possible to predict the behavior of weapons systems in hazardous situations based solely on the results of small scale tests. Hazard tests only produce comparative data on the response of any given explosive with respect to others. There is no criterion which can be applied to a given explosive, which, if exceeded, results in a deflagration, or detonation, dependent upon its level. Thus, much of the process of choice of an energetic material for a particular role in a weapon system, particularly where the safety of the system is concerned, depends upon subjective assessment and full-scale weapon testing trials. It is an expensive mistake, if, by the time full-scale hazard trials, such as required by the Royal Navy, the US Navy, or several other world navies [DoD Mil-STD-2105B (1994); OB Proc. 42657 (1990)] before embarkation of a weapon on board ship, are carried out the weapon fails one of the tests. This situation is avoided at present by carrying out many tests at the development stage of the weapon, at significant cost. These might be reduced in number and, hence, cost less if small-scale hazard tests bore a closer relationship to real hazard stressing of weapons.

The concept of critical energy for initiation has been invoked as a means of ranking the sensitiveness to initiation of different explosives to a given stimulus. In some circumstances it works well enough, e.g., the Walker–Wasley Criterion for the initiation of explosives as a result of projectile impact, $p^2\tau$, or the v^2d criterion for the initiation of explosives by shaped charge jet, where τ is the time period over which the pressure p is imposed on the explosive, and v is the velocity of a shaped charge jet d in diameter. The reason why such data are internally consistent is because in each case of the tests the energy was applied over roughly the same time, so that a time element is implicit in the analysis, even though what is said to be compared are energy inputs to the target explosive necessary to cause initiation at a given frequency, usually 50%. However, this concept of critical energy input may not be a true reflection of the criteria governing initiation, since it is a nonsteady phenomenon and a measurement of energy input alone ignores the period over which that energy is deposited in the target explosive. Bridgeman and his coworkers (1947) and Afanasiev and Bobolev (1971) subjected small samples of explosive to slow pressurization well into the range of detonation pressures. Remarkably, no explosives ignited, exploded, or detonated. This effect cannot be readily explained by a "critical initiation energy" argument, since, manifestly a significant amount of energy per unit mass of explosive was introduced into the explosive by virtue of the work done on it by the compression process. The only difference between these tests and those involving the initiation of detonation, say, in a shock initiation test is that, in the latter, the pressure is reached in a fraction of a microsecond, rather than over a period of several hundred seconds.

The introduction of the time factor transforms the concept of critical energy input into one of critical power input to the target explosive, the shorter the time for the delivery of a given energy level, the greater the power input by the stimulus. Lee (1992) has developed a theory which integrates the initiation theories of deflagration and detonation under a new criterion pertinent to both regimes. It can be derived from disparate initiation processes, ranging from shock to hot wire initiation. The theory builds on the importance of the time factor in initiation, while acknowledging the need to invoke the concept of sufficient intensity in the stimulus to generate an effect in the explosive. It might be possible to generate a very high power input to a charge, but, if the total energy involved is insufficient to raise a sufficient number of molecules over the activation energy threshold for the process to become self-sustaining, no event can possibly ensue. In order to complete the concept it is necessary to include an extensive factor for the material being subjected to an initiation stimulus. A dimensional analysis of the problem pointed to the importance of the mass of energetic material in receipt of the stimulus. The criterion treats of the critical power density necessary to cause initiation, i.e., the critical power input per unit mass of explosive. Analysis of the initiation theories associated with a number of initiation processes, including powder impact, shock, jet, premature explosion of shell, etc., indicates that the grouping of terms

$$\frac{1}{\rho} \cdot \left(\frac{dp}{dt}\right)_{\text{crit}} \tag{6.26}$$

is relevant. It has the dimensions power per unit mass, i.e., W/kg. It includes the time element in initiation and has an adequate intensity factor.

Interpolation of data published on initiation of burning, deflagration, and detonation has shown that all initiation theories have a criticality concept which can be expressed in terms of equation (6.26), which expresses critical power density in terms of a mechanical input, or as

$$\frac{C_V}{\rho} \frac{dT}{dt} \tag{6.27}$$

which expresses it in thermal terms. Analysis of work on critical "energy" input from papers and reports on hazard enable some general conclusions to be drawn. Critical initiation power levels of common military explosives are found in three bands:

Initiation of deflagration	ca. 10^8 W/kg
Deflagration/detonation transition (DDT)	ca. 10^{11} W/kg
Initiation of detonation	ca. 10^{12} W/kg

These bands coincide with the levels of power output expected from energetic materials reacting either in a deflagrative or detonative mode. No confirmation of these findings has been forthcoming and this concept has not found wide acceptance.

6.7. Initiation Mechanisms

Initiation of explosives requires the injection of energy to the reactant at a rate commensurate with the level of output required from the material. Thus, explosives are initiated from detonative output generated from a detonator via a booster explosive. Rocket and gun propellants can also be made to detonate under certain conditions, but are required to deflagrate in their normal roles and so are initiated by deflagrative processes. Initiation processes for high explosives fall into two main categories, electrical and mechanical. There are relatively few, if any, thermal initiation techniques pertinent to high explosives, because deflagrative output of high explosives is not sought due to its instabilities and unpredictable nature.

Some commonly encountered initiatory compounds are listed in Table 2.3 (in Chapter 2). A similar list is found in an excellent introduction to explosives, propellants, and pyrotechnics by Bailey and Murray (1989), which is part of a series of books on weapons, and related subjects, published by Brassey's. Only a very few materials fall into the required range of high sensitiveness, without being unduly dangerous and high thermal stability. Mercury fulminate, used in detonators for many years, was extremely sensitive to the point of being dangerous. Even lead azide has its deficiencies, mainly in regard to the relative ease by which it is hydrolyzed. This provides the possibility for the formation of hydrazoic acid, which may react with copper or brass to produce copper azide, or tin in solders to form tin azide, both of which are unacceptably sensitive to friction and impact. Silver azide was introduced in the 1970s, because it is less susceptible to hydrolysis. It increases the service life of unsealed detonators, especially those used in naval service.

It is often not possible to use a single initiatory explosive in an initiatory role, because neither its properties nor those of its fellows alone are optimized for a specific initiation stimulus. Mixtures are used to tailor materials and sensitiveness properties to the input stimulus and the desired output. For example, lead azide is not especially sensitive to friction, which is the mode of initiation in a stab-sensitive detonator, where a needle is forced into the explosive as shown in Figure 6.7.

The igniferous composition in this case may be tetrazene or, more likely, lead azide doped with about 5% of tetrazene, which aids the initiation of lead azide by increasing its stab sensitivity. A layer of lead or silver azide may follow, with an additional increment of RDX if the output is required to be especially powerful.

Some detonators respond to flash input from a source some distance away. A flash-receptive composition involves lead azide, for detonative output, lead styphnate for thermal, i.e., flash, receptivity, and aluminum to enhance the likelihood of lead azide burning to detonation. It also requires a lead azide increment to enhance detonative output further and finally a booster explosive. This is usually RDX nowadays, although tetryl was used at one time to generate an adequately powerful detonative output to ensure detonation transfer through a septum. A flash-sensitive detonator looks very similar to the stab detonator shown in Figure 6.7, except for the needle.

Small arms caps are sensitive to percussion and require a very sensitive explosive, such as lead styphnate to respond to the relatively low-power

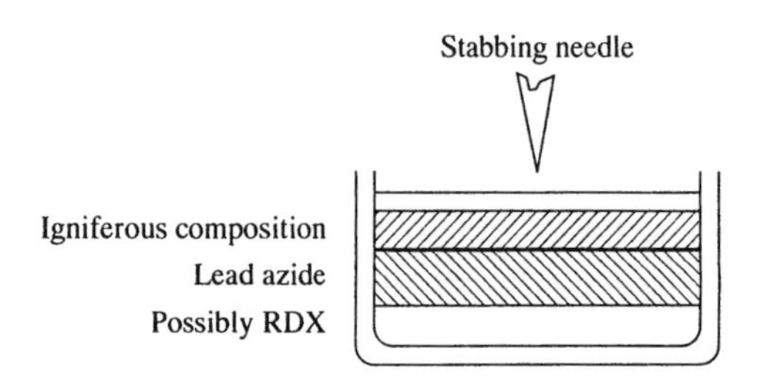

Figure 6.7. A stab detonator.

blow from the firing pin, but, since the output required is not detonative, some form of pyrotechnic composition is also necessary. This ensures that the output contains an adequate amount of hot particles which can be ejected from the flash holes in the primer on to the single-base propellant charge in the cartridge case.

Recent changes in the law in many countries have required the elimination of the use of tetryl from new weapons because of its mutagenic properties and the perceived hazard to workers exposed to it over long periods. This reduces the options for booster explosives to PETN and RDX. With the increase in demand for lower vulnerability weapons systems PETN has also fallen out of use, except for very specialized requirements, such as detonation transfer along thin channels in complex warhead initiation systems. RDX-based compositions are more favored as boosters nowadays, although in order to reduce their sensitiveness to thermal stimuli and mechanical impact, they may also contain TATB and some form of waxy or polymeric phlegmatizer. Propellant booster systems often employ black powder, since it burns with the generation of copious quantities of hot particles, which are ideal for the ignition of rocket or gun propellants. Alternatively, very fine particulate propellant is sometimes used and in very large rocket motors for space or strategic weapons, a small rocket motor with about 10% of the thrust of the main motor is sometimes employed and its output directed on to the main motor grain (Zeller, 1993).

6.7.1. Initiation by Heat

Strictly speaking, all explosive initiation processes require the input of heat in one form or another. For example, thermal decomposition of the energetic material in a slow cook-off is initiated by heat, as is the initiation of a bare explosive charge in the oblique impact test. In the former case the whole of the charge undergoes thermal decomposition in a controlled environment. In the latter, localized hot-spots associated with temperature spikes induce rapid local decomposition, which may, or may not, spread to engulf the whole of the charge. In neither case, however, is the process under sufficient control to be capable of being used as a reliable form of initiation of secondary high explosives or propellants directly. The problem with bulk thermal initiation is the uncertainty of the length of the induction period, since, while the theory is revealing, it consists of a number of approximations, which render the numerical prediction of thermal events insufficiently precise for the exact timing of the moment of initiation. Moreover, there is no guarantee that detonation will ensue from a thermal explosion initiated in this way and bulk thermal ignition is certainly not what is required in the initiation of rocket propellants especially, because motors are designed to burn from exposed surfaces, rather than uniformly from the center. In neither case is it possible to be sure what the effects of the casing of the weapon will be.

Hot-spots generated by nonshock, mechanical action are randomly distributed and there are uncertainties in their location and the temperatures at which they are formed. These factors render initiation of secondary explosives or propellants by this technique uncertain. However, hot-spot initiation is not unsuited for use in initiation trains, where especially friction-sensitive initiatory materials may be highly receptive to hot-spot initiation and reliable output can be obtained from relatively low-power inputs. Hot-spots play an important role in the initiation of some of the less sensitive commercial slurry explosives. The presence of glass microballoons increases local shock pressures and temperatures via shock focussing. This phenomenon was elegantly shown theoretically by Mader (1979) in his studies of the interaction of a shock wave with voids in nitromethane.

6.7.2. Friction or Stabbing

Friction forms an integral part of the initiation trains of a wide range of relatively low-technology weapons. It produces immense quantities of hot-spots in initiatory explosives or compositions where they can grow reliably to detonation, or to initiate further igniferous output materials from which propellant grains can be reliably ignited. The preferred form of friction input is via the action of "stabbing" a composition through a metal foil septum into a friction sensitive composition, which responds igniferously, as was shown in Figure 6.7.

6.7.3. Flash or Flame

Most energetic materials are sensitive to the application of a flame, some, especially those used in flash-receptive detonators and igniters, are so even if the flame lasts only for a few milliseconds, as in a flash. Flash-receptive materials, such as lead styphnate, in current use have igniferous, rather than detonative output, so the transition from igniferous to detonative output in an explosive munition must be via several intermediate compounds in addition to the flash-receptive material.

6.7.4. Percussion

Most initiatory materials will be initiated by a sharp blow and this is a relatively reliable method of initiation. In a small-arms cartridge, the impact of a striker pin against a specially designed anvil crushes the cap composition, as shown in Figure 6.8. Initiation occurs via crushing of crystalline particles and the generation of hot-spots from which the process grows reliably. Output is igniferous and is channelled as hot particles and gases through the flash holes to impinge on the propellant in the cartridge. Pressed powders are most sensitive because they allow the movement of particles against each other where the harder ones abrade the softer ones to generate heat.

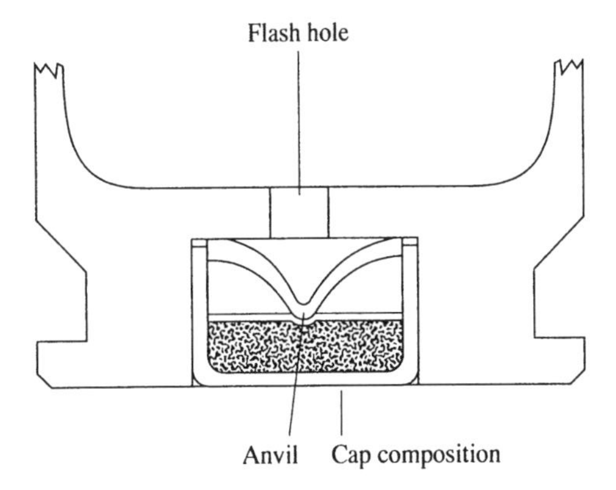

Figure 6.8. Ignition of the propellant in a small-arms cartridge.

6.7.5. Electrical

Electrical methods are very frequently used in weapon systems, the more so recently with the introduction of new types of separated electrolyte batteries, which only mix their electrolytes when the weapon is launched and armed. Electrical detonators and igniters have been designed with sensitivities from the order of microjoules to tens of joules of input electrical energy necessary for completely reliable initiation. The most common form of electrical device is the bridgewire, or foil, where the conductor is part of an electrical circuit. The conductor is stimulated either by a continuous electrical current, or the discharge from a capacitor, either of which deliver of the order of 10 mJ electrical energy. This energy is equivalent to a current necessary to ensure fully reliable initiation, an "all-fire current" of the order of 1–5 amp in a fuzehead device of the form shown in Figure 6.9. This is the basis for electrical demolition detonators, or blasting caps and microdetonators used in weapon systems. However, in some systems there is insufficient electrical energy available to initiate the fuzehead, for this type of detonator to be considered sufficiently reliable. Alternatively, a shorter time to detonation or ignition is required than the few milliseconds achievable by using a fuzehead detonator, which operates at a rate dependent on the burning of the chemicals covering the fuzehead and so is also likely to have wider time tolerances than can be permitted for optimum functioning of some systems.

There are devices in which the initiatory material itself is made part of the electrical circuit. This is achieved by adding graphite to the lead or silver azide in the first increment of the detonator or igniter, so as to render it electrically conducting. The graphite type and quantity incorporated into the composition can be used to control the resistance of the device and the amount of electrical energy deposited in it, so as to provide the best match for the energy available and the rise time of the delivery mechanism. Such

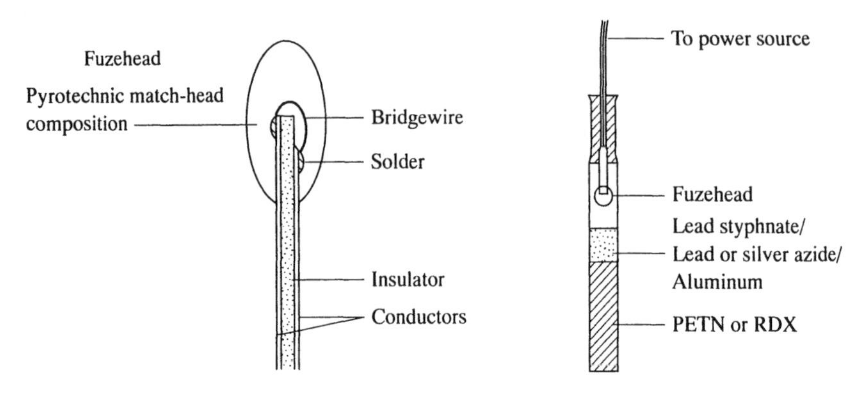

Figure 6.9. Fuzehead and bridgewire detonator.

devices can be extremely sensitive and ones with sensitivities of the order of 10 μJ have been used, especially in conjunction with piezo-electric crystals, which rely on relatively light impact to cause explosion. Piezo crystals are falling out of favor gradually because of their extreme sensitiveness and the requirement that very stringent electrical safety measures are taken with weapons with them on board. This is especially the case of air-carried systems and even more critically, on such systems flown from ships. A conducting composition detonator/igniter is shown diagrammatically in Figure 6.10. As with other detonator and igniter designs, the initial stimulus may need to be amplified by a series of other compositions to ensure that the booster or main charge or grain receives an adequate stimulus from the relatively low-energy input.

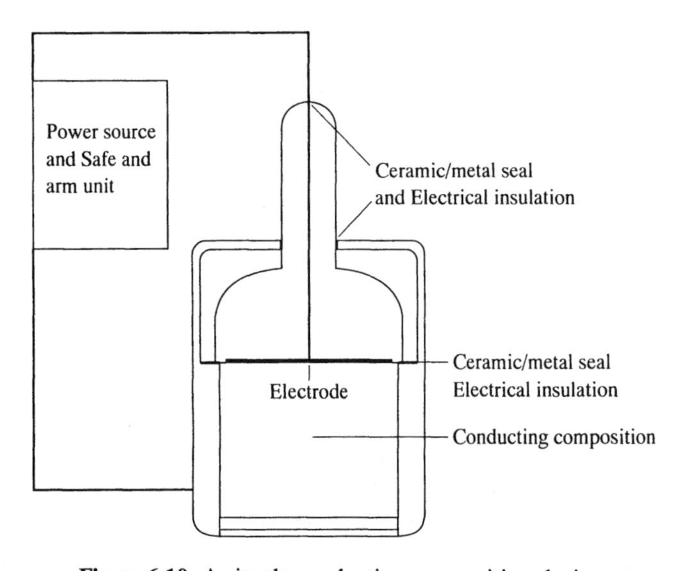

Figure 6.10. A simple conducting composition device.

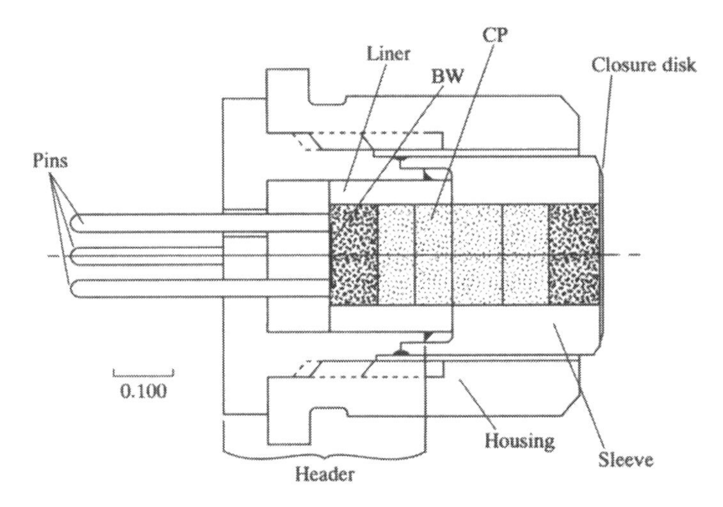

Figure 6.11. Low voltage DDT detonator using cobalt perchlorate explosive [after Lieberman et al. (1984)].

Low-voltage DDT detonators are now available for some devices where there is adequate electrical energy, see Figure 6.11. The explosive used in such devices is cobalt perchlorate, which can be induced to detonate in a relatively low-confinement system and there is no need for a sensitive initiatory explosive. Sandia National Laboratories in the United States have developed low-energy flyer plate detonators as shown in Figure 6.12, which

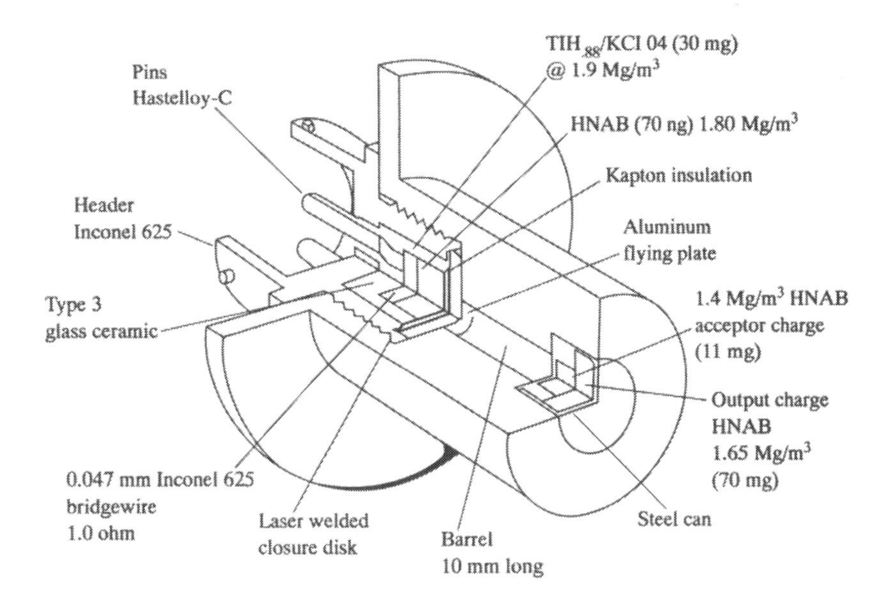

Figure 6.12. Low-energy flying plate detonator [after Jacobson (1981)].

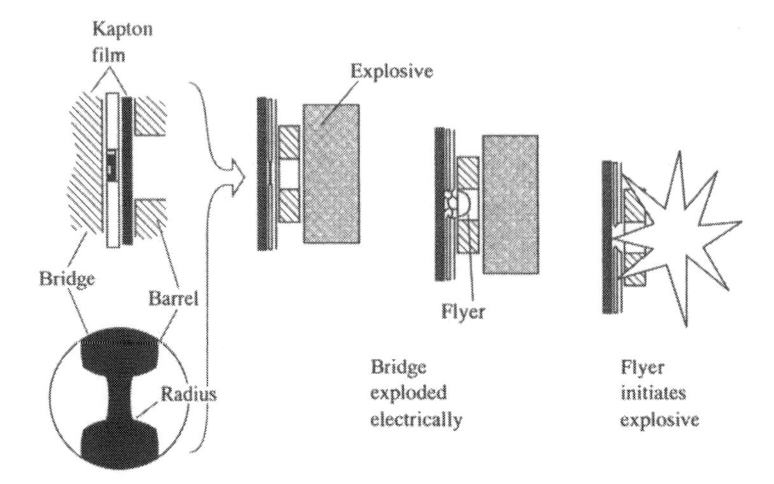

Figure 6.13. Principle of operation of an electrical slapper detonator [published with permission of Los Alamos National Laboratory, NM, USA].

use a 3.5 amp firing current to initiate a mixture of TiH/$KClO_4$ which detonates a small charge of HNAB explosive. This projects an aluminum flyer plate over a 10 mm range at up to 1.3 km/s to impact another two pellets of HNAB pressed to different densities, giving reliable full-order detonation output.

Slapper detonators require no intermediate explosives or pyrotechnics and induce full-order detonation in secondary explosives promptly and reliably (Figure 6.13). Unfortunately, they require relatively high-energy inputs and are expensive for use in anything other than the most expensive of weapon systems, requiring the highest level of reliability and safety. They use the very safe secondary explosive HNS IV, which is not in contact with the slapper plate and so can be used in line with the main explosive charge without a barrier in between. Operation is via a high-voltage discharge (up to 6.5 kV) delivering several joules of energy in nanoseconds to an aluminum foil. The bridge foil vaporizes, or bursts, under the intense electrical load and accelerates the plastic, usually "Kapton" flyer plate. The burst current of the aluminum foil determines the effective power the foil can deliver to drive the flyer. A 0.08 mm thick "Kapton" flyer may be accelerated to 4.1 km/s in 1.2 mm travel and, on impact against the target PETN explosive delivers a 185 kb shock with a 35 ns duration. This is sufficient to generate very reliable detonation "promptly" in the PETN.

6.7.6. Coherent Light

Flash initiation is one of the oldest initiation techniques and was used for gunpowder and was only discarded temporarily for explosive, as opposed to propulsive output, when it was found to be unsuitable to extract the maximum performance from nitroglycerine. Its resurgence came with the flash-receptive detonator and igniter, which have been mentioned already.

Laser (**L**ight **A**mplification by **S**timulated **E**mission of **R**adiation) light sources provide highly intense radiation levels in very precise time periods and are capable of being tuned to emit radiation at precisely predicted wavelengths. These attributes render them ideal for initiation of detonative or deflagrative systems, especially since some of the highly laser-sensitive compositions are also relatively insensitive to impact, friction, and other initiation stimuli. Table 6.2 shows a list of some of the capabilities of recently developed laser initiation systems.

Laser-initiated devices have safety advantages over electrically and mechanically initiated systems. They eliminate the possibility of initiation through electrostatic discharge. This is a very real problem on board ship and in all air-carried ordnance, especially with some types of rocket motor, which are also susceptible to lightning. Missile systems can be tested with greater safety than is the norm currently, where it is essential to input some electrical power to the systems to ensure that circuits are functioning and capable of operating at full power. With lasers, we can test at a different wavelength of light, i.e., one at which the energetic material, or the energetic material dopant is insensitive. Circuit functionality can also be tested at very low light power. Optical fiber connections are perfectly capable of transmitting the optical signals necessary to ensure reliable initiation and laser diodes can be coupled into 200 μm core fibers directly. Fibers of 400–600 μm diameter transmit power densities of up to about 3 GW/cm^2 which represents a power of about 2–4 MW in the fiber. Although this is near the limit of fiber power capacity it is routinely possible for a system such as this to operate with complete reliability. At present costs are relatively high, but will fall rapidly as the technology matures.

If an optical laser initiation system is used, it is possible to eliminate primary explosives from an ordnance device. There may be a need to use pyrotechnics, but lead styphnate, in particular can be eliminated. This implies safer manufacture and greater safety in deployment. The latter is likely to

Table 6.2. Laser sources available for the initiation of weapons.

Laser type	Response	Devices in which the laser can be used
High power laser diodes (1–2 W)	Deflagration	Actuators, e.g., explosive bolts and gas generators: DDT detonators
Solid-state lasers (10–100 W)	Deflagration	Actuators: DDT detonators
Q-switched rod lasers (< 1 J)	Shock	Prompt detonators

contribute significantly to the goal of designing and operating a truly "insensitive munition." It is possible to drive a slapper detonator foil from laser light impact. The flyer is accelerated as a result of laser ablation of a thin disk of aluminum at the end of an optical fiber. A thin foil can easily be accelerated to more than 2 km/s in 40 ns in such a system.

6.8. Initiation Trains

Initiation trains are the means by which energetic materials are initiated reliably in weapon systems. Discussion on initiation so far has emphasized the response of energetic materials to inadvertent stimuli. Understanding of this aspect of explosive behavior enables the explosive engineer to design safer weapons, using less hazardous energetic materials. However, such materials need to be brought to ignition reliably using systems which are intrinsically safe, despite the requirement for them to use sensitive energetic materials to ensure an adequate output from a relatively small initiation energy input. The twin demands of reliability and safety of weapon systems place a significant burden on initiation systems, especially with the steady decrease in the sensitiveness of the main warhead or gun propellant charge or rocket motor propellant grain. Initiation trains often involve a series of explosives, each of slightly lower intrinsic hazard and increasing charge size to translate a relatively low-power input to the full system output.

Main warhead explosives and gun or rocket propellants are relatively insensitive themselves and, in order to be brought directly to full energy output, may require relatively high-power input. Since, by definition, the mass of an initiating explosive device needs to be rather small, if initiation were in a single step, the energetic material in such a device would need to be very powerful and/or sensitive. There would be no guarantee that initiation would be achieved reliably without a relatively large charge of the sensitive material, which is dangerous. Initiation train design seeks to eliminate such problems. It does it by using a carefully graded set of explosives from a very sensitive primary explosive, used very sparingly to accept reliably the relatively weak system initiation stimulus, through intermediate explosive(s) to the main charge. Some initiation systems might have five explosive elements in the train before the main charge is initiated. The relatively large number and degree of complexity of initiation trains are due in part to the need for safety and the requirement to be able to build in a break in the chain for safety purposes, and also because some of the very sensitive materials which accept the first stimuli have relatively weak explosive outputs which need to be enhanced gradually before the main charge is initiated. A safety break in the initiation train can be between the initiator and the booster, or between the booster and the main charge. It is the purpose of booster, or intermediate explosives to increase the output of the train sufficiently that the main charge can be initiated reliably. In many instances, the first part of the initiation train may be a few tens of milligrams of initiatory composition, which may initiate a gram or so of enhancing material before the booster of up to about 50 g is initiated.

Initiation trains are as relevant to propellant initiation in either guns or rockets as they are to the initiation of high explosive charges, since similar security against inadvertent ignition is required. The main difference between propellant and explosive initiation trains is that in the former a final igniferous output is required from the ignition system, in the latter it must be a detonative output. Both ignition trains may start from igniferous input, but in the propellant ignition train amplification of the stimulus is achieved via the use of very vigorously burning materials such as gunpowder, which transmit burning particles and flame to the receptive main grain. In both cases reliability is achieved through the sensitivity of the original primary explosive material, which it is easy to cause to ignite from a moderate stimulus. Currently one of the major problems in the search for reduced vulnerability of weapon systems involves reduction in the hazard associated with the use of relatively large gunpowder or powdered propellant charges as propellant igniters. The problem resides in the requirement for an intense spray of incandescent particles to initiate the propellant. The advantages from the use of gunpowder still just outweigh the disadvantages associated with its sensitivity and unpredictableness, the latter a result of the fact that it relies on a natural product, charcoal, for its performance.

The safe and arm unit prevents the main charge receiving the transmitted stimulus because of a break in the initiation train, as shown diagrammatically in Figure 6.14.

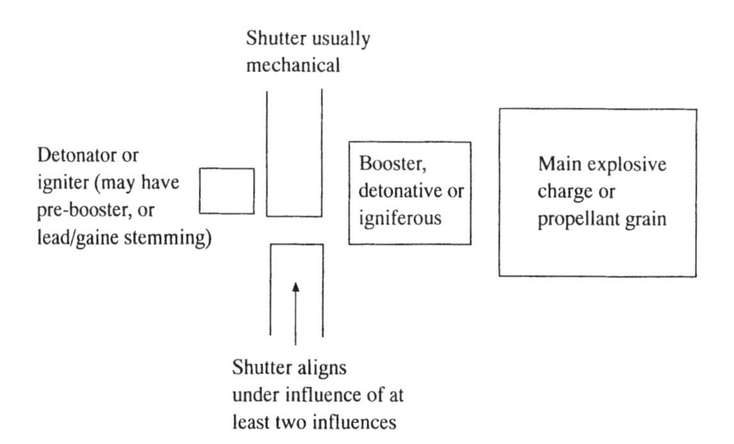

Figure 6.14. Diagrammatic representation of a safe and arm unit.

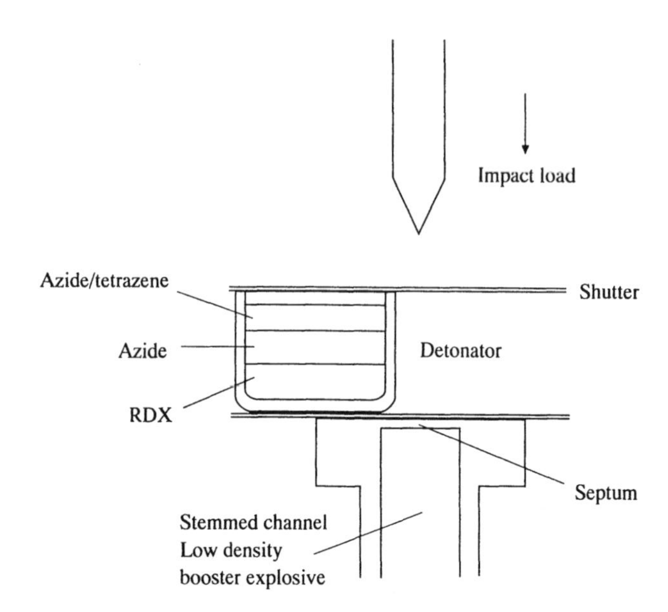

Figure 6.15. Arming unit illustrated in "Safe" position with the detonator out of line with the booster stemming.

An actual detonator arrangement with the presence of the potential break in the chain is shown in Figure 6.15, for a stab-sensitive shuttered system. The shutter is a mechanically operated "gate" which moves at some time during the weapon's arming sequence from a position where the initiation train is out of line physically to one where it is in line. If the train is out of line, a detonation cannot pass the location of the shutter because the receptor charge "below" the shutter is too far from the initiating explosive for the impulse, or thermal output from it to be sufficient to bring the receptor to useful reaction. In weapon design and development testing this assertion is required to be proved by many tests to establish the principle of "shutter sealing," which is of vital importance in assuring the potential user of the safety of his purchase.

Surprisingly, in this electronic age, it is only now, in the mid-1990s, that successful electronic safe and arm units are becoming available. This is because of the requirement that the software should be completely reliable and that it is thoroughly testable. Even the presence of a few electronic "go/no-go gates" in a microcircuit require an immense number of individual certifying tests and, until recently, the time involved in carrying them out was unreasonably long. It is probable that, before the year A.D. 2000 there will be more electronic safe and arm units available, but, until then many systems will remain pieces of precision mechanical engineering, which, in the case of artillery fuzes has been likened to mechanical watch movements.

6.9. Conclusions

The trends in explosives development demanded by weapon designers are greater output power associated with improved safety characteristics. It is probable that new, more powerful explosives molecules will be synthesized in the near future, but they may not be intrinsically safer, or less sensitive. The two requirements in an individual molecule may be mutually exclusive, they certainly seem to be so at present. However, our understanding of the mechanisms of initiation of explosives leads us to believe that we should be able to control the explosive response of energetic materials by managing their thermomechanical responses to initiatory stimuli. This implies that future energetic materials are likely to be composite materials, based on the incorporation of energetic material crystals as fillers in energetic polymeric matrices, possibly plasticized by energetic liquids.

Responses of energetic materials in weapons to bulk or local thermal initiation stimuli, e.g., fuel fires, or slow cook-off, or mechanically induced hot-spots, can be minimized through physical and chemical protection of energetic crystals in the explosive. There are several means by which this can be achieved. Thermal protection might be achieved through the use of energetic material crystal coatings which are both adhesives to fix the crystals in the polymeric matrix and endothermic and so absorb some of the energy from hot-spots otherwise likely to induce ignition of the energetic material. Endothermic polymeric matrices would be able to provide both protection and heat abstraction, but would be likely to reduce the overall output of the energetic material.

Greater exploitation of advances in the generation of electrical power and the continued reduction in the size of power sources may be expected to be exploited in the future to power laser and slapper initiators, since both offer a potential route to the reduction in the use of initiatory compositions with all their safety implications.

References

Bailey, A. and Murray, S.G. (1989). *Explosives, Propellants and Pyrotechnics.* Brassey, London.

Becker, R. (1921). *Z. Phys.*, **4**, 393.

Boggs, T.L. and Derr, R.L. (1990). *Hazard Studies for Solid Propellant Rocket Motors.* AGARDograph No. 316 AGARD-AG-316, NATO.

Bridgeman, P.W. (1947). *J. Chem. Phys.*, **15**, 311.

Cachia, G.P. and Whitbread, E.G. (1958). *Proc. Roy. Soc. London Ser. A*, **246**, 286.

Campbell, A.W., Davis, W.C., and Travis, J.R. (1961). *Phys. Fluids*, **4**, 288.

Campbell, A.W., Davis, W.C., Ramsay, J.B., and Travis, J.R. (1961). *Phys. Fluids*, **4**, 511.

Chapman, D.L. (1899). On the rate of explosion in gases. *Phil. Mag.*, **47**, 90.

Cook, M.A., Pack, D.H., Cosner, L.N., and Gey, W.A. (1959). *J. Appl. Phys.*, **30**, 1579.

Cowan, R.D. and Fickett, W. (1956). *J. Chem. Phys.*, **24**, 932.

Döring, W. (1943). *Ann. Physik*, **43**, 421.

Döring, W. and Burkhardt, G.I. Technical Report No. F-TS-1227-1A (GDAM A9-T-46).

Duff, R.E. and Houston, E. (1955). *ibid.*, **23**, 1268.

Field, J.E., Swallow, G.M., and Heavens, S.N. (1982). *Proc. Roy. Soc. London Ser. A*, **382**, 231.

Frank-Kamenetskii, D.A. (1955). *Diffusion and Heat Exchange in Chemical Kinetics*. Translated by N. Thon. Princeton University Press, Princeton, NJ.

Frank-Kamenetskii, D.A. (1938). *Dokl. Nauk SSSR*, **18**, 413.

Frank-Kamenetskii, D.A. (1939). *Acta Physicochim. URSS*, **10**, 365.

Frank-Kamenetskii, D.A. (1942). *ibid.*, **16**, 357.

van't Hoff, J.H. (1884). *Études de Dynamique Chimique*. Amsterdam, p. 161.

Gray, P. and Harper, M.J. (1959). *Trans. Faraday Soc.*, **55**, 581.

Gray, P. and Harper, M.J. (1959). *Seventh International Symposium on Combustion*. Butterworth, London, p. 425.

Gray, P. and Harper, M.J. (1960). Reactivity of solids. *Proceedings of the Fourth Symposium on the Reactivity of Solids*. (J.H. de Boer et al., eds.). Elsevier, Amsterdam, p. 283.

Gray, P. and Lee, P.R. (1967). Thermal explosion theory. In *Oxidation and Combustion Reviews* (C.F.H. Tipper, ed.). Elsevier, Amsterdam, **2**, 1.

Groocock, J.M. (1958). *Trans. Faraday Soc.*, **54**, 1526.

Heavens, S.N. (1976). D.Ph. Thesis. Cavendish Laboratory, University of Cambridge.

Heavens, S.N. and Field, J.E. (1974). *Proc. Roy. Soc. London Ser. A*, **338**, 77.

Hertzberg, G. and Walker, R. (1948). *Nature*, **161**, 647.

Hubbard, A.W. and Johnson, M.H. (1959). *J. Appl. Phys.*, **30**, 765.

Jacobs, S.J. (1960). *ARS J.*, 151.

Jacobson, A.K. (1981). Low energy flying plate detonator. Sandia National Laboratories Report, Sandia, Albuquerque, NM. Report Number SAND-81-0487.

Jacobson, A.K. (1981). *11th Symposium on Explosives and Pyrotechnics*. Franklin Institute, Franklin, Philadelphia, PA.

Jones, H. (1949). *Third Symposium on Combustion*, p. 590.

Jouguet, E. (1906). On the propagation of chemical reactions in gases. *J. Math. Pures Appl.*, **1**, 347 and **2**, 5.

Kistiakowsky, G.B. and Wilson, E.B. (1941). Report on the prediction of the detonation velocities of solid explosives. US Office of Scientific Research and Development Report, OSRD-69.

Kistiakowsky, G.B. and Wilson, E.B. (1941). Hydrodynamic theory of detonation and shock waves. US Office of Scientific Research and Development Report, OSRD Report OSRD-114.

Kistiakowsky, G.B. and Kydd, P.H. (1954). *J. Chem. Phys.*, **22**, 1940.

Lee, P.R. (1992). *Fourth International Symposium on Explosives Technology and Ballistics*. Pretoria, RSA, National Institute for Explosives Technology, p. 99.

Lieberman, M.L., Fleming, N.A., and Begeal, D.R. (1984). CP DDT Detonators III: Powder column effects. Sandia National Laboratories Report, Sandia, Albuquerque, NM. Report Number SAND-84-0143C.

Lieberman, M.L., Fleming, N.A., and Begeal, D.R. (1984). 9th *International Pyrotechnics Seminar, IIT Research Institute*, Colorado Springs, CO.
Mader, C.L. (1963). *Phys. Fluids*, **6**, 375.
Mader, C.L. (1979). *Numerical Modeling of Detonations*. University of California Press, Berkeley, CA.
Merzhanov, A.G., Abramov, V.G., and Dubovitskii, F.I. (1959). *Dokl. Akad. Nauk SSSR*, **128**, 889.
Merzhanov, A.G. and Dubovitskii, F.I. (1960). *Russian J. Phys. Chem.*, **34**, 1062.
Merzhanov, A.G., Barzykin, V.V., Abramov, V.G., and Dubovitskii, F.I. (1961). *Russian J. Phys. Chem.*, **35**, 1024.
Merzhanov, A.G., Abramov, V.G., and Gontkovskaya, V.T. (1963). *Dokl. Akad. Nauk SSSR*, **148**, 156.
Merzhanov, A.G. and Abramov, V.G. (1981). *Propellants, Explosives, and Pyrotechnics*, **6**, 130.
von Neumann, J. (1942). Office of Scientific Research and Development. OSRD Report 541.
von Neumann, J. (1990). *Collected Works* (1952–1963), vol. 16, p. 203. Pergamon Press, Oxford. Ordnance Board: Insensitive Munitions: Pillar Proceeding 42657.
Semenov, N.N. (1928). *Z. Phys.*, **48**, 571.
Semenov, N.N. (1930). *Z. Phys. Chem.*, **11B**, 464.
Semenov, N.N. (1932). *Phys. Z. Sowjet.*, **1**, 607.
Semenov, N.N. (1955). *Chemical Kinetics and Chain Reactions*. Clarendon Press, Oxford.
Simchen, A.E. (1964). *Israel J. Chem.*, **2**, 33.
Skidmore, I.C. and Hart, S. (1965). *Fourth Symposium on Detonation*, p. 1.
Taylor, G.I. (1950). *Proc. Roy. Soc. London Ser. A*, **200**, 235.
Thomas, P.H. (1958). *Trans. Faraday Soc.*, **54**, 60.
US Military Standard: Hazard Assessment Tests for Non-nuclear Munitions, (1994). DoD-STD-2105B (Navy).
Wake, G.C. and Walker, I.K. (1964). *New Zealand J. Sci.*, **7**, 227.
Walker, I.K. (1961). *ibid.*, **4**, 309.
Walker, I.K. and Harrison, W.J. (1960). *J. Appl. Chem.*, **10**, 266.
Walker, I.K. and Harrison, W.J. (1965). *New Zealand J. Sci.*, **8**, 106.
Zeldovich, Ya.B. (1940). *J. Exptl. Theoret. Phys. USSR*, **10**, 542.
Zeldovich, Ya.B. (1950). NACA Tech. Memo, p. 1261.
Zeller, B. (1993). Solid propellant grain design. In *Solid Rocket Propellant Technology* (Davenas, A., ed.). Pergamon Press, Oxford.
Zinn, J. and Mader, C.L. (1960). *J. Appl. Phys.*, **31**, 323.

CHAPTER 7

The Gurney Model of Explosive Output for Driving Metal

James E. Kennedy

7.1. Introduction

Measures of the output of explosives in most sources are based upon the strength of the shock wave or the chemical energy content of the explosive. These are most commonly the detonation wave velocity, D, the detonation pressure P_{CJ}, or the heat of detonation ΔH_d, which is derived from detonation calorimetry experiments or thermochemical equilibrium computations. These quantities in themselves provide a correct understanding of the relative output of one explosive in comparison to another (although density is also a factor), but they do not provide a direct measure of how fast an explosive can drive metal or other materials, which is the subject of interest in many applications and safety considerations. In this chapter we shall present a measure of explosive output which does permit the estimation of the velocity or impulse imparted to drive materials.

During the 1940s, Ronald W. Gurney, a solid-state physicist from Princeton, NJ, was pressed into wartime service at the US Army Ballistic Research Laboratories in Aberdeen, MD. He developed a couple of simple ideas into a way to estimate the velocity of explosively driven fragments (Gurney, 1943). These ideas form the basis of a method we now call the Gurney model, which has been used in various ways since Gurney returned to physics and contributed to the understanding of tunneling in semiconductors (Barut and Bartlett, 1978; Condon, 1978). The Gurney model and its extensions and applications are the subject of this chapter.

The method, based on energy and momentum balances, was devised to correlate fragment velocities from explosive/metal systems of widely varying sizes and proportions. Although shock waves played a very important part in the transfer of energy from the detonation of confined explosives to the surrounding metal munition cases, the assumptions Gurney made in his model to provide mathematical tractability had nothing to do with shock mechanics. Gurney assumed that:

(1) detonation of a given explosive releases a fixed amount of energy per unit mass which winds up as kinetic energy of the driven inert material (often metal) and the detonation product gases; and

(2) those product gases have a uniform density and a linear one-dimensional velocity profile in the spatial coordinates of the system.

The physical justification of these assumptions may be thought of as follows. The first assumption is equivalent to an expectation that the efficiency of energy transfer to the metal will be consistent (less than 100%), regardless of the geometry or massiveness of the confinement of the explosive. This assumption turns out to be a good one as long as there are no significant "end losses" of gases, which cause the gases to expand in a direction not contemplated in the one-dimensional model. The second assumption corresponds to a condition where there is opportunity for multiple shock reverberations in the gas space while the confinement is still intact, in which case the gas state inside the case tends toward constant density and a linear velocity profile. Both assumptions break down when the case mass is relatively light, because there is insufficient time for reverberations within the gas to drive toward steady expansion within the gas space; the result is that the case is then driven faster than the Gurney model predicts. More detail will be given later about limitations to the Gurney model, and how we may compensate for them.

Discussion in this chapter is presented in terms of explosives driving metals. While the equations apply for any driven material, metals have the advantage of staying together and containing the detonation product gases longer than other materials such as earth, and metals are relevant for many applications. Experimental data which have been taken to quantify the Gurney model have involved explosives driving metals.

This chapter presents the principal results obtained by use of the Gurney model in Section 7.3, including the terminal velocity formulas for various simple geometries, and impulse evaluation. The reduction of experimental data to provide values of the characteristic Gurney velocity, $\sqrt{2E}$, is described. Gurney energy values are compared with chemical energy values for explosives, and the efficiency of high explosives expressed in these terms is observed to be quite high. In Section 7.4 some direct applications of the Gurney model are presented. These provide insight on how parameter studies can be done, for example, the magnitude of the effect of tamping in increasing metal velocity in asymmetric sandwiches. The Taylor turning angle for metal plates driven by grazing detonation is analyzed, with emphasis on the direction of projection of the metal. Extensions of the Gurney model are discussed in Section 7.5. Methods are presented for estimating the value of $\sqrt{2E}$ from the chemical composition and density of explosives containing carbon, hydrogen, nitrogen, and oxygen (C–H–N–O explosives). Acceleration formulas linked to the gas expansion ratio are presented for slab and cylindrical geometries. Applications of the Gurney model for high-pressure-gas power sources other than detonating explosives, specifically exploding metal foils and laser-ablated plasmas, are discussed briefly. In Section 7.6, limitations of the Gurney model are discussed. These relate to inaccuracy which sets in when the metal mass to explosive charge mass ratio becomes too small, and accounting for side losses from bare or lightly confined charges. Finally in Section 7.7, the Gurney analytical treatment describing the velocity attained by driven metal is combined with analytical treatments of other physics to model some applications of engineering interest. These include momentum transfer by inelastic collision and detonation transfer by flyer plate impact. The results in these models, and in Gurney calculations in general, are enriched by their parametric nature, which provides insight concerning design choices.

7.2. Nomenclature

Subscripts are defined at the time of use in the text. A few subscripted variables are defined below for the sake of clarity.

A — $(2M/C+1)/(2N/C+1)$, Eq. (7.9).
A^* — v_O/v_I in Section 7.3.3.
a — Constant of proportionality, taken to be 1 (Section 7.3.3).
B — Parameter used in Section 7.5.3.1.
C — Explosive charge mass.
D — Detonation velocity.
D_ϕ — Phase velocity of detonation along the surface of a liner.
E — Gurney energy.
E_G — Gurney energy, treated as variable with gas expansion.
E_{eg} — Electrical Gurney energy.
$\sqrt{2E}$ — Gurney velocity.
e — Potential energy in explosive gases.
e_0 — Chemical energy content in unreacted explosive.
F — Parameter used in Section 7.5.3.1.
G — Term related to velocity gradient in liner, Section 7.3.3.
h — Initial thickness of explosive.
ΔH_d — Heat of detonation.
I_{SP} — Specific impulse.
i — Electrical current.
J_b — Burst current density.
K — A constant in Eq. (7.13).
L — Reduced donor length, l_d/l_{total}.
l — Length.
$\mathbf{M}$ — Momentum.
M — Metal mass.
m — Total metal mass for symmetric sandwich in Section 7.5.3.1.
N — Tamper mass.
n — Symmetry parameter, Section 7.5.3.2.
P — Pressure.
P_0 — Initial pressure for acceleration calculations, Section 7.5.3.3.

R Lagrangian radius, Eqs. (7.11) to (7.15).
r Eulerian radius.
s Axial position along the metal liner, Section 7.4.2.
$\mathbf{T}$ Kinetic energy per unit area of symmetric sandwich.
T Time of arrival of detonation at a given position along the liner.
t Time.
t_M Thickness of metal liner.
$\mathbf{U}$ Potential energy of gas.
U Shock wave velocity.
u Particle velocity.
v Metal velocity.
W Lagrangian, Section 7.5.3.3.
x Position in detonation product gases.
x_M Position of surface of plate M.
y Lagrangian position, Figure 7.2.
Z Ratio of impedances of acceptor to flyer, z_a/z_f.
z Shock impedance, $\rho_0 U$.
α Arbitrary angle, Eq. (7.11).
β R_O/R_I.
χ Total displacement of plates M and N in asymmetric sandwich, Section 7.5.3.3.
δ Metal projection angle.
ε Ratio of kinetic energy of driven metal to chemical energy of explosive.
ε_m Ratio of kinetic energy of driven metal to Gurney energy of explosive.
Φ Kamlet parameter, Section 7.5.1.
γ Ratio of specific heats, C_p/C_v.
η Parameter in Eq. (7.12).
φ Parameter in Section 7.5.3.1.
κ Discounting angle, Section 7.6.2.
μ Parameter in Section 7.5.3.3.
θ Metal deflection angle.
ρ Density.
ρ_0 Initial density of explosive.
τ Shock duration.
τ^* Characteristic acceleration time.
$\bar{\tau}$ Characteristic acceleration time, Section 7.5.3.3.
Ψ Impulse integral, Section 7.3.3.

7.3. Results of the Gurney Model

Following Kennedy (1970), the Gurney model may be applied to any explosive/metal system with a cross-section admitting one-dimensional translational motion of the metal typically normal to its surface, regardless

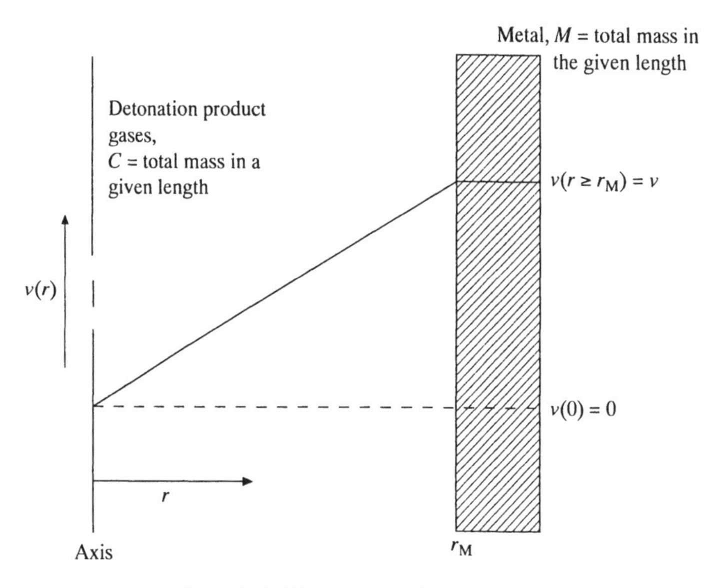

Figure 7.1. Linear gas velocity profile.

of the direction of detonation propagation. A specific Gurney energy E, in kcal/g, with a characteristic value for each explosive at a given density, is assumed to be converted from chemical energy in the initial state to kinetic energy in the final state. This final kinetic energy is partitioned between the metal and the detonation product gases in a manner dictated by an assumed linear velocity profile in the gases. Figure 7.1 illustrates such a velocity profile for a cylindrical charge in a metal tube. The velocity of the metal is assumed to be constant throughout its thickness, and the velocity of each gas and metal element is taken to be normal to the axis. These assumptions usually are not strictly true, but they greatly simplify the mathematics with little loss in accuracy.

Based on these assumptions, we can write an energy balance for any simple symmetric geometry which can be easily integrated to provide an analytical expression for the final metal velocity v as a function of the Gurney specific energy E and the ratio of the total metal mass to the total explosive mass, M/C. In addition, asymmetric one-dimensional configurations such as a sandwich of explosive between layers of metal of differing thickness require a momentum balance, which is solved simultaneously with the energy balance.

7.3.1. Terminal Velocity Formulas for Symmetric and Asymmetric Configurations

For purposes of calculating the dimensionless quantity M/C, it will often be convenient to consider a representative volume element of the explosive/metal assembly. This could be the volume normal to a unit area for a flat

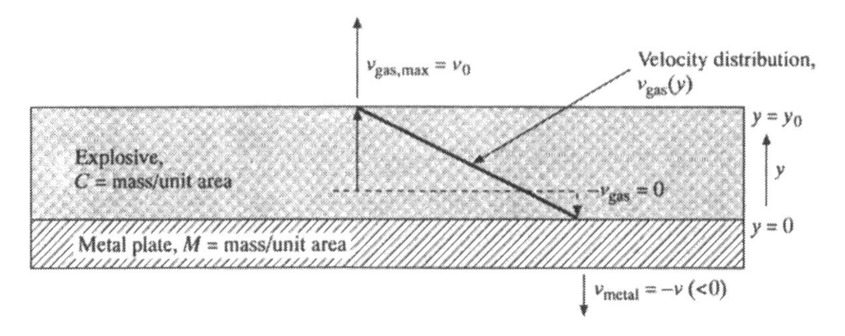

Figure 7.2. Assumed velocity distribution in open-face sandwich.

configuration, a unit length or a solid sector for a cylindrical configuration, or a spherical sector or unit for a spherical configuration.

Let us derive the equations which describe an asymmetric geometry, the open-face sandwich shown in Figure 7.2, as an example of the method. The velocity distribution in the detonation product gases as a function of Lagrangian position y is

$$v_{gas}(y) = (v_0 + v)\frac{y}{y_0} - v \tag{7.1}$$

and the velocity of the metal is v across its entire thickness. The initial thickness and density of the explosive are y_0 and ρ_0, respectively. Coordinate positions are identified with particles rather than spatial positions. This permits expressions to be written in terms of initial positions. A given particle is assumed to move at constant speed at all times in constructing energy and momentum balances for terminal velocity determination. Energy and momentum balances for a unit area then may be written, respectively, as

$$CE = \tfrac{1}{2}Mv^2 + \tfrac{1}{2}\rho_0 \int_0^{y_0} \left[(v_0 + v)\frac{y}{y_0} - v\right]^2 dy \tag{7.2}$$

and

$$0 = -Mv + \rho_0 \int_0^{y_0} \left[(v_0 + v)\frac{y}{y_0} - v\right] dy. \tag{7.3}$$

Integration of these equations and substitution of M/C for the equivalent ratio $M/\rho_0 y_0$ yields a result which can be expressed as

$$v = \sqrt{2E}\left[\frac{\left(1 + 2\dfrac{M}{C}\right)^3 + 1}{6\left(1 + \dfrac{M}{C}\right)} + \frac{M}{C}\right]^{-1/2} \tag{7.4}$$

and

$$\frac{v_0}{v} = 2\frac{M}{C} + 1. \tag{7.5}$$

7.3.2. Gurney Equations

Henry (1967) provided a comprehensive review of the Gurney method and derivation of many formulas. Gurney equations for common symmetric and asymmetric configurations are presented in Figures 7.3 and 7.4. The quantity $\sqrt{2E}$ occurs in each expression; its units are velocity and it is known as the Gurney characteristic velocity for a given explosive. The ratio of final metal velocity v to the characteristic velocity $\sqrt{2E}$ is an explicit function of the metal to charge mass ratio M/C, and, in the case of the asymmetric sandwich where the second plate may be considered to be a tamper of mass N, is an explicit function also of the tamper-mass-to-charge-mass ratio N/C. In all cases the velocity derived is that of plate M.

Figure 7.5 presents a plot of the dimensionless velocity, $v/\sqrt{2E}$, of metal M as a function of the loading factor M/C for configurations shown in Figures 7.3 and 7.4. Values of the Gurney velocity $\sqrt{2E}$ for a number of explosives are tabulated in Section 7.3.4, hence Figure 7.5 can be used directly to estimate the metal velocity to be expected from explosive/metal systems of simple geometry.

The partitioning of gas flow in the symmetric sandwich is obvious—half the detonation product gases flows in the direction of each metal plate. However, for asymmetric sandwich configurations, the proportion of gas flowing in the direction of plate M varies with M/C and N/C. In view of the linear velocity profile assumed for uniform-density gases in the Gurney method, the proportion of the total detonation product gases which flows in the direction of plate M may be seen to be $v_M/(v_N + v_M)$. The two velocities, v_M and v_N, may be evaluated by individual applications of (7.9). For an open-face sandwich, the proportion of gas flowing in the direction of plate M is found to be $v_M/(v_0 + v_M)$, which may be evaluated from (7.4) and (7.5).

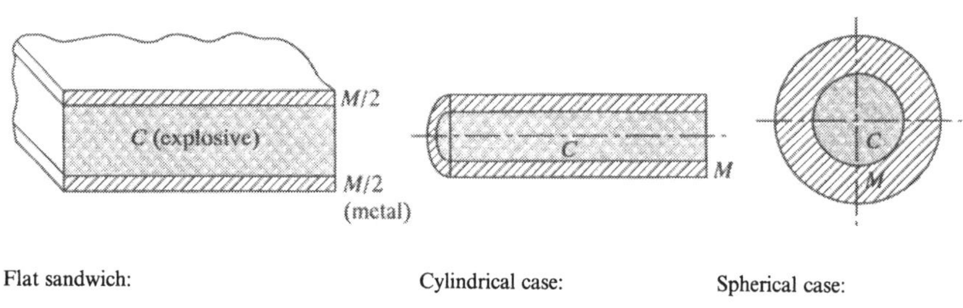

Flat sandwich:

$$\frac{v}{\sqrt{2E}} = \left[\frac{M}{C} + \frac{1}{3}\right]^{-1/2} \quad (7.6)$$

Cylindrical case:

$$\frac{v}{\sqrt{2E}} = \left[\frac{M}{C} + \frac{1}{2}\right]^{-1/2} \quad (7.7)$$

Spherical case:

$$\frac{v}{\sqrt{2E}} = \left[\frac{M}{C} + \frac{3}{5}\right]^{-1/2} \quad (7.8)$$

Figure 7.3. Symmetric explosive/metal configurations.

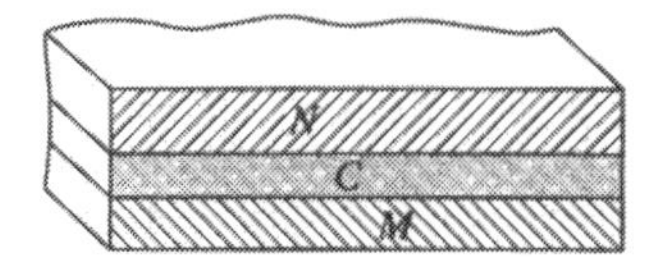

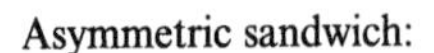

Asymmetric sandwich:

$$\frac{v_M}{\sqrt{2E}} = \left[\frac{1+A^3}{3(1+A)} + \frac{N}{C}A^2 + \frac{M}{C}\right]^{-1/2} \qquad (7.9)$$

where

$$\frac{N}{C} = \frac{\text{total tamper mass}}{\text{total explosive mass}}$$

and we define

$$A = \frac{1+2\frac{M}{C}}{1+2\frac{N}{C}}.$$

Open-face sandwich:

$$\frac{v}{\sqrt{2E}} = \left[\frac{\left(1+2\frac{M}{C}\right)^3+1}{6\left(1+\frac{M}{C}\right)} + \frac{M}{C}\right]^{-1/2} \qquad (7.10)$$

Figure 7.4. Asymmetric explosive/metal configurations.

7.3.3. Imploding Geometries

The principal motivation for analyzing imploding geometries is to model the acceleration of the liner in shaped charge operation. For imploding devices, according to Chou et al. (1981, 1983) a proper global momentum balance for a curved geometry contains an extra term that represents an outward impulse applied to the explosive gases. Following Chou, the control volume

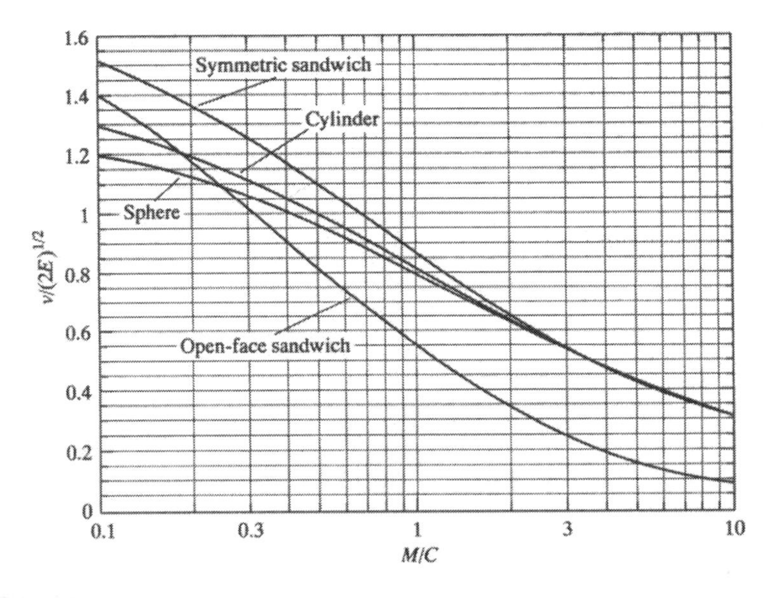

Figure 7.5. Dimensionless velocity of metal as a function of loading factor M/C.

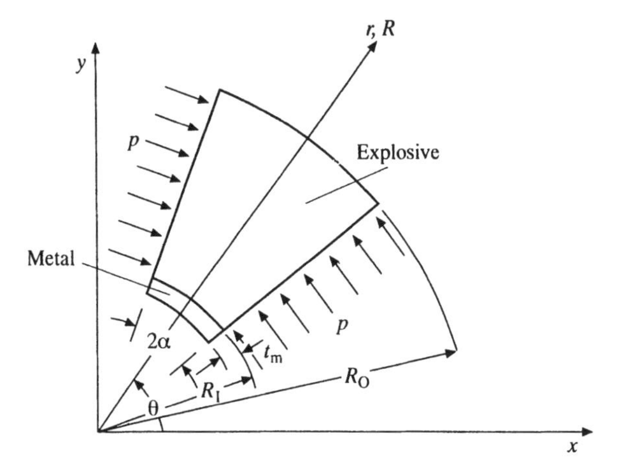

Figure 7.6. Control volume used in deriving modified Gurney formulas for imploding geometries.

used to derive modified Gurney formulas for imploding geometries is shown in Figure 7.6. Conservation of linear momentum in the pie-shaped sector shown in the figure includes the net pressure acting outward on the gas from the sides:

$$-\int_{-\alpha}^{\alpha} M_\ell v_l \cos\alpha \, d\alpha + \int_{-\alpha}^{\alpha}\int_{R_I}^{R_O} v_{gas}\rho_0 R \, dR \cos\alpha \, d\alpha = 2\int_0^\infty \int_{r_I}^{r_O} P(r, t) \, dr \, dt \sin\alpha, \quad (7.11)$$

where liner segment mass $M_\ell = \rho_M t_M(2R_I - t_M)/2$, $v_l =$ velocity of metal liner (taken to be positive), $R =$ Lagrangian radial coordinate, $r =$ Eulerian radial coordinate, and ρ_M, $\rho_0 =$ initial densities of metal, explosive. Subscripts I and O denote the inside (liner interface) and outside surfaces of the explosive.

The velocity of the explosive detonation product gas is assumed to be linear with respect to the Lagrangian coordinate, and setting the final kinetic energy of the system equal to the Gurney energy leads to a solution for the final velocity of the metal liner:

$$v_l = \frac{1}{1+\dfrac{M}{C}}\left\{-\frac{\Psi}{C} + \left[\frac{2E\left(\dfrac{M}{C}+1\right)}{\eta\left(\dfrac{M}{C}+1\right)-1} - \frac{\Psi^2}{C^2}\left(\frac{1}{\eta\left(\dfrac{M}{C}+1\right)-1}\right)\right]^{1/2}\right\}, \quad (7.12)$$

where

$$\eta = \frac{3}{2}\left[\frac{3\beta^2+4\beta+1}{4\beta^2+4\beta+1}\right], \qquad \beta = R_O/R_I, \qquad \text{and} \qquad \Psi = \int_0^\infty \int_{r_I}^{r_O} P(r, t) \, dr \, dt.$$

Note that as R_I and R_O both approach infinity, this formula approaches the classical formula for the open-face sandwich.

Since no exact solution is available, Chou et al. (1981, 1983) offer an approximate form for Ψ:

$$\Psi \cong KP_{CJ}\tau^*(R_O - R_I)\left(\frac{R_O}{R_I} - 1\right), \tag{7.13}$$

over the range $1.25 \le R_O/R_I \le 1.72$, K is a constant to be determined by fitting the results of wave-code computations, and τ^* is a characteristic time for exponential acceleration of the liner, discussed in Section 7.5.3.2.

For imploding cylinders with exterior confinement, Chanteret (1983) developed an expression to locate the zero-radial-velocity position, R_x, in the linear velocity profile, and made use of it in devising a formula for the imploding liner velocity. R_x is evaluated by solving a cubic equation,

$$R_x^3 + 3R_x\left[(R_O + R_I)\frac{\rho_0}{\rho_{CJ}}\left(\frac{M_I}{C}R_O + \frac{M_O}{C}R_I\right) + R_I R_O\right] - 3(R_I + R_O)R_I R_O\left[\frac{2}{3} + \frac{\rho_0}{\rho_{CJ}}\left(\frac{M_I}{C} + \frac{M_O}{C}\right)\right] = 0. \tag{7.14}$$

The imploding liner velocity is then given in terms of R_x as

$$v = \sqrt{2E_G}\left[G\left(\frac{R_O^2 - R_I^2}{R_x^2 - R_I^2}\right)\frac{M}{C} + \frac{1}{6}\right]^{-1/2}, \tag{7.15}$$

where G is a term which takes into account a possible velocity gradient in the liner thickness, and $G = 1$ for a uniform velocity profile. The velocity profile is strongly influenced by wave mechanics early in the acceleration. Here the Gurney energy term is E_G rather than E because E_G is taken to be a function of gas expansion as discussed in Section 7.5.3.2. The $1/6$ term represents the kinetic energy of the inward-moving gas, as derived by Chanteret.

Hirsch (1986) addressed the same problem as Chanteret, i.e., cylindrical implosion with exterior confinement. For the terminal inward velocity of the liner he derived

$$v_I = \sqrt{2E}\left\{A^*\left[\left(\frac{M}{C} + \frac{\beta + 3}{6(\beta + 1)}\right)\Big/ A^* + A^*\left(\frac{N}{C} + \frac{3\beta + 1}{6(\beta + 1)}\right) - \frac{1}{3}\right]\right\}^{-1/2}, \tag{7.16}$$

where

$$A^* = v_O/v_I = \left[\frac{M}{C} + a\left(\frac{M}{C}\right)(\beta - 1) + \frac{\beta + 2}{3(\beta + 1)}\right]\Big/\left[\frac{N}{C} + \frac{2\beta + 1}{3(\beta + 1)}\right] \quad \text{and} \quad \beta = R_O/R_I.$$

The quantity a is a constant of proportionality which arises in evaluating the impulse integral due to gas pressure in imploding geometries. Hirsch approximates the integral as $aM_O v_I(R_O - R_I)$, and the value of a must be estimated on the basis of wave-code computations. He notes that by use of $a = 1$ he obtained good agreement with the predictions of Chou et al. (1981, 1983). He argues further that a will not change rapidly as a function of C, and thus he expects that $a = 1$ for all values of M/C, N/C, and β.

Hirsch also derived an equation for implosion velocity of a spherical liner:

$$v_I = \sqrt{2E}\left\{A^*\left[\left(\frac{M}{C}+\frac{\beta^2+3\beta+6}{10(\beta^2+\beta+1)}\right)\bigg/A^*+A^*\left(\frac{N}{C}+\frac{6\beta^2+3\beta+1}{10(\beta^2+\beta+1)}\right)\right.\right.$$
$$\left.\left.-\frac{3\beta^2+4\beta+3}{10(\beta^2+\beta+1)}\right]\right\}^{-1/2}, \tag{7.17}$$

where

$$A^* = v_O/v_I$$
$$= \left[\frac{M}{C}+a\left(\frac{M}{C}\right)(\beta^2-1)+\frac{\beta^2+2\beta+3}{4(\beta^2+\beta+1)}\right]\bigg/\left[\frac{N}{C}+\frac{3\beta^2+2\beta+1}{4(\beta^2+\beta+1)}\right].$$

7.3.4. Impulse Estimation

Since the Gurney method provides an estimate of the final velocity imparted to an explosively loaded body, the total momentum $\mathbf{M}$ of the body can be readily calculated as the product of mass and velocity. We divide by the total charge mass C to derive the specific impulse delivered by the explosive

$$I_{SP} = \frac{\mathbf{M}}{C} = \frac{Mv}{C}. \tag{7.18}$$

7.3.4.1. Specific Impulse of Unconfined Surface Charges

Explosives may be detonated directly on the surface of a very heavy body in order to deliver a desired impulse for testing purposes. If we consider the loaded body to be rigid, all detonation product gases will flow away from the surface and maximum specific impulse will be delivered to the body (Defourneaux and Jacques, 1970). For a very large body-mass-to-explosive-mass ratio, $M/C \gg 1$, (7.4) reduces to

$$v \cong \sqrt{2E}\cdot\sqrt{\frac{3}{4}\frac{C}{M}}. \tag{7.19}$$

By utilizing this expression for v in (7.18), the relationship for specific impulse is found to be

$$I_{SP} = \sqrt{1.5E}. \tag{7.20}$$

Upon detonation of an unconfined charge as described above, essentially all the gas must flow away from the plate. Because the gas is perfectly confined on the metal side of this configuration, the peak backward velocity v_0 along the linear gas velocity profile will assume a maximum value. By equating the specific momentum of this flowing gas to the specific impulse driven into the metal according to (7.20), we derive a relation for the maximum free gas velocity:

$$v_{0,\,\max} = \sqrt{6E}. \qquad (7.21)$$

7.3.4.2. Impulse Increase by Tamping

The impulse imparted to a body by detonation of explosive at a given areal density (grams explosive/cm^2 of body surface) can be increased by tamping the explosive with, for example, an outer layer of metal. The Gurney method can be used to estimate the change in impulse as a function of tamper areal density. If the heavy body is considered to be rigid, its surface may be taken to be the symmetry plane in a symmetric sandwich configuration. The velocity of the tamper plate will then be given by (7.9), where the tamper-to-explosive mass ratio N/C is taken to represent M/C. The impulse imparted to the heavy body per unit mass of explosive will be equal to the momentum of the tamper and the gas following it. The effective specific impulse of the explosive when the tamping ratio is N/C can be shown to be

$$I_{\mathrm{SP,\,tamped}} = \sqrt{2E}\left(\frac{N}{C}+\frac{1}{2}\right)\bigg/\left(\frac{N}{C}+\frac{1}{3}\right)^{1/2}. \qquad (7.22)$$

Note that when no tamper is present ($N/C = 0$), this expression reduces to that for an untamped charge as given in (7.20).

7.3.5. Gurney Energy of Explosives

Table 7.1 lists Gurney velocities calculated from data from carefully conducted experiments in which there were no end losses or gas leakage

Table 7.1. Output of selected explosives.

Explosive	Density, g/cm^3	ΔH_d,* kcal/g	E, kcal/g	$E/\Delta H_d$	$\sqrt{2E}$,‡ km/s	I_{SP}, dyne-s/g
TNT	1.63	1.09	0.67	0.61	2.37	2.05×10^5
RDX	1.77	1.51	0.96	0.64	2.83	2.45×10^5
PETN	1.76	1.49	1.03	0.69	2.93	2.54×10^5
HMX	1.89	1.48	1.06	0.72	2.97	2.57×10^5
Comp B†	1.717	1.20	0.87	0.72	2.71	2.35×10^5

* ΔH_d data from Ornellas (1968, 1970).

† Composition: 63% RDX, 36% TNT, 1% wax by weight.

‡ $\sqrt{2E}$ data from Hardesty and Kennedy (1977) except for RDX data from Henry (1967).

through metal walls until after their acceleration was completed. The best known of these experiments were LLNL cylinder tests (Kury et al., 1965). Occurrence of end losses (e.g., charges in short tubes with $L/D < 6$, with open ends) or early fracturing of driven metal may decrease the effective value of $\sqrt{2E}$ by 10% to 20% or more (Weinland, 1969). Values of specific impulse calculated from Gurney velocities are also tabulated in Table 7.1.

Other forms may be used to express the efficiency of energy transfer from the explosive to the metal. Let us define efficiency, ε, as the ratio of the kinetic energy of plate M to the chemical energy in the explosive. By introducing the Gurney energy E into the numerator and denominator and rearranging, we derive the following expression in "Gurney quantities" (Hoskin et al., 1965):

$$\varepsilon = \frac{Mv^2}{2C\Delta H_d} \cdot \frac{E}{E} = \left[\left(\frac{v}{\sqrt{2E}}\right)^2 \cdot \frac{M}{C}\right] \frac{E}{\Delta H_d}. \tag{7.23}$$

The first term of the result is the fraction of the Gurney energy which winds up in the form of kinetic energy of plate M, and this term is solely a function of M/C (and N/C where applicable). We can evaluate this term immediately for any given configuration, using the appropriate form for $v/\sqrt{2E}$ for the configuration of interest. This term, defined as ε_m, is plotted in Figure 7.7 for asymmetric sandwiches with various tamping factors. In Figure 7.7, notice that the Gurney solution (solid curves) is compared with wave-propagation code solutions which employed the gamma-law ideal-gas equation of state. Agreement between the code results and the Gurney results is

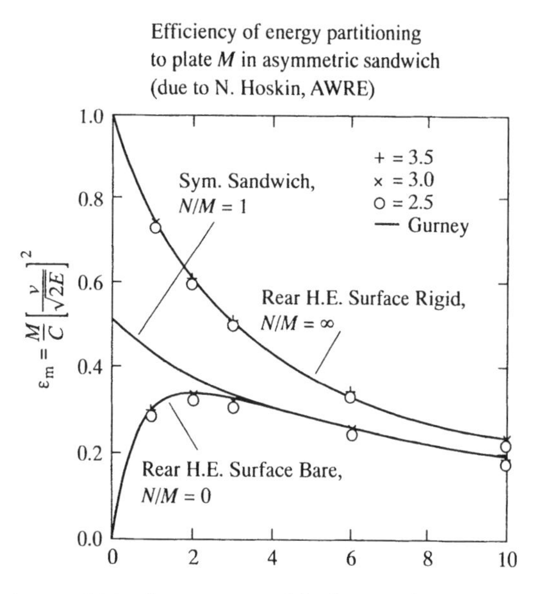

Figure 7.7. Efficiency of kinetic energy partitioning to plate M in asymmetric sandwich. Symbols at upper right denote values of γ used in wave-code runs.

quite good. The Gurney solution for the open-faced sandwich ($N/M = 0$) exhibits a maximum efficiency at a value of $C/M = 2$.

The second term, $E/\Delta H_d$, is the ratio of Gurney energy to chemical energy of the explosive. For chemical energy we have used the heat of detonation (Ornellas, 1968, 1970). Values of Gurney energy E and chemical energy ΔH_d are listed in Table 7.1 for several high explosives, and it is noted that the ratio $E/\Delta H_d$ usually is in the range of 0.6–0.7. This is quite a high efficiency value. It is high because explosives produce a large amount of gas (approaching 100%) and the high pressures which are present in detonation drive very rapid energy transfer. By comparison, the value of $\sqrt{2E}$ for a propellant is lower; $\sqrt{2E} = 1.7$ km/s for smokeless powder and 0.94 km/s for black powder (Henry, 1967). The efficiency of propellants is typically in the range of 0.2–0.3. For one pyrotechnic material, a titanium subhydride/potassium perchlorate mixture which has been used for the production of mechanical work, the efficiency is about 0.07 (Lieberman and Haskell, 1978); this is low because pyrotechnics usually produce little gas, and some of the gas they do produce is temporary gas (condensable) such as KCl.

Why is it that explosives are not 100% efficient? What happens to the energy that does not seem to appear as kinetic energy? In open configurations such as the open-face sandwich, the gas can lose communication with the metal surface during expansion. This occurs when the backward expansion velocity of certain parcels of the gas becomes higher than the sound velocity in the gas, and pressure signals from that gas can no longer be transmitted forward to the metal surface. In closed configurations such as the Lawrence Livermore National Laboratory (LLNL) cylinder test, some of the energy remains as potential energy (due to the pressure of the gas) at the time that the driven metal fragments and lets the gas expand through it, or when the gas eventually escapes out of the end of the configuration. In the Fickett–Jacobs cycle (Fickett and Davis, 1979) presented in Figure 7.8, the point is that expansion of the detonation product gases along the isen-

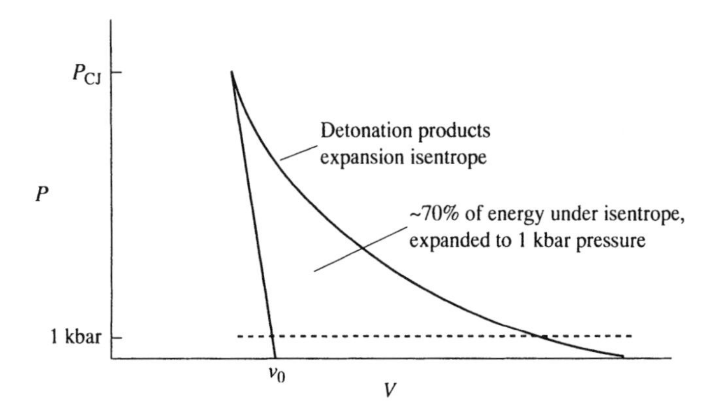

Figure 7.8. Fickett–Jacobs cycle diagram.

Table 7.2. Gurney velocities of additional explosive materials.*

Explosive	Density, g/cm^3	Composition by weight	$\sqrt{2E}$, km/s	References
Comp A-3	1.59	RDX/wax: 91/9	2.63	Dobratz and Crawford (1985)
Comp C-3	1.60	RDX 71%	2.68	Kennedy (1970)
		TNT 4%		
		DNT 10%		
		MNT 5%		
		NC 1%		
		Tetryl 3%		
Cyclotol 75/25	1.754	RDX/TNT: 75/25	2.79	Hardesty and Kennedy (1977)
Detasheet C	1.48	PETN 70%	2.1–2.3	Walters and Zukas (1989)
		Plasticizers 30%		
LX-14	1.835	HMX 95.5%	2.80	Dobratz and Crawford (1985)
		Estane 4.5%		
NM	1.14		2.41	Kennedy (1970)
Octol 75/25	1.821	HMX/TNT: 75/25	2.83	Hardesty and Kennedy (1977)
PBX-9404	1.84	HMX 94%	2.90	Kennedy (1970)
		NC 3%		
		CEF 3%		
PBX-9502	1.885	TATB 95%	2.38	Dobratz and Crawford (1985)
		Kel-F 800 5%		
TACOT	1.61		2.12	Kennedy (1970)
Tetryl	1.62		2.50	Hardesty and Kennedy (1977)

* Most experimental data are from LLNL cylinders tests (e.g., Kury et al., 1965).

trope to a pressure of 1 kbar corresponds to the state when 30% of the energy remains in the gas. At this pressure the rate of energy transfer is much lower than it was at the detonation state, and the rate of energy transfer continues to go down with further expansion. The reason that so much of the energy still remains in the gas at 1 kbar is that the value of the polytropic exponent has dropped to near the ideal gas value of about 1.4 at this low pressure. A simplistic way to view the energy transfer could then be to assume that it is perfectly efficient down to a pressure of 1 kbar, and not at all efficient at lower pressures. While not strictly correct, the smoothly varying behavior tends in that direction.

Values of $\sqrt{2E}$ for some other explosives are listed in Table 7.2.

7.4. Applications of Gurney Analysis

The appeal of simple Gurney analysis is that it is algebraic and lends itself to parameter studies. This can provide insight into a problem which is superior to that gained even by running a set of wave-code solutions to the same problem. The examples which follow represent problems in which Gurney analysis alone can provide such physical insight.

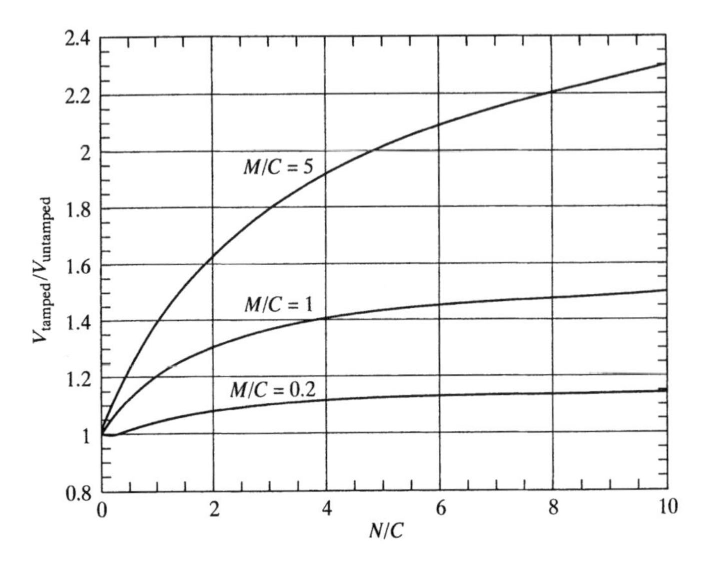

Figure 7.9. Gain in velocity of plate M due to tamping factor N/C.

7.4.1. Tamping Effectiveness

The form of the Gurney equations (explicit in $v/\sqrt{2E}$) is well suited to directly reveal the effect of a change in configuration (e.g., M/C or N/C) upon the velocity imparted to metal. The effectiveness of a metal tamper plate N in an asymmetric sandwich (7.9) in increasing the velocity of plate M is calculated as an example. Figure 7.9 is a plot of the proportionate velocity increase of plate M due to tamping for various loading ratios M/C. The figure illustrates that tamping a relatively heavy charge ($M/C = 0.2$) increases the velocity of plate M very little, while adding tamping to a light charge ($M/C = 5$) increases the velocity considerably, particularly in the range $N/C < 5$.

7.4.2. Direction of Metal Projection

In the derivation of Gurney equations, it is usually assumed that the metal moves in a direcion normal to its surface. This is true when the the detonation wave encounters the metal at a normal angle of incidence, but is not true when the metal is driven by detonation approaching from another angle.

Let us consider the limiting case of "grazing detonation," which propagates parallel to the metal surface as shown in Figure 7.10. Taylor (1941, 1963) shows that, for steady flow, the angle at which a metal plate is driven by grazing detonation is halfway between the normal to the original plane of the plate and the normal to the deflection angle, θ, of the plate as detonation progresses along the charge. Both the deflection angle and the angle

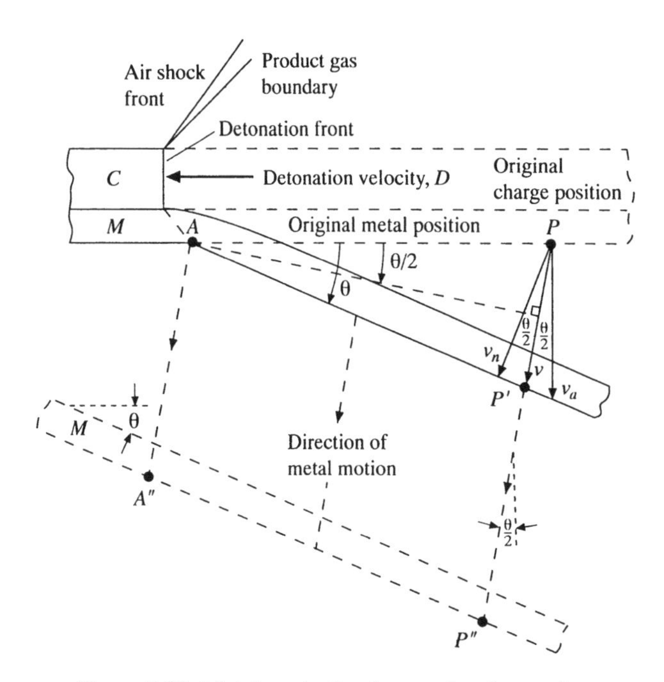

Figure 7.10. Metal projection by grazing detonation.

of projection, $\delta = \theta/2$, can be calculated from knowledge of the detonation velocity D and the plate velocity v (see (7.4)). The angle of projection is

$$\delta = \sin^{-1}\left(\frac{v}{2D}\right) \tag{7.24}$$

off the normal to the original metal plate surface, in the direction of detonation propagation (Taylor, 1941, 1963). When detonation strikes the metal surface obliquely at an angle between 0° and 90°, the phase velocity of detonation progress along the plate surface should be substituted for D in (7.24), and the angle of projection still applies.

Chou et al. (1981, 1983) derived an expression for the turning angle for the unsteady case when the plate is still accelerating and has not reached terminal velocity. This circumstance is important in the analysis of shaped-charge liner collapse. When v is a function of time and position s along the liner, $v(s, t)$, and prime denotes differentiation with respect to s, the general expression for the liner projection angle is

$$\delta = -\int_T^t v'\,dt + \frac{1}{2v}\int_T^t (v^2)'\,dt. \tag{7.25}$$

Here T denotes the time of arrival of detonation at the liner position s. To implement this equation, they recommend use of an exponential form to describe liner acceleration (a topic that is discussed in Section 7.5.3.2),

$$v = v_l \left\{ 1 - \exp\left[-\frac{t - T}{\tau^*} \right] \right\}, \tag{7.26}$$

where v_l, the terminal velocity, and τ^*, the characteristic acceleration time, are evaluated as functions of s alone. Integration of (7.25) with the use of (7.26) yields

$$\delta = \frac{v_l'}{2D_\phi} - \tfrac{1}{2}\tau^* v_l' + \tfrac{1}{4}\tau^{*\prime} v_l, \tag{7.27}$$

where D_ϕ indicates the phase velocity of the detonation wave along the liner surface. Note that the first term of this equation corresponds to the steady Taylor angle for small angles. Chou et al. utilize two-dimensional wave-propagation code calculations to evaluate v_l, v_l', τ^* and $\tau^{*\prime}$. Comparisons of computed values of δ with calculations per (7.27) show good agreement.

7.5. Extensions of Gurney Analysis

Traditional Gurney analysis is limited to predictions of the terminal velocity of metal driven by detonating explosives, based upon experimental data gathered on the output of explosives of interest. In this section we extend that foundation by discussing methods for the prediction of the effective output of explosives, expressed in terms of the Gurney velocity, based upon the chemistry and density of the explosive. We also describe analysis that permits estimation of the acceleration of the metal. The use of Gurney analysis to model problems with alternative sources of power, specifically the rapid electrical vaporization of conductors and laser ablation of solid materials rather than detonating explosives, is then described.

7.5.1. Estimation of Gurney Velocity from Chemistry, Density, and Detonation Parameters

Kamlet and Jacobs (1968) developed an approximate method to estimate the output of explosives in terms of their detonation velocity and detonation pressure, which is described in Chapter 4. Their method involved the hypothesis of simplistic chemistry, from which the chemical energy release and the product gases associated with detonation could be calculated. Recognizing that the energy release and the amount of product gas contribute to the output of an explosive, Kamlet then postulated a parameter Φ which was a function of these quantities. Kamlet's correlations to detonation velocity and detonation pressure were functions of Φ and density, ρ_0.

Hardesty and Kennedy (1977) developed an expression for $\sqrt{2E}$ as a function of the same parameters, Φ and ρ_0,

$$\sqrt{2E}, \text{km/s} = 0.60 + 0.54\sqrt{1.44\rho_0\Phi}. \tag{7.28}$$

This expression was compared with experimental results for 21 pure C—H—N—O explosives and mixtures, and the data were within $\pm 5\%$ error bars of this correlation. The average deviation was 0.04 km/s.

Kamlet and Finger (1979) later developed another correlation for $\sqrt{2E}$ involving the same parameters. Their result was

$$\sqrt{2E}, \text{km/s} = 0.887\Phi^{0.5}\rho_0^{0.4} \tag{7.29}$$

and their average deviation was 0.02 km/s, better than the correlation of Hardesty and Kennedy.

Cooper (1996) developed a correlation which applies to a larger set of explosives because it involves only the detonation velocity. His result was

$$\sqrt{2E} = \frac{D}{2.97}, \tag{7.30}$$

with a standard deviation of 3.7% (see Chapter 4). A result in this form is useful, particularly under circumstances where data on detonation velocity are available or can be obtained, but data on the chemical composition of the explosive are not readily available.

Roth (1971) performed gas dynamic analysis which led him to this expression for $\sqrt{2E}$ as a function of detonation parameters:

$$\frac{\sqrt{2E}}{D} = \frac{0.605}{\gamma - 1}. \tag{7.31}$$

Here γ is the ratio of specific heats at the detonation state.

7.5.2. Effects of Gaps

Computational work by Kennedy and Chou (1990) addressed the condition where there were gaps in explosive metal systems. They found that, for gaps between the explosive and driven metal or within the explosive assembly, the velocity of the metal was decreased in a manner that could be predicted by considering the explosive to be operating at a lower density, equal to the mass of the explosive divided by the total volume of the explosive plus the gaps. Formulas such as (7.28) and (7.29) could then be applied to evaluate the effective value of $\sqrt{2E}$.

7.5.3. Acceleration Solutions

7.5.3.1. Jones Analysis for Slab Geometries

The ingredient which must be added to the Gurney model to permit calculations of the acceleration profile for the driven metal is an equation of state for the detonation product gases. This enables calculations of the pressure in the gas during gas expansion, which in turn enables calculation of accelera-

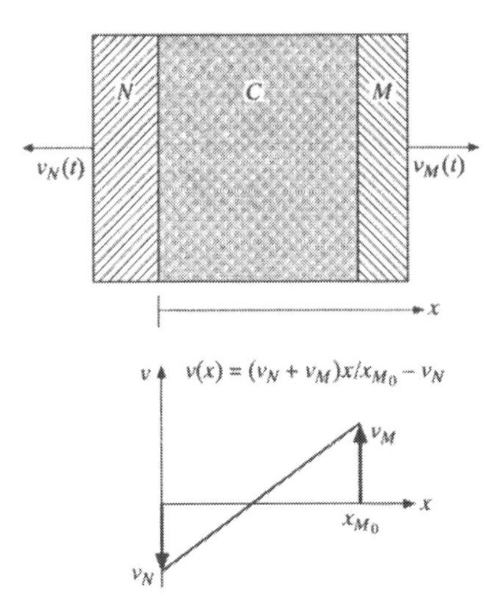

Figure 7.11. Asymmetric sandwich configuration. A linear velocity distribution in the gas is illustrated and described by the equation.

tion. The energy equation contains an added term, representing the potential energy in the gas due to the pressure.

Jones et al. (1980) published a solution for slab geometries using these principles and the ideal-gas equation of state for the detonation products. The general configuration for this derivation is that of the asymmetric sandwich, illustrated in Figure 7.11. The linear gas velocity profile assumption allows us to write the principle of conservation of energy as

$$Ce_0 = Ce + \tfrac{1}{2}Nv_N^2 + \tfrac{1}{2}Mv_M^2 + \tfrac{1}{2}\rho_0 \int_0^{x_{M0}} \left[(v_N + v_M)\frac{x}{x_{M0}} - v_N\right]^2 dx, \qquad (7.32)$$

which reduces to

$$Ce_0 = Ce + \tfrac{1}{2}Nv_N^2 + \tfrac{1}{2}Mv_M^2 + \frac{C\,(v_N^3 + v_M^3)}{6\,(v_N + v_M)}.$$

where e_0 is the chemical energy per unit mass of the explosive, e is the internal energy per unit mass remaining in the explosive, v_N and v_M are instantaneous velocities of the plates, and other symbols have their usual meanings. The momentum balance equation is written in the usual form, with the result

$$0 = -v_N\left(N + \frac{C}{2}\right) + v_M\left(M + \frac{C}{2}\right). \qquad (7.33)$$

With the definition of $A = v_N/v_M = (2M/C + 1)/(2N/C + 1)$ per (7.10),

(7.32) may be written as

$$Ce_0 = Ce + \tfrac{1}{2}v_M^2\left[NA^2 + M + \frac{C}{3}\frac{(1+A^3)}{(1+A)}\right]. \tag{7.34}$$

From this equation we can see that for $e = 0$ (when all of the chemical energy has been converted to kinetic energy and v_M reaches its maximum value) the final velocity of plate M is given by the terminal velocity formula of (7.9), with $\sqrt{2e_0}$ in place of $\sqrt{2E}$. This makes it clear that we should replace e_0 by E in this analysis, where $E < e_0$, in order to be consistent with the terminal velocity results.

Following (7.21), we use the ideal gas equation

$$e = \frac{P}{\rho(\gamma - 1)}, \tag{7.35}$$

where P is the pressure in the detonation product gases and γ is the ratio of specific heats, to eliminate e in (7.34). Let us define B and $\dot{x}_M$ as follows:

$$B \equiv \frac{N}{C}A^2 + \frac{M}{C} + \frac{1}{3}\frac{(1+A^3)}{(1+A)} \tag{7.36}$$

and

$$\dot{x}_M \equiv v_M, \tag{7.37}$$

where x_M is the Eulerian position coordinate of the surface of plate M and the superscript dot denotes time differentiation. Then, using the ideal gas and acceleration relations, we obtain an ordinary differential equation which describes the motion of the flyer

$$\rho E(\gamma - 1) = M\ddot{x}_M + \tfrac{1}{2}B\dot{x}_M^2\rho(\gamma - 1). \tag{7.38}$$

Use of the kinematics of the flyer and the density expressed in terms of x_M, and the initial conditions allows us to eliminate ρ from (7.38) so that the only unknowns are x_M and its derivatives:

$$2E = B\dot{x}_M^2 + \frac{2M/C}{(\gamma - 1)}\ddot{x}_M[x_M(1+A) - x_{M0}A]. \tag{7.39}$$

This equation is solved through use of the identity

$$\ddot{x}_M = \frac{1}{2}\frac{d\dot{x}_M^2}{dx_M}. \tag{7.40}$$

and variable changes to yield

$$2E = B\dot{\varphi}^2 + F\varphi\frac{d\dot{\varphi}^2}{d\varphi}, \tag{7.41}$$

where

$$\varphi = x_M - \frac{A}{A+1}x_{M0} \qquad \text{and} \qquad F = \frac{(M/C)(A+1)}{(\gamma - 1)}.$$

This is a first-order linear differential equation for $\dot{\varphi}^2(\varphi)$ whose solution (for the initial conditions $\dot{\varphi} = 0$ and $\varphi = \varphi_0$ at $t = 0$) is

$$\dot{\varphi}^2 = \frac{2E}{B}\left[1 - \left(\frac{\varphi}{\varphi_0}\right)^{-B/F}\right]. \tag{7.42}$$

Translating back to Gurney parameters, the solutions for the three slab geometies shown in Figures 7.3 and 7.4 are as follows:

Asymmetric sandwich:

$$v_M^2 = \frac{2E}{B}\left\{1 - \left[\left(\frac{x_M}{x_{M0}}\right)(A+1) - A\right]^{-B(\gamma-1)/(M/C)(A+1)}\right\},$$

where

$$A = \frac{2M/C + 1}{2N/C + 1},$$

$$B = \frac{N}{C}A^2 + \frac{M}{C} + \frac{1}{3}\frac{1 + A^3}{1 + A},$$

and

$$x_{M0} = \frac{C}{\rho_0}. \tag{7.43}$$

Open-face sandwich:

$$A = \frac{2M}{C} + 1,$$

$$B = \frac{M}{C} + \frac{1}{3}\frac{(1 + A^3)}{(1 + A)},$$

and

$$x_{M0} = \frac{C}{\rho_0}. \tag{7.44}$$

Symmetric sandwich:

$$v_M^2 = \frac{2E}{(m/C + 1/3)}\left[1 - \left(\frac{x_m}{x_{m0}}\right)^{-(\gamma-1)(m/C+1/3)/(m/C)}\right],$$

where

$$m = 2M$$

and

$$x_{m0} = \frac{x_{M0}}{2} = \frac{C}{2\rho_0}. \tag{7.45}$$

Equation (7.45) follows from (7.43) by using $x_m = x_M - x_M/2$ and $N = M = m/2$. This is consistent with the notation used in Figure 7.3.

7.5.3.2. Chanteret Analysis for Symmetric Slabs and Exploding Cylinders

Chanteret (1983) analyzed the acceleration of the metal by grazing detonation in plane (symmetric sandwich) and exploding cylindrical geometry.

Following Taylor (1941, 1963), he included the axial component of detonation product gas velocity in his energy balances, and began with a Bernoulli equation based upon a coordinate system moving with detonation velocity:

$$e_0 + \frac{D^2}{2} = e + \frac{P}{\rho} + \frac{(D-u)^2}{2}. \tag{7.46}$$

He expressed the results in terms on the gas expansion ratio, $(R/R_0)^n$, where R is the explosive-metal interface radius and R_0 is its undetonated value. The velocity as a function of gas expansion is calculated through the Gurney equation, generalized to

$$v = \sqrt{2E_G}\left[\frac{M}{C} + \frac{n}{n+2}\right]^{-1/2}, \tag{7.47}$$

where E_G varies with expansion and $n = 1$ for planar and $n = 2$ for cylindrical geometry. Chanteret shows that E_G can be expressed in several ways:

$$\begin{aligned} E_G &= e_0 - e - \frac{u^2}{2} + \frac{Pu}{\rho_0 D}\left(\frac{R}{R_0}\right)^n, \\ E_G &= \left(\frac{D}{D-u}\right)\left(e_0 - e + \frac{u^2}{2}\right), \\ E_G &= D\left[\frac{P}{\rho(D-u)} - u\right]. \end{aligned} \tag{7.48}$$

Evaluation of E_G requires use of an equation of state and isentrope description.

Chanteret compared various acceleration models for $\gamma = 3$. He showed that his acceleration solution agreed well with the results of wave-code computations for values of $M/C = 1$ or larger, and that his model slightly underpredicted the acceleration for smaller values of M/C. He also showed that the Jones model (Jones et al., 1980) overpredicted the acceleration for $M/C < 1$.

7.5.3.3. Flis Analysis by Lagrange's Principle

Following Flis (1994), the kinetic energy, **T**, per unit area of a symmetric sandwich system in which the gas velocity distribution is always linear in Lagrangian coordinates is

$$\mathbf{T} = (\tfrac{1}{2}M + \tfrac{1}{6}C)\dot{x}^2, \tag{7.49}$$

where $\dot{x}$ is the velocity of the metal plates. Using the ideal-gas law, the potential energy of the gas is found to be

$$\mathbf{U} = \frac{P_0 C}{\rho_0(\gamma - 1)}\left(\frac{x_0}{x}\right)^{\gamma-1}. \tag{7.50}$$

Forming the Lagrangian $W = \mathbf{T} - \mathbf{U}$, and applying Lagrange's equation

$$\frac{d}{dt}\left(\frac{\partial W}{\partial \dot{x}}\right) - \frac{\partial W}{\partial x} = 0 \tag{7.51}$$

yields the equation of motion

$$\left(M + \frac{C}{3}\right)\ddot{x} = P_0\left(\frac{x_0}{x}\right)^{\gamma}. \tag{7.52}$$

Integration indicates the velocity of the metal as a function of gas expansion:

$$\dot{x} = \left\{\frac{2P_0}{\rho_0(\gamma - 1)\left(\dfrac{M}{C} + \dfrac{1}{3}\right)}\left[1 - \left(\frac{x_0}{x}\right)^{\gamma-1}\right]\right\}^{1/2}, \tag{7.53}$$

where $P_0 = \rho_0(\gamma - 1)E$. For the integral value of $\gamma = 3$, this may be integrated to produce

$$x = x_0\sqrt{\left(\frac{t}{\bar{\tau}}\right)^2 + 1}, \tag{7.54}$$

where $\bar{\tau} = x_0/v$, and v is the terminal velocity of the metal per (7.6). Flis shows that the velocity history may now be written as

$$\dot{x} = v\frac{t}{\bar{\tau}}\left[\left(\frac{t}{\bar{\tau}}\right)^2 + 1\right]^{-1/2}. \tag{7.55}$$

Flis shows that the general solution for symmetric configurations, per Figure 7.3, the solution is

$$\dot{x} = \left\{\frac{2P_0}{\rho_0(\gamma - 1)\left(\dfrac{M}{C} + \dfrac{n}{n+2}\right)}\left[1 - \left(\frac{x_0}{x}\right)^{n(\gamma-1)}\right]\right\}^{1/2}, \tag{7.56}$$

where $n = 1$ for slab, $n = 2$ for cylindrical, and $n = 3$ for spherical geometry. Flis applied the same kind of approach to the asymmetric sandwich and obtained

$$\dot{\chi} = \left\{\frac{2P_0C}{\mu\rho_0(\gamma - 1)} - \left[1 - \left(\frac{h}{h + \chi}\right)^{\gamma-1}\right]\right\}^{1/2}, \tag{7.57}$$

where

$$\mu = \frac{MN + \dfrac{C}{3}(M + N) + \dfrac{C^2}{12}}{M + N + C}, \qquad \chi = x_1 + x_2,$$

x_1 and x_2 are the displacements of plates M and N, respectively, and h is the initial thickness of the explosive. Equation (7.57) can be integrated for the case of $\gamma = 3$ to yield

$$\chi = h\left\{\left[\left(\frac{t}{\bar{\tau}}\right)^2 + 1\right]^{1/2} - 1\right\}, \tag{7.58}$$

where $\bar{\tau} = \sqrt{h\mu/P_0} = h/(v_M + v_N)$ in this case.

7.5.4. Electrical Gurney Energy

There are other sources besides explosives which can produce impulsive acceleration of solids. One of these is the physical explosion of a thin metal foil by passage of a large, fast-rising electric current pulse through it. Exploding foils actuated in this way have been used for a number of purposes. One purpose is to power slapper detonators (Ornellas and Stroud, 1988), in which the exploding foil accelerates a thin dielectric film (typically polyimide) across a short gap to impact an explosive pellet and shock-initiate it. A large version of this type of device was called the electric gun by researchers at LLNL (Bloom et al., 1989), who used it to study the initiation of explosives by short-duration impacts of thin, large-diameter dielectric flyer plates.

Tucker and Stanton (1975) correlated the effective energy content of an exploding foil, for purposes of accelerating a dielectric film, with the current density flowing through the foil at the time of its vaporization (called the burst current density, J_b). They defined an electrical Gurney energy, E_{eg}, which they found to be a power function of J_b. For copper foils,

$$E_{eg} = 5.86 \cdot 10^{-10} J_b^{1.41}. \tag{7.59}$$

The units in (7.23) are kJ/g for E_{eg} and A/cm^2 for J_b.

The velocity to which the dielectric film was driven could then be calculated by use of traditional Gurney formulas, e.g., (7.6). The metal foil in this instance served as the driving material, and its mass represented the "charge" mass C; the dielectric film was the driven mass, and its mass was M.

The burst current was calculated through the use of circuit equations, integrating to evaluate the time and current, i, which flowed when the "action," $\int i^2\, dt$ (Tucker and Stanton, 1975) reached a characteristic value for burst of the foil material. Burst current density was then determined simply by use of the cross-sectional area of the foil.

It is interesting to note that the energy density of exploding metal foils can be larger than that of detonating explosives.

7.5.5. Laser Ablation Gurney Energy

Deposition of sufficiently large laser energy in a very short time can ablate a thin layer of any solid target material, and form a plasma which drives the remaining portion of the solid target layer. This mechanism has been shown by Paisley et al. (1991) to be capable of driving thin films of metals to velocities of several km/s. These films could be used, for example, to conduct equation of state or spall strength measurements with miniature samples of material (Warnes et al., 1995).

The energy content of the plasma formed by laser ablation has been described by Lawrence and Trott (1993) using a Gurney energy analogy. The laser ablation Gurney energy, a function of laser fluence and absorptance and thermal diffusivity of the target (flyer) material, was correlated to experimental data. The depth of target material ablated by the laser pulse was estimated analytically, and results of the calculation in terms of integral quantities such as velocity or momentum of the flyer were found to be insensitive to the value estimated for blow-off depth. Trends such as the effect of laser fluence and pulse duration on flyer velocity compared well with the data.

7.6. Limitations and Corrections in Gurney Analysis

The simple gas-dynamic assumptions made in the Gurney model do not apply for certain circumstances, and then Gurney analysis either should not be applied or should be applied with corrections to account for the deviations. In this section we shall describe deviation of results from the predictions of the Gurney model at low values of M/C, and the ineffectiveness of portions of an explosive charge wherein the product gases flow in a direction that does not contribute to the momentum of driven metal.

7.6.1. Comparison with Gas-Dynamic Solution for Open Sandwich

Aziz et al. (1961) performed an analysis for an open-faced sandwich configuration in which the driven metal was treated as a rigid body but the gases were treated with an ideal gas equation of state. There was an analytic result when the value of γ for the product gases was an integer. The piston velocity was found to be a function of time, M/C, and sound velocity in the product gases, and the result for the terminal velocity v of the plate was

$$\frac{v}{D} = 1 - \frac{27}{16}\frac{M}{C}\left[\left(1 + \frac{32}{27}\frac{C}{M}\right)^{1/2} - 1\right]. \tag{7.60}$$

A special feature of this solution is that because it treats the gas-dynamics fully, the result should be applicable for all values of M/C. Gurney solutions, on the other hand, cannot be trusted at low values of M/C where gas-dynamic behavior dominates because Gurney solutions are oversimplified for this regime.

In Figure 7.12, we compare the values of terminal velocity of the plate driven by Composition B explosive, for which $D = 8.03$ km/s, taking the polytropic exponent to be 3. For large values of M/C, we see that the solu-

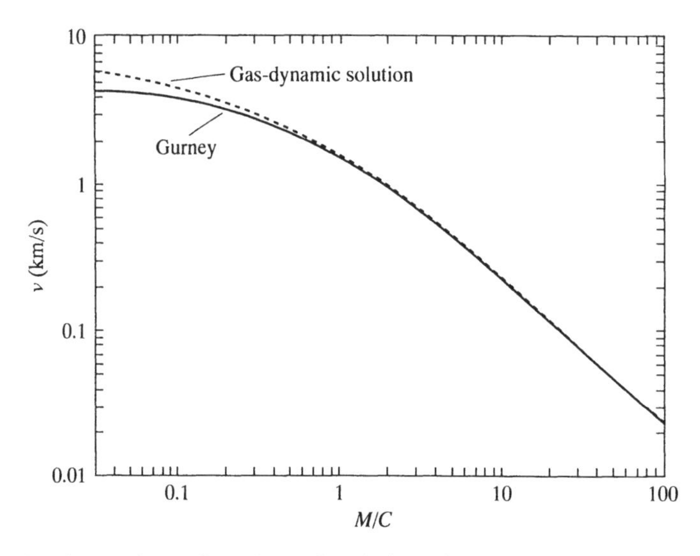

Figure 7.12. Comparison of gas-dynamic solution with Gurney solution for open-face sandwich.

tions overlay very well. At small values of M/C the solutions deviate, with the gas-dynamic model (Aziz et al., 1961) predicting higher velocities than the Gurney model. This leads us to recommend that the Gurney model not be used at values of $M/C < \frac{1}{3}$. If it is used, it should be recognized that the result represents a lower bound for the velocity, and the true value is expected to be higher.

For some configurations, however, good correlation of observed velocity with Gurney equation predictions continues to values of M/C much less than 1/3. Butz et al. (1982) show that symmetric sandwich configurations follow the Gurney model very well down to $M/C = 0.05$, the lowest value they measured.

7.6.2. Correction for Side Losses

Consider the case of an unconfined cylindrical charge of explosive resting on a plate, and detonated from the center of the rear surface. Baum et al. (1959) indicated that the explosive which was effective in driving the plate was the core of the explosive nearest to the plate. Detonation product gases from explosive which was in a zone affected by a lateral rarefaction wave moved in a lateral rather than axial direction and did not contribute to the axial momentum of the plate. For a free lateral surface, Baum estimated that the rarefaction wave intruded into the explosive at about half the detonation velocity. For the configuration shown in Figure 7.13, the explosive which could be "discounted" (neglected), then, was that outside the boundary of an angle whose tangent was $D/2D = 0.5$. We round off the resulting

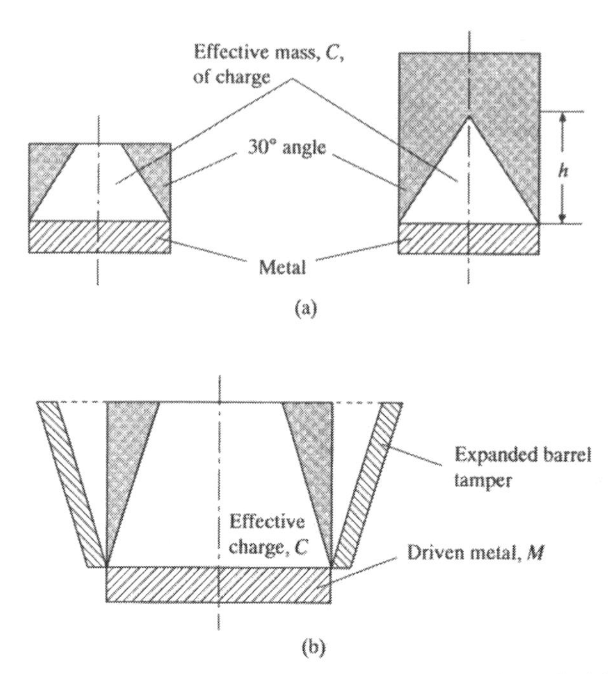

Figure 7.13. Discounting angle to account for side losses in cylindrical charges. (a) For unconfined cylinders, explosive outside the 30° discounting angle (shaded region) is disregarded in evaluating C. Thus material above height h is entirely discounted. (b) For cylinders with barrel tamping, the discounting angle corresponds to the velocity of the tamper. See (7.64) in text. Shaded region of charge is discounted.

estimate of the discounting angle to 30° to a normal to the plate, as shown in Figure 7.13. As can be seen in the figure, the limiting amount of unconfined cylindrical explosive which can be effective in driving the plate is that within a cone with a 30° half-angle.

The plate mass driven by the plate will usually be that which is directly under the explosive, because the detonation will typically shear out that part from the surrounding plate. If the plate is thick enough or cushioned so that the plate does not shear off or spall, it would be expected that the portion of the plate below the charge should be considered to be the portion driven by the charge; the remainder of the plate would then be carried along as a result of momentum sharing as described in Section 7.7.1.

Benham (1979) analyzed the effect of a finite amount of tubular confinement upon the velocity of the driven plate by using the discounting angle approach put forward by Baum. His result was that the discounting angle κ was scaled to the unconfined discounting angle by the ratio of the velocity of the tube wall, v_t, divided by the expansion velocity of a free surface of cylindrical charge, $v_{t,0}$; thus

$$\kappa = 30° \frac{v_t}{v_{t,0}}. \tag{7.61}$$

The values of both v_t and $v_{t,0}$ were derived from (7.7). For a barrel tamper as shown in Figure 7.13, described by the ratio N/C, the velocity v_t (see Figure 7.3) is

$$v_t = \frac{\sqrt{2E}}{\left(\frac{N}{C} + \frac{1}{2}\right)^{1/2}}. \tag{7.62}$$

It can be seen that for the unconfined case ($N/C = 0$), the result is

$$v_{t,0} = \sqrt{2}\sqrt{2E}. \tag{7.63}$$

Thus Benham's result for the discounting angle was

$$\kappa = \frac{30^\circ}{\sqrt{2}\left(\frac{N}{C} + \frac{1}{2}\right)^{1/2}}. \tag{7.64}$$

It should be noted that the entire charge mass C may be used in calculating the value of κ.

Calculated results were compared with experimental results over a range from $N/C = 0.06$ to 4.4, and close agreement was found by Benham (1979).

7.6.3. Scaling

When experimental data on velocity are available for a particular explosive/metal configuration, it is appropriate to compare those results with results of Gurney calculations. When the results agree, we can be comfortable using the Gurney approach to calculate design modifications to satisfy design goals. But the results may not agree for geometries with substantial side losses or which do not conform to the classical geometries for which Gurney equations have been derived in Section 7.3.1. In that case it is appropriate to use the experimental data to evaluate an effective value of $\sqrt{2E}$ for the energetic material formulation and density, using a Gurney equation for a geometry which approximates the configuration of interest. When the value of M/C is large, it makes little difference which Gurney equation is used among the symmetric configurations; this may be seen in Figure 7.5. But the value of N still makes a difference for slab geometries.

When explosive is used at a density other than that for which experimental data exist, (7.19) and (7.20) could be used as a basis for scaling from the experimental value of $\sqrt{2E}$ at the other density.

7.7. Combination of Gurney with Other Physics

The explicit and closed-form nature of Gurney model solutions makes it easy to combine this model with other physics to work applied problems.

7.7.1. Inelastic Collision Momentum Transfer

Sometimes the metal configuration being driven by explosives includes substantial gaps between parts, so that part of the metal is driven to terminal velocity before it collides with other metal parts. When the shape of parts undergoing collision is irregular (i.e., other than planar impact of flat plates), the parts often will become entangled and move on together. This is a circumstance in which the collision may be considered to be inelastic, and the momentum of the first flyer or fragments is shared with the impacted metal so that the final velocity of the total mass is reduced in a way which corresponds to the increase in mass after impact (Kennedy et al., 1996). The velocity of the initially driven metal M_1 is calculated using a standard Gurney formula, (7.3) through (7.10), and then momentum sharing is calculated as

$$M_1 v_1 = (M_1 + M_2) v_{1+2}, \tag{7.65}$$

where subscript 1 refers to the initially driven metal, subscript 2 refers to the impacted part(s), and v_{1+2} refers to their collective final velocity.

Inelastic collision is quite a lossy process in comparison with directly driving the mass $M_1 + M_2$ with the same explosive. Thus a gap in the assembly of driven metal can substantially reduce the velocity to which the assembly is driven.

What conditions make it appropriate to apply this relation, as opposed to calculating the velocity of the total mass $M_1 + M_2$ using Gurney equations? If M_1 has the chance to approach terminal velocity before impact with M_2, it is appropriate to use (7.65). We may address this question through consideration of either the shock reverberations in the driven plate, or the degree of expansion of the detonation product gases. "Plate" M_1 approaches terminal velocity after about three reverberations in the metal of the shock wave from the detonating explosive if the detonation is normally incident against the metal. However, the acceleration history is substantially affected by the angle of incidence. Kury et al. (1965) found that metal has been accelerated essentially to its final velocity by the time the detonation products have expanded to twice their original volume whenever detonation is normally incident upon the metal surface; whenever detonation propagates in a direction tangential to the metal surface, the metal approaches its final velocity only after the product gases have expanded sevenfold. Lee and Pfeifer (1969) found in two-dimensional computations that a tube wall was driven to the same final velocity (within a few percent) regardless of whether the detonation was normally incident or tangentially incident on the wall.

Use of momentum sharing calculations as outlined above has been found to be appropriate in testing scenarios where range safety is an issue, particularly whenever small fragments impact and drive larger metal bodies which have lower proportional air drag and therefore can travel further.

7.7.2. Detonation Transfer by Flyer Plate Impact

Detonation is often transferred from a detonating element (called the donor) to a second explosive charge (called the acceptor) by use of the donor charge to drive a flying plate which impacts the acceptor charge and shock-initiates it. This section presents an analytical model for this process, utilizing the Gurney model to describe the flyer plate velocity and a shock initiation criterion as the explosive response function.

7.7.2.1. Shock-Initiation Criteria

In applications involving detonation by flyer plate impact, it is typically desired that a minimal amount of donor charge be required and that efficient use be made of the flyer plate. These conditions lead to the use of thin flyer plates, which can be driven to high velocities by small donor charges. Gittings (1965) was the first to report on the response of explosives to the impact of thin flyer plates. Her experimental work was analyzed by Walker and Wasley (1969), who proposed shock-wave energy fluence, $P^2\tau/\rho_0 U$, as an initiation criterion. Here P is the shock pressure driven into the explosive, τ is the shock duration, and U is the shock velocity. This criterion was alternatively expressed as simply $P^2\tau$, and was generalized (Hayes, 1977) as a $P^n\tau$ initiation criterion; here n is an arbitrary value, fit to experimental data. Note that this criterion is for "large-diameter" flyers; it makes no reference to diameter of the flyer plate, and this is a weakness in this form of criterion.

Criteria of this form are useful for engineering design purposes, for example, the design of an interface to function reliably. But they are not conservative when it comes to safety analysis, because mechanisms other than shock initiation may operate as the limiting case in the response of explosive materials to hazard environments.

7.7.2.2. Analysis of Impact Interaction

Impact of the flyer material against the explosive material drives shock waves of equal pressure into both, such that both materials move with the same velocity and contact is maintained at the interface. For impact of a thin plate (the flyer) against a thick explosive, this initial shock response persists for the time interval of one shock reverberation through the flyer plate. This condition may be written in terms of the conservation of momentum, which in general is

$$P - P_0 = \rho_0 U(u - u_0) \equiv z(u - u_0), \tag{7.66}$$

where $z = \rho_0 U$ and velocities are taken to be positive in the direction of shock propagation. In this case P_0 is zero in both the flyer and the acceptor explosive just prior to impact, and u_0 is zero in the acceptor. The flyer plate impact velocity is denoted u_f. Let u denote the final particle velocity in both materials. The pressure is identical in both, so

$$P = z_a u = z_f(u_f - u). \tag{7.67}$$

Here subscripts $_a$ and $_f$ refer to acceptor and flyer, respectively. From this equation we find

$$u = \frac{u_f}{Z+1}, \tag{7.68}$$

where $Z = z_a/z_f$.

The shock duration may be approximated as the time required for a double transit of a shock wave through the flyer material thickness, ℓ_f, thus

$$\tau = \frac{2\ell_f}{U_f} = \frac{2\rho_f \ell_f}{z_f}. \tag{7.69}$$

In terms of material properties and impact parameters, from (7.67), (7.68), and (7.69) we may now write

$$P^2\tau = \frac{2\rho_f \ell_f z_a Z u_f^2}{(Z+1)^2}. \tag{7.70}$$

To apply this analysis, we may use any method to evaluate the flyer plate velocity, u_f, as a function of design parameters. In this section, we shall use the Gurney equations for this purpose.

7.7.2.3. Gurney Analysis to Evaluate Initiation Criterion

The analysis that follows was developed by Kennedy and Schwarz (1974). Consider the donor–acceptor–flyer configuration sketched in Figure 7.14. For simplicity, let us consider the tamper material behind the donor to be sufficiently thick and stiff that it approximates a rigid surface which moves very slowly when the donor charge is fired. This allows use of the symmetric sandwich formula, (7.6), to describe the velocity of the flyer plate as a function of the ratio between the masses of the flyer plate and the donor explosive. Subscript $_d$ shall be used to denote donor. The quantity u_f^2 required to implement (7.70) then can be written as

$$u_f^2 = 2E \Big/ \left(\frac{\rho_f \ell_f}{\rho_d \ell_d} + \frac{1}{3}\right) \tag{7.71}$$

and the expression for $P^2\tau$ becomes

$$P^2\tau = \frac{4E\rho_f \ell_f z_a Z}{(Z+1)^2} \left(\frac{\rho_f \ell_f}{\rho_d \ell_d} + \frac{1}{3}\right). \tag{7.72}$$

With the total length of donor plus flyer fixed as ℓ_{total} in Figure 7.14, we find it convenient to work with a "reduced donor length" L, defined as

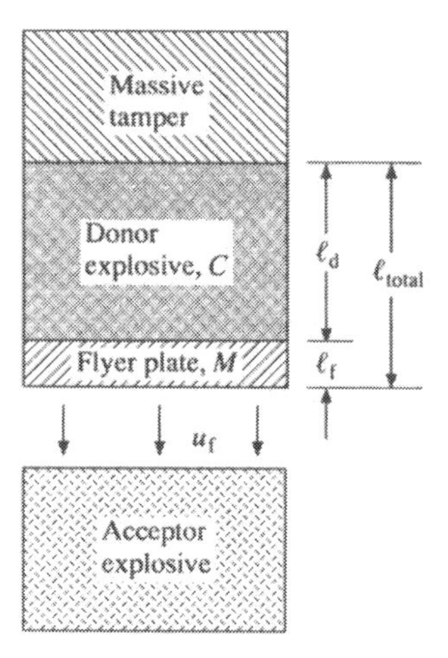

Figure 7.14. Detonation transfer configuration which is modeled. Gap between flyer and acceptor is assumed to be long enough that flyer attains velocity u_f per (7.6).

ℓ_d/ℓ_{total}. We rewrite (7.72) in terms of L, and express the value of $P^2\tau$ also per unit total length:

$$\frac{P^2\tau}{\ell_{total}} = \frac{4E\rho_f z_a Z(1-L)}{(Z+1)^2}\left[\frac{\rho_f(1-L)}{\rho_d L} + \frac{1}{3}\right]^{-1}. \tag{7.73}$$

If we consider the flyer plate impact to be the only source of shock loading of the acceptor, we can recognize that there must be a maximum in the function represented in (7.73). Consider the two limits in the function, where $L = 0$ and $L = 1$. At $L = 0$ there is no charge to drive the flyer, so $P^2\tau$ must be zero; at $L = 1$, there is a driving charge but no flyer plate so again $P^2\tau$ must be zero. Observing that the value of the function is nonzero at intermediate values of L leads us to the correct conclusion that there must be a maximum in the function.

A plot of $P^2\tau/\ell_{total}$ versus L is presented as Figure 7.15. In this figure, we have selected real materials to serve as the donor, flyer, and acceptor so that we would have values for the material properties, and these material choices are noted in the caption. It is noted that the curve for each of three flyer materials—(1) stainless steel, (2) aluminum, and (3) nylon—has a maximum and that the highest maximum is for the highest impedance material, stainless steel. Also note that the optimal fraction of the total height ℓ_{total} which is occupied by the flyer decreases as density and impedance of the flyer increases, i.e., optimal length of a stainless steel flyer is 0.1 ℓ_{total}; optimum for aluminum is 0.2 ℓ_{total}; and optimum for nylon is about 0.35 ℓ_{total}. (The optimal reduced flyer lengths would be shorter if the initiation criterion were, e.g., $P^3\tau$ or $P^4\tau$ rather than $P^2\tau$, because the premium on pressure (at

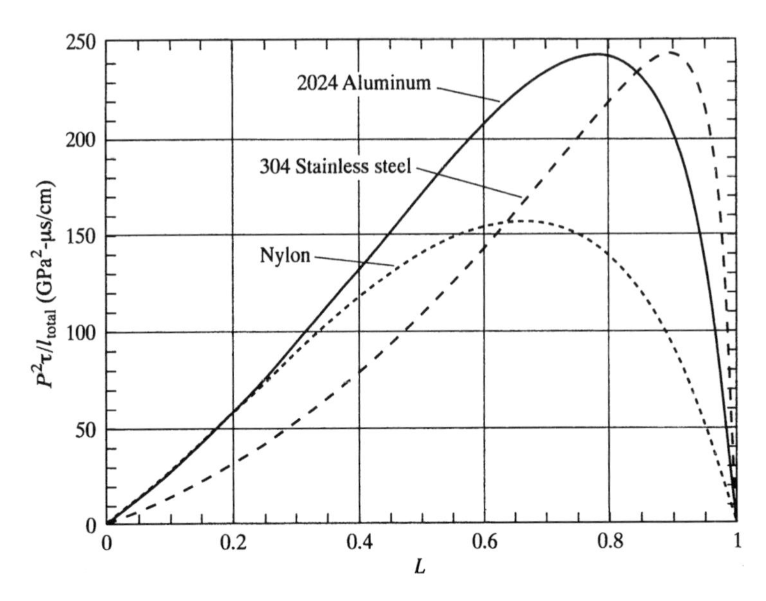

Figure 7.15. Initiation parameter as a function of reduced donor length, L. HNS was used as the donor and PBX-9404 was used as the acceptor explosive in this calculation.

the expense of shock pulse duration) would call for higher flyer velocities for all flyer materials.) The value of $P^2\tau$, required for initiation of PBX-9404, the acceptor in this calculation, is 5.5 GPa^2-μs, so it can be seen that the length required for initation of PBX-9404 would be a small fraction of a millimeter.

In some cases, it is desired that the amount of explosive be minimized to reduce collateral damage. In such a case it would be best to operate at a condition which produced the greatest amount of $P^2\tau/\ell_{total}$ per unit of L, the donor charge. This condition corresponds to the value of L at which the slope from the origin to a point on a curve is a maximum; in Figure 7.15 this would call for a tangent to the curve. This case is discussed by Kennedy and Schwarz.

If space were at an absolute premium, the total length modeled would have to include free space to allow for acceleration of the flyer plate. Then the acceleration form of the Gurney relations could be used as shown, for example, by Kennedy and Schwarz.

Acknowledgment

The help of Dr. William J. Flis of Dyna East Corporation in providing an extensive bibliography of work related to the Gurney method is gratefully acknowledged.

References

Aziz, A.K., Hurwitz, H., and Sternberg, H.M. (1961). Energy transfer to a rigid piston under detonation loading. *Phys. Fluids*, **4**, 380–384.

Barut, A.O. and Bartlett, A.A. (1978). Introductory note. *Amer. J. Phys.*, **46**, 319.

Baum, F.A., Stanyukovich, K., and Shekhter, B.I. (1959). *Physics of an Explosion*, pp. 505–507. Moscow (English translation from Federal Clearinghouse AD 400151).

Benham, R.A. (1979). Terminal velocity and rotation rate of a flyer plate propelled by a tube-confined explosive charge. *Shock and Vibration Bull.*, **49**, 193–201.

Bloom, G., Lee, R., and Von Holle, W. (1989). Thin pulse shock initiation characterization of extrusion-cast explosives. *Third Intl. Symp. on Behavior of Media Under High Dynamic Pressure*, La Grande Motte, France, June.

Butz, D.J., Backofen, J.E. Jr., and Petrousky, J.A. (1982). Fragment terminal velocities from low metal to explosive mass ratio symmetric sandwich charges. *J. Ballistics*, **6**, 1304–1322.

Chanteret, P.Y. (1983). An analytical model for metal acceleration by grazing detonation. *Proc. Seventh Intl. Symp. on Ballistics*, The Hague, Netherlands, April.

Chou, P.C., Carleone, J., Hirsch, E., Flis, W.J., and Ciccarelli, R.D. (1981). Improved formulas for velocity, acceleration, and projection angle of explosively driven liners. *Proc. 6th Intl. Symp. on Ballistics*, Orlando, FL. Also in *Propellants, Explosives, and Pyrotechnics*, **8** (1983), 175–183.

Condon, E.U. (1978). Tunneling—How it all started. *Amer. J. Phys.*, **46**, 319–323.

Cooper, P.W. (1996). *Basics of Explosives Engineering*. VCH, New York.

Defourneaux, M. and Jacques, L. (1970). Explosive deflection of a liner as a diagnostic of detonation flows. *Proc. Fifth Symp. (Intl.) on Detonation*, ONR ACR-184, pp. 457–466, August.

Dobratz, B.M. and Crawford, P.C. (1985). *LLNL Explosives Handbook*. Properties of Chemical Explosives and Explosive Simulants. Lawrence Livermore National Laboratory Report UCRL-52997 Change 2, January.

Fickett, W. and Davis, W.C. (1979). *Detonation*. University of California Press, Berkeley, CA, pp. 35–38.

Flis, W.J. (1994). A Lagrangian approach to modeling the acceleration of metal by explosives. *Developments in Theoretical and Applied Mechanics*, Vol. XVII, pp. 190–203.

Gittings, E.F. (1965). Initiation of a solid explosive by a short-duration shock. *Proc. Fourth Symp. (Intl.) on Detonation*. ONR ACR-126, pp. 373–380, October.

Gurney, R.W. (1943). The initial velocities of fragments from bombs, shells, and grenades. Army Ballistic Research Laboratory Report BRL 405.

Hardesty, D.R. and Kennedy, J.E. (1977). Thermochemical estimation of explosive energy output. *Combust. Flame*, **28**, 45–59.

Hayes, D.B. (1977). Optimizing the initiation of detonation by flying plates. Sandia National Laboratories Report SAND 77-0268, Albuquerque, NM.

Henry, I.G. (1967). The Gurney formula and related approximations for high-explosive deployment of fragments. AD813398, Hughes Aircraft Co. Report PUB-189.

Hirsch, E. (1986). Simplified and extended Gurney formulas for imploding cylinders and spheres. *Propellants, Explosives, and Pyrotechnics*, **11**, 6–9.

Hoskin, N.E., Allan, J.W.S., Bailey, W.A., Lethaby, J.W., and Skidmore, I.C. (1965). The motion of plates and cylinders driven by detonation waves at tangential incidence. *Proc. Fourth Symp. (Intl.) on Detonation*. ONR ACR-126, pp. 14–26, October.

Jones, G.E., Kennedy, J.E., and Bertholf, L.D. (1980). Ballistic calculations of R.W. Gurney. *Amer J. Phys.*, **48**, 264–269.

Kamlet, M.J. and Finger, M. (1979). An alternative method for calculating Gurney velocities. *Combust. Flame*, **34**, 213–214.

Kamlet, M.J. and Jacobs, S.J. (1968). Chemistry of detonations. I. A simple method for calculating detonation properties of C–H–N–O explosives. *J. Chem. Phys.*, **48**, 23–35.

Kennedy, J.E. (1970). Gurney energy of explosives: Estimation of the velocity and impulse imparted to driven metal. Sandia National Laboratories Report SC-RR-70-90, Albuquerque, NM, December.

Kennedy, J.E., Cherry, C.R., Cherry, C.R. Jr., Warnes, R.H., and Fischer, S.H. (1996). Momentum transfer in indirect explosive drive. *Proc. 22nd Intl. Pyrotechnics Seminar*, Fort Collins, CO. Published by IIT Research Institute, Chicago, July.

Kennedy, J.E. and Chou, T.S. (1990). Effects of gaps in explosive/metal systems. Presented at *14th Intl. Pyrotechnics Seminar*, Boulder, CO (unpublished), July.

Kennedy, J.E. and Schwarz, A.C. (1974). Detonation transfer by flyer plate impact. *Proc. Eighth Symposium on Explosives and Pyrotechnics*. Franklin Institute, Philadelphia, PA. Also Sandia National Laboratories Report SLA 74-5073, Albuquerque, NM.

Kury, J.W., Hornig, H.C., Lee, E.L., McDonnel, J.L., Ornellas, D.L., Finger, M., Strange, F.M., and Wilkins, M.L. (1965). Metal acceleration by chemical explosives. *Proc. Fourth Symp. (Intl.) on Detonation*. ONR ACR-126, pp. 3–13, October.

Lawrence, R.J. and Trott, W.M. (1993). Theoretical analysis of a pulsed-laser-driven hypervelocity flyer launcher. *Internal. J. Impact Engng.*, **14**, 439–449.

Lee, E.L. and Pfeifer, H. (1969). Velocities of fragments from exploding metal cylinders. Lawrence Livermore National Laboratory Report UCRL 50545, January.

Lieberman, M.L. and Haskell, K.H. (1978). Pyrotechnic output of $TiH_x/KClO_4$ actuators from velocity measurements. *Proc. 6th Intl. Pyrotechnics Seminar*, Estes Park, CO. Published by IIT Research Institute, Chicago, July.

Ornellas, D.L. (1968). The heat and products of detonation of HMX, TNT, NM, and FEFO. *J. Phys. Chem.*, **72**, 2390–2394. Also Private communication (1970).

Ornellas, D.L. and Stroud, J.R. (1988). Flying-plate detonator using a high-density high explosive. U.S. Patent 4778913.

Paisley, D.L., Warnes, R.H., and Kopp, R.A. (1991). Laser-driven flat plate impacts to 100 GPa with sub-nanosecond pulse duration and resolution for material property studies. *Shock Compression in Condensed Matter. Proc. APS Topical Conference*, Williamsburg, VA, pp. 825–828.

Roth, J. (1971). Private communication.

Taylor, G.I. (1941). Analysis of the explosion of a long cylindrical bomb detonated at one end. In *Scientific Papers of G.I. Taylor*, Vol. III (G.K. Batchelor, ed.). Cambridge University Press, Cambridge, 1963, pp. 277–286.

Tucker, T.J. and Stanton, P.L. (1975). Electrical Gurney energy: A new concept in modeling of energy transfer from electrically exploded conductors. Sandia National Laboratories Report SAND 75-0244, Albuquerque, NM.

Walker, F.E. and Wasley, R.J. (1969). Critical energy for shock initiation of heterogeneous explosives. *Explosivstoffe*, **17**, 9.

Walters, W.P. and Zukas, J.A. (1989). *Fundamentals of Shaped Charges*. Wiley, New York, p. 51.

Warnes, R.H., Paisley, D.L., and Tonks, D.L. (1995). Hugoniot and spall data from the laser-driven miniflyer. *Shock Compression of Condensed Matter. Proc. APS Topical Conference*, Seattle, WA, August, pp. 495–498, *AIP Conference Proc.* 370, Part 1 (S.C. Schmidt and W.C. Tao, eds.).

Weinland, C.E. (1969). A scaling law for fragmenting cylindrical warheads. NWC TP 4735, April.

CHAPTER 8

Hazard Assessment of Explosives and Propellants

Peter R. Lee

8.1. Introduction

8.1.1. Aims

This chapter is intended to provide a background to hazard assessment of explosives, in which, conventionally, we include propellants and pyrotechnics. It will discuss:

- the bases of testing regimes;
- what we expect the data to reveal;
- the machine-dependence of data and the significance of it;
- statistical methods for the analysis of results;
- the role of judgment in hazard assessment;
- differences between hazard and reliability testing and the role of explosive confinement;
- the terms sensitivity, sensitiveness, and explosiveness;
- hazard assessment stages, i.e., powder, charge, and system tests; and
- insensitive munitions tests.

It is not intended to be a catalogue of testing methods or a workbook for carrying them out, although some tests are described in detail in order to exemplify techniques considered to be of general interest. The reader is also referred to the following standard works on hazard assessment:

- *US Military Standard*, DoD-STD-2015B (Navy) (1994);
- *UK Manual of Tests*, Sensitiveness Collaboration Committee (SCC No. 3) (1988);
- *LASL Explosives Property Data* (Gibbs and Popolato, 1980); and
- *LLNL Explosives Handbook* (Dobratz, 1981).

Most national military explosives and weapons research and development agencies produce similar documents for internal use and readers are referred to ones pertinent to their own national testing authorities for such detail. In addition, *AGARDograph 316*, by Boggs and Derr (1990) is a fine example of a commentary on methods of testing as applied specifically to testing propellants and rocket motors for compliance with insensitive munitions requirements, although it is pertinent to the testing of all types of weapon system containing energetic materials. The development of the technologies with time is likely to be charted by these documents and their successors. They provide a comprehensive reference to the mechanics of the full range of tests and the outcomes expected from them. The NATO Insensitive Munitions Information Centre (NIMIC) in Brussels also has a role to play in the future of the development and standardization of new tests and assessment of the outcomes from them, presently for those nations involved in setting up the database, but also, in the future, for other nations as they join it. The future will undoubtedly see the development of cheaper tests, which will be more directly related to the expected hazard and these will be supplemented by better predictive modeling of the responses of materials and systems to tests and realistic hazard conditions.

8.1.2. Nomenclature

N	Number of trials carried out.
N_i	Number of ignitions (n_c number of contradictory results).
h_i	Height at which an ignition takes place.
H	Rockwell hardness of steel used in plate dent tests, or median height.
c	Lowest drop height in a series of tests.
d	Test variable interval (depth of dent in a dent test, or D).
σ	Standard Deviation.
T	Parametrical value of assumed 50% point in a test.
F_T	Summation over all contradictory test results based on the assumed 50% point.
α	Biot Number ($= \chi r/\lambda$).
χ (cal/cm^2 · °C · s)	Surface Heat Transfer Coefficient from the explosive to the surroundings.
r or r_0 (cm)	Characteristic linear dimension of the charge.
λ (cal/cm.°C · s)	Thermal conductivity of the explosive.
δ	Frank-Kamenetskii Criticality Criterion given by $\left(\frac{Q}{\lambda}\right)\left(\frac{E}{RT_0^2}\right) \cdot r_0^2 \cdot A \cdot c_0 \cdot \exp\left(-\frac{E}{RT_0}\right)$.
Q (cal/mole)	Heat of exothermic decomposition reaction: also (kJ/kg).
q (cal/mole · s)	A rate of heat generation by a chemical reaction given by $q = F(Q)\exp(-E/RT)$: also (kJ/kg · s).
E (cal/mole)	Activation energy of the exothermic decomposition reaction: also (kJ/kg).
R (cal/°C · mole)	Universal Gas Constant: also (kJ/K · kg).

T_0 (K)	Temperature: sometimes expressed in °C.
$A(s^{-1})$	Arrhenius Frequency Factor.
c_0 or ρ (g/cm^3)	Initial density or concentration of reactant.
U (m/s)	Shock and particle velocity; also in km/s or mm/μs.

8.1.3. Background

The science of hazard assessment of explosives began as a consequence of investigations into accidents with explosives, rather than as a result of a conscious effort by manufacturers and users to make a systematic study of accidental initiation. Accident investigators in different parts of the world, not unsurprisingly, came to similar conclusions concerning those stimuli to which explosives were especially susceptible. Consequently, roughly similar hazard tests were devised independently by investigators at different locations to assess the response of explosive materials to a range of accidental initiation mechanisms. In crude terms these early tests were associated with assessment of the effects of shock from another, detonating, explosive on a small bare charge and of a range of thermomechanical stimuli, including:

- mechanical impact;
- friction;
- heat;
- flame; and
- electrostatic discharges;

on small samples of powdered explosive. Different groups of scientists and engineers involved in the testing of explosives had slightly different perceptions of the mechanisms involved in each type of initiation and these perceptions affected the detailed design of tests. Consequently, tests designed and carried out at two different laboratories to give information on, say, the sensitiveness to impact of a sample of powder of the same explosive, might give different results. The reasons for this lie in the design of the test equipment, the test regime and the sample characteristics, e.g., mass, particle size, etc., of the explosive and the fact that most explosives do not consist of pure chemicals.

During the era when armed forces were strictly national, as opposed to operating as components of international forces, i.e., up to the end of World War II, each nation, and sometimes each armed service of some nations, developed their own idiosyncratic testing regimes. There was relatively little communication between the military–industrial complex of one country with that of another, whether they were allies or not, since most countries purchased weapons from their own manufacturers. Sometimes the lack of standardization was evident between different laboratories in the same country. If weapons were purchased from abroad, the vendor was either a former colonial "master" or an ally with a larger or better production base and it was usually the case that the purchaser adopted the safety mores of the supplier. For example, during World War II, the United Kingdom purchased large quantities of weapons from the United States, but carried out relatively little independent testing. It was recognized in the United Kingdom that United States hazard tests differed from United Kingdom ones, but the United States had a similar high regard for the safety of its operational armed forces and its civilian workforce as the United Kingdom, so it was possible to accept United States data unless the material or weapon was to be used for purposes other than those for which it had originally been designed.

Hazard testing began to develop as a technology in its own right as a result of several influences which began to emerge after World War II. Explosives technologists in allied countries were exposed to the results of German and Japanese technology via the outcomes of technical debriefing of former explosives technologists and through reading captured technical documents. The establishment of the North Atlantic Treaty Organization (NATO) emphasized the need for a greater unification of approach to a range of military matters by the member nations, which would have to include the exchange of certain basic technologies and experience connected with the storage, transportation, and use of high explosives. During the Korean War in the early 1950s there was little history of exchange of hazard testing information, because allied armed forces tended to operate as individual national units, linked by a joint command. Where foreign equipment was used it was likely to be of United States or United Kingdom origin, and therefore, acknowledged to have had adequate hazard assessment carried out on its energetic materials.

The stimulus for increased awareness of the importance of the understanding of hazard assessment started with the United States nuclear weapon programme and, it is assumed, those of the United Kingdom, the Soviet Union, France, and China. United Kingdom scientists were closely involved with the wartime work on the atom bomb and after the war continued close association with their United States colleagues. It was soon apparent that in order to gain maximum benefit from hazard work in the other country, some unification of testing regimes needed to be carried out. United States and United Kingdom nuclear weapon explosives hazard studies were given greater emphasis by the accidental conventional explosions associated with crashes of United Stated aircraft at Palomares in Spain and Thule in Greenland while carrying nuclear weapons. Wider international collaborative hazard assessment has developed more recently in the conventional weapon field, again mainly as a result of accidents, particularly those aboard United States aircraft carriers in the Gulf of Tonkin during the Vietnam War. In these accidents, fire, and explosions consequent on "Cook-off," i.e., the unplanned thermal initiation of explosive fillings in warheads and rocket motors, emphasized to armed forces around the world the level of vulnerability of weapons platforms to thermomechanical attack of the weapons they carry. While the major impetus for renewed interest in hazard assessment came from the navies of the world, problems associated with the response to thermomechanical stimuli of weapons on other platforms, e.g., tanks, as well as storage and transportation hazards in general, gradually drew in the other services.

The highly publicized and extremely damaging accidents on board the United States aircraft carriers highlighted the need to assess the hazards to explosives and weapons from stimuli more closely related to those likely to initiate a damaging event in the field than had been used as the bases for earlier tests. Work on lower vulnerability, sometimes referred to as "insensitive," munitions is still providing the stimuli for increased efforts to understand fundamental mechanisms governing the initiation of explosives. As a result of the leadership given by the US Navy in this field and the willingness of the United Stated and other nations to exchange testing data and details of test methods, the general level of the understanding of energetic materials hazards is increasing, as is the movement toward standardization of testing methods and goals to be achieved by future low vulnerability energetic materials.

Major constraints on the development of hazard assessment and the acquisition of data on current and new explosives and weapons are:

- the costs of testing full-size weapons; and
- the environmental impact of such testing, especially in densely populated countries.

There are two possible approaches which might alleviate these constraints and become more important in the immediate future. They are:

- increased use of numerical modeling; and
- the development of realistic miniature versions of tests, which will not only give results which can be related to those expected from full-scale tests, but which can also be conducted inside test cells.

Both techniques require increased understanding of:

- the mechanisms involved in the initiation of a given material in a given container by a given stimulus or stimuli;
- the thermal and thermomechanical properties of the energetic and non-energetic materials used in the weapons at relevant strain rates; and
- the thermal and pressure dependencies of the chemical reactions involved in the decomposition of the energetic materials.

8.2. Basic Precepts of Hazard Testing

8.2.1. Absolute and Relative Sensitivities

It is of the greatest importance to be aware from the outset that there is no such thing as an absolute measure of the hazard from any given explosive. All hazard testing to date results in comparative data. Thus, explosive A may be more or less sensitive than explosive B to a carefully defined set of test criteria. If explosives A and B are subjected to another set of test criteria they may not bear the same relative sensitivities to each other. No two test equipments, operating to ostensibly the same test regimes on the same materials, give exactly the same hazard data, for two reasons. Most tests involve subtleties we still do not fully comprehend and the nature of both accidental and simulated accidental explosive initiation is statistical. Confidence in the relationship between the results from a given test carried out at one site and the same test carried out at another can be gained only from a sufficient number of tests and a statistical analysis of their outcomes. If a test is expensive to carry out, there is a reluctance to conduct sufficient replicates to enable a statistical analysis to be made. It is under these circumstances that the "experience" and "judgment" of the scientists and engineers responsible for such tests are invoked.

Drop weight impact testing of small powdered explosive samples is carried out in most countries often at several different locations, in armed service and civilian establishments. In each case the general aims of the testing can be said to be the same; to determine the level of test stimulus which will cause a 50%, or some other percentage, probability that the explosive will ignite, burn, explode, etc. A range of essentially similar equipments is used in different countries and also within some countries to give ostensibly the same basic stimulus to the powdered explosive. They differ however in the detailed designs of the impacting weight, the quantity of the explosive tested and the criteria upon which the sample is adjudged to have exploded, etc. Consequently, different relative sensitivenesses of different explosives are determined dependent upon which machine is used. This is not unexpected, since each machine subjects the sample to slightly different stimuli. However, it has meant in the past that slightly different perceptions were gained by different sites of the relative powder impact sensitivenesses of the main explosives in common use. This is not a problem of itself, since it did not juxtapose, say, lead azide and TNT, when tested by one machine relative to another, but there were found to be slight differences, especially in the relative sensitivenesses of RDX, HMX, PETN, and tetryl. A similar and somewhat more obscure problem occurred in the United Kingdom in the 1970s. Three different establishments each operated seemingly identical "Rotter" powder impact testing machines. As a result of a carefully controlled comparative test, using the same samples of explosive, some differences in relative sensitivenesses of several different explosives were recorded. The differences in output data were attributed to minute differences in materials properties associated with the machines themselves and the mechanical responses of the concrete plinths upon which the machines were mounted and the subsoils beneath them.

These anecdotes refer specifically to powder impact testing of explosives, but are typical of the problems which began to emerge in the 1960s and 1970s when improved communications both within countries and between countries began to open up testing regimes to more critical scrutiny. The fact that there are differences between the data recorded at different sites does not render some data invalid compared with others, since, manifestly accurate recording of observed phenomena has been made at all locations. Differences emerge from differences in equipment and sample preparation and size. Provided results from a given equipment are reproducible and self-consistent, its users can be assured that they have a valid data set pertinent to the specific conditions operating in their test regime. What the opening up of test regimes to scrutiny by other laboratories has shown is that, while there is general agreement, with only minor variations from site to site and machine to machine, a few techniques and machines produce data which are significantly different from others. The users of such equipments had the option to redesign their equipment, or to acknowledge that they were not measuring exactly the same data as other testing centers. Moreover, the other test centers knew where differences lay in each others' records. It is likely that there will be less scope in future for such individualistic generation of hazard data, since the recent reduction in defence budgets across the world and the increased emphasis on multinational projects require much greater standardization of testing regimes and a greater need to ensure that the results from testing are intelligible to all participants in a programme.

8.2.2. Differences Between Hazard and Reliability Testing

Accidents are unforseen contingencies. They may involve the untimely initiation of the fuzing train of an explosive warhead or rocket motor because, as in the case of an abandoned land mine, the "wrong" person or animal activated it. However, the largest proportion of accidents involving explosives occur in such a manner that initiation is other than through the normal explosive train. Consequently, weapons behave very differently in accidents compared with when they are initiated through their explosive trains. Explosive trains are designed to deliver a precisely calculated stimulus to the motor or the warhead, as a result of which the propellant or explosive ignites and burns or detonates in a predictable fashion. The stimulus is designed to be sufficiently powerful to ensure reliable initiation, without imparting risk to the weapon during normal storage and handling when the fuze is unarmed. Part of the weapon development programme will have addressed the statistical probabilities of correct functioning under all conceivable service conditions and near 100% reliability of response is expected within the normal operating envelope of the weapon.

Reliability testing is not part of this chapter, but it is important to understand how it differs from hazard testing. In reliability testing we seek to identify the edge of the envelope of conditions under which a device will function reliably. In the main, the devices of interest in reliability testing are the detonators or igniters of warheads or rocket motors, since they control output and are more susceptible to deterioration from the effects of ageing, rough usage, or adverse climatic conditions than the main charge explosives or propellant grains. Although the functioning and output of the device depend on the properties of the energetic materials in it, the system characteristics of the device are also important. Weapon reliability depends upon the correct functioning of all its components over a range of prescribed conditions. In order to generate confidence that the weapon will function over this range of conditions, samples of initiation devices are subjected to thermal, mechanical, and environmental stressing deemed to represent the accumulated worst conditions likely to be encountered during the life of the system. The devices are then tested at a range of temperatures and the operating envelope is confirmed if all tests involve sensibly 100% correct functioning.

The significance of the noncompliance of those devices which fail to function is somewhat difficult to assess, since it may only be of the order of less than 1% of the total. Probit Analysis (Finney, 1971) is used to assess the significance of the failure level, but is outside the scope of this chapter.

In an accident, the stimuli to which the warhead or motor or bare explosive is subjected are unpredictable by their very nature and they are not usually delivered via the explosive train. The stimulus may be marginal and result in an explosive burn or deflagration, rather than a detonation. The statistical nature of initiation may disguise the hazard for an appreciable period of time when a system is being used, apparently safely, in service. An example of such a possibility is the in-service premature initiation of high explosive filled shell in gun barrels. An event might occur at a frequency of one in several hundreds, or even thousands and yet a responsible user would not merely wish, but require the system designers to eliminate the problem in order to maintain the confidence of the user in his weapon. The first step toward understanding the mechanism of an accidental event lies in the ability to generate it at will. It is not sound economics merely to continue to fire the shell under precautions until another event occurs and compare the history and characteristics of the shell involved. It is simpler, cheaper, and technically more sound to design a test in which the probability that an event will occur is increased sufficiently for initiation to be made to occur at will. The design of the experiments by which reliable initiation can occur is usually relatively simple, since there is usually one component of the stimulus which can be increased significantly relatively easily. For example, an event initiated when a weapon was dropped on to a protruding stanchion on a ship led to the Spigot Test. Events are generated in representative portions of warhead casing and explosive by dropping them (or the spigot) from significantly greater altitudes than was the case in the actual accident. Even so, it is not possible to carry out a statistically meaningful sample of tests involving actual weapons, because of costs. More importantly, because this is a simulation, the result of any individual test cannot be regarded as an absolute indication of the probability of another event occurring. All the tests can do is to show the relative importance of the different potential variables involved in the original accident. Referring back to our spigot intrusion accident and subsequent test development, all that we can expect is to be able to compare several explosives one with another when all are in identical casings, or the influence of casing type and, say, thickness, on the response of a given explosive.

Reliability testing requires a method which assesses the probability that the system will function perfectly, i.e., it seeks to identify criteria which would reduce initiation probability to less than 100%. On the other hand, hazard testing seeks to determine where the borderline between "safe" and "unsafe" conditions may be set. The natural borderline between an event and no event, or the 50% initiation point is the preferred one to seek. A statistical approach is required because accidental initiation is a statistical phenomenon.

8.2.3. Analysis of Test Results

Explosives either react, or they do not. This characteristic is very convenient as a basis for the statistical treatment of the response of a material to an initiation stimulus. In hazard testing we are interested in determining the stimulus likely to have a 50% probability of initiating the explosive, or explosive system, because it is convenient and easy to handle statistically. It is not directly relevant to predicting the stimulus required to give 100% probability of either initiation, or noninitiation. This is determined from reliability testing. Hazard testing is based upon an approximation to a true initiation stimulus so only relative data are generated.

8.2.3.1. The Bruceton Staircase Technique

The most widely used means to calculate the median and other data from "go–no go" explosives testing was developed at the Explosive Research Laboratory, Bruceton, PA, USA, during the years 1941 to 1945 (Mallory: OSRD Report 4040). The procedure can be applied to a wide range of test procedures which rely on assessing the response of an explosive system to the effects of a single test variable, e.g., the height from which a weight is dropped on to the sample, or the thickness of a barrier between a standard detonating donor and identical samples of a receptor explosive. The test variable is adjusted from test to test so as to straddle the level at which 50% of test specimens explode. After an event is recorded, the next test is carried out so that the level of stimulus received by the test sample is slightly reduced, e.g., the weight is dropped from a slightly lower height or the barrier is increased slightly in thickness and the test is repeated. After a non-ignition, the procedure is reversed. During test development Dixon and Wood (1948) found that certain experimental and system requirements had to be satisfied. These are:

- the test variable, or some function of it, should be normally distributed; the logarithm of the test variable is often used;

- the step between test variable values should be predetermined and based on the function of the test variable used, e.g., the logarithm;
- each test should be carried out on a new sample to eliminate the "explosive memory effect," where an explosive which apparently shows no sign of initiation may have decomposed sufficiently to have generated autocatalytic species, which assist in the initiation at a lower level of stimulus the next time it is applied;
- a statistically significant number of tests needs to be conducted; and
- a criterion for the judgment of what constitutes a "go" or a "no go" must be applied which is consistent with the requirements of the test, e.g., in an impact initiation test it should be an event which consumes, or at least changes, the larger proportion of the sample; in a gap test, the criterion should be that the acceptor detonates, and evidence in the form of an indentation on a steel witness block is usually taken as proof.

If N tests were carried out, resulting in N_x events and N_0 non-events, of whichever type, at a number of different test variables which form a predetermined set, the median value of the test variable can be calculated as follows. It is assumed that a number of preliminary tests have been carried out to determine the approximate location of the median value of the test variable. The first point counted in the total N tests is the one immediately before the first change in result, i.e., from all "goes" to a "no go," or vice versa. If $N_x \leq N_0$ the calculation should employ N_x, if $N_0 < N_x$, N_0 should be used. The median value of the test variable is given by

$$H = c + d\left(\frac{2A - 1}{2N_\alpha}\right), \tag{8.1}$$

where α represents either x or 0,

$$A = \sum \frac{(h_i - c)n_i}{d}, \tag{8.2}$$

and h_i is the height at which an ignition takes place, n_i is the number of ignitions, c the lowest drop height used for the explosive under test, and d is the test variable interval, or the function of it, such as the logarithm, already used to determine the set of intervals in the range of test variables. The standard deviation of the results is determined by finding M from

$$M = \frac{B}{N_\alpha} - \left(\frac{A}{N_\alpha}\right)^2, \tag{8.3}$$

where

$$B = \sum \frac{(h_i - c)^2 n_i}{d^2}, \tag{8.4}$$

and then using M to find s from tables. The Standard Deviation, σ, is given by

$$\sigma = ds. \tag{8.5}$$

The value of s depends on the position of the median relative to the actual values of the test variable. It is especially important when $M <$ ca. 0.35. Curves for finding s from M when M is small, according to whether the median corresponds with a testing point ($m - h = 0$), or is at 0.1, 0.25, or 0.5 of an increment from it, and a table of interpolation are included in the US Navy reports cited, which also show data for each M up to about 0.7. At $M > 0.7$ linear interpolation is sufficiently accurate.

This technique is simple to use and is easy to program. It is relevant to a range of "go–no go" tests related to explosives and has been found to be of adequate reliability for as few as 20 trials. However, if less than about 20 tests are carried out, more exact methods of determining Standard Deviations are required (Mallory, 1945).

8.2.3.2. Method of Minimum Contradictoriness

Where very few trials are possible, owing to the high cost of individual tests, or restricted availability of samples, this method, devised by Professor E.A. Milne (Marshall, 1983) may be used. The method was developed during World War II by Milne, working for the Ordnance Board, to assess the performance of armor plate in resisting penetration by projectiles. He sought the best estimate of the striking velocity of a projectile which would result in it piercing a given plate in 50% of the trials, when only a limited number of trials had been carried out at several, unrelated, velocities and some had led to perforation and others not. This is clearly also applicable to any testing regime resulting in a "go–no go" outcome, which is solely dependent upon a single test parameter, e.g., velocity of shot, drop height of weight, or thickness of barrier, etc.

As an example, say a test variable (TV) gives either a "go" (G), or a "no go" (N) in a series of odd tests over a range of TV as shown in Table 8.1.

It is not possible by inspection to identify where the 50% point might lie, but we can guess. Let us suppose it lies somewhere between 4.6 and 4.8 on the TV scale. All "goes" (G) which occur at higher values of TV and all "no

Table 8.1. Method of minimum contradictoriness.

Test variable (TV)	Results: "Go" (G), or "No go" (N)
5.0	4G
4.8	2G
4.6	2G:2N
4.4	2G:2N
4.2	1G:2N
4.0	1G:2N
3.8	1N
3.6	1N

goes" (N) which occur at lower values of TV than this assumed median point are deemed to be "contradictory." We only consider these contradictory results.

Let T be the parametrical value of the assumed 50% point, T_c is the same value for a contradictory result and, $F_T = \sum (T_c - T)^2$, where the summation is over all contradictory results. Milne assumed that T_0, the best estimate of the median point, was that value of T which made F_T a minimum. Nowadays F_T can be calculated very simply with an hand calculator, but during World War II, the effort required was more considerable and Milne sought to simplify the problem further, by averaging contradictory results. The method requires that we assume that T_0 lies in some particular interval on the TV scale, we have already chosen, say, 4.6–4.8, and then calculate $\sum T_c/n_c$ where n_c is the number of contradictory results arising from the assumption. The true value of T_0 has been found if the calculated mean value lies within the same interval as was assumed at the outset.

We guessed that T_0 lies between 4.6 and 4.8. This gives contradictions as follows:

$$(4.6 \times 2), (4.4 \times 2), (4.2 \times 1), (4.0 \times 1),$$

$$n_c = 6 \qquad \text{and} \qquad T_c/n_c = 4.36.$$

The value of T_c/n_c does not lie in the range in which we assumed the median would lie, therefore it is incorrect. If we try again and assume that the median lies between 4.2 and 4.4 the contradictions are as follows:

$$(4.6 \times 2), (4.4 \times 2), (4.2 \times 1), (4.0 \times 1),$$

$$n_c = 6 \qquad \text{and} \qquad T_c/n_c = 4.36$$

coincidentally the same as before. Now the calculated value of T_c/n_c lies in the same range as it was assumed to do so and this is the median value.

8.2.3.3. The Role of Judgment in Hazard Testing

Unfortunately, as a result of the cost and environmental impact of hazard testing, fewer experiments can be afforded and more judgment-based interpretation is required of the results conducted. When judgment is required to be exercised, it need not be exercised in an unscientific manner and there are useful procedures which provide frameworks for the prediction of the behavior of systems, even when it has not been possible to test them specifically. A useful checklist of activities and research into the archives can often provide as much information pertinent to a test which has not been carried out as one which has been carried out with insufficient instrumentation, or which suffers partial instrumental malfunction.

Activities involve mainly the assembly of data from the archived records of other tests. Data should be assembled from:

- all tests already carried out using the technique, whether or not they involved the same energetic material and/or case materials;

- all tests of a similar nature, e.g., bullet impact tests and spigot intrusion tests are linked relatively closely together, but less closely with Susan tests; oblique impact and friction tests have a somewhat tenuous relationship;
- basic powder tests and shock sensitivity (gap tests);
- thermal analysis, e.g., differential scanning calorimetry (DSC), differential thermal analysis (DTA), thermogravimetric analysis (TGA), of the energetic material will yield data on its decomposition energetics; and
- characteristics of the casing material, if any, e.g., thin aluminum melts in a fuel fire, releasing confinement relatively early, but steel does not.

These data need to be readily accessible. An electronic database is preferable, but it requires an enormous commitment of resources to be established, especially since an important component should be photographic evidence from trials. Such a database is an essential in an environment where staff responsible for hazard assessment are changed frequently. Technical staff management systems which require movement of personnel regularly,"for career purposes," are not necessarily consistent with the highest levels of safety. This is an area of research where expertise and experience can be of value to an organization far beyond the ability of the staff to carry out and interpret tests correctly. As experience mounts, the effectiveness of a true hazard assessment expert increases rapidly and the right person in such a role, for which a career track can be generated within this discipline, is an investment in safety for the whole organization which will bring rich rewards. In the past, experts in the field have had at their fingertips, or in their memories, the checklist shown above and, those items they had forgotten could easily be retrieved from a paper database. Such experts are becoming rarer and so are the data on experiments carried out even only 25 years ago.

Unfortunately, this holistic approach to hazard assessment is not capable of undergoing audit and objective assessment, but it can bring important benefits, provided the data are available and can be assembled. The NIMIC in Brussels could provide an invaluable international database to its users, since the results of hazard assessment tests should be non-project-specific and, therefore, noncommercial and not subject to company and national censorship.

8.3. Sensitiveness, Sensitivity, and Explosiveness

In Section 8.2, we discussed the differences between reliability testing, where we seek to ensure 100% reliability of a system, and hazard testing where we seek to compare data on the response of one explosive, or one case structure, with those of another at the 50% point. Conventionally, in the United Kingdom a distinction is drawn between these two characteristics of explosives and/or systems, by referring to the comparison in the likelihood of an accidental initiation occurring in a given explosive or system as governed by its *Sensitiveness* and the reliability which we are able to ascribe to an explosive, or system as its *Sensitivity*, rather than refer to both by the same term, "sensitivity." The value perceived in the use of this distinguishing nomenclature is in the manner in which it forces the user to be clear what characteristic is under discussion and, therefore, what testing philosophy is required.

A simple way to remember which term refers to which phenomenon, is to associate *Sensitiveness*, which ends with the letter "s," with *s*afety, and *Sensitivity*, which ends in the sound "i," with *i*nitiation reliability. In this way, hazard tests are *Sensitiveness Tests* and reliability tests are *Sensitivity Tests*. Examples of sensitiveness tests are powder impact, spark sensitiveness, oblique impact, spigot intrusion, etc. Sensitivity tests are exemplified by low-amplitude shock initiation, wedge tests, and gap tests.

A type of test which will not be referred to in detail in this chapter, is the performance test, in which we seek to assess the output of an explosive or device. These tests are essential in assessing whether explosives or propellants are capable of performing the roles required of them in weapons. They are also the bases for determining whether a weapon performs to specification, i.e., if a shaped-charge jet defeats a certain armor thickness, or a fragmenting warhead produces the correct pattern of fragments at the required velocities. Such tests as applied to materials are the cylinder expansion test, the strand burning test, and the closed vessel test. Arena trials for blast and fragment warheads, motor thrust trials, and armor defeat firings are the equivalent weapon tests.

Sensitiveness and sensitivity tests provide information on the response of materials to stimuli. The materials may be cased, but the casing, as part of the system, is included in the assessment of whether the material will either survive under conditions related to those involved in the sensitiveness test, e.g., proof firing of artillery shell, or will perform reliably when subjected to a particular stimulus, e.g., a detonation take-over, or a shutter sealing test, where, respectively, either 100% or 0% reliability is sought. In hazard tests we are interested only in the threshold for initiation. The response thereafter is not important. In sensitivity testing we require that the system operate as designed, i.e., an explosive should detonate and a propellant should burn reliably as a result of a given stimulus.

With the increased emphasis on the need to assess the behavior of weapons in order to reduce the effects of accidental initiations, it became obvious that the response of the whole system was important, not just that of the explosive filling. The significance of confinement in the response of explosives to unplanned stimuli began to be assessed in a controlled fashion in the 1960s in the West, accelerated by the testing and research associated with the premature initiation of explosives in shell in gun barrels. A rash of accidents had occurred world-wide as a result of the need to improve range and payload. Guns were being designed with increased chamber sizes and more energetic propellants, which resulted in shell being subjected to higher linear and angular accelerations and higher forces, such as driving band pinch. Prior to this time, when an occasional explosion occurred in the gun barrel, it was usually attributed to a fuze malfunction and more money was spent on improving fuze reliability. However, several events were observed in shell fired at proof without fuzes, and this led to the study of the behavior of explosives initiated by nondetonative stimuli and their subsequent burning under confinement.

In most cases, the solution to the problem lay in improved quality control during manufacture, but, as a spin-off it became possible to assess where the ignition of the explosive occurred from the pattern and extent of damage caused to the shell casing and gun barrel. The fact that a given explosive ignited in a given case can produce different damage patterns depending upon where initiation occurs, gave rise to the use of the term *Explosiveness* in conventional weapon studies. It was already part of nuclear weapons explosives technology. It refers to the *system response* to initiation of an energetic material contained within it. This concept is important in lower vulnerability, or "insensitive" munitions technology, since responses of weapons systems to stimuli can be controlled more easily than new low-vulnerability explosives can be discovered.

Insensitive munitions work, following the United States aircraft carrier accidents in the Gulf of Tonkin in the late 1960s and early 1970s, began with a search for improved explosives with "lower hazard than TNT." Several were identified, most notable of which was TATB, but, unfortunately, these materials either were of insufficient performance, or had other characteristics which rendered them unsuitable for use in high-volume production, low-cost weapon systems. At the same time the military explosives industry was on the verge of entering the age of the plastic bonded explosive (PBX) and it was found that even powerful PBXs responded less severely than traditional TNT-based explosives such as RDX/TNT (60/40), Composition B, or HMX/TNT (65/35) or (75/25), known generically as Octols, to similar stimuli in similar casings. A substantial research programme revealed differences between TNT-based explosives and PBXs and shown in Table 8.2.

The key to the differences in behavior between these explosives lie in differences in their responses to nondetonative stimuli under confinement. TNT-

Table 8.2. Mechanical and explosive differences between TNT-based explosives and PBXs.

Characteristic	TNT-based explosives	PBXs
Mechanical strength	Brittle, relatively weak	Relatively tough and strong
Closed vessel burning	Rapid	Relatively slow
Powder sensitiveness	Relatively insensitive	More sensitive
Charge hazard (explosiveness)	Relatively high	Relatively low
Shock sensitivity	Relatively high	Lower

based explosives are relatively brittle and weak. When subjected to mechanical damage, as a result of an impact, friction, or intrusion, the pressure pulse associated with the damage mechanism is often sufficient to crack the explosive ahead of any burning front initiated by the impact. This provides an increase in surface area for burning which accelerates the burning process in the manner experienced with damaged rocket propellants (Boggs and Derr, 1990). Thus, even under moderate confinement, TNT-based explosives burn rapidly and result in significant case damage and hazard from flying fragments. High explosive shell and bombs are heavily confined and burning velocities of TNT (700–800 m/s) and RDX/TNT (ca. 1600–1800 m/s) have been recorded under less confinement (thickness of steel casing). By comparison, even the relatively unsophisticated PBXs of the 1960s were tougher and stronger than the TNT-based explosives. Although the initial mechanical damage from impact, intrusion, or friction may initiate deflagration in them as with the TNT-based explosives, the deflagration does not propagate as readily and sometimes not at all. The reasons for this are greater toughness, i.e., resistance to cracking, of the PBX under the effects of the pressure pulse and the contribution of thermal insulation due to charring of the polymeric matrix. Some polymers decompose endothermically, which extracts energy from the burning front during the period it is accelerating after initiation.

The differences between PBXs of that era and TNT-based explosives lay in the differences in their responses to the initiating stimulus, which seems, at the outset, to be purely a sensitiveness difference between the explosives. However, the observer does not see these processes and could not analyze them if they were visible. The perceived differences between events initiated in TNT-based explosives and PBXs under similar confinement by a similar mechanism, e.g., a propellant igniter, are observed in the fragmentation patterns of the cases. TNT-based explosives cause many more fragments to be generated and for their velocities to be generally higher than those generated by the PBXs and all the TNT-based explosive is usually consumed, compared with only a proportion in the case of the PBXs. The differences in response depend strongly on the differences in interaction between the burning explosives and the casing. The sensitiveness of the explosive to ignition is unimportant since differences are observed in tests where all charges are ignited by a small propellant charge. Where a rapid acceleration of deflagration is possible because of high confinement, the casing remains in situ as the burning front passes it by. It retains confinement, partly because of its inertia, even after fragmenting, usually in a brittle mode and partly because the higher strain rate increases the strength of the material before it fails. Thus, the burning front can accelerate to high velocity, driven both by thermal and pressure effects on the kinetics of chemical decomposition of the explosive. If the processes associated with the acceleration of the deflagration after ignition are relatively slow, i.e., there is relatively low confinement, it is possible for plastic strains in the casing to cause it to fail at a given point as the process is passing it. This immediately reduces confinement and reduces the rate of the decomposition process. In the extreme, it may actually extinguish it. As a corollary, if the case is relatively thin, or weak, even deflagration of TNT-based explosives can be extinguished, since the casing can fail very rapidly and so eliminate confinement at a speed similar to the initiation of the burning process. A thin case is an advantage here because of its reduced inertia compared with a thicker one.

Once it became clear that the response of the system to a stimulus could be controlled to a large extent by the materials and design of the case, an alternative route to "insensitive" munitions other than through the synthesis of new explosive molecules with lower intrinsic vulnerabilities became available. The search continues for more energetic explosives, but it may be that the route to more powerful, lower vulnerability weapons lies in improved understanding of the control of the explosiveness of the systems. This approach would allow the deployment of even higher energy explosive materials. Already technology exists to reduce confinement of rocket motors engulfed in fire by means of an explosive line cutting charge known as a Thermally Initiated (Case) Venting System (TIVS). There is no reason why this and related technology should not be applied to protect warheads in fires and to manage the response of both motors and warheads to slow increases of temperature over many hours as required in slow cook-off or fragment attack tests. It is likely that the search for improvements in hazard control of systems could lead to the pre-eminence of mechanical engineering, rather than production chemistry in the search for safer weapons for the armed forces of the world.

Explosiveness of an explosive system has been said to be a function of the burning velocity that the energetic material can attain in it. While this is somewhat superficial, it does emphasize the most important feature of the response of a system to an accidental initiation. The greater the burning velocity the system can sustain as a combination of the explosive and casing properties, the greater will be the perceived response.

8.4. Stages in Hazard Assessment

Hazard assessment tests depend upon the level of understanding of the system or the material under test. In the early assessment of the hazards associated with a new explosive molecule, only relatively small quantities if the material will have been made and, consequently, only powder sensitiveness tests can be carried out. These are sufficient to identify and specify hazards, compared with other explosives. In many countries the results of such tests indicate whether further work can be carried out on the material without special precautions and whether it can be moved from one experimental site to another. The results of such powder tests are also used to assess the hazards in scale-up of manufacturing processes. Once sufficient quantities of the material have been made to experiment with compositions involving it, powder sensitiveness testing of them is required. This gives confidence that it would be possible to scale-up the manufacturing of the composition sufficiently to test fill in into cases and conduct performance testing on it. Should bare charges be required, certain charge tests are required to assess hazards from mishandling them. Finally, once a new composition has been deemed to be adequate for a given requirement, further charge hazard tests are required with the composition filled into the casing of the weapon system for which it is destined. In the current climate, these tests would almost certainly constitute insensitive munitions tests.

In summary, the sequence of events in the hazard and reliability testing of a new energetic material is as follows:

- powder sensitiveness tests to determine whether special precautions in handling are required, e.g., conducting floors and bench tops, controlled humidity, etc.;
- powder sensitiveness tests on the new molecule to permit scale-up manufacture and the movement of the material from one laboratory to another;
- powder sensitiveness testing of any new composition involving a new or current explosive molecule to permit manufacture of charges;
- bare charge sensitiveness testing to ensure that they can be handled without undue hazard, or that hazards are identified if they exist;
- shock sensitivity testing of bare samples, especially of explosives (and of propellants in the United States and other countries where shock sensitivity plays a part in the determination of the United Nations Classification); and
- weapon, or weapon simulant testing of the composition in a representation of the final warhead or motor environment in which it will be used, or in the actual hardware, if available, carried out to so-called Insensitive Munitions guidelines, as described in US Navy 2105B (1994) or a national or international equivalent set of tests.

8.4.1. Powder Tests

There are at least five major types of powder sensitiveness test:

- impact;
- friction;
- flame;
- flash; and
- electrical spark.

In the United Kingdom it is necessary to obtain data on each of these tests in order to enable a Safety Certificate to be issued, which permits the free transport of the explosive in question, under appropriate precautions between sites via the public highways. Data from these tests are required prior to scale-up in production, the filling of munitions, or the manufacture of bare charges. In some cases, particularly when the energetic material is especially critical, e.g., black powder, it may be necessary to conduct tests on each individual production batch as a preliminary check on suitability for the particular role required of it. There are no fixed responses which prevent production of a given material. Hazard data enable potential hazards to be identified and proper precautions taken. Even very sensitive materials can be transported and handled as long as they are acknowledged as such and so long as only relatively small amounts are involved. Thus, it is perfectly possible to transport primary explosives, such as lead azide and lead styphnate on roads in the United Kingdom, provided they are packed into individual 25 g tare containers, separated from each other by blast absorbing packaging, and no more than an approved number of these containers is carried at one time in a stout steel tank, which is capable of withstanding the effects of the explosion of one of the containers.

8.4.2. Impact Tests

The principle of operation of these tests is to simulate the effects on the sample of a quantity of the explosive being nipped and crushed between two metal surfaces. This is an entirely plausible initiation scenario, even in the case of a monolithic charge of the explosive, since most types of mechanical impact damage the explosive as well as the casing. Moreover, during rough usage, particularly if it involves vibration, the charge may rub against the charge container or some of the internal weapon furniture and fret, producing fine powder, which may, later, find itself in screw threads elsewhere in the warhead, or motor, when the weapon receives a sharp mechanical blow.

Powder impact testing is done routinely, using machines in which a steel mass of a few kilograms is caused to impact, either directly, or indirectly via a mechanical linkage, a small powdered sample of the explosive under test. A sample of a few milligrams of the explosive is subjected to impact in each individual experiment, but many, up to 200, replicate tests may be carried out to provide adequate statistical analysis of the response of the material to these stimuli. The apparatus and procedure vary from laboratory to laboratory. No two machines are precisely the same. Even those nominally the same can be different in subtle ways connected with the manner in which the large concrete, or steel block used as a support, is itself supported on foundations or the subsoil beneath the test building. Further differences reside in the manner in which the sample is prepared and held between two surfaces prior to impact. Some machine designs specify hardened steel dies for the purpose, others require that one or other surface is roughened and one even stipulates that sandpaper be placed between the sample and the dies. Different criteria also exist for the determination of whether or not an event has occurred.

Differences in terminology applied to different parts of the impact machines complicate interpretation of designs between different countries. The various parts are referred to in English as follows:

- Energy source — drop weight, drop hammer;
- Intermediate energy transfer link — plunger, striker, intermediate pin;
- Final energy transfer link — striker, striking pin, plunger, piston, drift;
- Sample confinement — cup, die cup, sample cup (or none);
- Test sample support — anvil, base; and
- Machine support — support block.

The mechanical links in the energy transfer system are referred to as "tools."

In principle, all machines carry out a similar test procedure in which the weight is dropped successively from different heights on to a different sample of the same explosive. In most cases this sample is powdered, but in some machines, especially in the United States, it was, at one time, customary to drop the weight on to a solid sample, the sizes, charge preparation, and geometry of which varied from machine to machine. A "go," or "no go" is recorded and the height increased if a "no go" occurred and decreased if a "go" occurred. A statistical analysis of the data along the lines of the Bruceton Staircase, or some other proven method, is used to determine either the 50% (median) point, or in some cases the 10% or 90% points, dependent upon the history of the laboratory. These levels of event are levels of probability that the explosive will react to give a positive result according to criteria pertinent to each machine type.

The are variations in the criteria used to judge whether or not an event has occurred. In many cases the judgment of the operator is relied upon to detect the smell from or sound of an event, or whether or not a flash occurred. This is highly subjective and in this modern age of high safety consciousness staff are not encouraged to sniff the products of combustion of explosives, not are they encouraged to be in a position to see the explosion of even a few milligrams of explosive. In many machines, attempts have been made to reduce the subjectivity of interpretation of the results of impact by using sound metering, or collecting the gases produced and measuring them. In the latter case a minimum volume of gas is associated with the explosion of the sample. If more than this quantity of gas is recorded an event is assumed to have taken place. This works well for most explosives, but is not appropriate for use with some pyrotechnics, which generate little or no gas on explosion.

8.4.3. Design Details of Some Powder Impact Test Machines

In principle, all impact testing machines are similar. An electromagnet is attached via a cable to a winding drum. A steel drop hammer of a given mass is raised to some predetermined height by means of the electromagnet

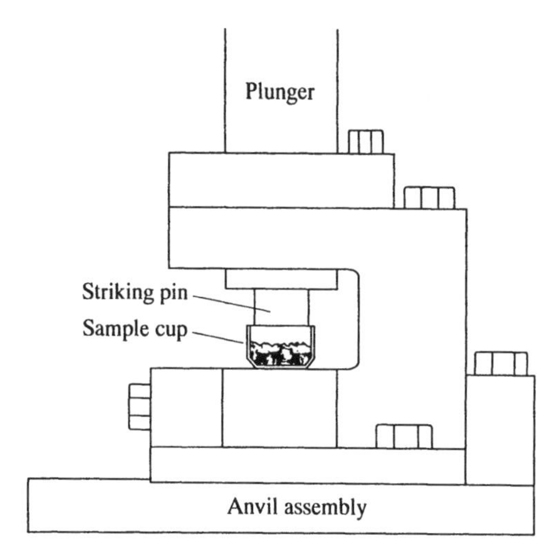

Figure 8.1. Detail of the impact assembly of the US Bureau of Mines machine.

and dropped by interruption of the current to the magnet by means of a solenoid switch. An event, or non-event is recorded and, usually, a Bruceton Staircase Method of determining the median impact height is employed. Important design characteristics of the equipments, which have a bearing on the results of the tests, are:

- explosive mass tested;
- state of granulation of the explosive;
- degree of confinement of the sample;
- mass of the impacting hammer;
- means of determining whether or not an event has occurred;
- conditions of the surfaces of the hammer and the anvil; and
- the supporting structure of the equipment.

The distinguishing characteristics of a few equipments are described and illustrated in the next section.

8.4.3.1. US Bureau of Mines, Pittsburgh (Walker, G.R., 1966)

The drop weight is 5 kg and can fall a maximum of 3.3 m. The weight falls, after release from an electromagnet via a solenoid switch, down a steel "T" beam frame on to a plunger which transmits the forces to a small striker pin that fits into a steel cup containing the sample. The striking pin is 12.7 × 19.1 mm long, of 440-C stainless steel, hardened to 55–60 on the Rockwell Scale and is ground and polished to give a freely sliding, but snug fit in the sample cup. The powdered sample is approximately 35 mg in mass. Explosion is judged by sound and sight; any charring, burning, smoke, flash, flame, or explosion is considered a "go."

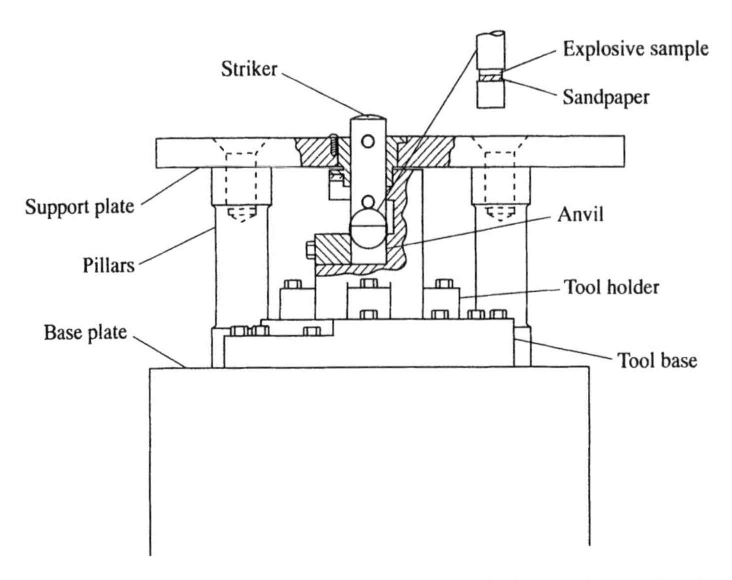

Figure 8.2. US NOL (Bruceton) impact machine; tool holder and assembly (Walker, G.R., 1966).

The important characteristics of this equipment are that the sample is quite well confined in the cup, and that it is probably expensive to maintain the striking pin in good condition. Dixon and Massey's (1951) and Brownlee's et al. (1953) methods are employed to give a 50% point, which is shown in Table 8.3 at the end of Section 8.4.3.

8.4.3.2. US Naval Ordnance Laboratory, White Oak, MD.: Laboratory Scale Test

This was the original "Bruceton" design (Mallory, 1945). Several weights are available, 2 kg, 2.5 kg, and 5 kg and the maximum drop height is 3.2 m. The 2.5 kg weight is usually used and is released through operation of a solenoid switch. The weight impacts a striker 31.75 mm in diameter and 88.9 mm long made of hardened tool steel (60–63 Rockwell "C" Scale), which slides freely in a guide. The striker surface is ground to a 41 $m \times 10^{-6}$ m finish. The explosive rests without restraint on a 25 mm square piece of 5/0 grade flint sandpaper, which itself is supported by the 31.75 mm diameter $\times$ 31.75 mm long anvil, also of tool steel hardened to 60 Rockwell "C." Samples tested are 35 mg $\pm$ 2 mg in mass and traditional TNT-based materials are prepared to a strictly controlled particle size range. Ground material is screened through US No. 16, 30, and 50 sieves and equal weights of material retained on Nos. 30 and 50 sieves are mixed for testing. However, other granular materials are tested "as received." Problems occur with PBXs, which necessitate them being cut up into small samples or mechanically ground.

A ceramic microphone is mounted in the horizontal plane of the anvil face 0.86 m from the center of the anvil, as part of a "noisemeter," which determines whether or not an event has occurred. The microphone is calibrated with known explosives, so that the difference between an event and a non-event is indicated by the illumination of a lamp in the operator's line of sight if an event occurs. A Bruceton Staircase technique is used and a 50% point is calculated. Results are shown in Table 8.3.

8.4.3.3. Los Alamos Laboratory Scale Impact Test and Navy Weapon Center, China Lake (Walker, G.R., 1966)

These tests and machines are very similar to each other and to the NOL test and machine referred to above. Data are said to be similar. Examples of results are recorded in Table 8.3.

8.4.3.4. Picatinny Arsenal Laboratory Scale Test (Walker, G.R., 1966)

This equipment is relatively simple compared with others. The steel anvil, which has a recessed opening to support the die cup, is supported on a steel sheet embedded in a large concrete block. The impact conditions between the striker and the sample in the sample cup are similar to those shown in Figure 8.1. Brass guides are used for the falling weight, which may be 0.5 kg, 1 kg, 1.5 kg, 2 kg, 2.5 kg, or 3 kg. The lighter masses are used for testing sensitive materials, such as initiatories. In this case, only the die cups are made of hardened tool steel (Colonial No. 6) and the cavity is formed by swaging. The cup is fitted with a vented plug, which is made from a high-carbon drill rod, which has been drawn and hardened. Only explosive material which passes through a US No. 50 sieve is used. Tests were carried out on at least 25 samples and about 30 are prepared to enable the approximate 10% point to be determined prior to the 25 tests. Later, the testing regime was changed to determine 50% points from 50 samples. The mass of each sample is recorded so that the average density of the explosive tested can be recorded. An "explosion" is recorded after audible or visual evidence of the event. Examples of results are recorded in Table 8.3.

8.4.3.5. Bureau of Mines Test (Walker, G.R., 1966)

In this test one of three weights may be used at 0.5 kg, 1.0 kg, or 2 kg over a 1 m maximum drop height. The impact block is fitted with an hardened steel anvil (Rockwell C-63 $\pm$ 2) with a mirror finish (5×10^{-6} m) supported on a steel block embedded in a concrete base. The plunger, which is equipped with a 10 mm diameter tip is also hardened to Rockwell C-63 $\pm$ 2 and has a similar mirror finish. No sample cup is used in this test, but impact conditions are similar to those shown in Figure 8.1. An electromagnetic release mechanism is employed to retain and drop the weight as with other tests. Granulated explosive used in this test must pass fully through a US No. 50

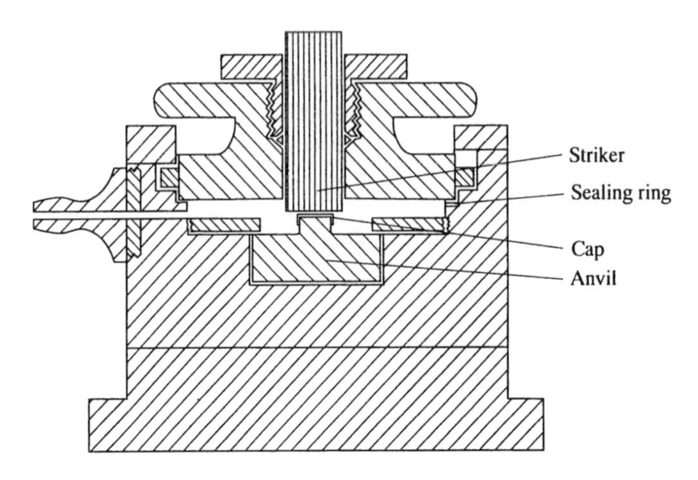

Figure 8.3. Explosion chamber, Rotter machine.

sieve and be retained on a US No. 100 sieve. Samples of about 20 mg are tested and the temperature of the test is adjusted to $25 \pm 2\,^{\circ}C$. An explosion is indicated by noise, smoke, or flame, and results from about 25 tests are analyzed by a Bruceton Staircase Technique. Examples of results are included in Table 8.3.

8.4.3.6. Rotter Test [SCC, No. 3, 1988: AOC DI(G) LOG 07-10]

The UK Rotter Test equipment uses the now-familiar technique of electromagnetic release of the weight via a solenoid switch. Drop weights may be of 2.5 kg, 5 kg, or 10 kg and may be dropped up to 3 m on to the sample. Drifts and anvils are made from hardened tool steel, heat treated to the "Y" condition and the 20–25 mg sample of explosive is tested in a brass cup. The Rotter machine is peculiar in that the cap containing the explosive is inverted over the anvil, as shown in Figure 8.3. The drift is 19 mm in diameter and the cup is 10 mm in diameter. There is relatively little control over the particle size, since so many materials are difficult to grind, or cut up.

The most important distinguishing feature of the Rotter Test is in the manner in which an event is detected. The entire impact assembly is sealed from the atmosphere with a rubber seal and the gases produced by the event, if any, are collected in a glass manometer (see Figure 8.4). If the volume of gas collected exceeds 1 ml, an event is assumed to have taken place. The only significant problem with this occurs if a so-called gasless pyrotechnic is being tested, in which case the test conditions are altered to note the appearance of smoke, flame, flash, or evidence of combustion when the spent cap is disposed of. The Bruceton Staircase Technique is used to give the 50% point. The result is called the "Figure of Insensitiveness" (F of I) in the United Kingdom and requires that 400 tests take place with alternately the sample and then "standard" RDX material impacted. When less formal

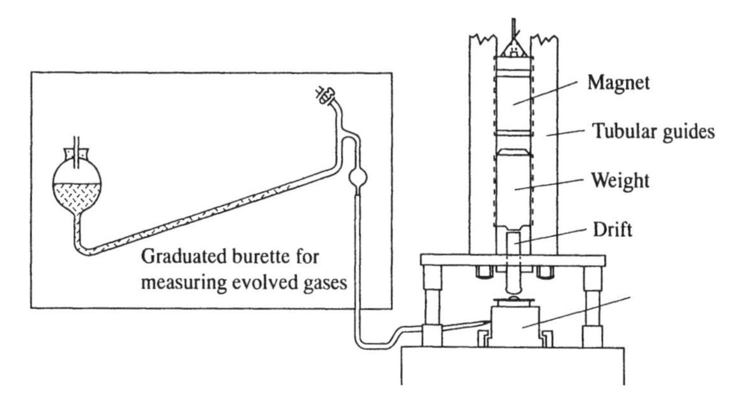

Figure 8.4. Rotter machine with gas evolution burette.

testing is required 50 interlaced samples are impacted to ensure that there are adequate data for a proper statistical analysis to be carried out. Interlacing of tests between the unknown sample and the "standard" RDX, for which the Figure of Insensitiveness is 80 is intended to minimize the effects of changes in the anvil during a test run, since the changes are experienced in the RDX tests as well. Whatever the test results for the RDX, its Figure of Insensitiveness is normalized to 80 and the Figure of Insensitiveness of the unknown adjusted pro-rata. Figure of Insensitiveness data are shown in Table 8.3 in parentheses.

8.4.3.7. Bureau of Explosives (NY) Laboratory Scale Tests (Walker, G.R., 1966)

In this test an 8 lb weight is dropped from predetermined heights on to the target sample up to a maximum height of 760 mm. An anvil is placed in the anvil housing with about 10 mg of a solid explosive. In many respects the procedure is the same as for other machines, although there is an increased confinement is this equipment in comparison with, say, the NOL, or Bruceton-type machine. However, results are essentially similar to those obtained from these machines and the Rotter machine. This tests is somewhat more expensive to carry out compared with others because of the complexity of the equipment.

8.4.3.8. German BAM Technique (Köhler and Meyer, 1993)

The German BAM Test and equipment are similar to those already discussed above. However, it is important to note some remarks made by Köhler and Meyer (1993). They reported that the BAM equipment had been modified to improve reproducibility of its results. This appears to have been achieved by increasing confinement in a different way compared with other impact test machines. This renders the results from the BAM machine dif-

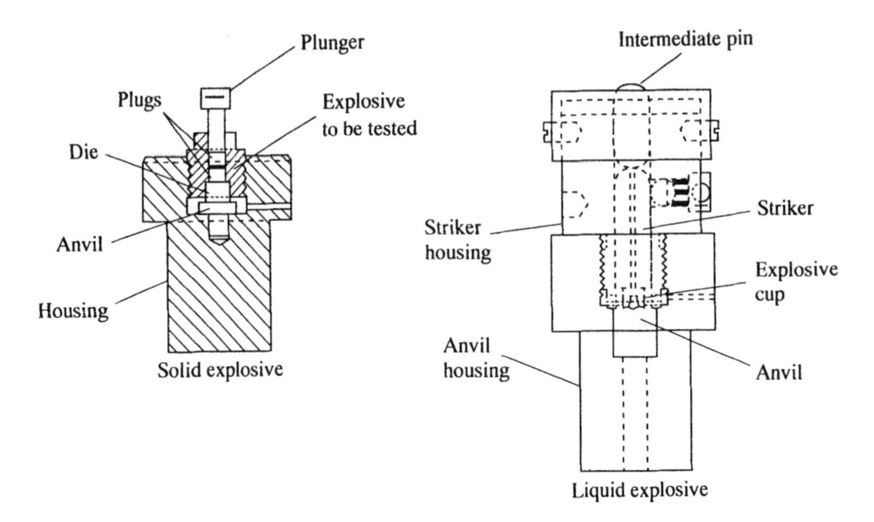

Figure 8.5. Bureau of explosives equipment.

ferent from all others reported (see Table 8.3), and suggests that this technique has some major differences from those elsewhere in the world. What these differences are cannot be determined from the simple drawings of equipment available, but they might repay an in-depth study, since it seems unlikely that one test could be so far out of step with the others without a major difference in conditions.

8.4.3.9. Test Results

Table 8.3 shows a set of test results from a range of powder impact machines. It is not exhaustive and it would have been improved by data from other national sources. It is not possible to convert all results to the same scale, hence there is no significance in sideways comparison of apparent levels of hazard. However, what can be compared from machine to machine is the order of relative sensitivenesses for each machine.

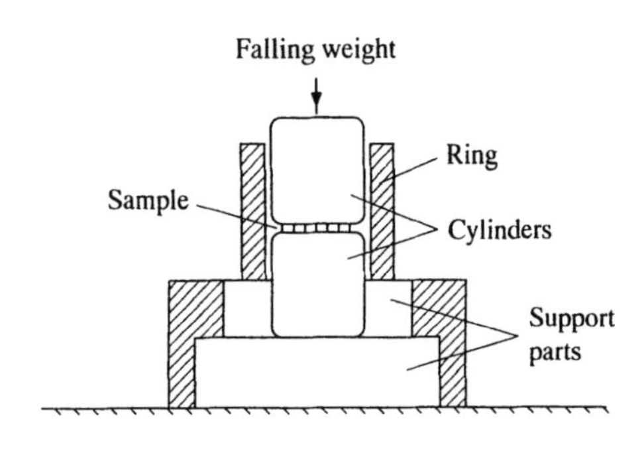

Figure 8.6. The BAM test impact site.

Table 8.3. Relative impact sensitivenesses of different explosives on a range of impact machines.

US Navy NOL, White Oak	Picatinny Arsenal, Dover, NJ	US Bureau of Mines	NY Bureau of Explosives	BAM Germany	Rotter UK (F of I)
Explosive D	Explosive D		TNT	TNT	TNT (160)
TNT	TNT	TNT	Explosive D		RDX/TNT (120)
Comp. B	Comp. B	Comp. B	Comp. B		Comp. B (110)
Tetryl	HMX	HMX/RDX	Tetryl		Tetryl (100)
RDX	RDX/Tetryl		RDX	RDX	RDX (80)
HMX		Tetryl			HMX (60)
PETN	PETN	PETN/PbAz	PETN	PbAz	PETN (50)
PbAz	PbAz		PbAz	PETN/Tetryl	PbAz (30)

RDX/TNT refers to United Kingdom material at 60/40 ratio by weight or w/w. This is not the same as US Composition B, although the latter also contains 60% RDX and 40% TNT (w/w). There are two reasons for this. The first is that UK RDX consists of only that molecule, whereas US RDX contains up to 12% HMX "contaminant." US Composition B is, thus, somewhat more powerful than UK RDX/TNT (60/40), by a slightly variable amount. The other difference is in the wax added to desensitize the explosive. UK RDX/TNT (60/40) contains 1% beeswax, whereas Composition B contains 1% of a synthetic wax. US RDX is made by the Bachmann Process and UK RDX by the Woolwich Method.

There seems to be general agreement between the numerical results (not quoted here, due to their different bases) from five of these six testing regimes that there are three groups of explosives, which can be distinguished by their general levels of hazard:

- Explosive D and TNT, Comp. B (RDX/TNT), Tetryl;
- RDX, HMX; and
- PETN and PbAz.

In the United Kingdom they are distinguished by the classifications, Comparatively Insensitive," "Sensitive," and "Very Sensitive." Only "Comparatively Insensitive" materials are permitted below the shutter in fuzes. Five of the tests identify these three groups, but they sometimes differ on the minor details of the order of hazard. The significant difference is in Köhler and Meyer's book (1993). The relative sensitivenesses of lead azide and both PETN and tetryl, if reproduced in the United Kingdom or the United States, would prevent the use of tetryl as a booster explosive. In fact, tetryl is no longer used as a booster explosive, but its hazard characteristics are used as a guideline in determining whether a given explosive might or might not be too sensitive to use as a booster. On the evidence of the BAM results, tetryl is more sensitive than lead azide, which means that it would not be possible to use it in the United Kingdom or United States beyond a shutter in a fuze. This is undoubtedly the result of the differences in testing conditions in the BAM test as compared with the others. The other differences in order amongst the rest of the explosives are relatively less significant and suggest that the conditions in the tests are reasonably similar.

The definitive book on the science of the responses of explosives to impact is by Afanasiev and Bobolev (1971). It does not confine itself to the effects of impact on powders, and extends well beyond the phenomenology of the tests referred to here. Afanasiev and Bobolev stress the need to understand the effects on an explosive of mechanical action, confinement, explosive burning characteristics, etc., in order to be able fully to predict initiation phenomena and materials responses and, by implication, reduce the cost of testing. Response characteristics including intensity, i.e., explosiveness, depend upon these properties, which are further influenced by the degree of confinement of a charge and the thermomechanical properties of any admixtures. A follow-up to this work would be very valuable, but pragmatics probably rule it out at present.

8.4.4. Powder Fricition Tests

It is undoubtedly the case that friction plays a large part in accidental initiations, but the types of surfaces between which an energetic material may find itself and the form of stressing can vary very widely. Hence, it is appropriate to consider the friction between a surface and powder on the one hand and a surface and a monolithic charge on the other. In this section we consider powder friction testing, which refers to the possible fretting of an explosive or propellant in a casing and the consequent risk of ignition as the powder is further chafed against the casing, or of the effects of a tool scraping against a loose powder during explosives processing. Friction testing is not regarded as of fundamental importance nowadays in the assessment of the hazard from explosives mainly because most of the tests were devised for materials which are more sensitive than those currently in use. Relatively few events are recorded and it is considered to be difficult to achieve a pure friction stimulus, without concomitant impact. The tests described below are, therefore, relatively infrequently used.

8.4.4.1. UK Mallet Friction Test (SCC No. 3, 1988)

In this test an operator swings a wooden mallet between his legs to deliver a glancing blow to a line of powdered explosive set on a target material. The wooden mallet can be used to impact wood or stone surfaces, or it may be fitted with either a steel or brass insert and impact aluminum bronze, brass, or steel targets. Ten "traces," or lines of explosive powder are struck once each by the operator, with another operator watching to determine whether an event occurred or not by the observation of flame or smoke or hearing an explosion.

There is no doubt that this is a crude test, but is was used in the United Kingdom for many years and is still used as a quick screening test, because results are reliable and reproducible. It also represents the sort of stimulus that a powder is likely to receive if it is handled without care during processing, weapon filling, cleaning equipment, or refurbishing ammunition or other weapons. Table 8.4 shows that the data from the mallet test are not sensibly different from those obtained using sophisticated friction machines in the United States and Germany. The reason for this is probably that all the tests merely produce a modified form of impact, between which the data recording and analysis procedures are incapable of distinguishing, because they are all relatively crude.

8.4.4.2. US Bureau of Mines Pendulum Friction Test (Walker, G.R., 1966)

The machine consists of a fixed mild steel anvil (Rockwell hardness B71) and a weighted pendulum, with a rigid arm fitted with a similar mild steel, or a hard fiber (ASTM Specification D-710-43T, colored red) shoe. The pendulum arm is adjusted so as to deliver a glancing blow to the anvil as it swings through its arc (Figure 8.7).

The anvil is a mild steel plate (Rockwell hardness B71) 83 mm × 305 mm with three parallel grooves machined in it. Each groove is 59 mm long by 3 mm deep and they are separated from each other by about 10 mm. Their purpose is to prevent the powdered explosive being swept away from the impact/friction site by the slipstream from the friction shoe. The pendulum

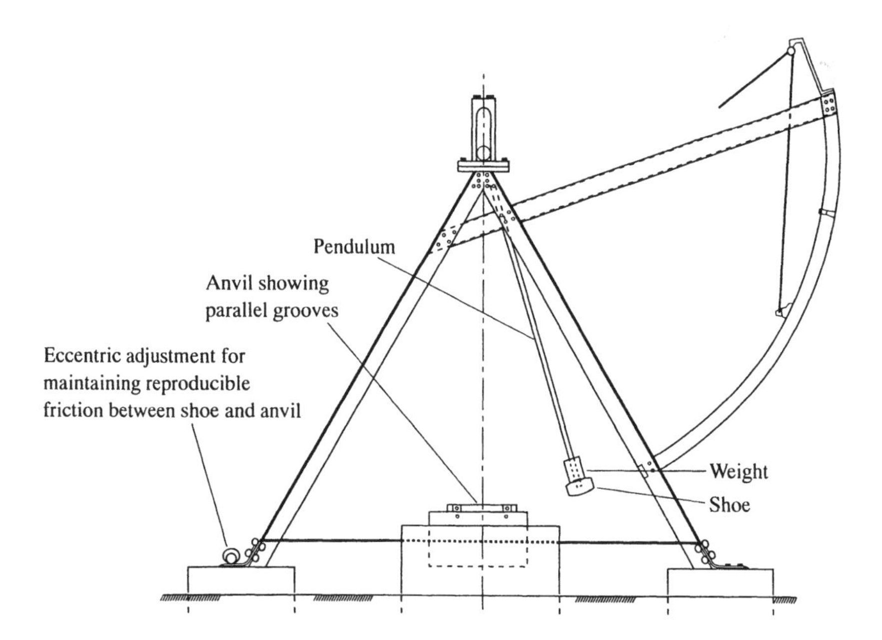

Figure 8.7. US Bureau of Mines friction pendulum.

arm is 2 m long and it is raised to a height of 1.5 m before release. The radius of curvature of the face of the shoes is 327 mm. The anvil can be heated to a temperature of $25 \pm 5\,^{\circ}\mathrm{C}$. The anvil and shoe are adjusted so that, after release, the pendulum arm swings 18 ± 1 times before coming to rest without a sample present. The sample mass is 7 ± 0.1 g, which is spread evenly in and around the grooves on the anvil before the test. The first tests are made with the steel shoe, and, if there are no indications of burning, smoke, explosion, or fire in ten tests, no more tests are carried out. When there has been an event using the steel shoe, the anvil is cleaned and repolished and the hard fiber shoe is used. If the results with the fiber shoe are no more than "slight local crackling" the material is deemed to have "passed" this test and be satisfactory for use as a secondary explosive. Results are shown in Table 8.4.

8.4.4.3. US Navy Weapons Center, Friction Pendulum Test (Walker, G.R., 1966)

This equipment is similar in principle to that of the Bureau of Mines. In this case the striker is a wheel set in the end of the pendulum arm. The striker weighs 1.8 kg and can be adjusted within its housing from test to test to minimize local wear on the striker wheel. The striker also has a calibrated prepositioning cam, which can be adjusted to give different heights of fall of the arm on to the sample. The striker plate or anvil, can be positioned in the two horizontal axes. The pendulum and the anvil can be adjusted relatively in the vertical plane. Prior to a test the pendulum is checked so that its impact point with the striker plate occurs 2° before bottom dead center and the pendulum is checked so that it can swing through its arc and contact all the 25 mg sample. The sample is spread over an area 3 mm × 6 mm and the striker released from a predetermined height. The result is recorded and a preliminary set of tests is carried out to determine the rough 50% point and then a 25 sample run is carried out using the up-and-down method. Events are recorded by observing for explosion, flame, flash, burning, charring, or smoke, and are listed in Table 8.4.

8.4.4.4. German BAM Friction Test (Köhler and Meyer, 1993)

In the German BAM Test, the powdered sample is placed between two surfaces, which are then loaded by a given force before they are caused suddenly to slide past each other. Tests are conducted with different forces between the sliding plates and the minimum force between them at which fires occur is recorded. This test has no impact component and is one of the purest friction tests available. It is also highly versatile, since it is relatively simple to replace one, or both surface materials. Results are shown in Table 8.4.

8.4.4.5. UK Rotary Friction Test (SCC No. 3, 1988)

This test has replaced the UK Mallet Friction Test. It has no element of impact and shares this characteristic with the BAM Test. It is shown in

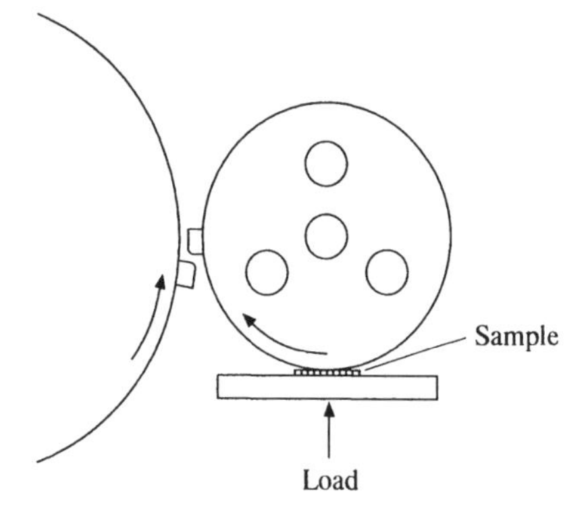

Figure 8.8. UK Rotary Friction Test (diagrammatic).

Figure 8.8. The sample is subjected to friction from a steel flywheel prepared to a standard roughness and under a preset load between it and a stationary block upon which the sample is placed. Energy stored in the flywheel imparts an impulse to the wheel assembly causing it to rotate, which generates high shearing forces in the sample under load between it and the stationary block. Flywheel velocity is adjusted using a potentiometer and the pressure between the wheel and the block is adjusted by means of an air regulator on a pneumatic ram. Fifty separate specimens of the material under test are subjected to a Bruceton Staircase procedure to determine a median flywheel angular velocity for the test material. Results are expressed in "Figures of Friction" relative to RDX.

Each wheel is used for six trials. It is advanced 120° for each one of them. Stationary blocks are offset from the wheel centerline so as to permit the four equal area faces of the blocks to be used twice, giving eight trials per block. Explosive samples approximately 15 mm^3 in volume are prepared by grinding to a given particle size spectrum and then placed on an unused portion of the block. The pneumatic ram is actuated to give a loading pressure of 276 kPa (40 psi). The flywheel speed is set and the friction wheel is rotated until stopped by the pawl.

A fifty-shot Bruceton Staircase run is performed using a logarithmic scale of flywheel speeds with an increment of 0.1 passing through 100 revolutions per minute (rpm). The upper limit is 398 rpm. Results are shown as Figures of Friction in Table 8.4, compared with RDX standard as 3.0.

8.4.4.6. Test Results

Table 8.4 shows some comparative friction test results from the United Kingdom, United States, and Germany. These data are remarkably consistent. The work in the United Kingdom on initiatory explosives using the Rotary Friction Test gives a sharp insight into the differences between this class of explosives and the noninitiatories. The most sensitive material of all tested in this equipment is VH2, the small arms cap composition. HTPB composite propellant has an Figure of Friction of 1.6, while Extruded Dou-

Table 8.4. Comparison of some friction data.

Explosive	US Friction Pendulum (P = Pass; F = Fail)		UK Mallet Test (Results are expressed as percentage fires: 0–24% ≡ 0; 25–75 ≡ 50; 75–100 ≡ 100)			BAM Test (Force between surfaces)	UK Rotary Friction Test
	Fiber shoe	Steel shoe	Wood/wood (% fires)	Wood/stone (% fires)	Steel/steel (% fires)	Force (N)	Figure of Friction
Expl. D	P	P					
TNT	P	P	0	0	0		6.0
Comp. B	P	P	0	0	50–100	360	
Tetryl	P	F	0	50	50–100	120	
RDX	P	F	0	50	50–100		3.0
HMX			0	50–100	50–100	60	1.5
PETN	P	F	0	100	50–100		1.3
PbStyphnate			100	100	100		0.17
PbAz			100	100	100		0.07

ble Base propellant has a Figure of Friction of 2.4. Materials with a Figure of Friction ≥ 6 are considered to be "Comparatively Insensitive" to friction, those with $3 \leq$ Figure of Friction < 6 as "Sensitive," and those where the Figure of Friction < 3 as "Very Sensitive" in line with the divisions made in Rotter Impact test data.

8.4.5. Other Powder Sensitiveness Tests

Other powder sensitiveness tests include one designed to determine the responses of energetic materials to flash, flame, temperature, and electrostatic discharge. While the results of these tests appear on Safety Certificates issued to permit transport and other work on explosives in the United Kingdom and Australia, the tests are somewhat parochial. More detail can be obtained from the UK SCC No. 3 (1988). Most nations operate a similar system of testing prior to permission to transport.

Results from powder tests are compared with results of testing other energetic materials which are routinely processed in order to assess whether any special precautions need to be taken. Other powder sensitiveness tests may be carried out if the impact and friction tests indicate abnormally high hazard. This implies that the material is also likely to be electrostatically sensitive, and such data are vital in order to be able to carry out any further work on the material with adequate precautions against electrostatic build-up and initiation via spark.

8.5. Charge Hazard Tests

The powder tests generate vital information on the manner in which an unknown explosive, or explosive mixture should be handled. The material may then be processed and munitions, test charges, or grains can be manufactured. Once charges have been fabricated, it is essential to be able to predict their responses to impact, friction, thermal, electrostatic, and shock stimuli, both as part of the determination of their sensitivities, i.e., what stimuli are required to initiate them reliably and their sensitivenesses and intrinsic explosivenesses, as opposed to the explosivenesses of systems in which they find themselves.

8.5.1. Shock Initiation Tests

These tests are also referred to as Gap Tests, since they seek to determine the effects of different materials imposed between a detonating standard donor explosive and a receptor explosive. The tests described in this section are not intended to be reliability tests themselves, although they do give an indication of what conditions might be required in a weapon to ensure reliability of transmission of a detonation pulse. They are in the strict interpretation of the differences between sensitiveness and sensitivity, *shock sensitiveness* tests, since they are intended to predict the hazard from unintentional detonation of one explosive when exposed to the shock from another, detonating explosive. The tests are true "sympathetic detonation" tests, since the initiating stimulus is pure shock from the donor to the acceptor through the gap, which screens out fragments. In many tests connected with insensitive munitions studies, the term "sympathetic detonation" is used erroneously to describe the effects on a given munition of the detonation of another munition close by. In such cases the initiation mechanism is not pure shock, but a mixture of shock and fragments acting synergistically.

Gap tests, like the powder impact sensitiveness tests above, involve a wide range of procedures and equipment, since they too were devised at different laboratories, often in isolation. There are two types of test, uninstrumented and instrumented. The instrumented tests involve more explosive and, not surprisingly, yield more useful results.

Tetryl is often referred to as the donor explosive in these and other tests, because most tests were devised before the mutagenic and other undesirable dermatological characteristics of the explosive were fully understood. Tetryl is not used generally in NATO now unless there is no alternative. In fact, there are several RDX-based explosive materials which are used in different countries as boosters, such as Composition A3 (RDX/desensitizer (98.5/1.5)) and some lower sensitivity explosives, such as PBXN-7. The reference to tetryl does not constitute either an endorsement of the explosive, nor disagreement with the findings of others on its unsuitability for continued use. Many tests were written round the explosive, which is probably the most effective booster explosive available, other than for its highly undesirable effects on the human metabolism.

8.5.1.1. Uninstrumented Gap Tests

Very many tests exist, each of which with its own equipment and rationale. In the main, tests may be divided into small and large scale. The small-scale tests were the original ones, using relatively small donors and acceptors, because, at the time, most explosives were both relatively sensitive to shock and had small failure diameters. A typical size of both donor or acceptor pellets is between about 12.5 mm and 25–30 mm right circular cylinders. The donor and acceptor are bare, pressed, or cast and machined cylinders, although in one test the donor and acceptor pellets are confined in brass cylinders. The barrier materials might be cellulose acetate "card," duralumin, "Lucite," brass, wax, or air.

Tests were carried out in a similar manner at different sites. A standard donor explosive pellet provides a standard shock pressure to one side of a barrier of one or other of the materials referred to above. The shock is transmitted and attenuated by the barrier to a different extent, dependent upon the thickness of the barrier. A thickness of barrier is sought which either allows or prevents detonation of the acceptor explosive at some level of probability (easiest to deal with is the 50% or median point). The criterion for a "go" in this case is that the acceptor detonates. It is not advisable to rely on sound, since a "no go" still results in a rapid deflagration of the acceptor, with a considerable noise level. A "go" is indicated by an indentation with a sharp lip impressed in a mild steel witness block, or, in the instrumented tests, by reference to the output from a series of ionization probes embedded along the length of the receptor pellet.

One modification of the test in favor at some laboratories involves the determination of the 50% point for a given material and associating it with the actual pressure peak at the acceptor face in contact with the barrier. This may be achieved if the shock output from the donor and the shock propagation characteristics of the barrier are known. Thus, the shock pressure can be related to the barrier thickness. This is valid, provided the barrier does not interfere with the rise time of the pressure pulse received by the acceptor. When a shock front strikes an acceptor in a gap test it is assumed that the shock front is sensibly vertical. This is true for most gap materials and relatively small gap thicknesses. However, in the case of a large gap composed of many somewhat compressible "cards" which might also have microscopic irregularities in them, the barrier might not be approximated by a monolithic block of the barrier material, which is the intention. What might be created is a complex composite material consisting of a combination of air and the barrier material. This would induce many internal reflections of shock between the constituent layers of the barrier, with concomitant reflections of the original shock, and lead to shock broadening, which is then a function of the donor explosive, the barrier gap thickness, and the barrier material. Hence, such a means of approximating the pressure pulse on the acceptor may not take into account a most important

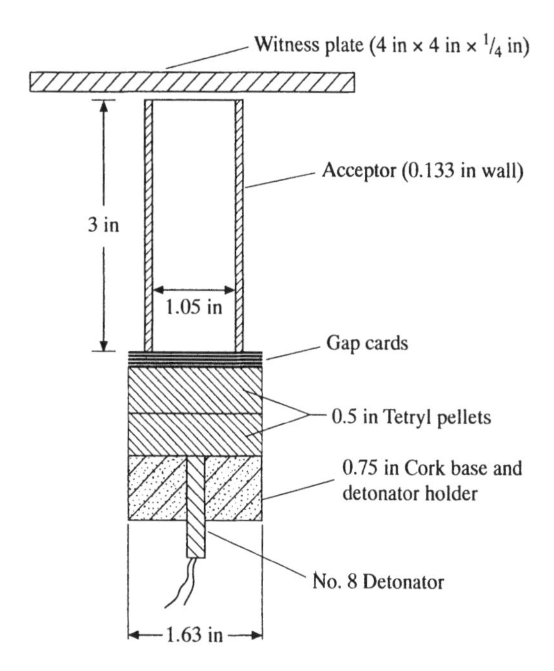

Figure 8.9. US Bureau of Mines shock sensitivity test.

factor, namely the rate at which the pressure rises. This, in turn, is important, since, as we may see in Chapter 6, rise time in pressure defines the rate of energy delivery, or power expended on the acceptor explosive, which is at least as important as the peak pressure. Some gap tests are detailed below.

8.5.1.1.1. Bureau of Mines, Pittsburgh, PA, USA (Walker, G.R., 1966)

The design of this test is definitive, even though results are not quoted from it in Table 8.6. The donor is a 41.3 mm × 25 mm long tetryl pellet of 50 g mass, pressed to $1.57 \pm 0.03 \times 10^3$ kg/m^3 the gap is made from cellulose acetate cards 2.54 mm × 10^{-4} m thick and 39.3 mm in diameter, unless large gaps are required in which case solid blocks of polymethyl methacrylate are used. The witness block for an event is a 6 mm thick sheet of cold rolled mild steel. The charge under test is contained in a nominal 25 mm diameter black steel (Schedule 40) with smooth finished ends. The Bruceton Staircase method is used to determine the 50% point from about 20 firings after the approximate median point has been determined. An event is adjudged to have occurred if a clean hole is punched in the witness plate. If a bulge or tear is observed a "no go" is recorded.

8.5.1.1.2. US Naval Ordnance Laboratory Test (Walker, G.R., 1966)

The aim in this test was to minimize the amount of explosive used, so the charge diameter was set at 5 mm. Some explosives, even at the time (1950) when this device was designed, had unconfined failure diameters greater than this and so the charge is encased in a thick brass sleeve, or case, 25.4 mm OD and 38.1 mm long, as is the donor, RDX. The RDX is carefully pressed from predried powder to a given density and charge weight and length. Some donors are fired directly against steel blocks of the type used as the recording medium for the tests and are required to produce dents of depths of a given range. The attenuator is made up of several "Lucite" disks of different thicknesses made up to the required overall thickness. An event is registered by a dent in a steel cylinder hardened to Rockwell B70-95 and the upper face in contact with the acceptor is machined to a surface finish of 1.6×10^{-6} m. The dent in each block is measured as is the hardness of the steel used. Dents are corrected to an equivalent dent, were the steel to have been of Rockwell hardness B83, using the formula

$$d = D_1 + 1.693 \cdot H, \tag{8.6}$$

where d is the corrected dent depth in millimeters, D_1 is the observed dent in millimeters, and H is the Rockwell hardness. The test explosive is granulated and dried in a similar way to the RDX donor and is also pressed according to a strict regime of increments and incremental loads. The acceptor density is recorded as the average of that of the 20 samples to be tested plus up to five used to determine the approximate position of the 50% point. The equipment is shown in Figure 8.10. The Bruceton Staircase method is used

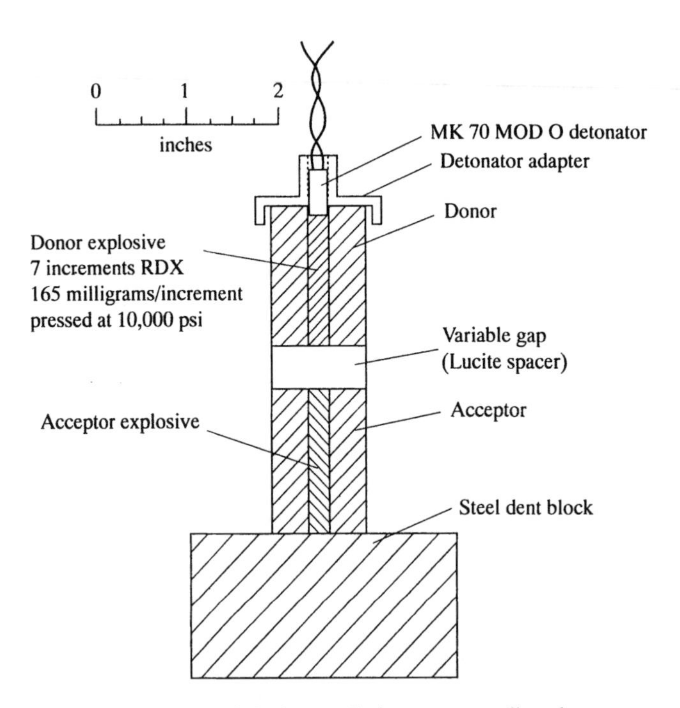

Figure 8.10. US Naval Ordnance Laboratory small scale gap test.

to determine the 50% point. Steps in thickness of the "Lucite" gaps are logarithmic, the increments of which are based on a concept known as the "decibang." This is a unit of initiation intensity analogous to the decibel. The expression of intensity is written

$$X = 10 \log \frac{RG}{GT}, \tag{8.7}$$

where RG is the "reference gap" and GT is the observed gap. The reference gap is chosen as 25.4 mm and the step between firing tests is taken as 0.125 decibangs. Data from this test are compared with other small-scale shock sensitivity tests in Table 8.6.

8.5.1.1.3. LANL, US Standard Scale Gap Test (Walker, G.R., 1966)

In this case both donor and acceptor are bare explosives, the donor is RDX/binder (92/8) pressed to a density of $1.688 \pm 0.002 \times 10^3$ kg/m^3 of 41.4 mm diameter and 101.6 mm long. The donor itself is initiated by a tetryl pellet and the detonation witness consists of two blocks of mild steel 50.8 mm thick. The acceptor is produced in the form of a pellet of the same dimensions as the donor and the attenuating gap is duralumin (2024 T4). A Bruceton Staircase method is used to determine the 50% point. Results are shown in Table 8.6.

8.5.1.1.4. LANL Small Scale Gap Test (Walker, G.R., 1966)

This test is related to the UK AWRE, Aldermaston test. It uses two donor pellets, a PETN booster, and an RDX/binder (94/6) donor. Both pellets are 7.6 mm in diameter, the PETN is 6 mm long and the latter is 5.3 mm long. The gap in this case consists of brass spacers 25.3 mm in diameter and of various thicknesses. Acceptor pellets are 12.7 mm $\times$ 38.1 mm long. The witness plate is mild steel and a "go" is recorded if a sharp-edged dent is produced in the face in contact with the acceptor explosive. A Bruceton Staircase technique is used to determine the 50% point and the results are shown in Table 8.6.

8.5.1.1.5. US Navy, China Lake Gap Test (Walker, G.R., 1966)

This test also operates at approximately 12.7 mm ($\frac{1}{2}$ in.) scale, but uses tetryl as the donor (12.7 mm $\times$ 12.7 mm pellet) and cellulose acetate cards as the gap. Acceptor pellets are 12.7 mm $\times$ 19.1 mm long and the witness block is a 152.4 mm long section of a 50.8 mm $\times$ 152.4 mm bar of 1018 cold rolled steel. A somewhat more complicated procedure than the Bruceton Staircase technique is used to determine a 50% point. No data are available to compare in Table 8.6.

8.5.1.1.6. Picatinny Arsenal Test (Walker, G.R., 1966)

This test operates using two bare tetryl pellets 39.8 mm $\times$ 40.6 mm long as donor, the attenuating gap is made from variable thickness cast and machined Aerowax B and the acceptor is the same diameter as the donor but 127 mm long. Some typical results are shown in Table 8.6.

8.5.1.1.7. UK Small Scale Gap Test (SCC No. 3, 1988)

This test is an amalgam of tests used at three different United Kingdom establishments and is now regarded as the small-scale national standard. The donor is a 12.7 mm × 12.7 mm long tetryl pellet and the acceptor is either a 12.7 mm square or circular cross-section pellet 25.4 mm long. Pressed or extruded charges are tested in circular cross-section charges and cast ones are machined to the square cross-section format. The gap is composed of brass shims to specification BS 2870 CZ 108, which are pre-prepared as 2.4 mm thick stacks made up of 48 individual foils soldered together. Gaps in excess of 2.4 mm are made up of complete soldered stacks and thinner ones are obtained by tearing off the requisite number of 0.05 mm thick shims. The witness block is a 25 mm or 25.4 mm block of mild steel. Detonation of the acceptor is adjudged to have occurred if the witness block has received a sharp-edged dent. The whole arrangement is held together by rubber bands. The initial gap is chosen by experience and the approximate 50% point reached. New or unknown materials are first tested at zero gap and the gap gradually increased by 2 mm increments. When, after a series of events a non-event occurs the gap is reduced by 0.1 mm and after a non-event it is increased by 0.2 mm. Once reversal occurs with a gap difference of 0.2 mm increments or decrements, single shim thicknesses of 0.05 mm are used until re-reversal occurs. This is the starting point for the investigation, which uses the Bruceton Staircase technique and about 25 tests to obtain results as compared in Table 8.6.

8.5.1.2. Instrumented Gap Tests

Several of the uninstrumented gap tests use acceptor samples of only 12.7 mm ($\frac{1}{2}$ in.) diameter. This was perfectly adequate to test sensitive explosives such as RDX/TNT and octol and their aluminized derivatives. Even the early plastic bonded explosives were sufficiently sensitive and had small enough failure diameters for testing to be possible on the small scale. However, as requirements to develop insensitive explosives and propellants have grown, some of the tests described above have ceased to be relevant. Early versions of the instrumented tests merely increased the diameter and masses of the donors and acceptors and placed ionization probes along the very considerable lengths of the acceptors. In this way, it was possible to show how the burning process accelerated in the acceptor to full detonation. The first tests of this type were developed in the United States and there has been some confusion outside the United States when the expression "pick-up" is used. In English, as opposed to American English, "pick-up" has the connotation that the process starts. In American English "pick-up" refers to acceleration.

The Bureau of Mines instrumented test is the same as the uninstrumented test shown in Figure 8.9 except that the acceptor is housed in a 406 mm long sample container fitted with T2 pressure switches. Wave velocities can be measured as a means of determining whether or not a true detonation, as opposed to a rapid deflagration has occurred. In the US Navy Large Scale Gap Test developed at the White Oak Laboratory the donor is two graphited tetryl pellets 50.8 mm × 25.4 mm long, pressed to a density of 1.51×10^3 kg/m^3. Two types of gap material are used, "Lucite" blocks and cellulose acetate cards. The test explosive is loaded into a cold drawn seamless mild steel pipe, of 36.6 mm ID and 47.6 mm OD, by whichever means is appropriate. The experimental arrangement is aligned within a rolled cardboard tube, with appropriate spacers. A "go" is adjudged to have occurred when the steel witness plate is completely holed.

The size and cost of large-scale gap tests precludes a Bruceton Staircase technique being used to determine the 50% point. An alternative approach adopted by the US Navy for an unknown material is to start at zero gap. If a detonation occurs the first thickness of spacer tried is 0.2 mm and if that, too, results in an event, the spacer thickness is doubled for each succeeding test until a non-event occurs. At this stage, each successive shot is made in an "up and down" manner, with a gap thickness half way between the nearest "go" and "no go" values. The procedure continues until the "go" and "no go" conditions are separated by a gap of 0.25 mm. The 50% point is determined by designating the gap thickness corresponding to the "go" as N and that corresponding to the "no go" as $N + 1$. A pattern of four further tests is carried out, two at each of levels N and $N + 1$. There are four possible outcomes of this testing and these are shown in Table 8.5. Two further tests are necessary if patterns III and IV are observed in the original four tests. Examples of the results of testing several common explosives are shown in Table 8.6.

Table 8.5. Determination of the 50% point: US Navy White Oak Large Scale Gap Test (Walker, G.R., 1966). (After detonation on first test at zero gap.)

Basic test pattern			Supplementary tests needed to establish 50% point	Supplementary test results		
Test	Gap size N	Gap size $N+1$		Gap size $N-1$	Gap size $N+2$	50% point
I	xx	oo	None	—	—	$N+\frac{1}{2}$
II	xo−	ox+	None	—	—	$N+\frac{1}{2}$
III	xo	oo	Two at $N-1$	oo		$N-1$
				xx	—	N
				ox		$N-\frac{1}{2}$
IV	xx	ox	Two at $N+2$		oo	$N+1$
				—	xx	$N+2$
					ox	$N+1\frac{1}{2}$

Table 8.6. Comparison of the results of some uninstrumented gap tests. (The gap or barrier material is referred to at the head of each column, and the 50% point is quoted in millimenters for all materials.)

US Navy White Oak; Large scale: "Lucite" and cellulose acetate		US Navy White Oak; Small scale "Lucite"		Los Alamos Standard scale Dural		UK Brass	
Expl. D	38.1			TNT	28.3	TNT(c)	Fails 0.0
TNT(c)	46.5	TNT (*p*)	6.4	Expl. D	43.0	R/T (UK)	0.64
Comp. B	60.5			Comp. B3	50.3	PE4*	1.70
						Pentolite	3.8
Tetryl	66.3	Tetryl	9.2	Tetryl	60.7	RDX/wax†	5.6
RDX	82.0	RDX	9.3	RDX	61.7	Comp B(US)	6.1
		HMX	10.3			Tetryl	7.3

* RDX/binder(88/12).

† RDX/wax(88/12), pressable.

8.5.1.3. Gap Test Results

Table 8.6 shows some uninstrumented gap test results. They are consistent in their ordering of the relative shock sensitivenesses of the common explosives shown, with only one exception. This is remarkable, bearing in mind their diversity. However, it would be quite difficult to relate the numerical data quantitatively, without having carried out some instrumented firings in order to determine the nature of the shock output from the acceptor-side face of the barrier in each case.

A particularly interesting feature of the results is the difference in shock sensitiveness recorded in the United Kingdom test between UK RDX/TNT (60/40) and US Composition B. The higher shock sensitiveness of the United States material is a result of the presence of a substantial "impurity" level of HMX in the RDX; in excess of 11% HMX has been recorded in Holston material. UK RDX/TNT (60/40) is more difficult to initiate reliably than US Composition B, which also has other useful properties. It flows better at the pouring temperature of ca. 80–85 °C used to pour the fillings of high explosive shell and the HMX appears to catalyze the processes associated with the volumetric expansion of HMX on thermal cycling.

8.5.1.4. Other Shock Initiation Tests

The data obtained from instrumented shock sensitiveness tests are useful, but their development represented a catching-up process by the conventional weapon laboratories of work done in the nuclear weapon laboratories much earlier. In the very precise explosive engineering associated with nuclear weapons it is important to understand the manner in which detonation waves develop and the speed in which they do so. The macroscopic effect of an initiation may well be explosion, but, if it occurs only after a substantial run-up distance in the explosive, the detonation wave may not have the requisite shape or arrive at the optimum time at some point in the warhead structure to carry out its designated task. The acceleration of a detonation wave to full output in a receptor explosive depends upon the initiation stimulus it receives from the donor via the barrier, as is well illustrated in the gap test work. However, prior to the late 1950s and early 1960s, little was known about this build-up phase. The nuclear weapons laboratories pioneered work on quantitative assessment of initiation through several tests which rely upon ultra high-speed photography and electrical/electronic detonation wave detection in test equipments.

8.5.1.4.1. The Wedge Test (Gibbs and Popolato, 1980; Majowicz and Jacobs 1958; and Campbell et al., 1961)

The test was developed in the United States to give time-dependent information on the run-up to detonation in a shocked explosive sample. It is not possible to give full weight to the importance of the work using the wedge test in a short chapter such as this, but it is described in detail by Gibbs and Popolato (1980). The Wedge Test arrangement is shown in Figure 8.11. The build up of a detonation in the receptor explosive is followed by means of a high-speed streak camera. The experiment consists of a planar detonation wave generator, which generates a detonation wave at a given peak pressure. This detonation wave is attenuated by means of plastic and/or metal

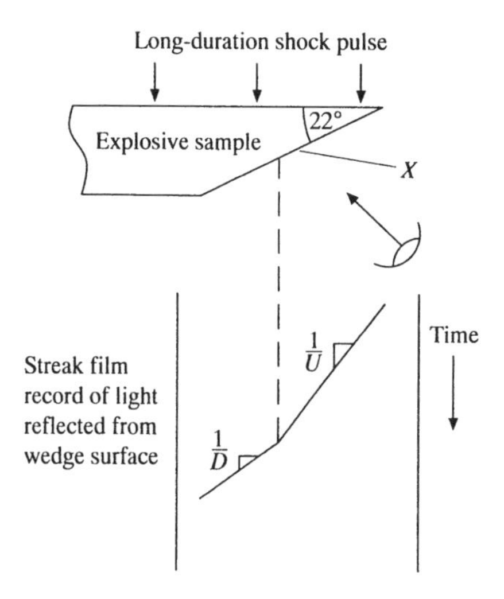

Figure 8.11. Diagrammatic representation of the wedge test and its streak record.

barriers, so that the pressure pulse received by the wedge of the explosive under test is of a precisely known intensity and duration. Understanding of the shock Hugoniot curves of the intervening attenuator materials enables the prediction of the pressure, P, and particle velocity, U_P, in the explosive sample. They are determined graphically from the intersection of the attenuator rarefaction locus and the explosive-state locus given by the conservation of momentum relation $P = \rho_0 U_P U_s$, where U_s is the shock velocity and ρ_0 is the original density of the explosive. The attenuator rarefaction locus is approximated by reflecting the attenuator Hugoniot about a line where the attenuator particle velocity is constant. The conservation of momentum relation, above, cannot hold during a non-steady-state process, such as initiation. However, it is possible to assume that near the interface between the explosive under test and the attenuator the effect of reaction is very small so that the assumption is very nearly correct.

The acceptor charge is wedge-shaped so that the locus of the arrival of the reaction front at its front face can be observed by a streak camera, from which film it is possible to interpret the velocity of the reaction in the explosive as a function of distance traveled in it and of elapsed time. Moreover, knowledge of the initiation pressure at the attenuator-sample face enables data to be obtained on the pressure dependence of the distance and run time to detonation in any explosive. A typical streak record is shown diagrammatically in Figure 8.11.

The streak record shows the onset of the initial shock in the explosive sample and its transition into a stable detonation wave after travel for a certain distance into the wedge. The initial shock may generate a more curved trace than that shown diagrammatically here as acceleration occurs, although there is a sharp transition at the point at which stable detonation starts. The logarithm of the distance, or time which the wave has run at the transition to detonation point, can be plotted against the logarithm of the initiating pressure to generate what are known as "Pop" plots after Alphonse Popolato who first devised them. Despite some scatter, these plots show the essential linearity of the log-log relationship. It has also been found to be useful to plot initial shock velocity against initial particle velocity in the explosive and this relationship, too, is linear. The reader is referred to Gibbs and Popolato, (1980) for examples of these important relationships.

8.5.1.4.2. Low Amplitude Shock Initiation (LASI) Test (SCC No. 3, 1988)

This is essentially a modified gap test, in which the gap material is polymethyl methacrylate (PMMA). The test was devised to determine the responses of different energetic materials to stimuli not expected to lead to detonation. It has much in common with the Wedge Test, since it provides a means by which the free surface velocity of an explosive subjected to a shock input can be related to the shock pressure stimulus received by the sample. It was devised at a time when there was great pressure on the laboratory where it was developed to reduce the noise from firings and prior to easy access to large plane wave generators. The equipment consists of a 50.8 mm gap test arrangement with the addition of a high-speed framing camera viewing the free surface of the explosive under test. Firings are made using a standard Tetryl donor, which is 50.8 × 50.8 mm long, pressed to a density of 1.5×10^3 kg/m^3. The acceptor charge is also 50.8 mm in diameter and 12.7 mm thick. The framing camera is focussed on the rear surface of the acceptor charge in order to determine its free surface velocity at several different thicknesses of the attenuating PMMA. A knowledge of the output pressure of the donor and the Hugoniot of PMMA allow the incident pressure on the acceptor explosive to be calculated as a function of the thickness of the attenuator.

Results are shown in Figure 8.12, which also shows the three characteristic regions associated with the development of detonation and a schematic inert response. The onset of reaction occurs when an explosive response curve starts to deviate from that of an inert. A steeply rising portion of the curve is associated with the acceleration of reaction and the final portion, which shows no increase in surface velocity as a function of the applied pressure, represents stable detonation. Some explosives exhibit pseudo-inert

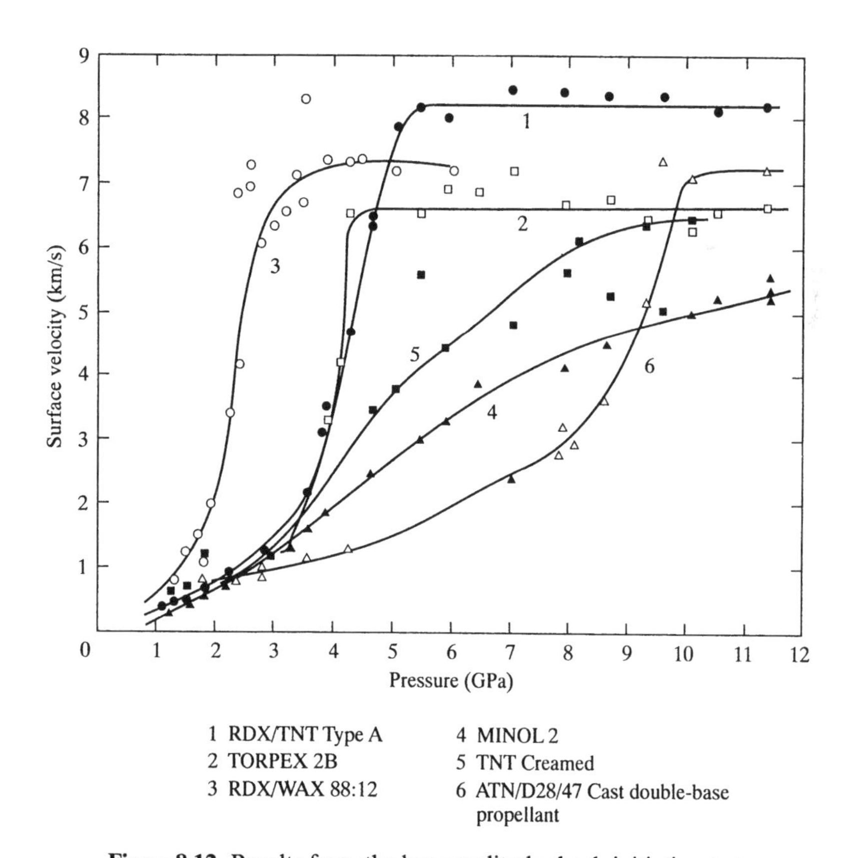

Figure 8.12. Results from the low amplitude shock initiation test.

responses, because they have long run distances to detonation in the geometry tested and are not detonating reliably after only 12.7 mm travel. This could be overcome by extending the length of the receptor pellet. Similar data to those shown in "Pop" plots are accessible from this test, provided different length receptor pellets are tested at fixed attenuator thicknesses. Low vulnerability explosives can be studied using this technique, provided charges with diameters in excess of the failure diameter are tested and the charge sizes associated with long run-up distances can be accommodated.

8.5.2. Charge Impact Tests

It is relatively rare that bare explosive charges are handled in connection with modern weapon design, since most are cast, pressed, or extruded into the case of the munition directly. However, some are, and in order to predict the likelihood of an event, two tests have been devised, which also enable an estimation to be made of the explosiveness of the bare explosive charge, as distinct from any system in which it might subsequently find itself.

8.5.2.1. Oblique Impact Test (SCC No. 3, 1988)

The test was devised at AWE, Aldermaston, in the United Kingdom, following an accident, in which a relatively large bare charge of an explosive slipped off an electric vehicle while being transported from one part of the site to another, fell about 400–500 mm on to an asphalted cleanway, and detonated. The explosive had been tested exhaustively as a powder and found to be acceptably insensitive for the purpose intended. However, no one had been prepared for the severity of the event, which caused casualties, and the test was devised to screen out other materials with this characteristic thereafter. There are several versions of the test, which was also used at Los Alamos and Livermore National Laboratories and the Pantex Company in the United States.

A 356 mm diameter hemisphere weighing approximately 21–22 kg is suspended by three wires from a gantry and allowed to swing so as to strike a prepared steel target plate at a grazing angle. It is a charge impact and friction test, rather than a pure impact test, because the pendulum effect of the suspension causes the explosive to be dragged across the target face after impact, so providing the friction component of the stimulus. The steel target plate is coated with a layer of sand (BSS 52–60, purified by acid and supplied by British Drug Houses), which is bonded to the steel by means of a thin layer of "Araldite" cement (Araldite MY753 and hardener HY951). The steel plate is screwed on to a Granolithic Paving slab 610 mm × 610 mm × 50.8 mm thick bedded into a concrete slab 711 mm × 711 mm × 127 mm thick. In the standard test the hemisphere impacts the target plate at 45°. If no events occur at the highest drop height, tests may be conducted at 76° incidence. Tests are carried out at 1, 2, 3, 5, and 10 feet (305, 610, 915, 1524,

Table 8.7. Classification of UK Oblique Impact Test Results (SCC No. 3, 1988).

Key	Description of event
H	Detonation, or a very energetic and severe deflagration. A crater is produced, no explosive is recovered, and the steel target plate is holed.
E	Explosion (Deflagration). Almost all the explosive is consumed, the steel target plate is distorted, and the concrete slab shattered.
P̲	Large partial explosion. The charge is broken, or badly damaged with 0.5 kg or more blown off in excess of 9 m. Open cracks may be found in the concrete slab.
p	Small partial explosion. Charge intact, except for a few tens of grams broken from the pole of the charge. The charge may be broken but none is thrown more than 9 m.
N	Fail. No sign of an explosive event on the charge or target and no puff of gas visible on the film record.

and 3050 mm). The first drop is made from 915 mm and the scale followed until three failures are recorded at a given height and where events occur at the next greater height. About five or six tests are conducted to determine the behavior of an explosive and each test is filmed at 3000–6000 pps.

The result from an individual test can vary from a puff of gas being seen on the film to a crater being caused in the test arena. Results are classified into five types in the United Kingdom as shown in Table 8.7.

It is not always possible to manufacture a hemisphere entirely out of explosive. An alternative is to construct a vehicle the bulk of which is made from "Jabroc" (a high density laminated wood). The explosive sample is prepared as a cylindrical insert 127 mm in diameter and 101.6 mm deep, with a 177.8 mm spherical radius at the open end. The total mass of the charge is similar to that for the complete hemisphere of explosive, and testing has confirmed that the composite charge performs in a reproducible and similar fashion to the bare hemisphere. Results are reported in Table 8.9.

8.5.2.2. LANL Oblique Impact Test (Gibbs and Popolato, 1980)

The Los Alamos version of the oblique impact, or skid test employs a technique where the charge is dropped vertically on to an inclined target plate, which may be inclined at either 45° or 76° in a similar fashion to the AWE test. In this test the hemisphere is only 245 mm in diameter, but up to fifteen drops are carried out following an up-and-down procedure to obtain a median height. In addition, drops can be carried out from higher than in the AWE oblique impact test and research has been carried out on the response of a range of explosives to a range of roughening agents on the steel target plate. Events are detected by the use of a pressure gauge to measure overpressure at 3.05 m from the target plate. Results are reported in Table 8.9, as the drop height that produces events in 50% of the trials and the average overpressure from events.

Table 8.8. Classification of the results of LLNL Oblique Impact (Skid) Tests.

Key	Description of event
H	Detonation.
$\underline{E}$	Violent deflagration; virtually all explosive consumed.
E	Medium order event with flame and light; major part of explosive consumed.
$\underline{P}$	Mild, low-order reaction with flame or light; charge broken up, or scattered.
p	Puff of smoke, but no flame or light visible on high-speed photographic record. Charge may be broken into large pieces.
B	Burn or scorch marks on explosive or target; charge retains integrity.
N	No reaction; charge retains integrity.

8.5.2.3. LLNL–Pantex Skid Test

The LLNL–Pantex version of the test uses an hemispherical billet of explosive 279.4 mm in diameter, but a similar pendulum device is used to generate the stimulus of impact and sliding friction on the target surface as occurs in the AWE test. The LLNL classification of test results (Dobratz, 1981) is more sophisticated than that devised by AWE as is shown in Table 8.8.

8.5.2.4. Oblique Impact (Skid) Test Results

Both the United States tests use smaller and lighter test vehicles than at AWE. This means that, in order to generate events with some of the less sensitive explosives, relatively high drop heights compared with those used at AWE are required. More subtlety in results is generated by the LLNL test and both the United States laboratories publish (Gibbs and Popolato, 1980; Dobratz, 1981) very large amounts of data on their own explosives, including the effects of surface and temperature variations. The importance of this test is that it is not only an impact and friction charge sensitiveness test, it is also an explosiveness test, since it gives an idea of the likelihood that an event, once initiated in a large, bare charge of a given explosive, will accelerate to high order.

The United States tests each use smaller charges than the AWE one, and in some cases the impacting material is not as abrasive as the relatively coarse sand used in the AWE tests. Consequently, the United States tests seem to indicate that the explosives can be dropped from relatively greater heights with relative impunity compared with the results of the AWE tests. All of the test results appear to indicate that even very high nitramine explosives can have relatively low explosiveness. This may be misleading for two reasons.

The explosives are not cased and so there is no contribution to the acceleration of the deflagration process from confinement. Only surface layers of the explosive are involved where there is little or no confinement apart from during the period of contact between the charge and the target. The fact that

Table 8.9. Comparison of Oblique Impact (Skid) Test results.*†

	LANL	LLNL				AWE			
		Heights (ft) and event types at 76° (Table 8.8)				Heights (ft) and event types at 45° (Table 8.7)			
Explosive	50% heights (ft) *76°* 45°	1.25	2.5	5	10	10	5	3	2
PBX9404	*1.8* 4.5	3N3H	HP2N	2H$_{(3.5)}$			H	HN	N
PBX9011	2.5	4N2H	HN$_{(1.75)}$						
R/T 75/25	*4*		P			Pp	3N		
Octol 75/25	75		N	P$_{(3.5)}$		2pN	2N		
Comp. B						3Np	N		
Comp. B3	*9.8*		3N3p	N5p			Pp	2Np	
PBX9501	26				P				
TNT	No tests				p	No tests			

* AWE uses US Composition B and B3: UK Octol has 71% HMX and 4% RDX in the 75% nitramine: PBX9404 contains 94% HMX: PBX 9011 contains 90% HMX: PBX 9501 contains 95% HMX.

† The small numbers in brackets after the event types in the LLNL test results refer to heights in feet at which intermediate tests were carried out.

some charges produce high-order events indicates that their intrinsic explosiveness is very large.

The reader should also be quite clear that impact surface type and finish and charge mass are extremely important in this test and that it is unwise to attempt to read anything into the actual drop heights, apart from general trends. LLNL used the test to assess the quality of the floor surface in some of its processing areas and conducted tests on a wide range of target finishes.

One surprising feature that the three test types seem to agree on is that Cyclotol 75/25 (RDX/TNT 75/25) appears to be more sensitive to this type of stimulus than is its direct relative Octol 75/25, despite the agreement elsewhere that HMX is more impact and shock sensitive than RDX. This factor alone emphazises the very considerable importance of carrying out tests which relate as closely as possible to the stimulus that is causing concern in practice.

8.5.3. High-Speed Impact Test: The Susan Test

This test was devised as a consequence of accidents involving nuclear weapons in air crashes at Palomares in Spain and Thule in Greenland. Accident investigations concluded that conventional explosives associated with nuclear weapons exploded as a result of forces induced by the crashing of the aircraft. Explosions scattered contamination over land and sea and the United States nuclear weapon laboratories were required to eliminate the possibility of this happening again. Their first task was to assess the vulnerability of explosives

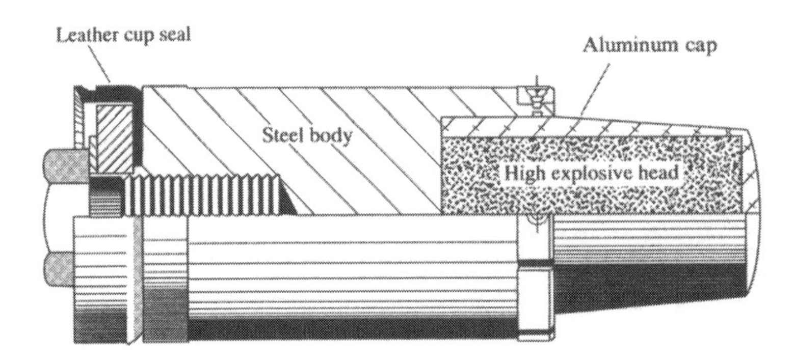

Figure 8.13. Susan Test vehicle.

to the stimuli they receive via their casings from effects associated with an air crash. The Susan* Test developed by Lawrence Livermore National Laboratory was the result. Tests have been carried out by LLNL (Dobratz, 1981) and, on behalf of various laboratories in the United Kingdom, by AWE.

This test uses a projectile, shown in Figure 8.13, which was designed specifically to collapse in a manner so as to squeeze and nip explosive between metal surfaces in the process (Green and Weston, 1970). The projectile carries an explosive charge of about 0.4 kg, dependent on explosive density and its all-up weight is about 5.8 kg. The test consists of firing the projectile at a predetermined muzzle velocity from a smooth-bore gun at a massive steel target so that it impacts nose-first at a known velocity. If a reaction in the explosive is induced, the overpressure from it is measured. Various masses of TNT are separately detonated at the target and the overpressures generated from each firing are used to produce a graph of TNT overpressure against charge mass. In addition, in the United Kingdom version of the test, two samples of the explosive under test are detonated in the arena to give an indication of the maximum level of event possible for any given explosive. Overpressures are measured by four pressure gauges set about 3 m from the impact point on the target in a special arena and high-speed cameras are used to record the events. The overpressure from the explosive under test is related to a mass of TNT required to be detonated at the same place to generate the same overpressure, and this mass of TNT is plotted as a function of the impact velocity of the projectile. In the LLNL test procedure (Dobratz, 1981) the explosive yield ranges from 0 kg to 0.34 kg of detonated TNT. This quantity of TNT is arbitrarily set as a maximum level, even though a larger value might be indicated from the results. Detonations occur very rarely in this test, but sometimes one or more of the pressure gauges records an extra large overpressure because of the asymme-

* The name derived from the name of the daughter of the originator of the test.

try of the collapse of the test projectile and its effects on the decomposition of the explosive in it. In the United Kingdom test (SCC No. 3, 1988) only about eight firings are usually carried out, the first at 91.4 m/s impact velocity, unless the explosive was expected to be very sensitive. If not, firings are carried out up to the maximum for the United Kingdom test of 305 m/s. In the LLNL test, higher velocities are possible and 1400 ft/s (426.7 m/s) was used in testing TATB.

8.5.3.1. Results of Susan Test Firings

Results are shown graphically in Figure 8.14 for some explosives. Dobratz (1981) cites a considerable amount of information on the Susan Test, since it was used extensively to assist in the early stages of the search for lower vulnerability explosives, especially those likely to be used for aircraft bomb fillings. A number of high-speed cine film studies were carried out to ascertain the mechanism for initiation of the explosive in this test and what happened to the test vehicle during impact. As a result of this extensive work, the emphasis shifted from the Susan Test being regarded as strictly a sensitiveness test to being a sensitiveness and explosiveness test, and comments on the processes occurring during the decomposition of several explosives give good insight into their responses to initiation under conditions of relatively low confinement. Details on the behavior of some well-known explosives follow in Table 8.10. It is useful to compare results obtained in this test with those from the Oblique Impact Test because both have the elements of sensitiveness and explosiveness. Threshold drop heights or impact velocities are a measure of sensitiveness, but the responses indicate the effects of different binders on the explosiveness. Even explosives containing more than

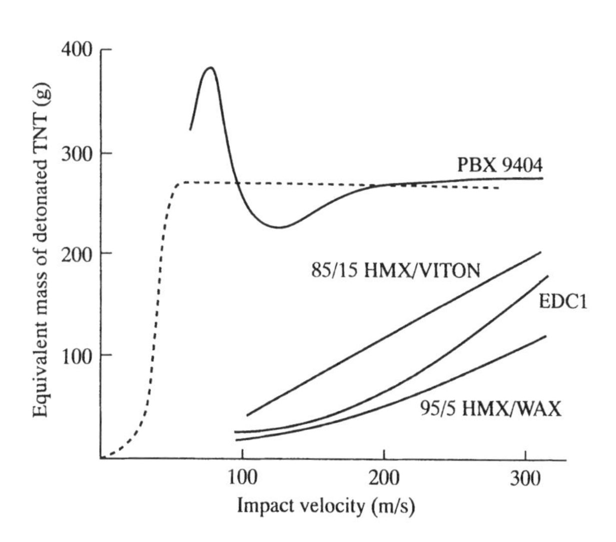

Figure 8.14. Susan Test results.

Table 8.10. Behavior of some explosives in the Susan Test (Dobratz, 1981).

Explosive	Characteristics
TATB	Almost indistinguishable from mock HE at 427 m/s; no evidence of accelerated burning at high-impact velocities.
Expl. D	Less reactive than TNT; very difficult to ignite mechanically; very low probability of violent event under low confinement.
TNT	Minor events at low velocities, due to pinch: no violent reactions above .366 m/s; very low probability of violent event under low confinement.
XTX-8003	Moderately difficult to ignite (49 m/s); low-reaction level; very small probability of building to violent reaction under low confinement.
Comp. B3	Threshold velocity 55 m/s; ignition only after extensive splitting; no detonation even at 427 m/s; low response only if confinement low.
Comp. A3	At low velocity like TNT; at higher velocities greater than TNT; difficult to ignite; low probability of high order if low confinement.
Cyclotol 75/25	Threshold velocity 55 m/s; moderately high responses at low velocities; difficult to ignite but large events possible.
PBX9011	Threshold 50 m/s; more energetic than Comp. B3; difficult to ignite; reaction enhancement at pinch stage; low probability violent reaction when low confinement.
PBX9501	Threshold 61 m/s; above this, reactions become violent over narrow velocity range; reactions do not always grow to very violent ones.
PBX9404	Threshold 32 m/s; reactions are violent without pinch stage enhancement; any mechanical initiation can build to a violent event.

90% of HMX may have relatively low-order responses if, as in the case of PBX 9501 (HMX/estane/BDNPA-F 95/2.5/2.5) the binder is plasticized, or in the case of PBX 9011 (HMX/estane 90/10) about 10% of relatively compliant binder is used. However, as with the explosiveness of compositions tested in the Oblique Impact Test, the confinement of these charges is quite low and these compositions exhibit higher energy responses in even thin steel casings. The Susan Test vehicle is designed to fail early and so affords little confinement to the charge.

Results such as these, taken with others from the United States, the United Kingdom, and other laboratories testing explosives in the early 1970s, improved the understanding of the effects of the thermomechanical properties of the explosives on their response to thermomechanical stimuli. The Susan Test showed very clearly also, how confinement aids the development of a damaging event, by allowing acceleration of the burning rate, eventually to a rate at which it is not possible for relief to be gained by loss of confinement, because of the inertia of the container walls. These concepts arose at an important time for hazard assessment, since they coincided with the need to improve the understanding of the responses of weapons to stimuli associated with fires and fragments as the importance of lower vulnerability weapons began to be realized.

8.5.4. Intrusion Tests

The tests described so far involve the interaction between either powdered explosive, or charges with stimuli which affect the surface of the charge only. However, there is a considerable database on the damaging effects of explosives charges and weapons impacting protrusions in factories, storage sites, or weapon platforms, or being impacted by projectiles. In either case, the explosive may be significantly damaged by the intrusion into it of a foreign body as well as being stimulated by the original impact.

There are two types of attack mechanism relevant here, Spigot Intrusion, which involves little, if any, shock input and occurs relatively slowly and bullet or fragment attack, where the attack mechanism may involve some shock element and takes place rather quicker. Although bullet and fragment tests are also carried out on bare explosive charges, their discussion is best left to later, for discussion as part of the discussion of weapons system testing.

Spigot Intrusion tests were developed as a result of the experience of several navies of weapons being dropped on to stanchions and other protrusions aboard ships, while ammunitioning or de-ammunitioning procedures were taking place. It was also recognized that the processes taking place during the intrusion of a spigot into an explosive charge involve impact and friction sensitiveness and charge explosiveness. This gives a comparison with charge impact sensitiveness data for the same explosives obtained from other sources and another means of assessing explosiveness.

Spigot Intrusion tests simulate the effects of a large charge in a weapon being impaled by some type of blunt-ended rod. This means that either the vehicle housing the charge to be tested, the charge itself, of the vehicle carrying the spigot, should be of the order of at least 20 kg in weight. Otherwise, the spigot may not penetrate even a lightweight casing around the test charge. Either the test vehicle containing the charge is dropped on to the spigot, or the spigot, housed in a weighted vehicle, is dropped on to the test charge.

8.5.4.1. LANL Spigot Intrusion Test (Gibbs and Popolato, 1980)

The initiation processes which this test seeks to simulate are those associated with mechanical impact of the charge, shearing of the explosive due to the entry of the spigot and, possibly, adiabatic heating through air compression in a cavity. All of these features are provided for in the test vehicle which is shown in Figure 8.15.

The explosive charge is machined or cast as a billet 152.4 mm × 101.6 mm high in the form of a right circular cylinder weighing between about 3.2 kg and 4.1 kg. The explosive charge is glued into the counterbore of an inert plastic bonded material, which has the same shock impedance characteristics as the explosive being tested. The charge support is in the form of a truncated cone 328.9 mm at its largest diameter and 222.3 mm at its lowest

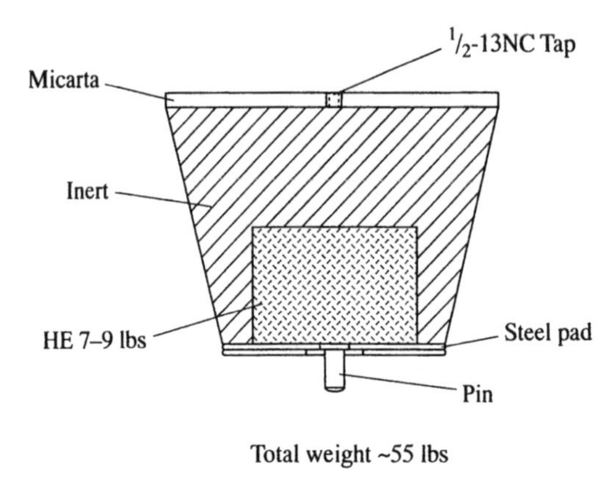

Figure 8.15. LANL Spigot Intrusion vehicle.

diameter. The cone is also 222.3 mm deep. A wire sling is bonded to the top of the support vehicle to enable it to be raised remotely prior to being dropped on to the target. The bottom of the explosive charge fits flush to the bottom of the support vehicle. The open end is covered with a 12.7 mm thick steel plate, which has a 19.1 mm diameter hole at its center. A 44.5 mm × 6.4 mm deep counterbore is cut in the steel cover plate on the side facing the explosive. A steel spigot, or pin, with a 28.6 mm diameter by 6.4 mm deep head and a 31.75 mm long by 19.1 mm diameter shaft is fitted through this hole so that the shaft protrudes from the bottom of the explosive. As assembled, the head of the steel pin is separated from the bottom of the explosive charge by between 0.35 mm and 0.5 mm gap and the shaft extends about 25 mm below the bottom of the steel cover plate.

Normally 20 drop tests are carried out and the median point is found using the up-and-down procedure. The results are reported as the drop height that produces events in half the trials. The magnitude of the event is important in these experiments, since it provides a measure of explosiveness. Test data are shown in Table 8.11.

8.5.4.2. AWE Spigot Test (SCC No. 3, 1988)

This test is based on the LANL test referred to above. There are a few minor differences, the main ones of which are that the test vehicle is manufactured from "Jabroc" and the lower portion of the vehicle has a steel ring fitted around it to maintain integrity of the vehicle for longer, during impact. The total weight of the filled vehicle is 26.8 kg, but variants have been tested at up to 136 kg in weight. The vehicle is lifted to its predrop height by a jib crane and released remotely. The vehicle is assisted in falling vertically and without yaw and pitch by the attachment of a cord to the upper face. This cord remains attached to the crane after release of the vehicle and guides it to the target. Each drop is filmed at 3000 pictures per second. The impact velocity is determined for each test from the cine record and the height from which the vehicle would have had to have been dropped to impact at this velocity is quoted as the drop height. This is to ensure that the unreeling cord used to guide the fall of the vehicle cannot be said to interfere with the impact conditions. The violence of the reaction of the explosive in the vehicle on impact is important and the same categories of event as described in Table 8.7 are used to classify spigot test events. A Bruceton Staircase technique is used to assess the 50% point, but, in those cases where the demonstration that no event occurs, the drop heights and responses are recorded without statistical analysis.

8.5.4.3. US Navy Spigot Test

The US Navy carried out Spigot testing at the Navy Surface Weapon Center, White Oak, using equipment which involved simulated weapon casings. The heights involved were considerably greater than those used in the Los Alamos and United Kingdom tests. Results are quoted but they are not consistent with those determined at LANL and AWE, because of significant differences in procedure and construction of the vehicles used.

8.5.4.4. Spigot Test Results

Allowing for differences in vehicles and instrumentation, there is good agreement between the LANL and AWE results. The US Navy data involve nearly an order of magnitude increase in height for events to occur compared with the other two results. There is no air gap between the spigot and the explosive in the US Navy tests and so no contribution from the effects of adiabatic compression of air. In both the United Kingdom and LANL tests there is an air gap built in to the test characteristics in order to sensitize the explosive. The original references in the United States contain many pages of the results of the Spigot Intrusion testing of a wide range of explosives and, while the 50% impact heights are important, of much greater utility is the type of reaction generated. Only PBX 9404 and Composition A3 produce high-order events, all others are relatively benign reactions. This is quite likely due, in part, to the loss of confinement of the explosive on impact due to the severe damage that is done to the vehicle, especially when it falls a great distance. Such damage is less likely in a real weapon and hence these results bear little relationship to the actual hazard likely in dropping, say, an uncased torpedo on to a spigot while loading or unloading a submarine.

8.5.5. Thermal Hazard Tests

Thermal hazard tests are different from thermal stability tests which refer to the determination, through a proven technique, e.g., the Abel Heat Test for gun propellants (Boggs and Derr, 1990), of the expected lifetime of a pro-

Table 8.11. Spigot Test results.

Explosive	LANL 50% point (ft) (event type, Table 8.8)	AWE (drop height in m) (event type, Table 8.7)			US Navy drop height (ft) (Event type, Table 8.8)		
					N	P	H
PBX 9404	(49)H	(25.9)H	(21.3)2H3N	(16.8)4N			
Comp. B3	(85)P	(21.3)P	(16.8)Pp	(12.2)N	519	610	
Octol 75/25	(45)P	(21.3)E	(16.8)E3N	(12.2)3N			
Comp. A3	(>150)N	(44.2)PN	(39.6)	(35.1)PN	244	285	396
TNT		(44.2)N	PN		610		

pellant composition in a weapon configuration. There are many other stability and compatibility tests, which generally involve accelerated ageing of the propellant at temperatures between 50 °C and 140 °C and the examination of distinct degradation phenomena, such as:

- rate of gas evolution;
- rate of stabilizer depletion;
- loss of weight under standard test conditions; and
- delay before the production of No_x gases.

Boggs and Derr (1990) carried out a comparative investigation on six of these tests for propellants and found that there was little correlation between the results from them. This is to be expected because different tests use different criteria to judge the main hazard perceived by the designers of the test, which may not be the same from test to test and site to site. These tests are required by the user to ensure that weapons have adequate life under expected storage and transport conditions and that they will remain safe and serviceable for an acceptable length of time in tactical deployment. They examine the depletion of stabilizers and the deterioration of relatively sensitive nitrate esters stored in conditions where deterioration is expected. Details from the results of the tests are used to assess the lives of rocket motors gun propelling charges and predicting when renewal is required.

8.5.5.1. Thermal Hazard Testing

8.5.5.1.1. Thermal Explosion Theory

The prediction of the thermal sensitiveness of energetic materials requires detailed knowledge of the thermal decomposition kinetics of the explosive or propellant materials involved. Kinetics data are incorporated into Thermal Explosion Theory, as reviewed by Gray and Lee (1967), Merzhanov and Abramov (1981) and by Lee in Chapter 6 of this book entitled "Theories and Techniques of Initiation," in order to generate equations relevant to the storage of weapons containing energetic materials at elevated temperature.

Thermal explosion theory is based upon the fact that energetic materials, explosives, propellants, and pyrotechnics decompose thermally with the evolution of heat. They are exothermic materials. The release of the heat of decomposition causes localized heating of the reactant. The temperature of the explosive tends to rise as a result of this heat evolution, but this may be nullified, or reduced, by the loss of heat from the locality of the decomposition to cooler parts of the mass of explosive decomposing, or its immediate surroundings. If, despite these losses, the rate of heat evolution exceeds the rate of heat loss, the temperature will rise out of control and explosion will occur. The theory first addresses the situation in symmetrically heated reactants.

The theories formulated by Semenov (1955) and Frank-Kamenetskii (1955) are reviewed in Chapter 6 of this book. It is shown there that what were regarded as separate and conflicting theories by their originators, are really the extremes of a continuous single theory, brilliantly connected by Gray and Harper (1959). They showed that the Frank-Kamenetskii theory refers to a system where the only resistance to heat flow resides in the reactant (the explosive), and the Semenov theory ascribes all resistance to heat flow at the surface of the reactant. This means that the Frank-Kamenetskii theory requires a distributed temperature in the reactant and the surface temperature of the reactant equal to that of the surroundings, whereas the Semenov theory requires an uniform reactant temperature and a temperature step at the surface of the reactant and the surroundings. Needless to say, in reality there is both a temperature distribution in the reactant and a temperature step at the reactant surface. Nevertheless, the Semenov and Frank-Kamenetskii theories also refer to actual physical conditions. The Semenov case refers to a well-stirred fluid in a thin-walled container, or a small, bare solid explosive, such as a single crystal, in air. The Frank-Kamenetskii case refers to a bare solid, or an unstirred fluid of low thermal conductivity in a thin metal case in a fluid environment. The Semenov, Frank-Kamenetskii, and intermediate conditions can be summarized by means of the value of a parameter known as a Biot number, given by

$$\text{Bi or } \alpha = \frac{\chi r}{\lambda}, \tag{8.8}$$

where χ is the surface heat transfer coefficient from the Semenov theory, r is the radius, or half-width of the infinite slab or cylinder or sphere, and λ is the thermal conductivity of the reactant. The Biot number is a measure of the relative importance between heat transfer at the surface of the reactant and within the reactant. If $\text{Bi} \to 0$ the thermal conductivity of the reactant is very high and the surface heat transfer coefficient is small. This is the Semenov condition, where all resistance to heat flow resides in the surface and we can simulate it by using a fluid as the reactant and ensuring that its temperature remains uniform by vigorous stirring. On the other hand, when $\text{Bi} \to \infty$ the thermal conductivity is small and the surface heat transfer coefficient is large. This is the Frank-Kamenetskii condition.

8.5.5.1.2. Hot-Spots

The thermal theory can be extended to cover the reaction of localized small quantities of heated explosives or foreign matter within the explosive matrix, through the theory of hot-spots. Here, the thermal field is very distorted locally, since the hot-spot may be of the order of 10^{-5} m in diameter and at a temperature of more than 1000 °C. Hot-spots are the points of initiation in many explosives tests involving the expenditure of mechanical energy on powders or charges, including impact, friction, skid, and intrusion. Any severe mechanical stressing of the reactant results in some damage to it. Impact, friction, or intrusion all cause break-up of the explosive and some of the tiny pieces may have their temperatures raised to very high values because of impact, friction, and/or shear. These small high-temperature regions may be sufficiently hot and have large enough bulk for their thermal energy to be sufficient for the burning process to spread throughout the charge. The equations of hot-spot initiation can be made similar to those of standard bulk thermal explosion so that they can be solved numerically and conditions for hot-spot initiation can be explored. There are two types of hot-spot. Reactive hot-spots, which consist of a limited quantity of the parent explosive which has been locally heated to very high temperatures, or inert hot-spots, which center on density discontinuities, such as air bubbles or dense inclusions, such as grit. Work on the system may cause the solid inclusion to move relative to the explosive matrix enough to generate heat as a result of friction, from which an explosion might grow. Alternatively, internal air cavities may be closed adiabatically as a result of mechanical action.

Hot-spots are the mechanism for the initiation of explosives by most of the mechanical methods described above, since each implies some form of shearing stress on the charge, either monolithic or powdered. However, while Merzhanov et al. (1963, 1966) were able to solve the relevant equations semi-analytically, and Lee (1992) produced numerical data describing criticality in compressed air, sand, and steel hot-spots in RDX and TNT, it is not possible, at present, to induce hot-spots exactly to a prescription and at a specific location in secondary explosives. However, hot-spots are not difficult to control or predict in certain initiatory materials and so are of great utility in the initiation of some initiatory materials in fuzes or igniters.

8.5.5.1.3. Fast and Slow Cook-Off Testing

Bulk thermal hazard testing began in earnest in the mid-1970s as part of the Insensitive Munitions Program launched in the United States, which started as an assessment of the response of a range of in-service systems to the effects of immersion in a fuel fire and impact of weapons by bullets or fragments. Later tests followed on, including slow cook-off.

In a Fuel Fire test, the thermal field is unlikely to be uniform and symmetrical, but less distorted than in a hot-spot. In slow cook-off tests it is possible that the thermal field could be relatively uniform. However, in either case, there are distortions to the conditions assumed in the simple thermal explosion theory, due to the time-dependence of the processes, the reactant consumption during decomposition, the casing around the charge, and the fact that the charge starts off cool and the temperature of the explosive must be raised to the stimulus temperature, either completely, or locally for reaction to start.

In a Fuel Fire, the heating of the explosive follows two distinct phases. The first is during the period when it is heated from ambient to the temperature of the flames, and the second is when self-heating due to the exothermic decomposition raises the temperature of the explosive above that of the flames. During the first phase the explosive acts as an inert because the temperature rise is relatively quick. In the second phase it reacts exothermically. Numerical solution of the non-steady-state equations of thermal explosion theory as outlined in the earlier chapter, first by Zinn and Mader (1960) and then Merzhanov et al. (1963), enables us to understand the phenomena associated with fuel fires and slow cook-off in qualitative terms at least.

Zinn and Mader sought to model the thermal explosion of 25.4 mm diameter spheres of RDX which were suddenly dropped into a Woods metal bath at several hundred degrees Celsius. Merzhanov et al. (1963) looked at thermal explosion in cylindrical symmetry. What they found is well illustrated by the curves reproduced by Boggs and Derr (1990) and shown in Figure 8.16. The curves illustrate two important phenomena of importance to thermal initiation of energetic materials, which have direct bearing on what happens in full-scale testing. Curves show how the temperature distribution changes as a function of time in an initially cool reactant dropped into a hot oven. The concept of the relatively inert, heating phase is well illustrated by the fact that almost the whole of the reactant passes through the temperature of the hot bath at about the same time.

Frank-Kamenetskii (1955) solved the approximate steady-state equation for the infinite slab, the infinite cylinder, and the sphere. Later, Gray and Lee (1967) obtained approximate solutions of the steady-state equation for the cube, the regular circular cylinder, several prisms, and the parallelepiped. When $\delta = \delta_{crit}$, the critical steady-state distribution of temperature in the reactant is defined. If $\delta < \delta_{crit}$ steady-state temperature distributions can also be obtained, each with a lower central temperature than that at criticality. Any increase of δ over δ_{crit} results in no solution being available for the steady-state equations. A new set of possibilities arises if the non-steady-state (time-dependent) equations are solved. Now, instead of producing one fixed temperature distribution for each subcritical state, a series of time-dependent temperature distributions is obtained for values of δ whether or not δ_{crit} is exceeded. Examples are shown in Figure 8.16 from the work of Merzhanov et al. (1963).

The families of curves have different characteristics according to the variation of the Biot number, Bi (Equation (8.8)), which is a measure of the

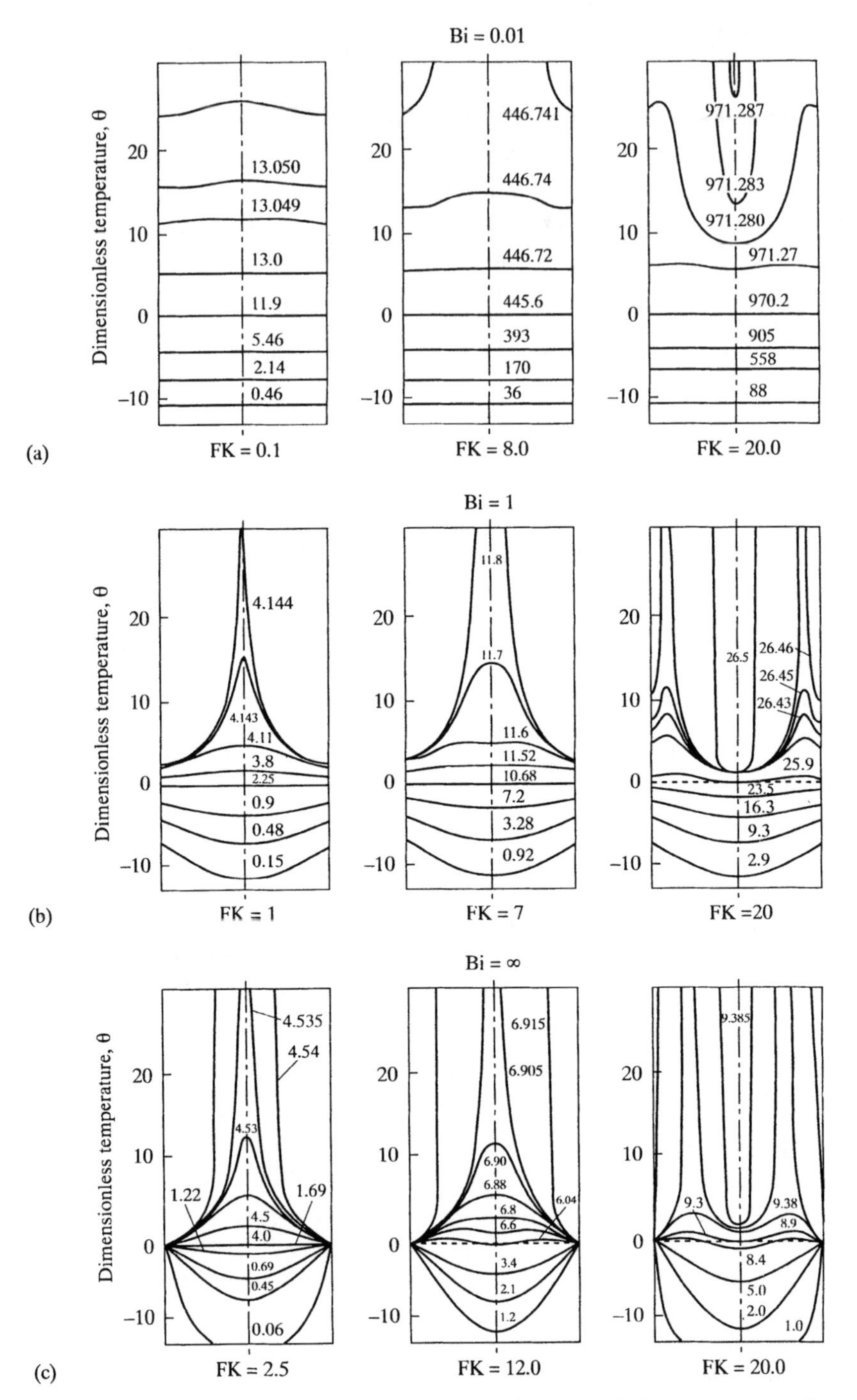

Figure 8.16. Non-steady-state temperature profiles for cylindrical reactants at various values of the Biot number and the Frank-Kamenetskii criticality criterion. (After Boggs and Derr: *AGARDograph 316*, 1990.) (a) Nonsteady temperature profiles at $\mathrm{Bi} = 0.01$, for several values of dimensionless time t/t_{ad}. (b) Nonsteady temperature profiles at $\mathrm{Bi} = 1$, for several values of dimensionless time t/T_{ad}. (c) Nonsteady temperature profiles at $\mathrm{Bi} = \infty$, for several values of dimensionless time t/t_{ad}.

relative importance of heat transmission through the surface of the reactant to the surroundings and the rate at which heat is produced in the reactant and arrives at its surface from the interior: it compares internal and external resistances to heat flow. As Bi tends to zero, so the physical conditions tend to those assumed by Semenov (1955), where resistance to heat flow is concentrated at the surface of the reactant or its container. As Bi tends to infinity, the physical conditions espoused by Frank-Kamenetskii (1955) become more nearly applicable and all resistance to heat flow resides in the reactant by virtue of its low thermal conductivity. Hence, those temperature distribution curves (profiles) associated with solutions of the non-steady-state equations where $\mathrm{Bi} \rightarrow 0$ tend to be flattish whereas those associated with $\mathrm{Bi} \rightarrow \infty$ tend to be significantly curved.

The second influence on the shape of the time-dependent temperature profiles is the value of δ, in relation to δ_{crit}, the Frank-Kamenetskii Criticality Criterion, referred to as "FK" in the diagrams. This dimensionless parameter encapsulates the main chemical kinetics characteristics of the explosive material and its size and shape. For values of $\delta < \delta_{\mathrm{crit}}$, which for a cylinder, as studied by Merzhanov et al. (1963), is about 2, the steady-state equations of heat conduction with distributed sources of heat can be solved to give steady-state temperature distributions across the reactant, i.e., no explosion occurs: the system self heats to give a peak central temperature, which then decays (see values of "FK" < 2.0 in Figure 8.16). In Figure 8.16, if δ just exceeds δ_{crit} the peak reaction temperature occurs in the center of the reactant and the explosive explodes or inflames from the center, more or less as an extension of the steady-state theory. However, as δ exceeds δ_{crit} by successively larger amounts, the point of ignition gradually moves outward toward the surface. In the limit, when $\delta \gg \delta_{\mathrm{crit}}$ ignition occurs at the surface of the reactant.

The importance of the curves in Figure 8.16 is that they help us to understand qualitatively the conditions in fast and slow cook-off tests and why the results of such tests are so different. The important parameter is the Biot number. In cases where the Biot number is small, i.e., 0.01, the temperature profiles appear to be relatively flat and there is a temperature step at the walls of the reactant, exactly as predicted by Gray and Harper (1959). This is an approximation to the Semenov condition. The nearly uniform temperature in the reactant need not necessarily be ascribed to its high thermal conductivity (or it being a being stirred fluid) in this non-steady-state analysis. It might also be due to a slow increase in the temperature of the environment. This would cause the wall temperature of the reactant to rise slowly as shown in Figure 8.16(a). This is exactly the situation in a slow cook-off. Hence initiation is expected to take place at the center of a reactant in a slow cook-off test. On the other hand, in a fast cook-off situation and especially when the weapon being tested has a relatively thin skin of high thermal conductivity metal, e.g., aluminum or steel, the major impediment to the transfer of heat through the system resides in the low thermal conductivity of the reactant, the explosive, or propellant in the weapon. This is approximated by the situation in Figure 8.16(c), where the Biot number is very large. In both cases there is also the complication of the degree to which the critical condition is exceeded. In the slow cook-off case, the criticality criterion is likely to be exceeded only very slightly, thus augmenting the conditions consistent with a central initiation. In the Fast Cook-Off, or Fuel Fire immersion, the criticality criterion is likely to be exceeded significantly simply due to the very high temperature to which the system is exposed, because the absolute temperature of the environment appears in the expression for δ as a negative exponential, which has an enormous effect on its value. This means that initiation is likely to occur near the surface of the system. In the former case of slow cook-off the explosive itself augments any confinement to the weapon as a whole and this tends to maximize the event. In the case of fast cook-off, if ignition occurs near the surface of the reactant where there is minimum confinement, the event is likely to be the minimum possible for the material/case combination being tested, because the case soon fails and reduces the confinement on the explosive.

In either case the presence of a case makes a difference, since it provides confinement and protection to the energetic material from the heating stimulus and it also defines the surface heat transfer coefficient between the system and the surroundings. However, these data indicate in principle why weapons tend to "pass" fuel fire test conditions and "fail" slow cook-off conditions, and why a system which meets fast cook-off requirements may not meet those for slow cook-off.

Thermal theory can also provide estimates of the so-called "safe" diameter of an explosive charge and the "safe" lifetime over which it can be stored at a given temperature. If we use the expression for δ_{crit} developed in the earlier chapter,

$$r_{safe} = \sqrt{\frac{\delta_{crit}\lambda R T_{crit}}{\rho E q_{T_{crit}}}} \quad \text{where} \quad q = F(Q)e^{-E/RT}. \tag{8.9}$$

where $q_{T_{crit}}$ is the heat generation at criticality. This is an expression of the intuitive assumption that larger charges cook-off quicker than smaller ones at a given supercritical temperature and may also explode at temperatures at which smaller charges are "safe." The theory can also be used to determine the period of "safe" life, if the "safe" diameter can be predicted for storage at ambient temperature. If we wished to determine the stability of an energetic material over, say, a 10 year period of storage at ambient temperature, we would need to simulate the temperature profile the charge might see during that time. The US Navy uses the following profile (Boggs and Derr, 1990):

10 years at 30 °C (countries in Northwest Europe use 20–25 °C) plus 14 days with the following daily schedule:

9 hours at 35 °C,
5 hours at 50 °C,
5 hours at 60 °C,
5 hours at 71 °C.

It is not reasonable to have to wait 10 years for data on a new system and so the effects of 10 years' storage at 30 °C have been simulated by one week's ageing at 85 °C. This temperature and time were established by experiments carried out in the 1980s by van der Mey and his coworkers (1984, 1985) and by van Geel (1971). Degradation of propellants after this short exposure at these elevated temperatures is in close agreement with the effects of long-term storage at the lower temperature.

These tests and others involving the thermal decomposition of explosives require data on the thermal decomposition kinetics of the explosive. Decomposition data can be determined using Differential Scanning Calorimetry (DSC). Definitive work in this field was done by Rogers (1972, 1975) and his results are quoted extensively by Gibbs and Popolato (1980) and Dobratz (1981). Details of decomposition kinetics of reactants are determined by heating a small sample in a Differential Scanning Calorimeter and subjecting it to a programmed temperature increase with time. The heat output from the decomposing explosive in one of two identical platinum alloy cups causes its temperature to rise more rapidly than the system programming would expect. The other cup which contains a sample of a reference material is heated electrically in order to maintain the same rate of temperature rise. The difference in the rates of energy supply to the two cups is proportional to the heat generation in the sample. Phase transitions of an exothermic or endothermic nature as well as the thermal decomposition process can be followed. Tests can be done on sealed or unsealed cups and there is provision for the use of inert atmospheres. Sealed cups can be used for testing at internal pressures of up to about 15 MPa from approximately 10 mg of sample material. Some decomposition kinetics and enthalpy data for a number of common explosives are recorded in the *LANL Explosive Data Manual*, and Table 8.12 is an amalgam of two tables from this volume and also includes some of the data already presented in Table 2.4 in Chapter 2 of this book.

8.6. Electrostatic Sensitiveness

An excellent digest of current thinking on electrostatic sensitiveness of propellants appears in Boggs and Derr's monograph (1990).

Table 8.12. Decomposition kinetics and enthalpy data for some common explosives (Dobratz, 1981; Gibbs and Popolato, 1980).*

Explosive	State solid (s) liquid (l)	Density (ρ) kg/m^3 × 10^{-3}	Heat of reaction (Q), kJ/kg	Frequency factor (A), s^{-1}	Activation energy (E), kJ/mole	Heat of combustion (ΔH_c^0), kJ/mole	Heat of formation (ΔH_f), kJ/mole
HMX	liquid	1.81	2093	5×10^{19}	220.58	−2765.4	47.3
	vapour			1.51×10^{20}			
HNS	liquid	1.65	2093	1.53×10^{9}	126.82	−6446.4	78.27
NQ	liquid	1.63	2093	2.84×10^{7}	87.48	−880.6	−84.9
PETN	liquid	1.74	1256	6.3×10^{19}	196.72	2589.6	−461.8
RDX	liquid	1.72	2093	2.02×10^{18}	197.14	−2100.3	61.53
TATB	vapour			3.14×10^{13}			
	solid	1.84	2511	3.18×10^{19}	250.6	−3080.1	−139.8
Tetryl	liquid	1.73		2.5×10^{15}	160.72	−3499.1	31.81
TNT	liquid	1.57	1256	2.51×10^{11}	143.98	−3420.4	−50.23

* Heat of Reaction is the exothermicity of the decomposition reaction, for which the activation energy and frequency factor are quoted. These are usually determined in Differential Scanning Calorimeter experiments, such as those referred to by Rogers (1972, 1975). Heats of Combustion and Formation refer to the standard state at 25 °C, determined in an oxygen-bomb calorimeter, fitted with an automatically controlled adiabatic jacket, calibrated by burning standard benzoic acid to determine its effective energy equivalent (Gibbs and Popolato, 1980).

8.6.1. Powder Tests

Most explosives laboratories carry out relatively unsophisticated electrostatic discharge testing of powdered explosives in order to determine any serious electrostatic hazard. The tests assess the hazard from the point of view of prediction of the precautions required in the handling of the materials. For example, some tests seek to determine whether the material is susceptible to discharge from people's clothes, in which case full second degree precautions are required, including earthing of personnel, bench tops and equipment, and imposition of minimum humidity conditions. Alternatively, the tests may indicate that the dry bulk material is susceptible to discharges in handling, or that certain types of electrical equipment might constitute an hazard. The tests involve discharging capacitors of known size into small quantities of the explosive and determining at what level (of no more than about three or four possible levels), if any, initiation occurs.

In the main, high explosives in weapons seem to be relatively immune from the effects of high-voltage electrostatic discharge, although propellants in rocket motors are not. This may change as explosives and propellants become more and more similar, chemically and physically, and advanced composite warheads begin to become more common. The difference, as perceived now, probably resides in the fact that the propellants are often encased in multiple layers of nonconducting materials and are able to build up significant levels of static electricity, which may discharge through the grain under certain circumstances. They also contain inherently more sensitive materials, such as ammonium perchlorate and nitroglycerine. The theory is currently being developed at China Lake, where Covino and her coworkers (1988, 1990) have identified:

- volume resistivity;
- dielectric breakdown; and
- dielectric constants;

as important parameters in the understanding of electrostatic discharge breakdown of propellants. Equipments have been built at China Lake to measure both surface and volume resistance of propellants over a temperature range from −30 °C to +100 °C. French work is quoted by Boggs and Derr (1990) in which it is claimed that the shape of the volume resistivity versus reciprocal of absolute temperature curve of a propellant can be used to predict whether or not it is susceptible to capacitative discharge.

The propellant in a motor is subject to electric stress if there is a significant voltage across the grain. If this dielectric fails, electrical energy leaks rapidly through the structure of the propellant to earth. In an attempt to understand why some propellants are electrostatic discharge sensitive and others are not, the Percolation Theory of Hammersley (1966) was developed. This predicts that the aluminum particle size and the electrical properties of the binder are major factors in the determination of the electrical properties of the propellant. Percolation theory enables us to determine a critical ratio between conducting and nonconducting particles above which the system is nonconducting. Somewhat complex algebraic expressions for defining a breakdown percolation coefficient have been produced, but they do not include many critical propellant parameters and so cannot describe the propellant system accurately as yet. In fact, while the research is not yet mature, our knowledge has advanced significantly with the work of Covino et al. (1988, 1990) and it is aimed at being capable of indicating whether an ESD hazard might be expected ab initio.

8.7. Assessment of the Results of the Tests on Energetic Materials

If we were unaware of any other hazard characteristics of explosives, the results so far presented would indicate that there are some types of stimulus, such as the impact of a bare charge with a very hard rough surface, which induce high-order events in some main charge explosives, especially those which contain more than 90% of HMX. However, that the commoner TNT-based octols and cyclotols and TNT itself are unlikely to react especially violently, even when relatively large bare charges are subjected to fairly energetic mechanical stimuli. Thermal stimuli, especially slow cook-off would be expected to generate high-order thermal explosions in large charges, because of the self-confinement, but, by implication, unless high confinement were present, fast cook-off would produce low-order events. The evidence from the accidents in the Gulf of Tonkin and the United States railroad freight wagon fires in the late 1960s and early 1970s, in which many weapons loaded with Composition B filled bombs exploded violently, is consistent with these general findings. Thus, for many acceptably safe explosives, with few major shortcomings, responses to stimuli depend markedly on the immediate environment, i.e., confinement and thermal insulation. The effects on the environment as a result of these responses depend on the explosiveness of each individual system rather than specifically the characteristics of the explosives alone.

The powder and bare charge tests reported above, especially the charge tests, were designed originally to assist in the screening of explosives' sensitiveness, without regard to the explosiveness of the system, and most tests prior to the 1970s were conducted without significant confinement. The accidents referred to gradually drew the attention of the explosives community to the fact that the systems responses of weapons to accidental initiation of their explosives fillings had not been addressed. This is not surprising because no hazard tests at the time imposed sufficient confinement on charges to cause other than the most sensitive and energetic explosives to detonate. Often relatively low-order events were recorded with octols and Cyclotol 75/25 in the Oblique Impact, Susan, and Spigot Tests, yet weapons filled with these explosives responded damagingly in weapon accidents.

Hubbard and Lee (1978) in the United Kingdom and Bernecker and Price (1975) in the United States attempted to follow the development of non-detonative reaction (deflagration) of explosives such as RDX/TNT (60/40) in relatively small charges. Both used steel tubes fitted with screwed end caps. While it was relatively easy to generate a very energetic deflagration event, traveling at up to 1800 m/s, most tubes failed before a higher-order event occurred. However, one of the outcomes of the tests, especially with large tubes containing up to 12 kg of RDX/TNT (60/40), was that even a rapid deflagration could be extremely damaging and generate fragments which were a serious hazard in themselves. This was proven later when, during trials to assess the response of high explosive shell to the effects of the fragments from donor shell an experiment was conducted in which the donor shell was caused to deflagrate, rather than detonate. Some of the several receptor shells arranged around the donor detonated, as shown by ground signatures and overpressure records. As a result of this and similar tests, it became clear that no hazard tests adequately predicted such behavior.

8.8. "Insensitive" High Explosives and Propellants

As a result of the United States aircraft carrier accidents, several nations initiated work to reduce the vulnerability of the energetic materials which were commonly used to fill warheads and rocket motors, and also to develop tests to assess which weapons in their inventories were most likely to respond in a manner which would contribute to the spreading of a conflagration or lead to mass explosion of weapon stores.

Much effort was devoted to the search for an energetic molecule which was both highly insensitive and powerful, yet could easily be processed into mass-produced warheads and rocket motors. It has not been found yet, although there are new energetic molecules being synthezised and assessed, especially in the United States, at China Lake, and in Xiang, in China. Materials such as TATB, nitroguanidine, and ammonium nitrate were assessed and many different compositions using them alone, or with other explosives, were designed, developed, and tested, without much success. The problems usually centered around relatively low detonative output of explosives and low specific energy of propellants, notwithstanding costs of manufacture of TATB, difficulties of filling mass-produced munitions with TATB, or nitroguanidine compositions, which are best handled as pressable compositions, and the perennial problems of the phase transition of ammonium nitrate at 38 °C and its hygroscopic nature.

Explosives scientists and engineers appeared to be in a cleft stick in the mid-1970s concerning "insensitive" explosives, since they could provide compositions which would fulfil weapons system designers' requirements for low vulnerability to fire and fragments, but the compositions did not perform adequately to meet performance requirements and were too expensive to make and fill into most munitions. Alternatively cheap, easy to process, high-performance explosives were available, which did not meet the requirements of lower vulnerability. The US Navy drove forward the work by insisting that certain low vulnerability requirements had to be met by new weapons systems before they would be permitted to be embarked on US Navy vessels. The US Department of Defense recognized the high strategic and tactical value of lower vulnerability systems and backed the programme wholeheartedly. Similar sentiments were expressed elsewhere and other nations began their own programmes. Cooperation and the free flow of information have characterized the search for lower vulnerability explosives and weapon systems for the past 20 years in the major Western democracies. Much of what has been achieved is due to the large sums of money invested by the United States, but other nations have generated their valuable inputs.

The breakthrough has come with the realization that, since the response of weapons to thermomechanical attack is a system response, i.e., it depends upon the explosive filling, the casing and packaging of the weapon, all these factors need to be addressed in efforts to reduce the response. No longer is the problem merely one of energetic material hazard, or *sensitiveness*, but it is now a system, or *explosiveness* problem. The search for lower vulnerability explosives now centers more on the understanding of the physico–chemical processes associated with the initiation and growth of reaction in a cased explosive, rather than on the search for the "Holy Grail" of a powerful, but insensitive explosive molecule.

Many more explosives than those referred to here as examples of the responses of energetic materials to hazard tests, have been tested than there is room to record in a document such as this, and many other tests have been carried out than those referred to here. Our understanding of initiation and growth of explosion has expanded significantly in the past 20 years beyond sensitiveness and sensitivity of materials. This is hinted at in the results of the Oblique Impact, Spigot, and Susan test results. Charge responses to stimuli depend on the materials properties of the charges and on the environments under which they were tested. Different versions of a given sensitiveness test induce different results because they are testing the explosive in different environments. Thus, it is likely that, to a greater or lesser extent, even those tests purporting to be sensitiveness tests are, at least in part, explosiveness tests, because the environment which the explosive finds itself on ignition has such a dramatic effect on the outcome of the test.

Once this is understood, differences between the results of hazard testing at different locations using different equipments become explicable, at least qualitatively. Moreover, as understanding of the effects of environment of explosives in weapons, e.g., confinement and thermal insulation afforded by the casings, explosives can be tailored to respond in the minimum fashion to stimuli as much by weapon structural engineering and explosives composition control as by choice of molecular type. There is insufficient space in a review of this nature to dwell upon the weapon structural design which has already been developed to reduce system responses. Indeed, much of it is classified for military or commercial security reasons, but trends in design are well documented, especially via NIMIC in Brussels.

This review will concentrate on the energetic materials and the means by which it is possible to reduce their responses to potentially hazardous thermomechanical stimuli. Of course, there are conditions where no explosive other than, say, TATB could possibly withstand the stressing levels imposed on a system without exploding, but there are worthwhile hazard versus performance trade-offs which can be made using sensitive explosives, such as RDX and HMX. The clues were available many years ago, when it was found that TNT-based explosives appeared to respond very energetically to nondetonative stimuli under confinement, yet were perfectly safe to manufacture and test, even to shoot from guns with accelerations of up to 50,000 g. These explosives rely on the TNT matrix to bind them together. TNT forms rather large, weak crystals with many parallel planes of weakness. Under compressive, tensile, or shear stress, especially at relatively high strain rates, TNT fractures in a brittle fashion, generating large extra surface area. A burning process develops faster in a charge with an high surface area than one with a low area. For example, pressed gunpowder burns relatively quietly as a pellet, but a train of loose gunpowder can burn at tens of metres per second.

One of the main features of an explosive composition which has a relatively low level response to initiating stimuli is a matrix which is resistant to cracking under the initiating conditions and the subsequent local pressurization of the explosive. Plastic explosive, such as UK PE4, or US C4, packed into a closed steel tube and ignited by a propellant charge will often do little more than blow the end closure cap of the tube off, leaving more than 95% of the explosive unconsumed. Even this damage is mainly ascribable to the energetics of the initiating propellant charge and its pressurization of the container. The same explosive materials cooled below the glass temperatures of their respective explosive binders will react very rapidly, causing the tube to fragment into many pieces, with less than 5% of the explosive unconsumed. In the first case, pressurization was unable to generate new surface, the explosive burned relatively slowly and overstrained the container, which burst and relieved the confinement in the vicinity of the burning area. The explosive was unable to burn sufficiently rapidly after the loss of confinement to maintain an adequate supply of chemical power to unburnt material and so the process was extinguished. The container burst as a result of excessive plastic strain due to relatively slow pressurization.

In the second case the explosive cracked extensively. This accelerated the burning process and increased the pressurization rate of the container above the stain rate at which plastic failure occurs in the metal and it would have failed in a brittle manner. However, despite the speed of the failure process, it would have been too slow to relieve the local pressure due to the very rapid decomposition of the explosive and the inertia of the fragmented casing. Thus, the confinement remained effectively in place until the explosive was consumed.

Polymeric bonded explosives also show high toughness and compliance at normal temperatures. Although, as has been shown above, it is important to ensure that the glass temperature of the polymeric material either is lower than the minimum expected operating temperature of the explosive, or some suitable plasticizing agent is added to depress it. In essence, this is the key to low vulnerability explosives. The correct choice of polymer, plasticizing agent and, possibly, some adhesive to bond the solid solution of polymer and plasticizing agent to the explosive crystals, usually RDX or HMX. A further refinement would be to use a polymeric matrix which decomposes thermally in an endothermic fashion. In the early stages of deflagration after initiation, when the process is still relatively slow and results in relatively low energy output per unit time, its decomposition would cause reduction of temperature locally and serve to reduce the reaction rate of the exothermically decomposing explosive. Extra energy in detonation may be obtained by the use of energetic polymers with energetic plasticizers, but, as yet no successful explosive or propellant composition of this type using both is available and little is known about the balance between exothermic and endothermic polymers required for optimized safety and performance.

The technology of lower vulnerability explosives has advanced more rapidly than that of propellants and pyrotechnics, because, in the main, explosives compositions are simpler than propellants and pyrotechnics and the interaction between the polymer and the explosive molecules does not affect unduly the behavior of the composition, when operating in the design mode. Propellants are more complex in their compositions and require many additives. Rocket propellants require platonizing agents, antismoke additives, burning rate modifiers, and antioxidants. Gun propellants need stabilizing agents, antiflash additives, antioxidants, etc. Pyrotechnics are notoriously difficult to design in plastic-bonded form because of the use of metal salts and metals and metalloid elements, which form highly energetic fuel–oxidizer mixtures. This environment reduces the life of many polymeric molecules and may generate either heat, or gas, or both in the matrix. Lower vulnerability propellants and pyrotechnics are about 10 years behind explosives in development.

8.9. System or Weapon Tests for IM

Following the accidents in the Gulf of Tonkin, the US Navy and other Western navies reconsidered the potential hazard from weapons carried on board ship by testing them for their response to the effects of immersion in a fire and to the effects of the impact of projectiles. Gradually the tests have been refined to include the effects of exposing weapons to a slowly ramped temperature as from a fire in an adjacent compartment on board a ship, or from more than one fragment from a detonating warhead close by. Currently many nations use the US Navy 2105B document of January 1994 as their guide to hazard testing of weapons systems and this review uses diagrams prepared by Dr. M. Held of Deutsche Aerospace, which he generously supplied. The tests employed throughout NATO and in Australia and New Zealand are essentially similar and involve assessments of:

- thermal response of a weapon exposed to a standard fuel fire (Fast "Cook-Off");
- thermal response of a weapon to a ramped temperature rise (Slow "Cook-Off);
- response of a weapon to attack by a single, or multiple bullets;
- response of a weapon to multiple fragment impacts;
- response of a weapon to dropping from an height related to that likely to be experienced by the weapon during ammunitioning or deammunitioning procedures on board ship or at a wharf; and
- response of a weapon to the effects of the impact of a shaped charge jet.

In each case the response of the weapon or weapon system is to be no greater than burning, with no projection of fragments, or debris more than a

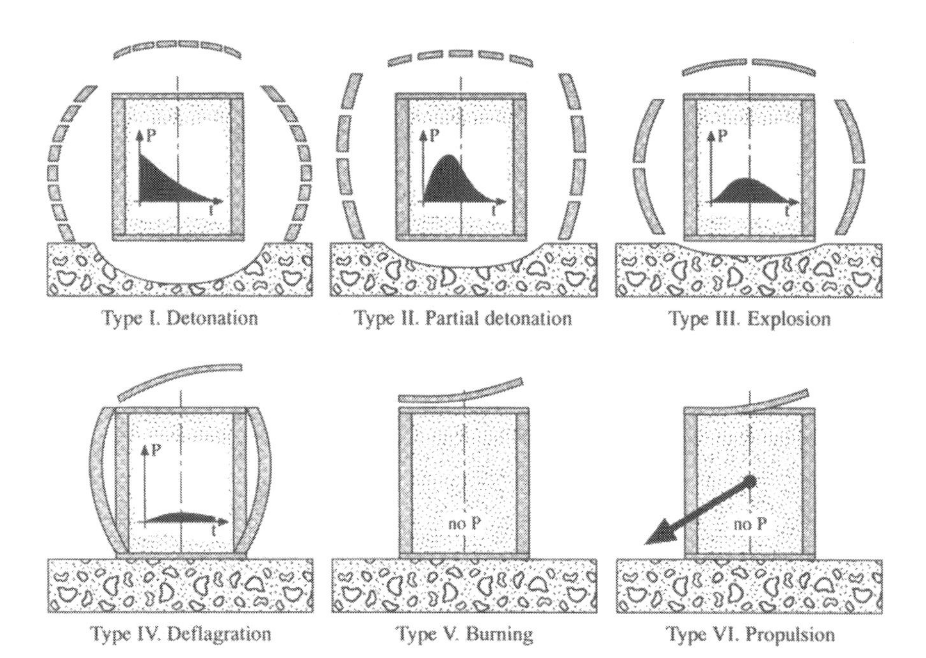

Figure 8.17. Explosive reaction levels. (With permission of Professor Dr. M. Held, Deutsche Aerospace.)

given distance. A useful illustration of the levels of severity of the results of tests is shown in Figure 8.17. The terminology is not exactly the same as in Tables 8.7 and 8.8, but it is a relatively simple matter to use the two tables and Figure 8.17 to gain a clear picture of the levels of damage resulting from the tests.

Results from these and related tests are weapon specific and therefore say little about the hazard characteristics of the explosives or propellants involved, because the responses of the systems are dominated by system characteristics. For example, it is possible to render even sensitive and highly energetic propellants, relatively innocuous in rocket motors in fuel fires, provided the motors are fitted with thermally sensitive devices which remove the confinement of the propellant if the temperature approaches the thermal explosion temperature of the composition used in the conformation tested. The response of weapon systems is, as a result, not only classified in the military sense, since it is undesirable to inform potential adversaries of any possible vulnerabilities of one's weapon systems, but also commercially, since competitors should be prevented from knowing what measures are available to reduce vulnerabilities and explosivenesses of systems in one's inventory.

8.9.1. Insensitive Munitions Thermal Tests

8.9.1.1. Slow Cook-Off Tests

The original purpose of slow cook-off testing was to obtain insight into the response of weapons to exposure to a slowly ramped temperature. The test was intended to simulate the effects on the weapon of heat generated in an adjacent compartment on a ship, from a fire there. A temperature rise rate of 6 °F (3.3 °C) per hour was chosen as typical of such an environment and the test consists of a disposable oven with a capacity for a controlled temperature increase at the rate specified as shown in Figure 8.18. The principle of the test is realistic, although doubt has been thrown on its validity in some quarters recently, because the temperature ramp is not limited and the temperature is programmed to rise monotonically until an event occurs. Those objecting to the test may not understand fully the principles which have been discussed above, namely that the so-called "Insensitive Munitions Tests" provide means by which the *explosiveness* of weapon systems is assessed as well as, in fewer of the tests, *sensitiveness*. An event has to be permitted to occur before an assessment can be made of its violence. In the mechanical stimuli tests some sort of event occurs. In the slow cook-off test, it is possible that, if the ramp was programmed to cease at a given temperature, some of the explosives tested might not explode or even react. This would waste the effort expended on the test, since a non-event gives no data on sensitiveness or explosiveness, moreover it provides a significant safety problem, since the weapon would need to be disposed of in situ. While it is true that all slow cook-off tests end in an event, there is discrimination

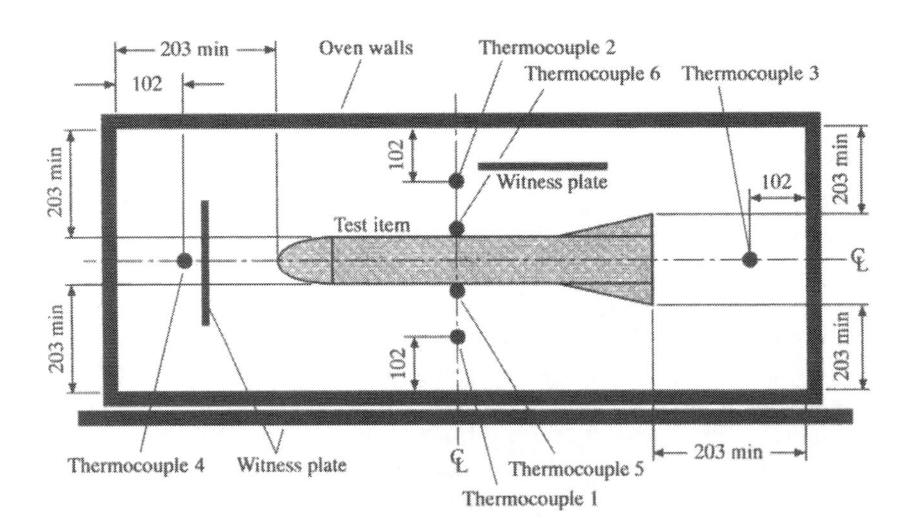

Figure 8.18. A typical slow cook-off disposable oven. (With permission of Professor Dr. M. Held, Deutsche Aerospace.)

between both materials and weapons via the explosiveness of the response to the stimulus. A response has to be generated in any explosiveness test, otherwise the effects of system characteristics, other than those of the energetic material, will not be readily available.

Slow cook-off tests are somewhat destructive, since no weapon system as yet has been able to meet the requirements that there should be no event more serious than burning. It is often the case that severe explosions occur. Therefore it is essential that the test be carried out in disposable equipment which uses the cheapest types of monitoring equipment. This usually consists of thermocouples set at various positions in the air-stream and at the surface of the weapon (see Figure 8.18), video cameras and air blast gauges around the test site. The weapon is heated by circulating hot air which has sufficient heat capacity and adequate thermal transfer characteristics because the heating rate is required to be very slow. An important adjunct to such tests is the need to be able to destroy the weapon after test, if it does not completely burn out or explode during it. The UK Test (SCC No. 3, 1988) uses a trolley with an explosive charge on it which can be set remotely against the side of the disposable oven if problems arise during the test.

United States tests, principally the one developed at the Navy Weapons Center, China Lake, are definitive (US Military Standard, DoD-STD-2105B (Navy), 1994). It is permitted to preheat the weapon uniformly to 55 °C below the predicted temperature at which cook-off is expected to occur and which is determined from thermal explosion theory and experience with other weapons with the same or similar explosives. In some United States tests an expendable video camera is used to monitor formation of exudate, if this is considered to be necessary.

8.9.1.2. Fast Cook-Off Test (Fuel Fire Test) (*US Military Standard*, DoD-STD-2105B, 1994; *Manual of Tests*, SCC No. 3, 1988)

The test is designed to ensure that the weapon is fully engulfed in the flames from a liquid aviation fuel fire and that the fire burns for a sufficient time to ensure that, should any reaction in the weapon be likely to occur, it will have the opportunity to do so. The United Kingdom and United States tests are very similar and an amalgam of the two will be described. The United States test specifies JP-4, -5, or -8 or Jet Al fuels, the United Kingdom test AVCAT or commercial kerosene to a British Standard Specification. The test is conducted in an open hearth, with a concrete base, into which sufficient fuel is poured to enable the fire to burn for about 20 min, or as long as is necessary for the munition to react fully. In the United Kingdom test the hearth is about 10 m square and is surrounded by a wall of concrete blocks with ventilation gaps between them. These act both as a windbreak and also protect the test site from fragments. Both tests use several thermocouples (with time constants of less than 0.1 s) to record temperatures in the flames from the fuel, at each end of the weapon and either side of its midpoint in

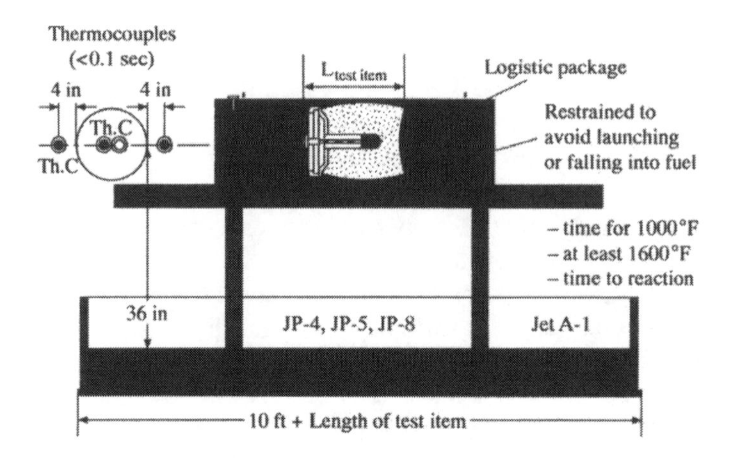

Figure 8.19. Fuel fire weapon support and restraint. (With permission of Professor Dr. M. Held, Deutsche Aerospace.)

the horizontal plane of its centerline about 100 mm from the skin of the weapon. Figure 8.19 shows a test arrangement.

It is necessary, for an adequate test to be conducted, that the temperature in the vicinity of the weapon should reach 538–550 °C between 30 s and 35 s from ignition. An average flame temperature of at least 870° is required to be measured on all thermocouples to generate a valid result. Reaction severity is determined from video and sound recordings and blast overpressure gauges. The US Navy has also added the capability of video scanning of the motor nozzle of a missile system, if it is being tested unpackaged and a 9 MeV X-ray facility to capture initiation and break-up of the weapon.

8.9.1.3. Results of Cook-Off Tests

United States and United Kingdom test results indicate that it is not possible, as yet, to predict the response of a given weapon to fuel fire testing based solely on the results of small-scale testing. The effects of the casing and packaging of weapon systems are of such overriding importance in influencing the outcome that, until full three-dimensional internal and external heat paths of any given munition can be modeled adequately, it will not be possible to predict results without full-scale testing.

Attempts have been made in the United Kingdom (Manners, 1992) and the United States (AGARD, 1992) to carry out tests on generic weapons, especially rocket motors. Results have given considerable insight into the relative sensitiveness of different propellant compositions in a given case design and material, based on assessment of minimum temperature of ignition. The tests have also provided comparative data on the behavior of different case materials filled with a given propellant. In United Kingdom tests "Kevlar" overwound and steel strip laminate cased motors filled with cast double base, elastomeric modified cast double base, and hydroxy-terminated polybutadiene propellants all gave 100% "pass" results in fast cook-off tests on model scale motors. Tests carried out by Hercules (AGARD, 1992) using their own propellants gave similar results for composite, strip laminate, and aluminum motor bodies. However, none of these tests enable us to generate the data for an unknown weapon, because the state of our knowledge of the thermal behavior of explosives and the thermal behavior of complex materials structures in a fluctuating and wide temperature range is too poor at present. Even if it were not, it is not certain that the fastest supercomputers available at present would be capable of delivering full three-dimensional output data cost-effectively.

8.9.2. Impact Tests

There are two types of impact test to discuss: bullet and fragment attack and drop test. The drop tests themselves are of two types, simple drop tests of packaged and unpackaged weapons to ensure that after dropping from the maximum height at which they are allowed to be slung from in a loading/unloading operation of a weapons platform, usually a ship, they are safe for disposal and a drop test on to a spigot. The latter test is a modification of the one described earlier. The spigot simulates a metallic protuberance, exemplified by, say, a stanchion on board a ship. Rarely are tests performed using complete weapons because of the cost. If doubts have been expressed on the case-energetic materials combination, it is customary to carry out tests using the technique described earlier for the Spigot Test.

8.9.2.1. Bullet and Fragment Attack Tests

Bullet attack tests are usually carried out by firing an armor-piercing (AP) 12.7 mm (0.5 in.) projectile at a warhead or motor at its muzzle velocity of about 875 m/s. Provision is made for high-speed camera observation of the events and for monitoring the overpressure and sound from the event, as shown in Figure 8.20, in which an arrangement for firing three closely separated projectiles is shown.

Few high-order events have been observed in firings against most rocket motors and warheads from the impact of a single bullet. This is not very surprising, since the AP bullet is designed to penetrate a target with the minimum loss of energy to the target. In this way it can penetrate deeper and, since it gives up its energy slowly to the target, the target does not need to respond rapidly and, even if it did respond, it would only need to accommodate the effects of a low-energy transfer rate. Consequently, the explosive would not necessarily be heavily damaged and cracked beyond the impact point. Thus, if it were to ignite, the ignition might not accelerate into a fast burning or detonation, because relatively little extra surface would have been generated to aid the acceleration of any burning initiated on impact. If,

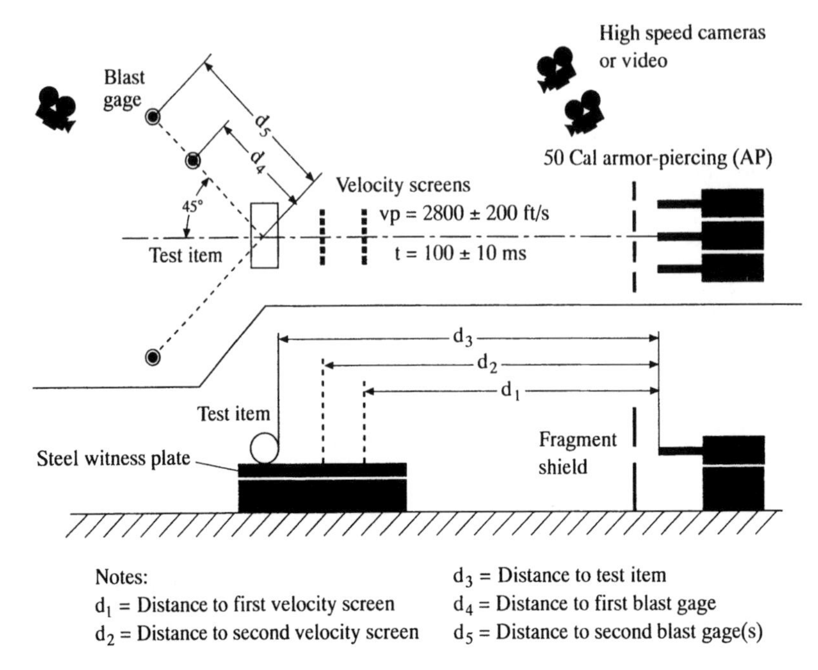

Figure 8.20. Three bullet impact test. (With permission of Professor Dr. M. Held, Deutsche Aerospace.)

on the other hand, the explosive or propellant is cracked by the energy input from the projectile, substantial new surface may be created. The effects of firing a burst of three bullets into the target is both more realistic and also more likely to initiate a significant event, since the second and third bullets, arriving closely after the first contribute their relatively low-energy inputs to a system under complex transient stressing. While the individual energy inputs are relatively small, the overall effect is to generate a longer pulse in the charge and the explosive or propellant is likely to respond more vigorously to it by cracking.

If the projectile impacting the target is nearly flat-fronted (standard fragments used in testing have 160° conical front faces), impact with the case and the energetic material, if the fragment is able to perforate the case, cause significantly more damage. A more intense shock wave is generated in the target by this type of fragment than an AP bullet and this may cause significant damage to both the casing and the energetic material.

Consequent upon the delivery of this large energy pulse, the energetic material may crack severely and generate much new surface. The initial pressure pulse will be in the form of a shock. Most energetic materials can be made to react under the effects of relatively modest shocks, although this might induce relatively long runs to detonation, should it even be possible. If there is adequate confinement, burning initiated as a result of the shock

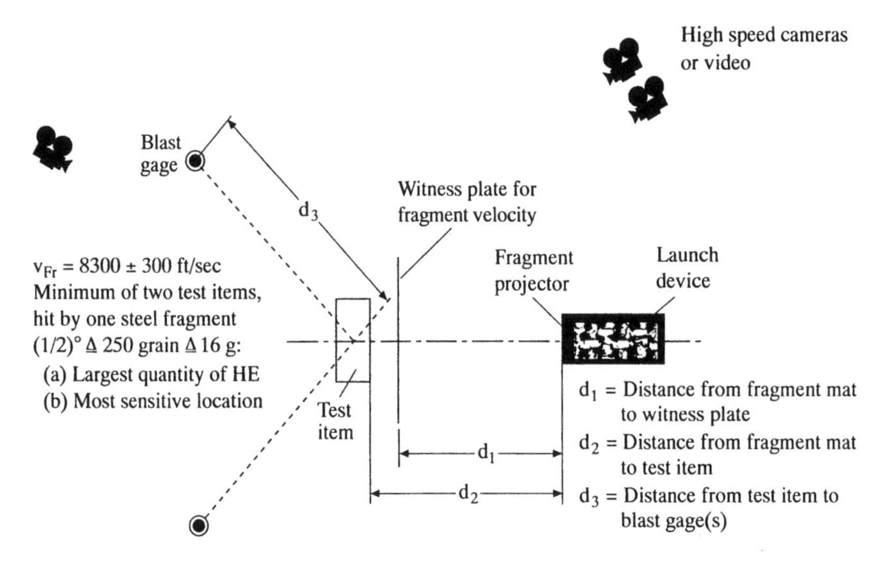

Figure 8.21. Arrangements for a fragment attack test. (With permission of Professor Dr. M. Held, Deutsche Aerospace.) *Note*: V_{Fr} is the velocity of the impacting fragment.

input from the projectile may accelerate and a significant event ensue because of the extra surface created. The event is likely to be greater than one from a single bullet of the same mass and velocity, because of the more intense initiating conditions. Moreover, at high-impact velocities, the shock content of the impact is likely to be high and, since fragment impact testing also involves multiple fragments, it is much more likely that a multiple fragment attack will engender a very rapid reaction than a multiple bullet attack. This is because some fragments will be impacting after others have cracked a given volume of explosive.

A fragment attack test is shown in Figure 8.21.

8.9.2.2. Shaped-Charge Jet Impact Tests

US Military Standard 2105B calls for attack of the weapon under test by the jet from a shaped charge and for it to be attacked by the spall from a shaped charge attacking an armor plate in the vicinity of the weapon. The test arrangements are shown in Figures 8.22 and 8.23. While both of these tests are realistic, they are not always used because in many cases the weapon is unlikely to see such stimuli. Naval Stores, unless they are specific marine items, are unlikely to include shaped charges. Consequently direct, or indirect attack by jets or jet spall is unlikely. However, in land service situations the tests have great relevance.

It is usual for the results of shaped-charge jet attack to be detonation of the energetic material in the target, since this is mode of attack delivers the greatest concentration of power to a receptor of any of the insensitive

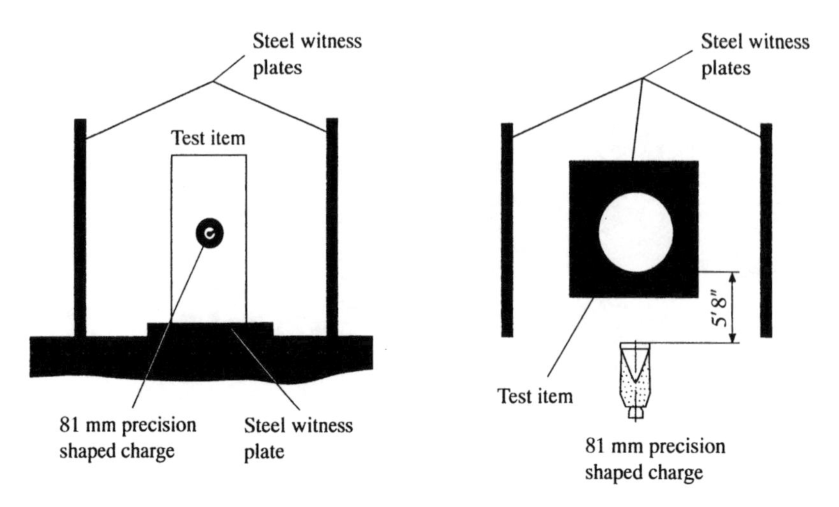

Figure 8.22. Attack by shaped charge jet. (With permission of Professor Dr. M. Held, Deutsche Aerospace.)

munitions tests. Some TATB/Kel-F charges have, on some occasions, withstood the effects of shaped-charge attack and merely broken up, but other examples are difficult to authenticate. Spall attack is similar to attack by fragments, and in some cases is expected to be slightly more benign because the fragments from the rear of the plate tend to be smaller than those from the fragment generator discussed in connection with the fragment attack test.

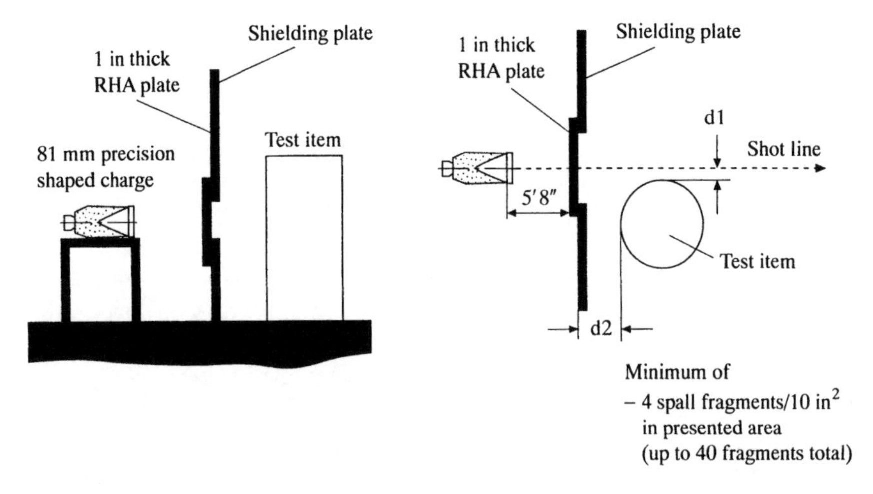

Figure 8.23. Attack by shaped charge jet spall. (With permission of Professor Dr. M. Held, Deutsche Aerospace.)

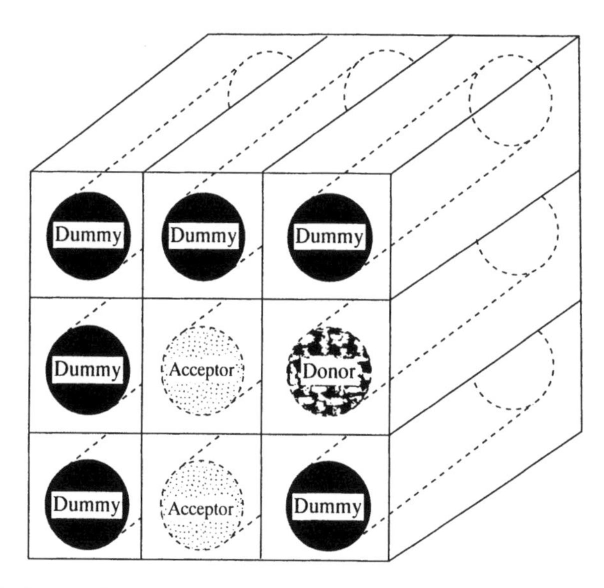

Figure 8.24. Sympathetic detonation test. (With permission of Professor Dr. M. Held, Deutsche Aerospace.)

8.9.2.3. Sympathetic Detonation Test

In this test a donor charge is detonated in a stack of charges with an arrangement as is shown in Figure 8.24. The term "Sympathetic Detonation" is a misnomer, because initiation occurs in acceptor charges as a result of both blast and fragment attack. In this test, which is especially important for bombs, shell, and other ordnance with high explosive content stacked in large quantities, the arrangement shown in Figure 8.24 maximizes the possibility of an event occurring in acceptor charges. Experience shows that it is the diagonal charges which are at most risk from the donor charge, because they are sufficiently far from the detonating donor for fragments from it to have developed much higher velocities on impact than is the case with nearest horizontal or vertical neighbors. In the latter cases, fragments may not be traveling fast enough to initiate detonation promptly and the receptor explosive charge is also protected from blast by the casing of the donor. One means of reducing the probability of events arising from transmission of detonation diagonally is to fill the region between the two or more layers of charges with a material with shock and fragment absorbing and deflecting capabilities. Aluminum tube is extremely effective, since it slows down fragments and acts as a shock-wave diffractor.

8.9.3. Procedures for Insensitive Munitions Testing

Figure 8.25 shows a schematic arrangement for the IM Testing of a weapon system. Complete testing, as scheduled here, is unlikely to be demanded by a

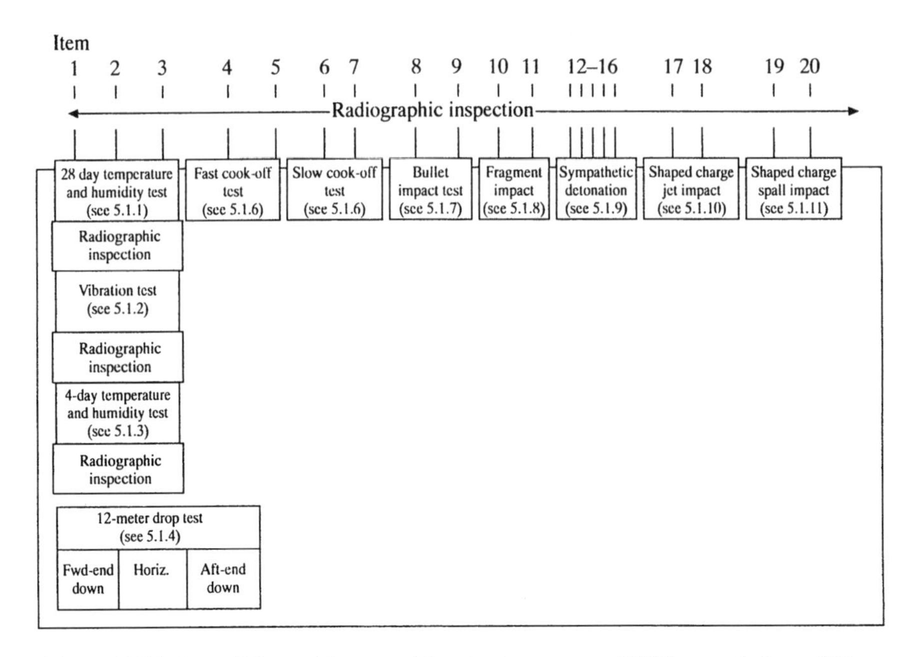

Figure 8.25. Overall insensitive munition test sequence. (With permission of Professor Dr. M. Held, Deutsche Aerospace.)

customer, mainly because not all the tests are strictly appropriate to the life cycle threats to all systems. This is convenient, since the tests are extremely expensive and demanding of range preparation and firing time. Under such circumstances considerable reliance is placed on the disinterested advice from experts with experience of the responses of other munitions to the same test.

8.9.4. Research Required to Improve Understanding of Insensitive Munitions and the Results of Insensitive Munitions Testing

In many developed countries, pressure from local populations is having its effects on the number and sizes of explosives tests which will be tolerated. Moreover, the end of the Cold War has led to significant reduction in weapons programmes and budgets worldwide. Consequently, development costs of new systems are being pared to the bone. A significant proportion of development costs in the past has been associated with weapon testing for performance and safety and there is pressure to reduce them.

However much is known about the explosive, or propellant, it is not yet possible to predict the outcome of Insensitive Munition, or any other hazard tests from numerical studies, because the initiation conditions are complex and the first and, sometimes, only response is deflagration. Deflagration of energetic materials, as opposed to the controlled burning of rocket and gun propellants and pyrotechnics has not been studied formally, because it has been argued in the past that it does not assist in the improvement of weapon performance. Proper analysis of the processes associated with unstable burning are required before predictions on the behavior of energetic materials in accidents can be analyzed and predictions made on the outcomes.

It is suggested that, in order to understand and predict responses of weapons to hazard tests, the following research on the deflagration process needs to be carried out:

- determination of the pressure dependence of the burning of a range of high explosives and propellants, in order to assess the effects of confinement on reaction acceleration;
- application of the theories of fracture mechanics to the complex composite structures of explosives and propellants, in order to determine the capacity for cracking of given materials in response to transient stresses; and
- determination of the mechanical properties of energetic and casing materials at rates of strain consistent with impact, pressurization due to deflagration, etc.

Equipped with data of this type and adequate computer codes, we should be able to predict the likely responses of casings and energetic materials in them to stimuli associated with impact and intrusion, and understand more clearly how to control the responses of energetic materials to unplanned initiations.

Acknowledgment

The author is indebted to Professor Dr. Manfred Held of TDW (previously known as Deutsche Aerospace, MBB), Schröbenhausen, Germany, for a set of diagrams illustrating the Insensitive Munitions Tests and the levels of explosive response from them.

References

Afanasiev, G.T. and Bobolev, V.K. (1971). *Initiation of Solid Explosives by Impact.* Translated by I. Shechtman (P. Greenberg, ed.). Israel Program for Scientific Translations.

AGARD Conference 511 Proceedings (1992). Insensitive munitions.

Bernecker R.R. and Price, D. (1975). Sensitivity of explosives to transition from deflagration to detonation. US Naval Ordnance Laboratory, White Oak, Silver Spring, MD, NOL TR-74-186.

Boggs, T.L. and Derr, R.L. (1990). Hazard studies for solid propellant rocket motors, *AGARDograph 316*, NATO, Brussels.

Brownlee, K.A., Hodges, J.L., and Rosenblatt, M. (1953). *J. Amer. Statist. Soc.*, **43**, 262.

Campbell, A.W., Davis, W.C., Ramsay, J.B., and Travis, J.R. (1961). *Phys. Fluids*, **4**, 511.

Covino, J. and Dreitzler, D.R. (1988). *19th ICT Conference, Combustion and Detonation Phenomena*, 80-1.

Covino, J. and Hudson, F.E. (1990). *AIAA J. Propulsion Power*.

Dixon, W.J. and Wood, A.M. (1948). *J. Amer. Statist Soc.*, **43**, 109.

Dixon, W.J. and Massey, F.M. (1951). *Introduction to Statistical Analysis*. McGraw-Hill, New York.

Dobratz, B.M. (ed.) (1981). *LLNL Explosives Handbook*. Lawrence Livermore National Laboratory, Livermore, CA, USA, UCRL-52997.

Finney, D.J. (1971). *Probit Analysis*. Cambridge University Press, Cambridge.

Frank-Kamenetskii, D.A. (1955). *Diffusion and Heat Exchange in Chemical Kinetics*. Translated by N. Thon. Princeton University Press, Princeton, NJ.

van Geel, J.L.C. (1971). *RVO/TNO*, Rijswijk, Netherlands, TL-1971-1.

Gibbs, T.R. and Popolato, A. (1980). (eds.). *LASL Explosive Property Data*. University of California Press, Berkeley, CA.

Gray, P. and Harper, M.J. (1959). *Trans Faraday Soc.*, **55**, 581.

Gray, P. and Lee, P.R. (1967). Thermal explosion theory. In *Oxidation and Combustion Reviews* (C.F.H. Tipper, ed.). Elsevier, Amsterdam, **2**, 1.

Green, L.G. and Weston, A.M. (1970). Data analysis of the reaction behavior of explosive materials subjected to Susan test impacts. Lawrence Livermore National Laboratory, Livermore, CA, UCRL-13480.

Hammersley, J.M. (1966). *J. Soc. Indust. Appl. Math.*, **11**, 894.

Hubbard, P.J. and Lee, P.R. (1978). RARDE Technical Report 2/78.

Köhler, J. and Meyer, R. (1993). *Explosives*. VCH, Weinheim.

Lee, P.R. (1992). *Fourth International Symposium on Explosives Technology and Ballistics, Pretoria, RSA, National Institute for Explosives Technology*, 99.

Majowicz, J.M. and Jacobs, S.J. (1958). *Amer. Phys. Soc. Bull.*, **3**, 293.

Mallory, H.D. (ed.). Bruceton Staircase: The development of impact sensitivity tests at the explosives research laboratory, Bruceton, PA, during the years 1941–1945. US Naval Ordnance Laboratory, White Oak, Silver Spring, MD. NAVORD Report 4236.

Manners, D.J. (1992). UK work on low vulnerability rocket motors (LOVUM Project). RARDE, Waltham Abbey, Essex, UK, Report.

Marshall, W. (1983). Private communication, re: Milne, E.A.

Merzhanov, A.G., Abramov, V.G., and Gontkovskaya, V.T. (1963). *Dokl. Akad. Nauk SSSR*, **148**, 156.

Merzhanov, A.G. (1966). *Combust. Flame*, **10**, 341.

Merzhanov, A.G. and Abramov, V.G. (1981). *Propellants, Explosives, and Pyrotechnics*, **6**, 130.

van der Mey, P. and Heemskerk, A.H. (1984). *Symposium Paul Vieille*, Le Bouchet, France.

van der Mey, P. and Heemskerk, A.H. (1985). *AGARD Symposium on Smokeless Propellants*, Florence, Italy.

OSRD Report Number 4040. Statistical analysis for a new procedure in sensitivity experiments. A Report by the Statistical Research Group, Princeton University, to the Applied Mathematics Panel NDRC. AMP Report No. 101. IR. SRG-P No. 40.

Rogers, R.N. (1972). *Thermochim. Acta*, **3**, 437.

Rogers, R.N. (1975). *ibid.*, **11**, 131.
Rogers, R.N. (1972). *Anal. Chem.*, **44**, 1336.
Semenov, N.N. (1955). *Chemical Kinetics and Chain Reactions*. Clarendon Press, Oxford.
Sensitiveness Collaboration Committee (SCC No. 3) (1988). *Manual of Tests*. RARDE NP1 Division, Fort Halstead.
US Military Standard, DoD-STD-2105B (Navy) (1994). Hazard Assessment Tests for Non-nuclear Munitions.
Walker, G.R. (1966). *Manual of Sensitiveness Tests*. The Technical Cooperation Program (TTCP), Panel O-2, Canadian Armament Research and Development Establishment.
Zinn, J. and Mader, C.L. (1960). *J. Appl. Phys.*, **31**, 323.

CHAPTER 9

Safe Handling of Explosives

Jimmie C. Oxley

9.1. Explosive Safety

There are four to five billion pounds of explosives produced annually in the United States [IME (1988)]. A relatively small portion is produced for military purposes (Figure 9.1). The vast majority of explosives are produced for industrial use: mining (55% coal, 16% nonmetal, 17% metal), civil and petroleum engineering; and agriculture. Considering the large quantities of explosives produced each year, there are surprisingly few catastrophic accidents. The manufacturers and handlers of explosives have developed effective codes of safety. Explosives manufacturing now results in fewer accidents than the chemical industry as a whole, which, in turn, has less accidents than the mining or lumber industry [Cook (1958)]. A likely explanation is the perceived hazards associated with explosives. Explosives manufacturers and users have established stringent procedures in anticipation of the potentially severe consequences of the misuse of explosives. The best compilation of explosives accidents to 1980 can be found in *History of Accidents in the Explosives Industry* by G.S. Biasutti. Unfortunately, this book is a private publication which makes it difficult to find as well as to cite. Below is a tabulation of the accidents reported by Biasuttii and an update drawn from data collected by the Department of Defense (DoD) Explosive Safety Board. The two columns of accident statistics illustrate the changing use of explosives. Nitroglycerin (NG) was involved in many early accidents; but as of this writing, only one commercial nitroglycerin plant survives in the United States. This reflects the change in commercial blasting agents away from NG-based dynamites toward ammonium nitrate based formulations. Ammonium nitrate (AN) blasting agents are significantly safer to manufacture and use than NG-based materials, even though the worst industrial accident in United States history involved AN (the Texas City disaster of 1947 took nearly 600 lives). The increase in accidents involving RDX/HMX, which came into use during World War II, is not surprising. The evident increase in accidents involving primary explosives may not be

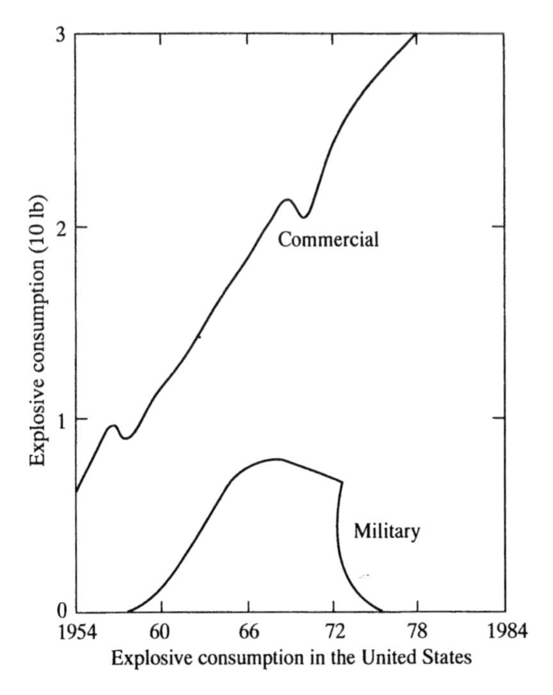

Figure 9.1. Commercial application of explosives is larger than military application.

significant. It probably reflects a difference the data gathering procedures used by Biasuttii and the Explosive Safety Board.

Number of accidents in the manufacture, storage, and transport of explosives		
	Prior to 1980	1980–1994
Black powder	60	0
Nitrocellulose	20	13
Nitroglycerin	156	5
NG-based blasting agent	188	4
Ammonium nitrate explosives	21	6
Detonators, primaries	31	68
PETN	10	1
TNT	26	15
RDX and HMX formulations	2	36

An energetic material, when provided with suitable stimulus, will undergo a self-sustaining reaction which can grow into violent deflagration or detonation. A variety of stimuli, including shock, friction, impact, spark, or heat can initiate an explosive event. Knowledge of kinds of stimulation and the magnitudes required to cause catastrophic reactions is essential in order to devise procedures for the safe handling of these materials. The purpose of this chapter is to discuss methods used to evaluate the sensitivity of an energetic material to energy-inputting stimuli.

9.2. Sensitivity Testing

9.2.1. Impact

Impact testing determines whether impact (i.e., sudden compression) will initiate rapid decomposition of an energetic material. There are various configurations, but basically, a weight is dropped on a sample or a sample is impacted against a rigid target. One of the most common type of impact testers is the ERL (Explosive Research Laboratory) machine. It consists of a free-fall weight (2.5 kg or 5.0 kg), a striker and anvil (1.25 in. in diameter), and a supporting frame. Samples (30–40 mg) are placed in a dimple in the center of a 6.5 cm^2 sheet of 5/0 garnet paper (Type 12 configuration). A Type 12B configuration, without garnet paper, is also used on solids, while a third configuration, Type 13, is used for liquids. A standard test set consists of 20–30 shots performed by following the Bruceton "up-and-down" testing technique. Results are reported in terms of the height at which an event is obtained 50% of the time (H_{50}). Although drop-weight impact testing is widely used for initial characterization of small amounts of energetic materials, results can sometimes be misleading or inconsistent and are often operator-dependent. For this reason, the explosives industry uses one material to allow cross-comparison of results; they standardize on TNT—212 cm (flake) and 154 cm (granular). Exemplary results on the ERL apparatus at Los Alamos National Laboratory (LANL) are given below for a Type 12 tool and a 2.5 kg weight [Gibbs and Popolato (1980)]. A more complete list can be found in Table 9.1. Impact sensitivity can be affected by a number of variables, most notably particle size and temperature. The latter effect is illustrated by the AN data given below [ME (1984)]. A number of scientific studies have been undertaken in an attempt to understand the chemistry occurring on the drop-weight impact machine. Both radiometric and spectroscopic methods have been investigated [Miller et al. (1985); Mullay (1987)], and drop weight impact results have been correlated with sensitivity of broad classes of energetic materials [Storm et al. (1990)].

	H_{50}(cm) [Gibbs and Popolato (1980)]	Ammonium nitrate [ME (1984)]	
		Temperature	H_{50}(cm)
PETN	12.0	°C	(Picatinny apparatus)
RDX	22.2	25	79
HMX	26.0	100	69
Octol	38.1	150	69
TNT	154.0	175	30

Table 9.1. Selected explosive properties.

Explosive	Melting point, mp, °C	Density, g/cc	Detonation		Drop weight, H50 cm 12	Vapor pressure	
			Velocity, km/s	Pressure, kbar		log P (mmHg)	Temperature °C
AN	169	1.73			149		
AN							
ANFO	169						
AP (200 μm)					47		
AP (90 μm)					44		
ADN	94						
CL-20							
HNS	318	1.74	7			1.00E − 09	100
HMX	285d	1.89	9.11	393	26	3.40E − 05	150
NC							
NG	13	1.6	7.58	230			
NQ	245d	1.7	8.28	260	> 320		
NTO					86		
PETN	143	1.7	7.98	300		8.00E − 05	100
RDX	204	1.77	8.64	347	22	0.013	150
TATB	448	1.85	7.66	275	> 320	1.00E − 04	150
TNT	81	1.64	6.94	190	212		
TNAZ	101						
Tetryl	129	1.69	7.7		49	$13.71 - 6776/T$	
Styphnic acid					46		
Picric acid		1.76	7.68	265	79	$12.024 - 5279T$	
Ammonium picrate					152		
PYX	460	1.75			63		
DATB	286	1.78	7.6		> 320	$13.73 - 7314/T$	
TACOT	410	1.85	7.2				
NONA	440–450					2.50E − 09	150
DIPAM	304	1.79					

Explosive	Isothermal kinetics			Moles gas/cmp	Differential Scanning Calorimetry (DSC) data			
	E kcal/mol	*A* s^{-1}	Temp. Range °C		Heat (DSC) cal/g	Exo (20°/min) Temp, °C	*E* kcal/mol	*A* s^{-1}
AN	26.8	1.6E+07	200–290	0.88	400	320	29.1	6.46E+08
AN	46.2	5.0E+14	290–380					
ANFO	35.2	2.40E+11	270–370		750	290, 370	47.3	3.2E+16
AP (200 μm)	21.3	4.81E+04	215–385	1.2	360	360	22.7	8.32E+05
AP (90 μm)					460	310	—	—
ADN	38.3	9.57E+15		1.4	—	194	31.3	1.66E+17
CL-20	42.4	4.00E+17	146–220, acetone		—	—		
HNS								
HMX	52.9	2.46E+18	230–270	5.4	850	280		
NC				0.00	760	220		
NG								
NQ					250	230		
NTO	38.4	1.05E+13	240–280	2.3	700	260		
PETN				6.3	1000	210		
RDX	37.8	1.99E+14	200–250	3.9	960	257		
TATB				3.4	930	397		
TNT	32.8	1.19E+10	240–280	3.2	610	330	34	2.5E+11
TNAZ	46.6	3.55E+17	220–250	3.8	—			
Tetryl							38	2.5E+15
Styphnic acid					800	280		
Picric acid				4.10	1000	320		
Ammonium picrate					1000	310		
PYX					1000	380		
DATB				3.40				
TACOT								
NONA								
DIPAM								

9.2.2. Friction

As with drop-weight impact testers, there are a number of friction sensitivity testing devices. In the ABL (Allegheny Ballistics Laboratory) sliding anvil test, a thin layer of powdered sample ($\sim$40 mg) is placed on a flat steel plate and slid by a nonrotating 0.3 mg wide steel wheel at a initial velocity of 240 cm/s. The force between plate and wheel is varied in 0.1 log (pound) intervals over the range 10–1000 lbs. The results are reported in terms of the force at which initiation of the sample is probable 50% of the time. Initiation is evidenced by production of a flash, smoke, or noise.

9.2.3. ESD

Sensitivity of an explosive to electrostatic discharge (ESD) is determined by subjecting a sample to a high-voltage discharge from a capacitor. The discharge energy is increased and decreased until the 50% probability point is determined. The commonly used ESD apparatus holds the sample (3/16 in. diameter, 1/4 in. high) on a steel dowel by means of a polystyrene sleeve. The sample is covered with lead foil. To induce a spark, a charged needle is moved down to penetrate the foil. Discharge takes place as the needle penetrates the foil. The discharge passes through the explosive to the grounded steel dowel. The needle, charged to varying degrees, is moved in and out of the sample until initiation is evidenced. Voltage, confinement, sample and particle size, temperature, and moisture content can be varied. Exemplary results from LANL [Gibbs and Popolato (1980)] using a 3 mil foil are shown below. Tests for electrostatic sensitivity have been devised which use significantly larger quantities of energetic material [Mellor and Boogs (1987)].

	Joule
PETN	0.19
RDX	0.21
HMX	0.23
TNT	0.46
TATB	4.25

9.3. Thermal Stability

Thermal stability describes the sensitivity of a given energetic material to thermal shock. Thermal stability of an energetic material determines its shelf-life, critical dimensions, and critical temperature. The first-step in evaluating the thermal stability of a new formulation is laboratory-scale thermal hazards analysis. Herein a number of common analytical techniques are discussed and their limitations considered. Since each analytical method has its own weakness, test results should always be confirmed by parallel methods. Furthermore, laboratory-scale results must be corroborated by large-scale tests. If the thermal stability of a new formulation is less than desired, a detailed study of the mechanism of decomposition is necessary so that an appropriate "fix" can be found.

Laboratory-scale thermal hazards testing can be subdivided into micro-, small-, and mid-scale test techniques. Micro-scale testing typically involves less than 5 mg of sample and includes such methods of analysis as DSC (Differential Scanning Calorimetry), DTA (Differential Thermal Analysis), TGA (Thermal Gravimetric Analysis), and a number of classical isothermal heating and analytical techniques. Henkin [Henkin and McGill (1952)], ODTX (One-Dimensional Time-to-Explosion test) [Catalano et al. (1976)], vacuum stability [MIL (1975a)], CRT (Chemical Reactivity Tests), and ARC (Accelerating Rate Calorimetry), trade name of Columbia Scientific, are herein classified as small-scale tests involving between 40 mg and 5 g of sample. The Koenen test [UN (1986)], SCB (Small-Scale Cook-Off Bomb) [UN (1986)], 1 liter cook-off, and sealed cook-off are mid-scale tests involving from 30 g to 1000 g samples.

9.3.1. Comparative Thermal Stabilities

Testing the thermal stability of energetic materials has long been of interest. The **Abel test** (1875) was developed to test the thermal stability of nitrocellulose, nitroglycerin, and nitroglycol. This test exploits the fact that nitrate esters decompose to produce NO_2. A 1 g sample is heated at 180 °F, and the time required to change starch/iodide paper from white to blue is noted. Like many of the thermal stability tests, it is only useful in comparing like materials, for example, the relative thermal stability of two different batches of nitrocellulose. One shortcoming of the test is that it allows no extrapolation of the results. Furthermore, for energetic materials which do not produce nitrogen oxides upon decomposition, the test is useless.

The **Taliani test** (1904) [MIL (1975b)], the vacuum stability test, and the gun propellant surveillance program (**65.5 °C test**) all test relative thermal stability by observing production of decomposition gases. Both the Taliani and the "65.5 °C" tests employ constant volume test containers and monitor build up of pressure. A modern version is the chemical reactivity test, where a sample is loaded at atmospheric pressure and heated at 120 °C for 22 h. Evolved gases are quantified and identified using gas chromatography. In the **vacuum stability test** [MIL (1975a)], 5 g samples are heated for 40 h at 90 °C (double-base propellants), 100 °C (single-base propellants), or 200 °C (high explosives). Volume of gas produced is monitored by a mercury manometer attached directly to the test tube. The test is cheap, is widely used, and can provide relative data for the purposes of quality control, with the proviso that the comparisons are limited to like samples.

The **ARC** (Accelerating Rate Calorimeter) is intended "to maintain a

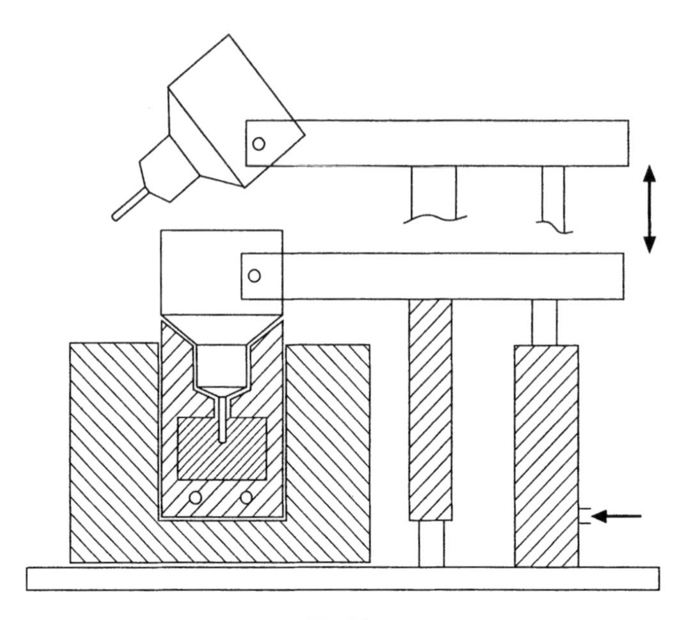

Figure 9.2. Henkin apparatus.

sample in an adiabatic state and permit it to undergo thermal decomposition due to self-heating while recording the time–temperature–pressure relationship of the runaway process." But in practice true adiabatic conditions cannot be maintained with solid samples, and analysis of data is somewhat complex. Nevertheless, researchers use this instrument to derive kinetic parameters as well as to do comparative thermal stability tests.

Henkin Time-to-Explosion. In the modified Henkin time-to-explosion test [Henkin and McGill (1952)], a 40 mg sample is sealed into a blasting cap under known pressure and heated in a Wood's metal bath (Wood's metal being an alloy: 50% Bi; 25% Pb; 12.5% Sn; 12.5% Cd) (Figure 9.2). Time-to-explosion (usually detected as a noisy pressure rupture) is measured, and a critical temperature is bracketed by "Go/No Go" runs at varying temperatures. Rogers has shown these correlate well with critical temperatures predicted from E and Z data obtained by DSC [Rogers (1975)]. Using his Arrhenius values it can be seen that the typical explosive exhibits "GO" Henkin results when its rate of decomposition is on the order of $10^{-2}\ s^{-1}$ to $10^{-3}\ s^{-1}$.

The **ODTX** [Catalano et al. (1976)] was developed at Lawrence Livermore National Laboratory to determine thermal stability and kinetics of thermal decomposition. A spherical charge (1.27 cm diameter) of explosive ($\sim$2.2 g) is confined between O-ring-sealed, heated anvils at pressures of 1500 atm. Time to confinement failure is measured at various temperatures. Kinetic parameters can be derived from this technique.

Various slow cook-off tests have been designed from which kinetics of thermal decomposition might be obtained, but usually they are used to satisfy set safety testing criteria. The United Nations "Orange" book [UN (1986)] lists two cook-off tests in the series 1 test plan:

Koenen Test. A steel cylinder of 24.2 mm inside diameter is filled to a height of 60 mm with the test material. The tube is sealed with the exception of an orifice, the diameter of which is varied from 20 mm down to 1 mm during the test sequence. The test is initiated by heating the sample with four propane burners. Two times are noted—the time from the start of heating until a reaction is evidenced and the time until the reaction terminates. In subsequent tests the orifice size is decreased until the reaction to heating is an explosion, which is defined by the number of pieces into which the tube fragments.

Small-Scale Cook-Off Bomb (SCB) Test. This test simulates transport and storage situations involving slow heating. A sample is loaded into a small steel cylinder of 400 cm^3 volume. The ends of the cylinder are capped with witness plates, and the vessel is heated from 25 °C to 400 °C at a rate of 3 °C per minute. Thermocouples in the walls of the vessel allow observation of self-heating. A positive test is one which either ruptures or fragments the vessel or punctures or deforms the witness plates.

9.3.2. Quantitative Thermal Stabilities

There are a number of the thermal stability tests which are useful in comparing the relative stability of like materials, for example, two different batches of nitrocellulose. Such tests usually lack sufficient information to extrapolate results to different conditions. Furthermore, tests which evaluated thermal stability by observing production of decomposition gases cannot be used to compare different classes of explosives, which have extremely different gas-producing capabilities (Table 9.2) [Oxley et al. (1998)].

To obtain thermal information that can compare unlike chemical compositions and to be able to utilize the predictive heat-flow models, quantitation of the thermal hazard is necessary, and the concept of critical temperature must be introduced. Critical temperature is the lowest constant surface temperature at which a specific material of a specific size and shape will self-heat catastrophically. A self-sustained reaction can occur when the rate of self-heating, resulting from decomposition reactions, exceeds the rate at which heat can be dissipated. If an energetic sample is kept below its critical temperature, no runaway self-heating will occur. (That is not to say no decomposition will occur; it does.) Thus, thermal safety of a charge depends on whether a balance can be maintained between the heat generated by decomposition reactions and the dissipation of heat to the surroundings. This heat flow in and out of the energetic material depends on the thermal

Table 9.2. Moles of gas produced per mole explosive at various temperatures, times, and percent decomposition.

	120 °C		220 °C		320 °C	
Compound	22 hr	> 99% decomp.	22 hr	> 99% decomp.	22 hr	> 99% decomp.
NTO	0		2.13	2.19	2.4	2.34
AN	0		0.88	0.87	0.89	0.88
AP	0		0.87	0.88	1.25	1.25
ADN	1.25	1.446	1.46	1.46	1.42	1.44
TNT	0		1.83	2.18	2.68	3.23
TNX	0		1.01	1.66	2.13	2.22
TNM	0		0		1.77	1.92
TNA	0		0.25	0.5	3.21	3.41
DATB	0		0.24	0.24	3.19	3.45
TATB	0		0		3.23	3.4
Picric acid	0		3.47	3.64	3.86	4.09
HMX	0		4.33	4.31	4.24	5.41
RDX	0		3.84	3.88	2.92	3.88
TNAZ	0		3.01	3.04	3.46	3.77
DMN	0		0.57	0.57	0.6	0.78
NG	2.54	3.15	3.55	3.5	3.96	4
EGDN	0	2.17	2.3	2.26	2.74	2.75
PETN	0		5.04	5.19	6.32	6.32
NC	0		0		0.001	0.00155

conductivity of the material, the reaction rates of the chemical decomposition producing heat, the size and shape of the sample, its heat capacity, phase transition energies, if any, and reaction energies (Figure 9.3).

The Frank-Kamenetskii (1969) equation models conductive heat transfer within a sample. The material is assumed to be unstirred and too viscous for convection to dissipate heat:

$$E/T_c = R \ln(r^2 pQAE)/(Tc^2 \tau s R), \tag{9.1a}$$

where r is radius (cm), s is shape factor (0.88 infinite slab, 2.00 infinite cylinder, 3.32 sphere), R is gas constant, (1.99 cal/mol deg), p is density (g/cm^2), τ is thermal conductivity (cal/deg cm), Q is heat of decomposition (cal/g), E is activation energy (cal/mol), A is pre-exponential factor (s^{-1}), and T_c is critical temperature (Kelvin). Since the Frank-Kamenetskii model allows for only one mode of heat removal, conduction, it predicts the worst case. The Semenov (1935) equation performs a similar calculation (where V is volume, S is surface area, and a is heat flow coefficent for sample), but allows for stirring of the sample as well as conduction to remove heat.

$$E/T_c = R \ln(VpQAE)/(Tc^2 SaR). \tag{9.1b}$$

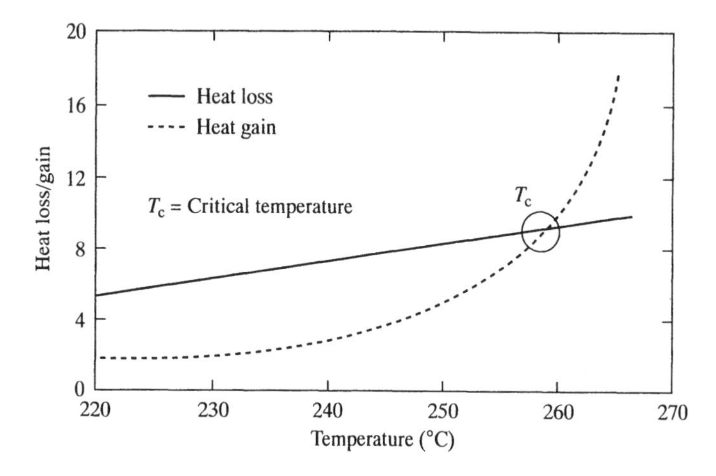

Figure 9.3. Heat loss and gain with increasing temperature.

In either equation T_c is found on both sides of the equation and, thus, is solved for iteratively. Simple mathematical analysis shows that since activation energy is outside the log term, it has the most dramatic effect on critical temperature. Typical activation energy values are between 20 kcal/mol and 40 kcal/mol, while pre-exponential factors are of the order of 10^{13} s^{-1} to 10^{16} s^{-1}. The values E and A are usually obtained assuming Arrhenius kinetics. The rate of x decomposition, dx/dt, is related to a function of the unreacted material, $f(x)$, and a constant, k, as follows:

$$-\frac{dx}{dt} = kf(x), \tag{9.2}$$

where

$$f(x) = x^n \tag{9.3}$$

and

$$\ln k = \ln A - \left(\frac{E}{R}\right)\left(\frac{1}{T}\right). \tag{9.4}$$

The order of reaction is n, and x is the fraction of unreacted material. The values E, activation energy, and A, the frequency factor, are typically found by plotting $\ln k$ versus $1/T$ to obtain a line for which the slope is $-E/R$ and the intercept is $\ln A$; thus, the values of E and A are interdependent.

Determination of the rate constant, k, as a function of temperature can be done a number of ways. Rate constants can be measured at each temperature of interest. At a given temperature (isothermally), the reaction progress is monitored as a function of time by observing either the formation of products, the disappearance of reactants, or the production of heat. For example, consider the thermal decomposition of ammonium nitrate:

$$NH_4NO_3 \rightarrow N_2O + 2H_2O + \text{heat}. \tag{9.5}$$

Nitrogen gas is also formed in this reaction, but the amount is relatively small in the temperature range of interest. The rate of the reaction can be followed by a number of techniques. Since the ratio of moles of nitrous oxide to moles of AN is nearly invariant over a wide temperature range, the rate of formation of products can be quantified by infrared spectroscopy [measuring production of nitrous oxide (N_2O)] or by gas manometry (measuring total volume of all gaseous products) or by observing the production of heat with some type of heat-measuring device (a thermal analyzer). Alternatively, the loss of reactant ammonium nitrate with time can be measured by monitoring nitrate or ammonium concentrations by ion chromatography.

There are many types of thermal analyzers or calorimeters; an insulated thermos with a thermometer would be the simplest. However, three types of thermal analyzers are commonly employed to measure thermal stability:

TGA Thermal Gravitimetric Analyzer;
DTA Differential Thermal Analyzer; and
DSC Differential Scanning Calorimetry.

Any of these instruments can be used to heat a sample at constant temperature (isothermally) or with constantly increasing temperature. The latter method is called "programmed" heating since the temperature is raised a set number of degrees per minute. In simplest comparative stability studies, if two samples are heated at the same rate, the sample with the weight loss (TGA) or the exotherm (DTA or DSC) at the lowest temperature is the least thermally stable.

In TGA the sample is placed on a delicate balance, and as it is heated a plot of its weight as a function of temperature rise is recorded. Weight loss is generally due to evolution of gas or moisture, but volatile samples may, themselves, sublime. The method suffers from the disadvantages of other gas measuring methods. Differential thermal instruments, DTA and DSC, record the enthalpy change between a sample and a reference. When a material undergoes a change, such as melting or crystal-phase transition or a chemical reaction, there is heat flow. The heat involved in these transitions can be quantified. DSC scans can be run in programmed or isothermal mode. The programmed mode is most commonly used; it continuously raises the temperature of the sample and reference at a programmed rate. An exothermic reaction may be a deviation up or down from the baseline; it is easily determined from the labeling of the ordinate axis. The position of the exothermic maximum moves to lower temperatures as the DSC programmed heating rate, B, decreases; this fact means that for thermograms to be comparable they must be run at the same programmed heating rate. This fact is also used in a number of variable heat-rate methods for attaining activation energies.

The ASTM (1995) method rests on the assumption that though the temperature of the exothermic maximum varies with scan rate, the fraction of material decomposed at the peak maximum remains the same regardless of scan rate. The heating rate, B, is plotted against $1/T$, resulting in a line with slope proportional to $-E/R$. Calculation of A involves evaluation of the complex term $d[\ln P(p)]/dp$. Commercial software is available for these calculations.

The kinetics of decomposition can also be determined from isothermal scans. The isothermal method holds the sample and reference at a constant set temperature. However, there is a finite amount of time necessary to bring these materials to the set temperature. If the temperature has been properly selected, as soon as the energetic sample reaches temperature it will start to evolve heat due to internal decomposition. For nth-order decomposition reactions, the rate of decomposition is a function of the amount of unreacted sample present (x). The highest decomposition rate occurs at the beginning of the decomposition, when the most reactive material is present. The rate decreases as the reaction goes forward since less reactant is available. The rate of decomposition, the decrease in concentration of reactant with time ($-dx/dt$), is equal to the rate constant, k, multiplied by the concentration of reactant remaining (x) to the nth power, where n is the order of the reaction. The mathematics are written

$$-\frac{dx}{dt} = kx^n, \tag{9.6}$$

$$\ln\left(-\frac{dx}{dt}\right) = n(\ln x) + \ln k. \tag{9.7}$$

For n about 1, a first-order reaction, the mathematics simplify to

$$\frac{dx}{x} = -k(dt), \tag{9.8}$$

$$\ln(x) = -kt - C \quad \text{where } C \text{ is a constant.} \tag{9.9}$$

Since in a first-order reaction rate ($-dx/dt$) is proportional to the fraction remaining x, the equation becomes

$$\ln\left(-\frac{dx}{dt}\right) = -kt - C. \tag{9.10}$$

The rate of decomposition or rather signal deflection from baseline (b) on the thermogram can be plotted versus time, producing a line of slope $-k$.

Isothermal determinations of E require that the rate constant k is determined at a number of temperatures. For normal ordered decompositions, the reaction initiates at its maximum rate (maximum concentration) and slows with time as the reactant is depleted. The most easily interpreted reaction orders are zero-, first-, or second-order. Zero-order is readily recognized because the rate of decomposition is constant or, at least, unrelated to the concentration of the reactant. First-order reactions will yield k as the slope of the plot of $\ln(x)$ versus t, while second-order reactions require the plotting of $[A]_t^{-1}$ versus t (Figures 9.4 and 9.5). While among these three orders, first-order is the most commonly observed in the decomposition of energetic materials, there are many times the order of the reaction is not

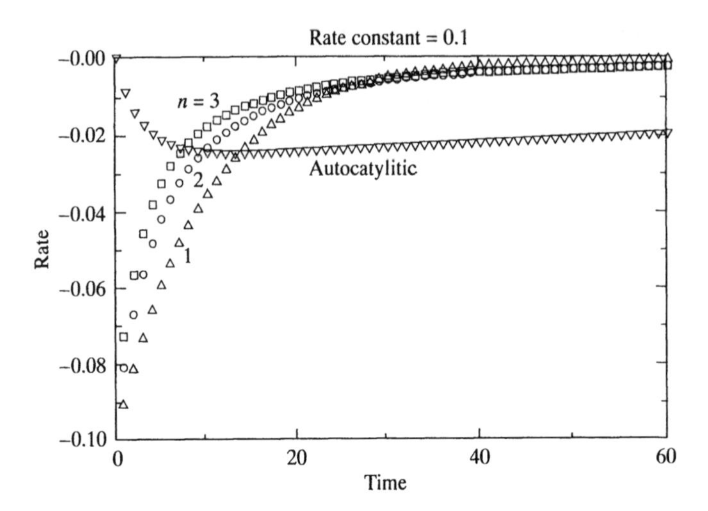

Figure 9.4. Simulated *n*th-order reactions.

open for straightforward interpretation. Furthermore, some energetic materials exhibit auto-catalytic behavior, i.e., there is an induction period in which the rate increases to a maximum followed by the normal decay of the reaction rate to zero. It is assumed that during the induction period decomposition products form which catalyze further decomposition. Isothermal scans which build up to maximum decomposition rates are also observed with solids which decompose immediately upon melting. For this case, melt with decomposition, raising the temperature of the experiment should cause the decomposition to return to normal-ordered behavior.

Several concerns arise when either DSC results or conventional kinetics derived from small-scale testing are used to predict the thermal stability of bulk material storage. First, Arrhenius plots are not usually straight over large temperature ranges; and, thus, the kinetics at high temperature may not be applicable at normal storage temperatures. Furthermore, the microsamples may not be representative of the overall material. Kinetic parameters determined from small-scale samples, therefore, must be used with caution in predicting the behavior of larger-scale systems. In order to determine if there is a problem with scaling, the kinetic parameters determined from small-scale experiments must be tested to see if they correctly predict the results of larger tests.

9.3.3. Summary of Thermal Analytical Tools

A number of laboratory thermal analysis tests have been devised over the years. Many are only useful to compare relative thermal stability among

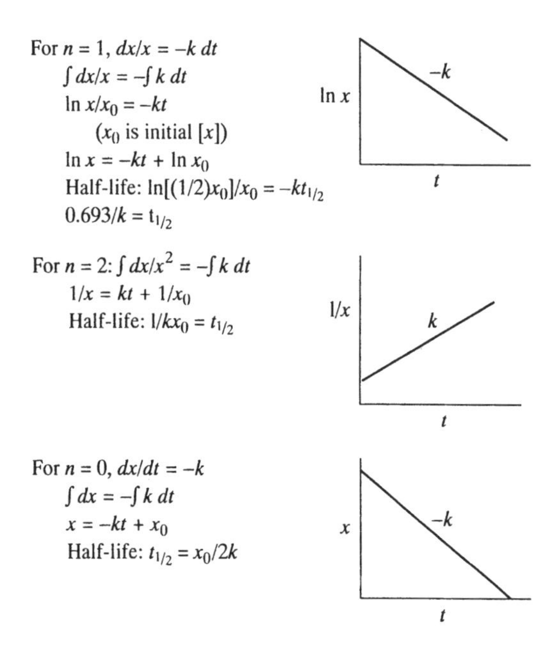

For $n = 1$, $dx/x = -k\,dt$
$\int dx/x = -\int k\,dt$
$\ln x/x_0 = -kt$
(x_0 is initial [x])
$\ln x = -kt + \ln x_0$
Half-life: $\ln[(1/2)x_0]/x_0 = -kt_{1/2}$
$0.693/k = t_{1/2}$
ln x
−k
t
For $n = 2$: $\int dx/x^2 = -\int k\,dt$
$1/x = kt + 1/x_0$
Half-life: $1/kx_0 = t_{1/2}$
1/x
k
t
For $n = 0$, $dx/dt = -k$
$\int dx = -\int k\,dt$
$x = -kt + x_0$
Half-life: $t_{1/2} = x_0/2k$
x
−k
t

Figure 9.5. Rate constant determinations. For *n*th-order reactions.

similar compounds. Most are based upon monitoring product formation, either gas formation (a cumulative event) or heat production (a real time event). Among the gas monitoring techniques are vacuum stability, Abel, Koenen, CRT, Henkin, ODTX, SCB, TGA. Many of these could be used to obtain kinetic rate constants at a fixed temperature, but that is not their common application; in fact, most are used in a dynamic heating mode. To measure rate of heat production a calorimeter of varying degrees of sophistication can be used. Most common are the DSC, DTA, or ARC. The more sensitive the instrument is to heat released, the sooner (i.e., lower temperature) exothermic decomposition can be detected. Unfortunately, the price is also usually proportional to heat sensitivity.

Table 9.3 compares the temperatures at which exothermic decomposition ($E_a = 20$ kcal/mol; $Z = 10^{-8}$ s^{-1}; zero-order) of an energetic material can be detected by various calorimeters or by a GO in Henkin or in a 1 liter cook-off. DTA, DSC, and Henkin are the least sensitive techniques. An exothermic event will not be detected until the rate of decomposition is quite fast, between 10^{-2} s^{-1} and 10^{-3} s^{-1}. ARC and C-80 (Setaram, Astra Scientific; San Jose, CA) are more sophisticated and expensive calorimeters; they will detect thermal decomposition on the order of 10^{-4} s^{-1} and 10^{-5} s^{-1}, respectively. As the sample size is increased (1 liter) or the calorimeter becomes more heat sensitive, even lower rates of reaction (i.e., decomposition at lower temperatures) can be observed.

Table 9.3. Rate constant of decomposition necessary for detection (and corresponding temperature) of compound decomposing zero-order (without depletion) with Arrhenius parameters: $E_a = 20$ kcal, $Z = 10^8\ s^{-1}$.

	Rate constant, s^{-1}	°C
DTA	8×10^{-3}	160
Henkin	6×10^{-3}	155
DSC	3×10^{-3}	140
ARC	2×10^{-4}	100
C80	7×10^{-6}	60
1 liter	2×10^{-6}	46
HART	1×10^{-7}	20

In addition to the heat-sensing limitations of a given instrument, most calorimetric results are complicated by the thermal constants of the sample and instrument and by multiple heat-change events. Furthermore, many of the thermal analysis techniques are used in a dynamic heating mode (temperature is constantly raised). Although dynamic heating saves operator time, it makes results difficult, and sometimes impossible, to extrapolate to real situations. The temperature over which a calculated activation energy is valid is completely unknown.

Due to all the complications mentioned above, as well as the possibility of side-reactions, the best way to assess the thermal stability of compound *X* at temperature *Y*, is to monitor the concentration of *X* over time at a constant temperature of *Y*. Since the temperature of interest (*Y*) is usually too low to allow observation, then the loss of *X* with time at temperatures as close to *Y* as practical should be observed. Techniques must be devised for each compound. For example, the thermal decomposition of RDX or PETN can be followed by high-performance liquid chromatography and a column such as a Alltech Econosphere C18 5U reverse phase column with mobile phase acetonitrile/water (1:1, PETN) methanol/water (1:1, RDX). With a flow rate of 1 ml/min retention times are 14 min and 8 min, respectively, for PETN and RDX. Detection can be accomplished with a UV detector—229 nm PETN and 216 nm for RDX [Oxley et al. (1994a, b)]. Most nitroarenes, such as TNT can be separated on a gas chromatograph [Oxley et al. (1995a, b)], and ionic species, such as AN, are best examined by ion chromatography.

9.3.4. Time-to-Explosion

Time-to-explosion is defined as the length of time a sample can be exposed to temperatures above critical before a thermal event occurs. Thus, thermal safety of an energetic material depends on whether a balance can be maintained between the heat generated by decomposition reactions and the dissipation of heat to the surroundings. Since critical temperature depends on how fast heat escapes to the surroundings, the surface-to-volume ratio of a sample is important. A small surface-to-volume ratio means generated heat cannot be readily dissipated from the sample; therefore, critical temperature is low. Yet, when a sample of small surface-to-volume ratio is subjected to external heat, it takes longer to heat through; thus, a small surface-to-volume ratio means a long time-to-explosion. As the size of an energetic sample increases, its surface-to-volume ratio decreases. Thus, critical temperature decreases as the size of the sample increases, while time-to-explosion increases with increasing sample size. The geometry of a sample also affects its surface-to-volume ratio. For geometric shapes of the same radius, an infinite slab has the smallest surface-to-volume ratio; a sphere has the largest; and an infinite cylinder is intermediate. Accordingly, the infinite slab has the lowest critical temperature; the sphere the highest (Figure 9.6).

While critical temperature is determined by the most rapid heat-producing decomposition pathway, time-to-explosion depends on the lowest temperature decomposition path available, making predictive modeling more dif-

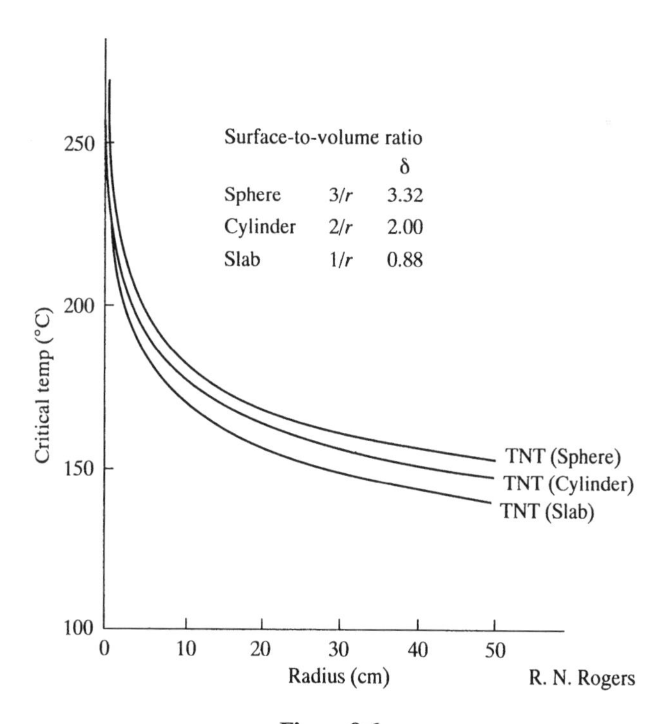

Figure 9.6.

ficult. Time-to-explosion can be computed from the adiabatic model

$$t = \left(\frac{CRT^2}{AQE}\right) \exp\left(\frac{E}{RT}\right),$$

where t is time, E is activation energy (cal/mol), A is pre-exponential factor (s^{-1}), C is heat capacity (cal/g deg), Q is heat of reaction (cal/g), and T is temperature in degrees Kelvin. If only experimental induction times (t) are known, E values are sometimes determined from this relationship. However, most time-to-explosion models are based on zero-order kinetics, which does not allow for depletion of the energetic material, and, thus, predicts times which are too short. These models are particularly inaccurate if the thermal decomposition process involves an induction time, such as a material that melts with decomposition or decomposes auto-catalytically. Like critical temperature, time-to-explosion is dependent on geometry. However, at very high temperatures, geometry is not important because the site where the explosion develops changes. With external temperatures only slightly higher than critical temperature, the induction time is long. The charge warms throughout; the hottest region prior to the explosion is the center; the explosion develops at the center. At external temperatures much higher than critical temperature, the explosion initiates near the surface, while the interior is still cool [Mader (1979)].

9.4. Case History of an Ammonium Nitrate (AN) Emulsion Accident

The general principles of thermal hazards analysis can be illustrated by a case history. In the course of a mining operation, a bore hole was loaded with an ammonium nitrate (AN) emulsion explosive (designated in Figure 9.7 as CETX-101) and allowed to sit for several hours. A premature detonation occurred. Based on earlier work by the Bureau of Mines [Miron et al. (1979); Torshey et al. (1968)] destabilization of the AN emulsion by iron pyrite ore (FeS_2) was suspected. A DSC thermogram (Figure 9.7) shows that the emulsion with the ore exhibits an exotherm at lower temperature than the emulsion alone. Since these thermograms were collected at the same scan rate 20°/min, they can be directly compared, allowing the conclusion that the presence of pyrite ore does indeed thermally destabilize the AN emulsion. Unfortunately, such an observation is merely qualitative; more rigorous experimentation is necessary to permit quantitative predictions. The thermal differences between neat AN and the AN and fuel-containing blasting agent needed to be examined.

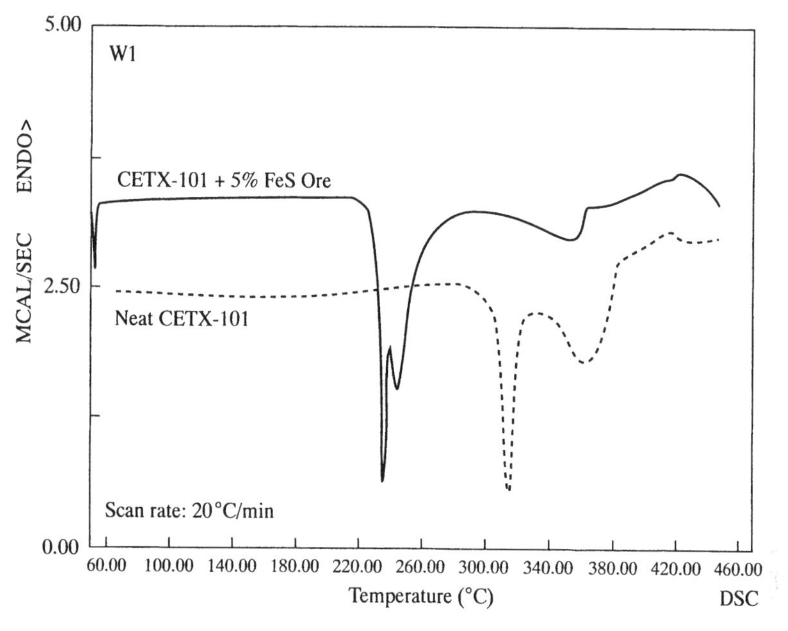

Figure 9.7.

9.4.1. Thermal Stability of AN Formulations by DSC

Figure 9.8 shows the DSC (Differential Scanning Calorimetric) thermograms of ammonium nitrate, AN slurried with 5 wt% mineral oil (ANFO), and an AN emulsion (CETX 101). There are several points of contrast. The emulsion exhibits no solid state II → I phase change nor melt because AN is dissolved and dispersed in small water droplets in the water-in-oil emulsion. Both the emulsion and ANFO have two exothermic regions. Since both the exothermic regions of the emulsion are at higher temperatures than neat AN (320 °C and 359 °C versus 317 °C), it should be more thermally stable than AN. But the first exothermic region of ANFO is somewhat lower than neat AN (304 °C versus 317 °C), which suggests some decomposition occurs more readily than in neat AN and some less readily.

In some laboratories, DSC is used as the sole method of analyzing thermal stability. If this is the case, it is essential for the user of these results to understand the degree of precision and reproducibility that can be expected from such data. Table 9.4 tabulates the DSC data for the thermal decomposition of ammonium nitrate, ammonium nitrate/fuel oil (ANFO), and AN emulsion (CETX 101). The temperatures of the exothermic maxima [for ANFO and emulsion two exotherms (exo 1 and exo 2)] and the 10% and 30% decomposition points show a standard deviation of less than 7% (first four columns). However, the error in total heat of reaction is twice as large (last column). For the AN system, where there is only a single exotherm, the variation in fraction reacted at the exothermic maximum is only 3%. How-

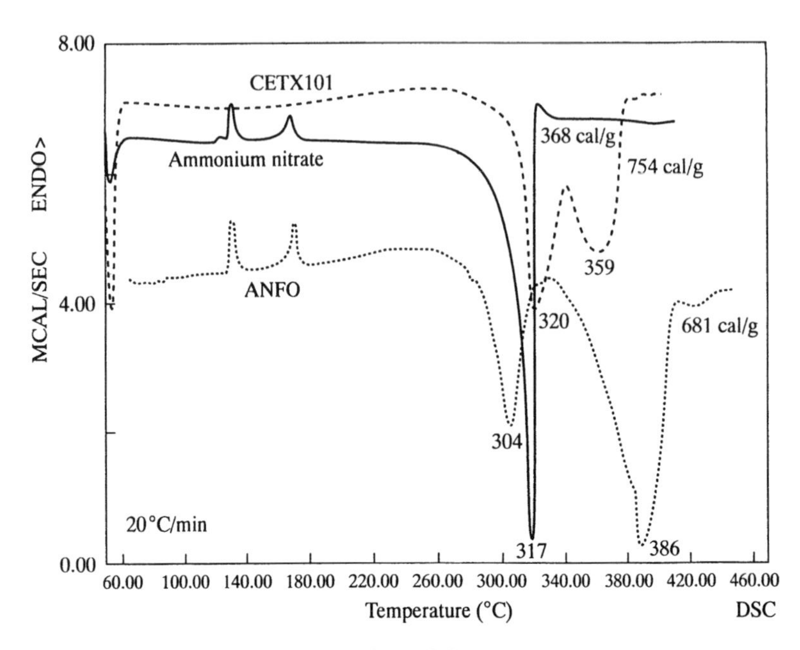

Figure 9.8.

ever, in the ANFO and AN emulsion systems, where there are two exothermic regions of somewhat variable shape, the error in the fraction reacted at the exothermic peak is greater than 25%.

Activation energies are frequently obtained from programmed DSC results. Programmed DSC (as opposed to isothermal DSC) records the heat-flow in or out of a sample while both the sample and the reference are ramped in temperature at a set rate (scan rate). The ASTM [ASTM (1995)] method exploits the fact that the temperature of the exothermic maximum varies with scan rate (which is the reason only thermograms obtained at the same scan rate can be compared). Plotting the temperature of the exothermic maximum versus the scan rate yields a straight line with a slope which is assumed proportional to activation energy. The corresponding frequency factor is obtained by certain assumptions and calculations [ASTM (1995)]. Since the ASTM method is included in most commercial DSC software, it is widely used to obtain Arrhenius parameters. Both ANFO and the AN emulsion exhibited two exotherms; this fact alone indicates the ASTM method is inappropriate for their analysis. Nevertheless, the exothermic maximum of both exotherms was plotted versus scan rate (B). Furthermore, the temperature of the 10% or 30% decomposition point versus scan rate was also plotted. Although straight lines were obtained for all four plots, the activation energies varied widely, from each other and from the values obtained by conventional isothermal techniques. This was true even in the case of AN, which had only a single exotherm. Table 9.5 shows the values obtained.

Table 9.4. Average of DSC scans at 20 deg/min.

# scans	% Decomposition 10%	% Decomposition 30%	Temperature, °C Minimum exo 1	Temperature, °C Minimum exo 2	Fraction reacted exo 1%	Fraction reacted exo 2%	Heat cal/g
Ammonium Nitrate							
26	292 °C	307 °C	318 °C	—	76%	—	350
Std. dev.	9	9	21		2		47
%	3%	3%	7%		3%		13%
ANFO							
29	294 °C	336 °C	297 °C	384 °C	13%	75%	737
Std. dev.	9	20	6	5	3	10	94
%	3%	6%	2%	1%	25%	13%	13%
CETX-101							
15	316 °C	329 °C	321 °C	364 °C	18%	72%	767
Std. dev.	5	8	5	8	6	6	82
%	2%	2%	2%	2%	33%	9%	10%

Table 9.5. Calculated Arrhenius parameters.

	by DSC ASTM [20] E_a kcal/mol	by DSC ASTM [20] Log A (1/s)	by Conventional Methods Temperature range	by Conventional Methods E_a kcal/mol	by Conventional Methods Log A (1/s)
Ammonium Nitrate					
10% decomposition	40.5	13.96	230–290 °C	22.9	6.09
30% decomposition	32.2	10.28	290–370 °C	41.9	13.50
exo minimum	32.0	9.89			
ANFO					
10% decomposition	32.0	8.17	230–370 °C (early)	35.2	11.38
30% decomposition	39.2	13.28	230–370 °C (late)	44.8	13.67
exo 1 minimum	26.2	8.11			
exo 2 minimum	43.1	12.51			
AN Emulsion CETX-101					
10% decomposition	38.1	12.34	210–370 °C	49.2	15.96
30% decomposition	68.3	23.60			
exo 1 minimum	40.3	13.09			
exo 2 minimum	38.7	11.49			

9.4.2. Thermal Stability of AN Formulations by Isothermal Techniques

Conventional isothermal thermal techniques were used to follow the decomposition of AN, ANFO, and the AN emulsion. For neat AN, rate constants could be followed satisfactorily by quantifying produced gas or monitoring nitrate remaining by ion chromotography (IC). However, for both ANFO and the AN emulsion, gas production was erratic, and rate of decomposition was assessed by IC monitoring of nitrate or ammonium ion remaining. In an AN emulsion, which contained calcium as well as AN, monitoring of the ammonium ion remaining was exclusively used to assess rate of decomposition. While it appears obvious that monitoring nitrate with an inert nitrate present would lead to confusing results, it is an unfortunate fact that once standard analytical procedures have been established, the user sometimes fails to question the validity of a technique for a specific problem.

The Arrhenius plot for the decomposition of AN over the temperature range 200 °C to 400 °C is shown in Figure 9.9. The break in the Arrhenius plot near 290 °C indicates that there is a shift in the dominent decomposition mechanism over this 200° temperature range; this change in mechanism has been confirmed by more detailed tests [Brower et al. (1989)]. DSC results give no indication of multiple decomposition modes; in fact, a general failing of DSC is that it is difficult to say to what temperature range DSC kinetics apply. For the purpose of using the Frank-Kamenetskii model, the temperature used to calculate activation energy and the pre-exponential factor should be that closest to the temperature of interest. However, the temperature of interest, such as the storage temperature, is often lower than can be measured directly. In such cases, Arrhenius parameters determined at higher temperatures are used. This is a common practice, but the users of such data must be aware that the mechanism of decomposition, and, hence, the Arrhenius parameters, may be quite different at storage temperatures than at the temperature measured.

We have shown that the decomposition of AN operates predominantly by an ionic mechanism in the temperature range 200 °C to 290 °C [Brower et al. (1989)]. Formation of nitronium is rate-determining. HX can be any source of acidity. Added nitric acid increases the rate of AN decomposition dramatically, while bases such as ammonia or water retard the decomposition.

$$NH_4NO_3 \Longleftrightarrow NH_3 + HNO_3, \tag{9.11}$$

$$HNO_3 + HX \underset{\text{slow}}{\overset{-X}{\Longleftrightarrow}} H_2ONO_2^+ \longrightarrow NO_2^+ + H_2O, \tag{9.12}$$

$$HX = NH_4^+,\ H_3O^+,\ HNO_3,$$

$$NO_2^+ + NH_3 \longrightarrow [NH_3NO_2^+] \longrightarrow N_2O + H_3O^+, \tag{9.13}$$

$$NH_4NO_3 \Longleftrightarrow N_2O + 2\,H_2O. \tag{9.14}$$

Scheme I

Above 290 °C, a free-radical decomposition mode is dominant. Homolysis of nitric acid is the rate-controlling step [Brower et al. (1989)].

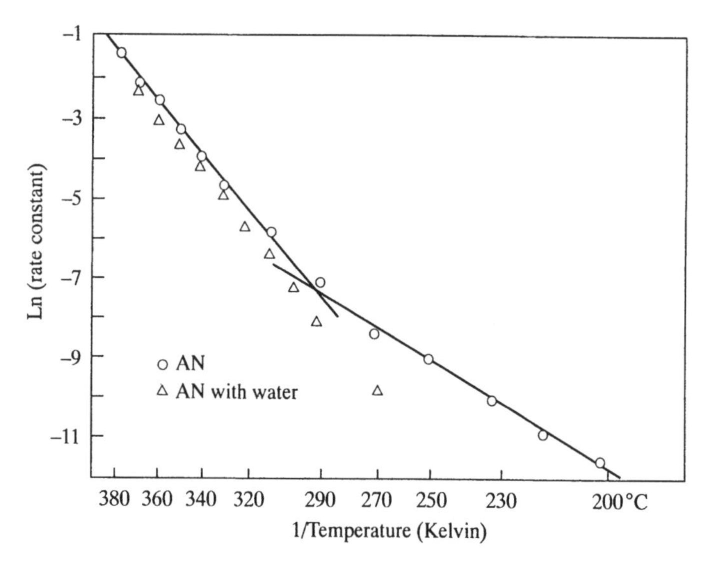

Figure 9.9. Ammonium nitrate. Arrhenius plot.

$$NH_4NO_3 \Longleftrightarrow NH_3 + HNO_3, \tag{9.15}$$

$$HNO_3 \xrightarrow{\text{slow}} NO_2 + HO, \tag{9.16}$$

$$HO^{\cdot} + NH_3 \longrightarrow NH_2 + H_2O, \tag{9.17}$$

$$NO_2 + NH_2 \longrightarrow NH_2NO_2, \tag{9.18}$$

$$NH_2NO_2 \longrightarrow N_2O + H_2O. \tag{9.19}$$

Scheme II

Formation of nitrogen may occur by several routes:

$$HNO_2 + NH_3 \longrightarrow N_2 + 2\,H_2O, \tag{9.20}$$

$$NO + {}^{\cdot}NH_2 \longrightarrow NH_2NO \longrightarrow N_2 + H_2O. \tag{9.21}$$

The first-order plot of remaining nitrate after isothermal heating at 270 °C is shown in Figure 9.10 for AN and for ANFO. As indicated by the DSC thermogram (Figure 9.8), ANFO decomposition initially is as fast or faster than that of AN, but it soon slows dramatically. Experiments have shown that the extent of the early fast decomposition depends on the thermolysis temperature and on the oil content of the emulsion. As temperature decreases or oil content increases, the period of fast decomposition decreases (Figure 9.10(a), (b)) [Oxley et al. (1989)]. We have also shown that the thermal scans of ANFO and the AN emulsion both become single high-temperature exotherms if the samples are sealed under an ammonia atmosphere (Figure 9.11) [Oxley et al. (1989)]. Build-up of ammonia retards AN

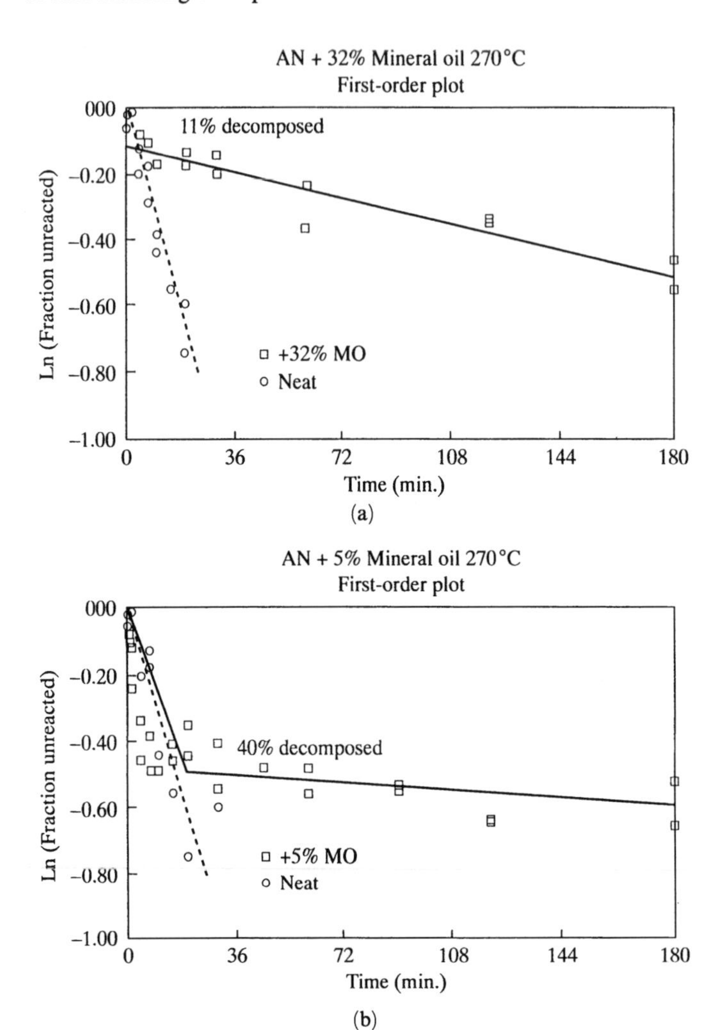

Figure 9.10.

decomposition. This build-up is the reason for the two exothermic regions. This evidence helps us to understand the decomposition mechanism of AN fuel-containing formulations. Nitric acid, which reacts with ammonia in fuel-free AN, is diverted into reaction with hydrocarbon. Ammonia builds up and retards further AN decomposition until the temperature rises (as in programmed DSC). This is the explanation for the two exothermic regions and the reason for the greater stability of the mineral-oil-containing formulations. It is also the reason more N_2 is found in the mineral-oil-containing formulations. In neat AN, the reaction of nitric acid and ammonia is responsible for the principle decomposition product nitrous oxide. When a hydrocarbon is present both nitrous acid and nitric oxide are more abundant, and dinitrogen becomes dominant in the decomposition gases.

Neat ammonium nitrate	Gas composition			from DSC
$HNO_3 + NH_3 \longrightarrow N_2O + 2\,H_2O$	N_2O	N_2		
Ammonium nitrate + Fuel	78%	22%		neat AN
$HNO_2 + NH_3 \longrightarrow N_2 + 2\,H_2O$	N_2O	N_2	CO_2	
or	23%	59%	18%	emulsion
$NO + {}^{\cdot}NH_2 \longrightarrow N_2 + H_2O$				
Hydrocarbon	39%	45%	16%	ANFO

$$RCH_3 + NO_2 \longrightarrow RCH_2^{\cdot} + HONO$$

$$RCH_2^{\cdot} + NO_2 \longrightarrow RCH_2NO_2$$

$$RCH_2^{\cdot} + HONO \xrightarrow{-NO} RCH_2OH$$

$$RCH_2NO_2 \longrightarrow RCH_2ONO \xrightarrow{-NO} RCH_2O^{\cdot} \xrightarrow{[O]} RCHO \xrightarrow{[O]} RCOOH \xrightarrow{[O]} CO_2$$

Nitrogen oxides equilibria

$$NO + 2\,HNO_3 \Longleftrightarrow 3\,NO_2 + H_2O,$$

$$HNO_3 + HNO_2 \Longleftrightarrow 2\,NO_2 + H_2O.$$

Scheme III

In an ammonium nitrate-fuel formulation, it is still the dissociation of AN forming nitric acid that triggers decomposition, but now the fuel is also involved in the decomposition offering new decomposition pathways. The complete oxidation of the hydrocarbon to CO_2 is one of these. However, Scheme III does not explain the reason the emulsified AN is more thermally stable than ANFO. The DSC thermograms shown in Figure 9.12 show that if all the emulsion ingredients are mixed together, but not emulsified, increased thermal stability is not obtained. Furthermore, a broken emulsion showed lower thermal stability than an intact one. We concluded that it is the emulsification process that provided the thermal stability [Oxley et al. (1989)]. If the emulsion broke as soon as heating commenced, it would not be any more stable than ANFO. However, we have observed two 1 g samples of emulsion held at 150 °C for 5 days: one, CETX-93, broke as soon as it reached 150 °C; but the other, CETX-101, after 5 days at 150 °C, separated only slightly, forming little pockets of AN. The thermal instability of CETX-93 was confirmed with mid-scale tests. A 1 liter cook-off of CETX-101 exhibited a critical temperature between 205 °C and 220 °C, while CETX-93 had a much lower critical temperature between 150 °C and 170 °C.

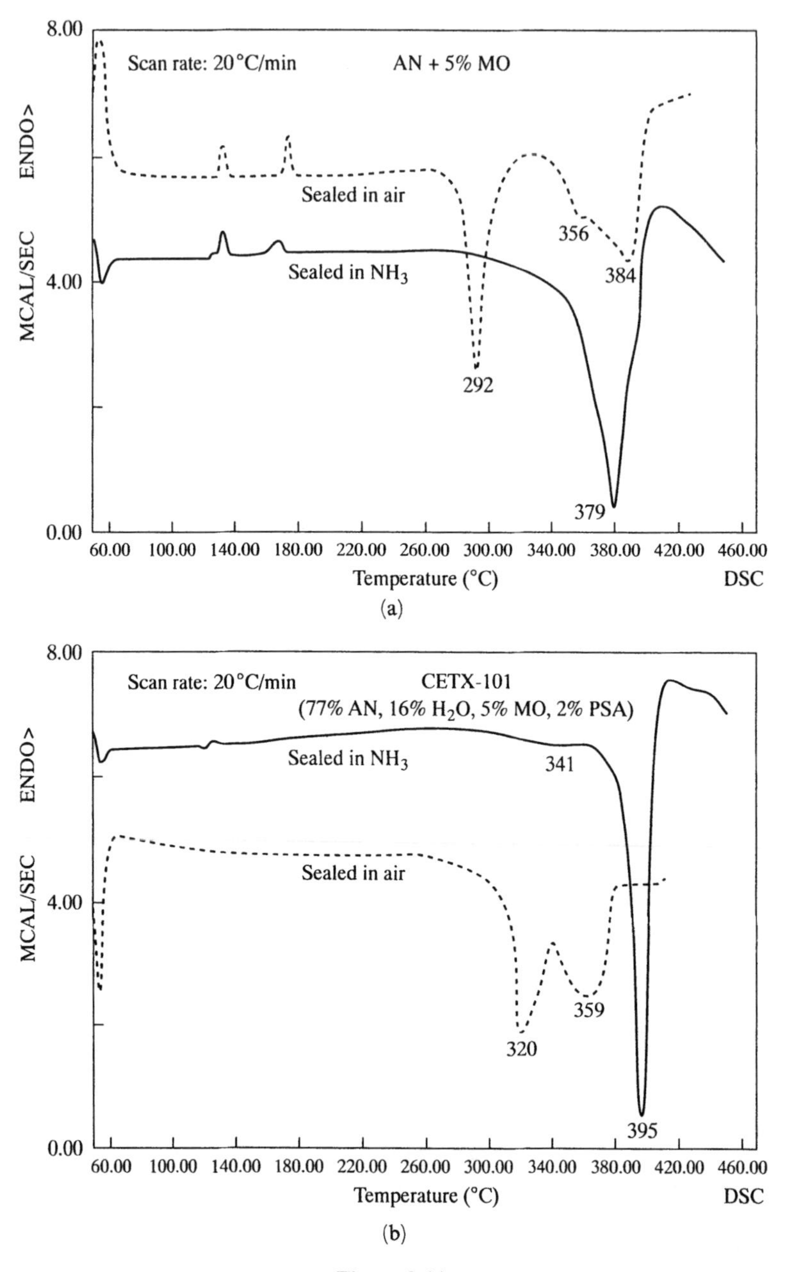

Figure 9.11.

9.4.3. Thermal Stability of AN with Additives

Our interest in the effect of additives centered around possible contaminants in mining operations. We were interested in the effect of ferric and ferrous ion in addressing the effect of pyrite ore. Bureau of Mines reports have noted that pyrite and its weathering product ferrous sulfate ($FeSO_4$) both

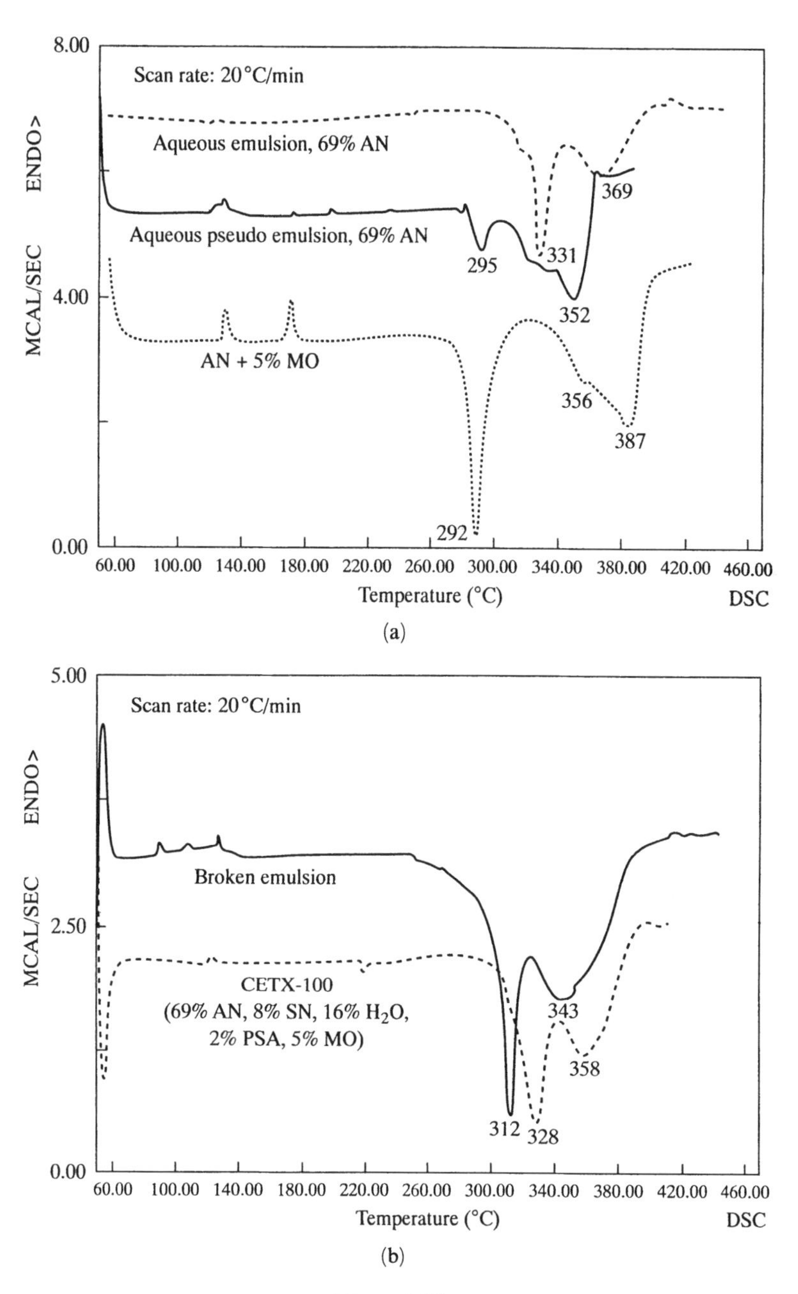

Figure 9.12.

accelerate the decomposition of AN and ANFO [Miron et al. (1979); Forshey et al. (1968)]. We reviewed this effect on thermal stability to see how well DSC predicted the instability and to determine whether it was the ferrous ion, the sulfide ion, or both, which cause accelerated decomposition. (Note: the iron pyrite encountered in mining is often mixed with chalcopyrite which contains copper, which itself could catalyze AN decomposition).

To determine the effect of various anions on the thermal stability of AN, it was necessary that the additives be introduced with inert counter ions. Our first thought was to use sodium ion or ammonium ion to insure compatibility with AN. In a survey of ammonium and sodium salts (SO_4^{2-}, BF_4^-, $B_4O_7^{2-}$, HCO_3^-), we observed the sodium salt-containing AN formulations were invariably slightly more stable than the ammonium salt-containing AN formulations. Furthermore, the sodium salts of these anions as well as the anions CO_3^{2-}, $CH_3CO_2^-$, CHO^-, $C_2O_4^{2-}$, and F^- dramatically stabilized AN thermal decomposition. Figure 9.13 shows that this increased thermal stability was confirmed by observing the temperature of the DSC exothermic maximum and the temperatures of the 10% and 30%

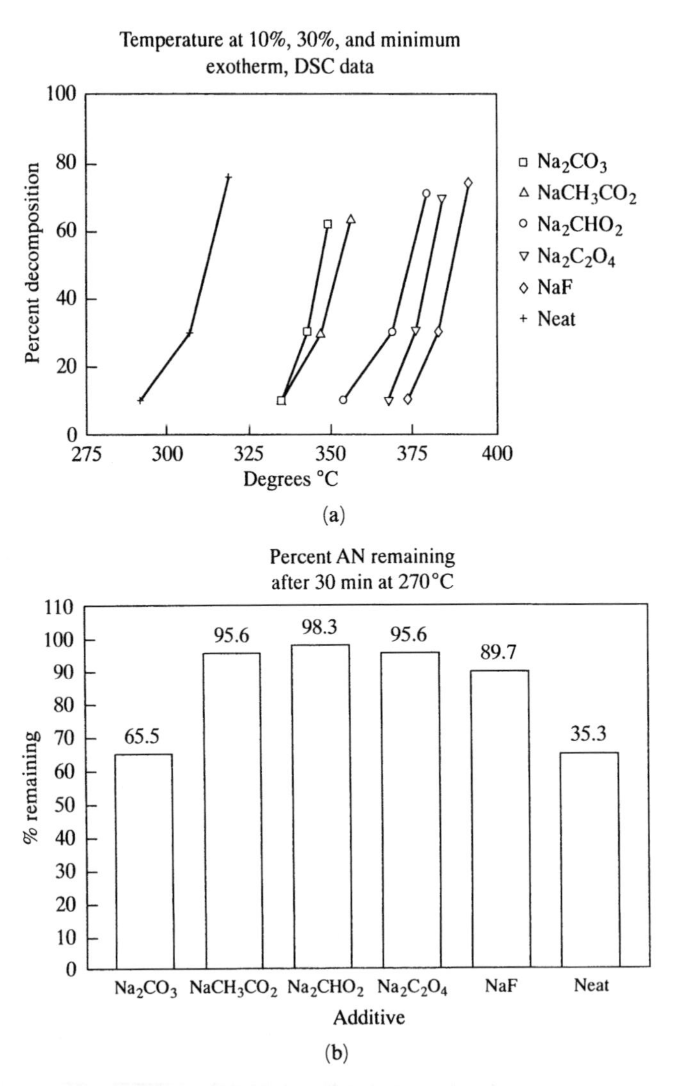

Figure 9.13. Thermal stability of AN with 5% sodium salts of weak acids.

decomposition points and by analyzing the percent nitrate remaining after heating the mixtures for 30 min at 270 °C. The methods agreed—the lower the observed DSC temperature, whether the exothermic maximum or the point of 10% or 30% decomposition (reading the top chart of Figure 9.13 horizontally across), the less AN remained after 30 min of isothermal heating [Oxley et al. (1992)].

The effect of the added salts can be directly related to the thermal decomposition mechanism of AN (Scheme I). Ammonium cation, the salt of a weak base, acts as an acid (HX) accelerating the decomposition of AN formulations.

$$NH_4^+ + H_2O \leftrightarrow NH_4OH + H^+.$$

The anions of weak acids make the mixture more basic, thus stabilizing AN.

$$CHO^- + H_2O \leftrightarrow CH_2O + OH^-$$

Urea also stabilized the AN formulations by lowering the acidity of the melt [Barclay and Crewe (1967)].

Having determined that the effect of added anions on AN was related to their effect on the acidity of aqueous solution, we began to study the effect of various metal cations on AN thermolysis. The thermal stability of AN with most common metals was assessed from DSC thermograms of their nitrate salts. The lower the temperature of 10% AN decomposition, the less stable the system was assumed to be. The destabilizing effect of chromium ion on AN thermal stability has been studied by a number of researchers. Rosser, Inami, and Wise [1964] found that all chromium compounds soluble in liquid AN would catalyze its decomposition. They suggested that this was due to an oxidation/reduction cycle where Cr (+6) is reduced to Cr (+3) by N (−3) and reoxidized to Cr (+6) by N (+5). If an oxidation/reduction cycle were important in promoting AN decomposition, then metals with multiple oxidation states should be most destabilizing to AN. This was not the case (Table 9.6). Cobalt and copper nitrate exhibited no strong destabilizing effect on AN, even though both commonly use two oxidation states (Co^{+2}/Co^{+3} and Cu^{+1}/Cu^{+2}). On the other hand, aluminum, which only has one oxidation state, Al^{+3}, had a strong destabilizing effect on AN. Chromium, aluminum, and iron nitrate, all of which strongly promote AN thermal decomposition, have two properties in common. They were the only +3 oxidation state metals examined, and they all have a rather small ionic radius. The results of these two properties is that each of these ions has a relatively high charge-to-radius ratio. When dissolved in water, these three nitrates produce acidic aqueous solutions. Charge-to-radius ratio is related to acidity because it determines how many water molecules can surround an ion and how a cation interacts with water. The scheme below shows the extremes:

$$\text{M with high charge-to-radius ratio} : M^{+n} + nH_2O \rightarrow M(OH)_n + nH^+,$$

$$\text{M with low charge-to-radius ratio} : M^{+n} + xH_2O \rightarrow M(OH_2)_x^{+n}.$$

Table 9.6. Thermal stability of AN + nitrate salts.

	DSC data (20 °C/min) temperature (°C)		Properties		
Nitrate additives (5 wt%)	1st exo.	10% decomp.	Oxidation state	ionic radius	pH aqueous solution
Cr	294	251	3	76	5
Fe	287	258	3	69	2
Al	298	262	3	68	3
Cu	304	280	2	87	4
AN	324	289			6
Ba	316	288	2	135	7
Mn	314	288	2	81	7
Co	306	293	2	79	6
K	329	293	1	152	7
Na	326	294	1	95	7
Pb	324	294	2	133	7
Mg	323	296	2	86	7
Zn	330	298	2	88	6
Cd	322	299	2	109	7
Ag	343	305	1	126	6
Ni	327	311	2	83	7
Ca	353	317	2	114	7

Figure 9.14 indicates that urea and various salts of weak acids provide sufficient thermal stability to counteract the effect of destabilizing metal salts. This suggested a "fix" to the premature detonation problem.

9.4.4. Verification of Small-Scale AN Kinetics by Larger-Scale Tests

Several concerns arise when either DSC results or conventional kinetics derived from small-scale testing are used to predict the thermal stability of bulk material. First, Arrhenius plots are not usually straight over large temperature ranges; and, thus, the kinetics at high temperature may not be applicable at normal storage temperatures. Furthermore, the microsamples may not be truly representative of the overall material. Therefore, kinetic parameters determined from small-scale samples must be used with caution in predicting the behavior of larger-scale systems. In order to determine if there is a problem with scaling, the kinetic parameters determined from small-scale experiments must be tested to see if they correctly predict the results of larger tests.

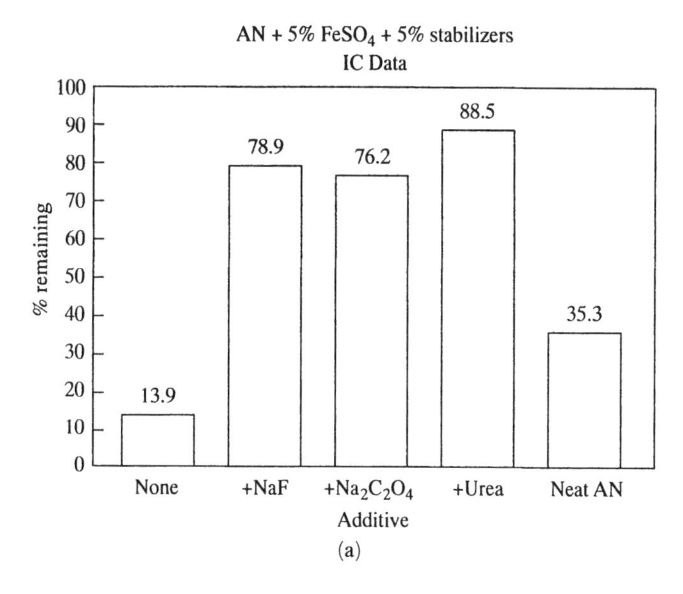

(a)

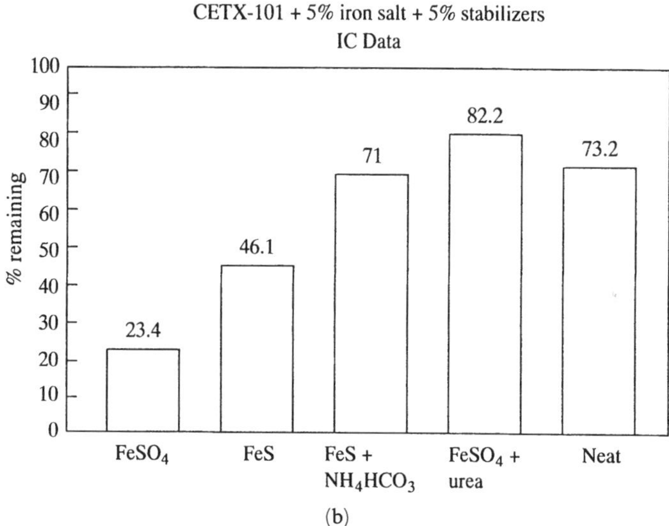

(b)

Figure 9.14. Stabilizers counteract destabilizers.

9.4.4.1. One-Liter Cook-Off

A 1 liter cook-off can be used to check the scale-up of a predictive model. The set-up designed at Eglin Air Force Base consists of a 1 liter round-bottom flask holding the sample immersed in a cylindrical jar full of low-viscosity silicone oil. Three thermocouples, one at the center, one horizontally at the edge, and one half-way between are placed into the sample flask to monitor temperature (Figure 9.15). One thermocouple is placed directly in the bath to monitor its temperature. There is a small vent in the top of the flask to ensure that the event observed is thermal runaway and not pressure

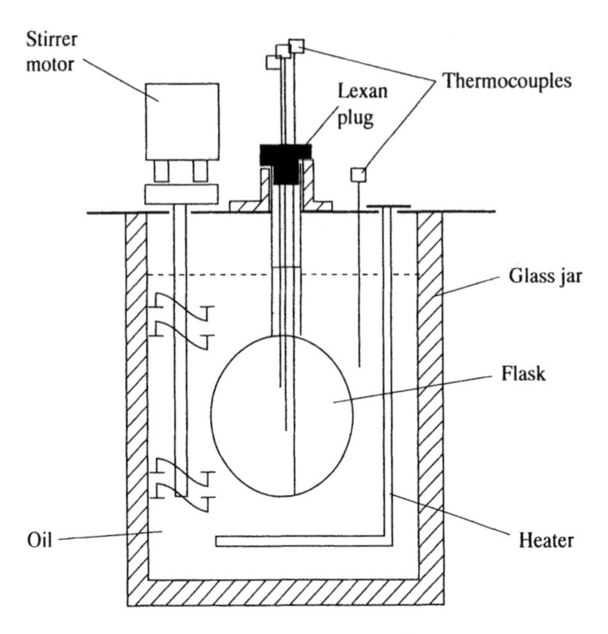

Figure 9.15. 1 liter cook-off setup.

rupture. The 1 liter of sample is heated isothermally until a thermal runaway occurs or the experimenter runs out of patience. Time-to-explosion is measured, and a critical temperature is bracketed by "Go/No Go" runs at varying temperatures. If the test results confirm the predicted critical temperature, the kinetic values used to make this predictions can be used with some confidence.

In a 1 liter cook-off, a sample of AN emulsion CETX-101 was heated for 44 hours at 190 °C and for 2.5 hours at 207 °C (the temperature change reflects the experimenters' impatience) before thermal runaway was observed. This is a reasonable bracketing of the Frank-Kamenetskii predicted critical temperature of 202 °C. In contrast, CETX-93, an emulsion which exhibited a greater tendency to break, underwent thermal runaway in the 1 liter apparatus between 150 °C and 170 °C.

9.4.4.2. Sealed Cook-Off

A vented cook-off, such as the one described above, may give erroneous results due to loss of volatiles, such as water. The kinetics of AN emulsion CETX-93 contaminated with 10 wt% iron pyrite predicted a 1 liter critical temperature of 98 °C. However, the 1 liter cook-off of that formulation at 110 °C gave no indication of thermal runaway. The sample bubbled violently and portions of it escaped from the flask. Both these actions allowed heat losses not accounted for in the Frank-Kamenetskii predictive model. Therefore, a new cook-off configuration was designed which would prevent boil-off of the volatile components during low-temperature (less than 150 °C)

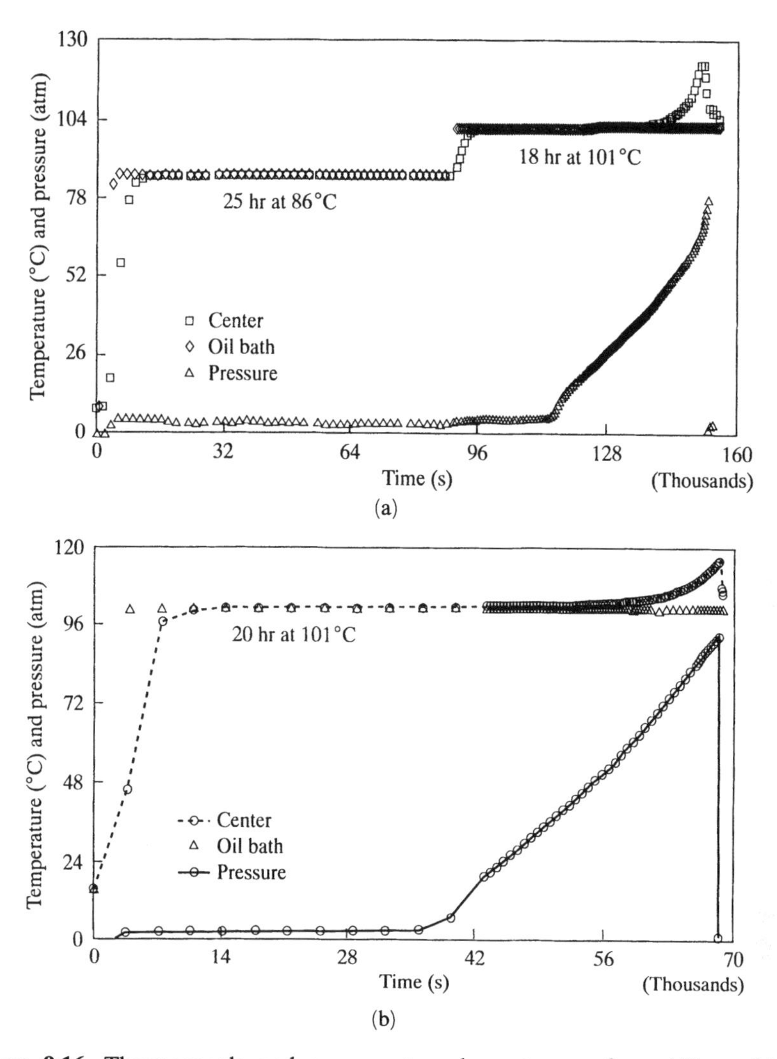

Figure 9.16. Thermocouple and pressure transducer traces of an AN emulsion (CETX-93) with 10% iron pyrite ore.

cook-off experiments. A number of configurations could be used. The objectives are to have a sufficiently sturdy container to hold pressure, to be relatively inexpensive since it may be sacrificed, and to be a shape easily modeled. A stainless-steel cylindrical chamber (6.35 cm diameter by 19.05 cm depth, about 600 ml total volume) was machined. It was equipped with a thermocouple set in the center of the sample, a pressure transducer, a pressure relief valve, and a pressure rupture disk rated to 540 atm. Two cook-off experiments were performed using this configuration with an AN emulsion (~300 ml), destabilized with 10 wt% ore; thermal runaway was observed between 86 °C and 101 °C. (Predicted critical temperature in the cylindrical configuration was 105 °C.) Figure 9.16 shows the trace of the thermocouples in the bath and in the center of the sample, as well as the readings from the pressure transducer for both cook-off experiments. For a sample of the AN emulsion CETX-93 spiked with 10 wt% ore and 3 wt% urea, the predicted 1 liter critical temperature of 187 °C was high enough so that the 1 liter configuration could be used. Runaway was observed between 159–174 °C. These results appear to indicate that the emulsion with urea was not only more stable than the AN emulsion destabilized with pyrite, but it was also more stable than the emulsion alone.

9.4.5. Visual Observations

With all the emphasis on high-tech instrumentation, we should not forget the value of visual observations and common sense. The thermolysis of the AN emulsion spiked with pyrite ore in the 1 liter and sealed cook-off configurations allowed examination of the thermal decomposition process at temperatures much lower than those examined by Henkin, DSC, or conventional isothermal techniques. In the low-temperature tests, visual observations suggested little AN was consumed in the exothermic process. The residue from the cook-off of AN emulsion CETX-93 with 10% pyrite ore was a brown mud, permeated with long white needles, and a yellow–green residue [identified as containing iron (III) and sulfate]. Since the white crystals melted around 169 °C, they were assumed to be AN. It appears that the observed thermal runaway was caused by the heat of oxidation of the iron sulfide to sulfite. AN participated in this reaction by providing the oxidant, nitric acid; but general AN dissociation did not occur

$$4H^+ + 2FeS_2 + 6\,NO_3^- \rightarrow Fe_2(SO_4)_3 + 3N_2 + SO_4^{-2} + 2H_2O.$$

These observations serve to emphasize the importance of confirming the prediction of small-scale techniques, with larger-scale tests. The exothermic decomposition of AN observed by DSC was not the important exothermic event at lower temperatures.

9.4.6. Summary

An overview of the present state of the art in laboratory-scale thermal analysis experiments is illustrated by a case history study of an ammonium nitrate (AN) composite blasting agent which underwent premature detonation in a bore hole. A full study of the AN mixture was made. DSC and isothermal techniques were used to assess the rate-determining mechanism in AN decomposition. The effect of added fuel and contaminates on AN was studied in detail. It was found that straight chain hydrocarbons, in the long-term, retard thermal decomposition, while acidic species, such as ammonium ion and ferric salts, enhance decomposition of AN formulations and basic species, such as urea and the salts of weak acids, stabilize them. Similar observations have been made for ammonium perchlorate [Oxley et al. (1995c)]. The final evidence was obtained from 1 liter cook-off experiments and micro heat-flow calorimetry. These showed an exothermic reaction at low temperature which involved the AN blasting agent only as a catalyst. An effective "fix" to the problem was devised based on the determined decomposition mechanism of AN. The importance of thermal stability studies near the temperature of interest is thus emphasized.

9.5. Thermal Stability of Organic Explosives

In general, nitramines [Oxley et al. (1992b, 1994c)] are more thermally stable than nitroarenes [Oxley et al. (1995a, b); Brill and James (1993); Minier et al. (1991)] and nitroarenes are more thermally stable than nitrate esters [Hiskey et al. (1991)]. All three share one common mechanism by which they can decompose—homolysis of the X—NO_2 bond. The relative bond energies for this bond are estimated as O—NO_2, 40 kcal/mol, N—NO_2 47 kcal/mol, and C—NO_2 70 kcal/mol for nitroarenes [Melius (1988)]. If homolytic cleavage of NO_2 were the only decomposition route, then nitroarenes should appear significantly more thermally stable than nitramines, in line with their bond energies.

	O—NO_2 PETN	N—NO_2 RDX	C—NO_2 TNT	Salt AN	(data from [Oxley et al. (1994a, b)])
200 °C	2.7E−2	5.9E−4	4.4E−6	8.6E−6	s^{-1} isothermal rate
240 °C	too fast	2.0E−2	9.1E−5	1.1E−4	s^{-1} isothermal rate
DSC	210	257	330	320 °C exothermic max 20 °C/min.	

However, nitroarenes are extremely sensitive to acidity [Oxley et al. (1995a, b)]. Acids attack the NO_2 group, lowering the energy necessary to dissociate it from the ring since it now leaves as HNO_2. Bases react with the acidic protons on nitrotoluenes forming Meisenheimer complexes [Terrier (1982)]. Furthermore, there are other decomposition pathways available to each type of energetic material. The dominant reaction path is dependent upon the temperature of the study.

9.6. Toxicity of Explosives

The toxicity of explosives, given in Table 9.7, is garnered from materials safety data sheets. Users are warned against touching, breathing, and ingesting these materials. Standard precautions are recommended. For most explosives, symptoms cited include headaches, nausea, and skin irritation.

Table 9.7. Environmental properties of explosives.

	MW (g/mol)	mp (°C)	Tex (°C)	Vapor pressure (torr, 20 °C)	bp (°C)	State	Color	TLV (mg/m^3) TWA	STEL	PEL	LD50 rat (oral) (mg/kg)	Irritant	Methemoglobin agent	Cancer	Fatal	Chronic and/or Cautions
NITRATE ESTERS																
NG, nitroglycerin	227	13	270	2.5E − 04	218	L	None → lt .y	0.05	0.1		105	Y	Y		Y	Nausea, hallucination; do not fight fire
80 ppm immediate danger to life and health; narcotic; 0.02 ppm cause headaches																
EGDN, EthyleneGlycolDinitrat	152		237	0.04		L	None → lt .y	1.0								Warnings as in NG
DEGN, DiethyleneGlycol Dinit	196			3.6E − 03	160											
Dynamite (NG/EGDN)						S										
NC, nitrocellulose C12H14N6022 (14.4% N)			160			S	Cream									Do not allow to dry
Collodion NC (NC/i-pro as most volatile component, alcohol may cause dizziness or irriation; estimated lethal dose to man 0.5–5 g/kg																
PETN	316	141	210	8E − 5*	200	S	White	10				N		N		Blood press. drop; 65 mg/d speeds up heart shipped wet 40% in water or in methanol; methanol toxicity 400 ppm
DETASHEET (PETN, acetyl tributyl citrate, NC) DuPo XTX-8003 PETN 80%, silicone 20%																
NITROARENES																
1,3-DNB, dinitrobenzene	168	89			297	S	Yellow	0.15			83	Y	Y		Y	Reprod. tract; somnolence, headache, coma
1,4-DNB	168	173				S	Yellow–brown	0.15		1		Y	Y		Y	Reprod. tract; somnolence, headache, coma
2,6-DNT, dinitrotoluene	182	65				S	Light yellow	1.5			177	Y	Y			Reprod. tract; somnolence, headache, coma
2,4-DNT	182	69	270	9E − 07	250d	S	Light yellow	1.5			268	Y	Y	Y		Reprod. tract; somnolence, headache, coma
2-methyl-4,6-dinitro-phen	198	84				S	Yellow	0.2		0.2	10	Y	Y		Y	Somnolence, headache, coma
HNAB, hexanitrobenzene	452	220	332	1E − 7*		S						Y		N		Lungs
HNS, hexanitrostibene	450	318d	320	1E − 9*		S	Cream	0.05				Y				Kidney/liver
Picric acid	229	122	300	1.0E − 09		S	Yellow	0.1	0.3	0.1		Y			Y	Kidney/liver; avoid metals
Tetryl	287	129	187	3.5e − 5*	187	S						Y	Y		Y	Eyes, digestion, lungs, liver, anemia
DATB, diaminotrinitroben	243	286	322	6E − 3**		S	Yellow					Y				
TATB triaminotrinitroben	258	320	347	1e − 4**		S	Yellow	1.5	3		5000	Y	N	N	N	None
PBX-9502 (TATB/KEL-F800) (95/5)																HF fumes at high temperature

TNT	227	81	288	0.1*		S	Cream	0.5				Y				Anorexia, liver, aplastic anemia; avoid NH^{4+}
Baratol ($Ba(NO_3)$2/TN Pentolite (TNT/PETN) (50/50)							Octol (HMX/TNT) (70/30-75/25)									
Cyclotol (RDX/TNT) (75/25—68/32) +1% wax or 0.3–0.6% calciu						Comp. B (RDX/TNT/wax) (50/50-60/40)										
NITRAMINES																
NQ, nitroguanidine	104	230d	230			S	White					Y		N		Avoid strong acids
Guanidine Nitrate	122					S	White				1028					Altered sleep, excitement; str oxid. avoid M
RDX	222	204d	217	4E – 5*		S	White	1.5	3	NE	100	Y			N	convulsions, systemic poison
Comp. A-3 (RDX-wax) PBX-9407 (RDX/polyvinyl chloride) (94/6)							PBX-9010 [RDX/polystyrene (KelF800)] (89/11–91/9)									
C-4 (RDX/polyisobutylene, dioctyl adipate or sebacate, pet oil) (90/2.1/5.3/1.4–92/2.5/5.9/1.6)																
HMX	296	280d	253	3E – 9*		S	White				300	Y				Ship with 10–15% H_2O/methanol
PBX-9404 (HMX/NC/trichlroethylpho PBX 9011 (HMX/Estane) (90/7); PBX 9501 (HMX/Estane/BDNPA-F) (95/2.5/2.5)															Y	1–2 g fatal ingested
LX-07 (HMX/Viton A) (90/10); LX-10 (HMX/Viton A) (95/5); LX-14 (HMX/Estane) (95.5/4.5)																
Miscellanous																
AN, NH_4NO_3	80	169	246			S	White				4820					Oxidizer
AP, NH_4ClO_4	117	450d	—	4.4e – 9*		S	White				4200					Oxidizer
HAN, $HONH_3NO_3$	96	48			102	S/L	White					Y	Y			Strong oxidizer
LP 1864 (HAN/TEAN/water)		−102		11.5	124	L	Colorless	TEAN = triethanolamine nitrate								
Ba $(NO_3)_2$	261	332	—			S	White		0.5							Oxidizer, poison
NaN_3	65	—	—			S	White				27			Y		Poison, mutigen
Nitromethane, CH_3NO_2	61	−29	315	37	101	L	Colorless		250			Y				Liquid flash point 95F
2-Nitropropane	89	−93		13	120	L	Colorless			90		Y		Y		Gastrointestinal/liver
NTO, nitrotriazole	130	270d	270			S	White			NE	5000	Y				Avoid oxiders
FEFO, C5H6N4)10F2	320	14	~195	2.1E – 04		L			1.5						Y	Highly toxic on contact, strong vesicant
DPA, diphenylamine (stabilize)	169					S	White	10								

TLV = Threshold Limit Value; TWA = Total Weight in Air; STEL = Short-Term Exposure Limit; PEL = Permissible Exposure Limit.
State = Solid, Liquid; Tex = Temperature of explosion.

In general, the greatest hazard is from the explosive characteristics rather than the toxicity. In fact, as noted on Table 9.3, the toxicity of several products is governed by the solvent (isopropanol, methanol) or remaining monomer (styrene) from the polymeric binder (KEL-F800) rather than the main ingredient, (i.e., the explosive fill).

In general, nitrate esters tend to be coronary vasodilators. For that reason, both nitroglycerin (NG) and PETN are used as heart medication. The onset time from exposure to headache and the persistence and severity of the headache appears to be proportional to the volatility of the nitrate ester. Continual exposure to NG can result in a temporary tolerance, but this is quickly lost after several days without exposure [Gotell (1976)]. Some early workers with NG reportedly wore bits of NG-soaked cloth on their days off so as to retain their NG-tolerance. NG is readily absorbed through the skin as well as by breathing.

Nitroarenes, as well as many other chemicals, increase blood formation of methemoglobin [Linch (1974)]. A methemoglobin value of 1.5% has been recommended as an index of overexposure, without judging the degree of adverse effects [Hyg]. Methemoglobin at 1% levels substantially reduces the capacity of blood to carry oxygen; at 15% methemoglobin cyanosis and headaches result [CDC (1988)]. A symptom of cyanosis is a purplish–blue color of the skin. In general nitrobenzenes are more toxic than nitrotoluene, possibly due to the greater ease with which the nitrotoluene can be oxidized [Davis (1943)].

RDX was patented in 1899 for possible medicinal use and in 1920 as an explosive [Davis (1943)]. Modern-day material safety data sheets list "convulsions" as the symptom of severe RDX or HMX poisoning.

References

Barclay, K.S. and Crewe, J.M. (1967). The thermal decomposition of ammonium nitrate in fused salt solution and in the presence of added salts. *J. Appl. Chem.*, **17**, 21.

Brill, T.B. and James, K.J. (1993). Kinetics and mechanism of thermal decomposition of nitroaromatic explosives. *Chem. Rev.*, **93**, 2667–2692.

Brower, K.R., Oxley, J.C., and Tewari, M.P. (1989). Homolytic decomposition of ammonium nitrate at high temperature. *J. Phys. Chem.*, **93**, 4029–4033.

Catalano, E., McGuire, R., Lee, E., Wrenn, E. Ornellas, D., and Walton, J. (1976). The thermal decomposition and reaction of confined explosives. *Sixth Symposium (International) on Detonation*, pp. 214–222.

Cook, M.A. (1958). *The Science of High Explosives*. American Chemical Society Monograph. R.E. Krieger, Florida, facsimile.

Davis, T.L. (1941, 1943 reprint of two volumes). *The Chemistry of Powder and Explosives.* Angriff Press, Hollywood, CA.

Forshey, D.R., Ruhe, T.C., and Mason, C.M. (1968). The reactivity of ammonium nitrate–fuel oil with pyrite bearing ores. Bureau of Mines Report, RI 7187.

Frank-Kamenetskii, F. (1969). *Diffusion and Heat Transfer in Chemical Kinetics*. Plenum Press, New York.

Gibbs, T.R. and Popolato, A. (1980). *LASL Explosive Property Data*. University of California Press, Berkeley, CA.

Gotell, P. (1976). Environmental and clinical aspects of nitroglycol and nitroglycerin exposure. *Occupational Health and Safety*, p. 50.

Henkin, H. and McGill, R. (1952). Rates of explosive decomposition of explosives: Experimental and theoretical kinetic study as a function of temperature. *Indust. Engng. Chem.*, **44**, 1391.

Hiskey, M.A., Brower, K.R., and Oxley, J.C. (1991). Thermal decomposition of nitrate esters. *J. Phys. Chem.*, **95**, 3955–3960.

Linch, A.L. (1974). Biological monitoring for industrial exposure to cyanogenic aromatic nitro and amino compounds. *Amer. Indust. Hyg. Assoc. J.*, **35**, 426–432.

Mader, C.L. (1979). *Numerical Modeling of Detonation*. University of California Press, Berkeley, CA, p. 145.

Melius, C.F. (1988). *Proceedings 25th JANNAF Combust. Meeting*. Huntsville, AL.

Mellor, A.M. and Boggs, T.L., eds. (1987). Energetic materials hazard initiation: DoD Assessment Team Final Report. US. Army.

Miller, P.J., Coffey, C.S., and DeVost, V.F. (1985). Infrared emission study of deformation and ignition of energetic materials. NSWC TR 85-452, NSWC, White Oak, MD.

Minier, L., Brower, K., and Oxley, J.C. (1991). The role of intermolecular reactions in thermolysis of aromatic nitro compounds in supercritical aromatic solvents. *J. Org. Chem.*, **56**, 3306–3314.

Miron, Y., Ruhe, T.C., and Watson, R.W. (1979). Reactivity of AN–FO with pyrite containing weathering products. Bureau of Mines Report, RI 8373.

Mullay, J. (1987). Relationships between impact sensitivity and molecular electronic structure. *Propellants, Explosives and Pyrotechnics*, **12**, 121–124.

Oxley, J.C., Kaushik, S.M., and Gilson, N.S. (1989). Thermal decomposition of ammonium nitrate-based composites. *Thermochem. Acta*, **153**, 269–286.

Oxley, J.C., Kaushik, S.M., and Gilson, N.S. (1992a). Ammonium nitrate explosives—Thermal stability and compatibility on small and large scale. *Thermochem. Acta*, **212**, 77–85.

Oxley, J.C., Hiskey, M.A., Naud, D., and Szeckeres, R. (1992b). Thermal decomposition of nitramines: Dimethylnitramine diisopropyl-nitramine, and N-nitropiperidine. *J. Phys. Chem.*, **96**, 2505–2509.

Oxley, J.C., Smith, J.L., and Wang, W. (1994a). Compatibility of ammonium nitrate with monomolecular explosives, Part I. *J. Phys. Chem.*, **98**, 3893–3900.

Oxley, J.C., Smith, J.L., and Wang, W. (1994b). Compatibility of ammonium nitrate with monomolecular explosives, Part II: Nitroarenes. *J. Phys. Chem.*, **98**, 3901–3907.

Oxley, J.C., Kooh, A., Szeckeres, R., and Zheng, W. (1994c). Mechanism of nitramines thermolysis. *J. Phys. Chem.*, **98**, 7004–7008.

Oxley, J.C., Smith, J.L., Ye, H., McKenney, R.L., and Bolduc, P.R. (1995a). Thermal stability studies on homologous series of nitroarenes. *J. Phys. Chem.*, **99** (24), 9593–9602.

Oxley, J.C., Smith, J.L., Zhou, Z., and McKenney, R.L. (1995b). Thermal decomposition studies on NTO and NTO/TNT. *J. Phys. Chem.*, **99** (25), 10383–10391.

Oxley, J.C., Smith, J.L., and Valenzuela, B. (1995c). Ammonium perchlorate decomposition: Neat and solution. *J. Energetic Mater.*, **13** (1&2), 57–91.

Oxley, J.C., Dong, X.X., Han, T., and Smith, J.L. (1998). Decomposition gas production by common explosives. *J. Energetic Mater.* (Submitted).

Rogers, R.N. (1975). Thermochemistry of explosives. *Thermochim. Acta*, **11**, 131–139.

Rosser, W.A., Inami, S.H., and Wise, H. (1963). The kinetics of decomposition of liquid ammonium nitrate. *Trans. Faraday Soc.*, **67**, 1753.

Rosser, W.A., Inami, S.H., and Wise, H. (1964). Decomposition of liquid ammonium nitrate catalyzed by chromium compounds. *Trans. Faraday Soc.*, **60**, 1618.

Semenov, N.N. (1935). *Chemical Kinetics and Chain Reactions.* Oxford University Press, London.

Storm, C.B., Stine, J.R., and Kramer, J.F. (1990). Sensitivity relationships in energetic materials. In *Chemistry and Physics of Energetic Materials* (S.N. Bulusu, ed.). Kluwer Academic, Dordrecht, pp. 605–639.

Terrier, F. (1982). Rate and equilibrium studies in Jackson–Meisenheimer complexes. *Chem. Rev.*, **82**, 77.

ASTM (1995). Method E 698-79. Standard test method for Arrhenius kinetic constants for thermally unstable materials. *1995 Annual Book of ASTM Standards*, vol. 14.02. Philadelphia, PA.

Hyg. Documentation of the threshold limit values and biological exposure indices. *Amer. Conf. of Gov. Industrial Hygienists.* Cincinnati, OH.

IME (1988). *Institute of Makers of Explosives 75th Anniversary Booklet* (from US Bureau of Mines Data).

CDC (1988). Methemoglobinemia due to occupational exposure to dinitrobenzene. Centers for Disease Control. *MMWR*, **37** (22), 353–355.

ME (1984). *Military Explosives.* Army Technical Manual TM9-1200-214, September, pp. 8–95.

MIL-STD-286B; method 403.1.3 (1975).

MIL-STD-286B; method 406.1.2 (1975).

UN (1986). *Recommendations on the Transport of Dangerous Goods Tests and Criteria*, 1st ed. United Nations, New York.

CHAPTER 10

Demolitions

Chris A. Weickert

10.1. Introduction

This chapter provides a broad overview of demolitions and demolition devices. Focus is on systems or techniques which use explosives to either launch a projectile against or to be placed in direct contact with a particular structure. Blast from an explosive charge, combustible fuel or dust, can cause significant damage to a structure, however this is not a conventional demolition technique. For a discussion of this topic, see the books by Baker (1973), Kinney and Graham (1985), Baker et al. (1983), and Glasstone and Dolan (1977). Although demolition can be performed with various mechanical techniques, they are outside the scope of this chapter.

Explosives suitable for demolition applications are available in numerous forms: liquid, slurry, foam, gel, pellet, plastic (sheet and block), cast and pressed. Such explosives find use in a wide variety of industrial and military applications: explosive welding, forming, powder compaction, rock shattering, jet or projectile formation in shaped charges or explosively formed projectiles (EFP). For demolitions these may be used in direct contact or as the filling for shaped charge or EFP devices which produce a high-velocity projectile used to cut or demolish a structure.

Historically, demolition techniques used bulk explosives in contact with a structure and/or its critical elements to produce the required effect. In the 1800s, it was discovered that a cavity or depression between a direct contact charge and a structure, "focussed" the effects of the blast. This gave rise to the term "shaped charge." Refinements to the technique followed whereby the depression (originally done by hand) was formed by pressing the explosive against an inert form. Lining the depression with metal further increased the performance. As shaped charges were developed with an efficient cutting capability, many of the old demolition techniques were replaced. Research on shaped charges led to the development of a charge which formed a high-velocity projectile and hence the term "explosively formed projectile." Explosively formed projectiles offer considerable future potential for demo-

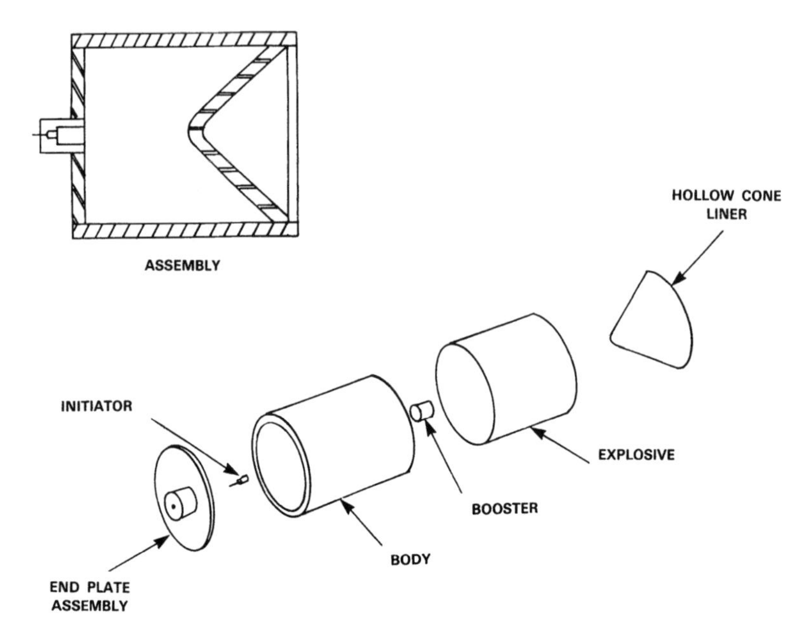

Figure 10.1. Components of a shaped charge device.

lition applications, since their long stand-off capability allows them to be used in situations where it is dangerous to use contact demolition techniques.

The basic components of a shaped charge device are illustrated in Figure 10.1. The device consists of an initiator or detonator, a booster, and explosive charge encased in a charge body with a plate at one end and a hollow-cone liner on the other end. Upon initiation, a detonation wave propagates through the booster and high-explosive charges. As this detonation wave impinges on the liner, the high pressures produced cause the liner to behave as a fluid and force it to collapse on the axis of symmetry. When the liner reaches the axis of symmetry the inner layer of material forms a high-velocity jet (approximately 15–20% of the liner mass), while the remainder of the mass forms a low-velocity slug. The jet and slug formation processes are illustrated in Figure 10.2.

As the shaped charge jet forms it continuously lengthens due to a velocity gradient. Typically, jet velocity ranges from 7 km/s to 10 km/s at the tip to 1 km/s to 2 km/s at the tail. The velocity gradient is a result of the variation of the explosive/liner mass ratio along the liner. Due to this gradient along the length of the jet, a stretching process occurs and eventually the jet breaks up. Sequential flash radiographs of the jet breakup are shown in Figure 10.3. At later times the jet particles tumble, thus reducing their penetrating effectiveness. Therefore, there is an optimum distance from the target in order to maximize penetration. This stand-off distance is typically 5–6 charge diameters (CD) for a conical shaped charge and results in a penetration depth

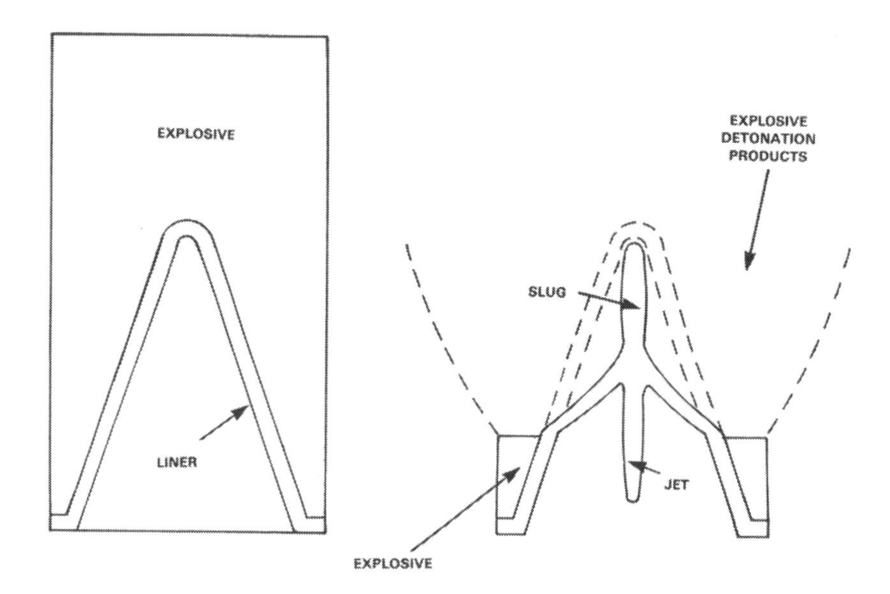

Figure 10.2. Shaped charge jet and slug formation.

of more than 6 CD in steel. Shown in Figure 10.4 is the penetration profile for a 75 mm diameter shaped charge in mild steel plates. Many different liner geometries and materials have been used to produce shaped charges for different applications. Shaped charges are treated in detail in the books by Walters and Zukas (1989) and Carleone (1993). The historical aspects of shaped charges are presented by Kennedy (1983) and Backofen (1978).

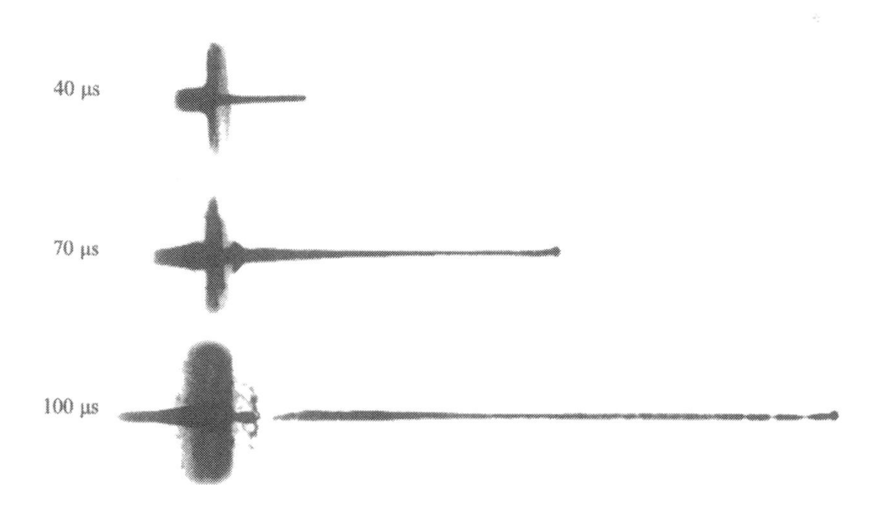

Figure 10.3. Flash radiographs of a 75 mm diameter shaped charge.

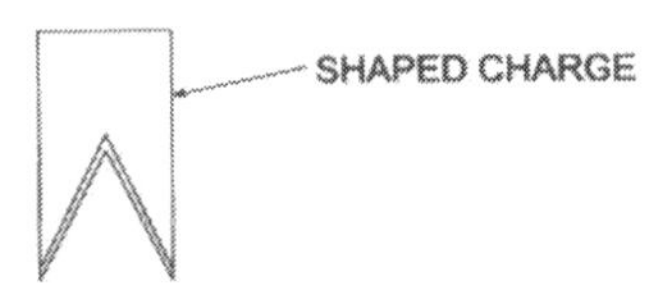

Figure 10.4. Experimental hole profile.

An explosively formed projectile device is similar to a shaped charge, except that the liner is saucer-shaped rather than conical (Figure 10.5). Detonation of the explosive collapses the entire liner into a single high-velocity projectile (Figure 10.6), typically 2–3 km/s. The stand-off distance for EFP devices can be well in excess of 100 m. In comparison to a shaped charge device, the penetration depth for a single piece EFP (0.8–1 CD) is much less, however the hole diameter in the target is considerably greater. Similar to shaped charges, many different liner designs and materials over the years have been used to produce EFPs for various applications. An introduction to EFP technology can be found in the books by Walters and Zukas (1989) and Carleone (1993).

The literature on shaped charges and explosively formed projectiles tends to be spread out over a wide range of government and industrial reports,

Figure 10.5. Comparison of shaped charge and EFP liners. Left: Shaped charge liner. Right: EFP liner.

journal papers, and symposia. For those readers wishing to pursue this topic in more depth, probably the largest source of information is contained in *The International Symposia on Ballistics.* A second source is *The Hypervelocity Impact Symposia* which are republished in the *International Journal of Impact Engineering.* Information on the behavior of materials at the high strains and strain rates that liner materials encounter, can be found in the *Shock Compression of Condensed Matter*, *EXPLOMET*, and *EURODYMAT* conferences. The most recent of each is listed in references (1994–1996).

This chapter covers three areas of demolition applications: (i) Wall Breaching; (ii) Bridge Demolition; and (iii) Explosive Ordnance Disposal (EOD). Since explosively formed projectiles represent the newest demolition tech-

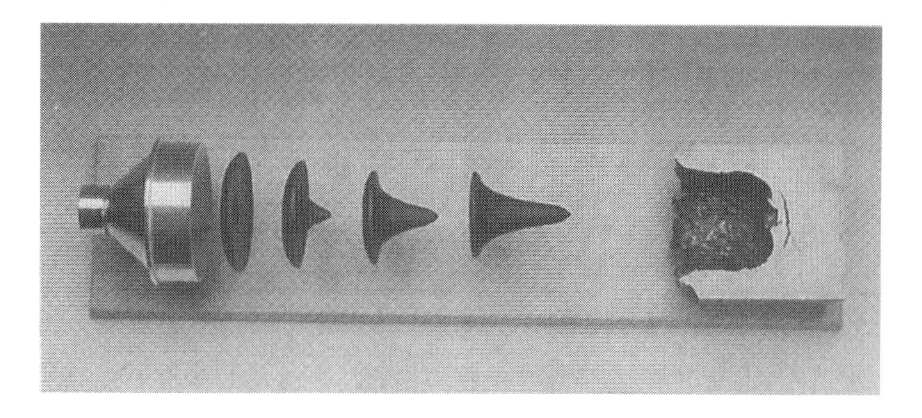

Figure 10.6. Formation of an explosively formed projectile.

nique, the first section covers the mechanics of EFPs in detail. Other technologies are introduced in the following sections on applications.

10.2. Explosively Formed Projectiles

The design of explosively formed projectile devices involves many parameters which affect the projectile characteristics. Physical properties and responses of the explosive, liner, and casing materials represent some of these parameters. Other parameters are geometric, such as liner contours, casing dimensions, charge diameter, etc. Successful EFP design will arrive at parameters which lead to an EFP capable of defeating a specific structure for a particular application. The purpose of this section is to illustrate the effects of the various parameters.

There are several techniques to produce EFPs of a particular shape. One technique is to fold the rim of the liner forward (Figure 10.7) or backward (Figure 10.8) with respect to the center of the liner. Alternatively, the liner can be folded into a "W" shape (Figure 10.9). Each of these types of projectile shapes can be attained by liner contour design and/or by the high explosive configuration. The type of the design to be used, depends on the particular application. In general, the forward-fold liners result in longer EFPs in comparison to the backward-fold liners and therefore have greater penetration capability, however the velocity gradient in the projectile usually causes them to break into several pieces, which limits their range. The backward-fold liners result in a projectile with better aerodynamic characteristics and therefore greater range. Since most applications for EFP technology require long stand-off and deep penetration into a structure, the majority of research has been conducted on backward-fold liner designs.

Once a designer has chosen the type of EFP for a particular application, it then becomes necessary to determine the design parameters. The design and optimization of EFPs require experimental tests. However, due to the large number of parameters controlling the EFP formation, a trial and error approach soon becomes very expensive. Furthermore, the speed at which events unfold makes it difficult to gain an in-depth understanding of the mechanics through analytical and experimental means alone. Hence, EFPs are designed using numerical modeling techniques in combination with results from experiments. These permit the designer to explore and narrow down the number of parameters prior to experiments, and provide insight into the mechanics governing the various stages of EFP formation and target penetration. Numerical modeling techniques are discussed in detail in the books by Zukas et al. (1982) and Zukas (1990). Perhaps the most difficult aspect of the numerical modeling is simulating the behavior of the liner materials at the very high rates of strain. Details of material behavior under these conditions can be found in Asay and Shahinpoor (1993), Meyers (1994), Bushman et al. (1993), and Blazynski (1987). Sophisticated constitutive

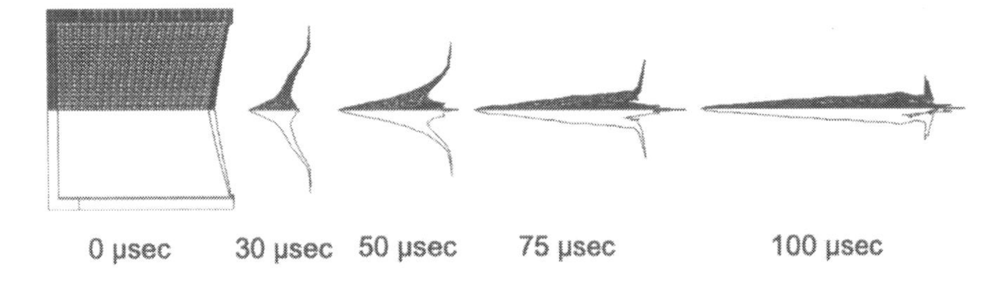

Figure 10.7. ZeuS simulation of a forward-fold EFP.

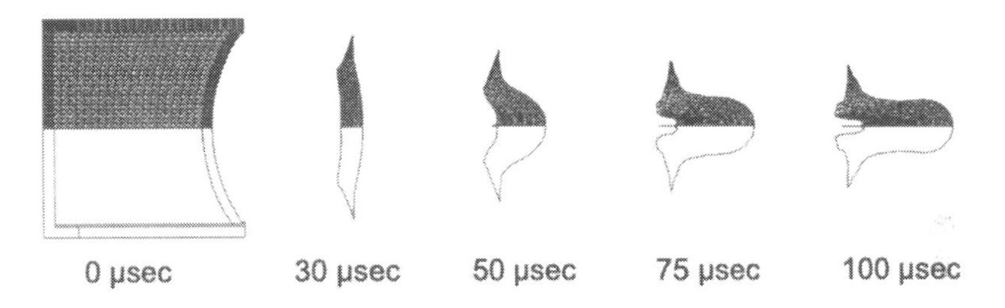

Figure 10.8. ZeuS simulation of a backward-fold EFP.

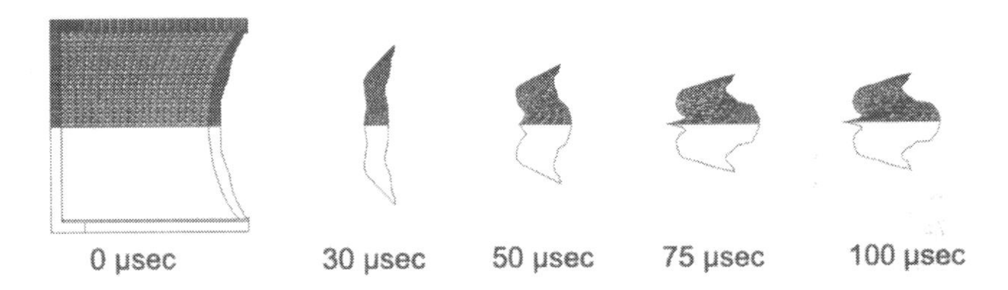

Figure 10.9. ZeuS simulation of a W-fold EFP.

models which include thermal softening, work hardening, strain rate effects, and equations of state have been incorporated into the computer codes. However, the criteria used for material failure and the subsequent action taken by the code, for example, zero strength elements or elimination of elements, still needs improvement. A given experiment can almost always be matched by the codes, but this generally requires adjustments specific to that problem. In spite of these limitations, numerical modeling has been used very successfully in EFP design.

The final outcome of an EFP device is a high-velocity projectile with particular shape and velocity characteristics. All design parameters, whether geometric or material, affect these characteristics. Within the geometric characteristics, the liner contour design can be used to produce an EFP with a particular shape. However, equally important is the casing or confinement

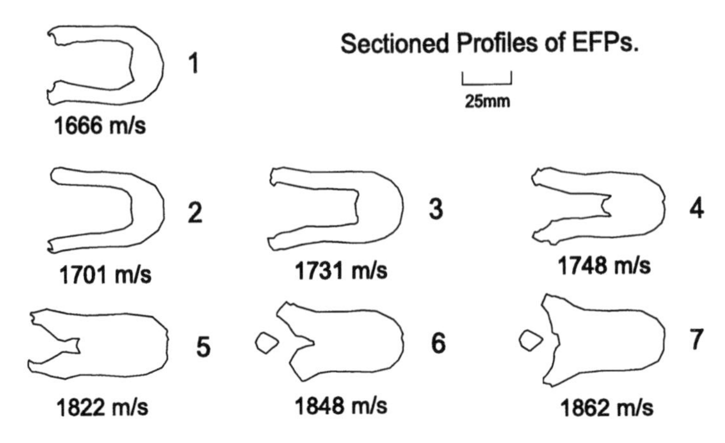

Figure 10.10. Experimental EFP profiles (1.0–7.0 mm case).

geometry. By simply changing the thickness of the casing of the EFP device, both the resulting projectile velocity and shape are changed significantly as shown in Figure 10.10. Similarly, changing from a cylindrical charge to a taper-backed charge (to reduce the quantity of explosive) and thinning the casing (to reduce the weight), significantly affects the characteristics of the projectile which is formed as shown in Figure 10.11. By redesigning the liner profile though, it is possible to produce the same shape of EFP but with a lower velocity as shown in Figure 10.12.

The materials used for the charge components also define the EFP characteristics. Figures 10.13 and 10.14 show four soft-recovered EFPs from experiments using 100 mm diameter charges with identical geometry (casing and liner). The broken EFP (Figure 10.14 (bottom)) was actually intact during flight but fractured during the soft-recovery process. The cast explosive filling for the charges was either Composition B or Octol. The liners were machined from Armco iron or tantalum. The more energetic explosive (Octol) drives the liner harder, resulting in a longer EFP. The tantalum liner material produces a longer and almost solid EFP in comparison to the Armco iron EFP which is hollow. Notice the velocity difference for the EFPs as a result of the material characteristics (specifically density).

In many explosively formed projectile designs, the objective is to produce a long rod-type projectile in order to maximize target penetration capability. However, a long projectile is inherently unstable in flight. This instability can be corrected by spin or drag stabilization [Weickert (1986)]. Spin stabilization is difficult to achieve with an EFP and hence most EFP designs use drag stabilization. Early designs using this technique had a flared tail on the projectile as shown in Figure 10.15 (top). Recently designs have been

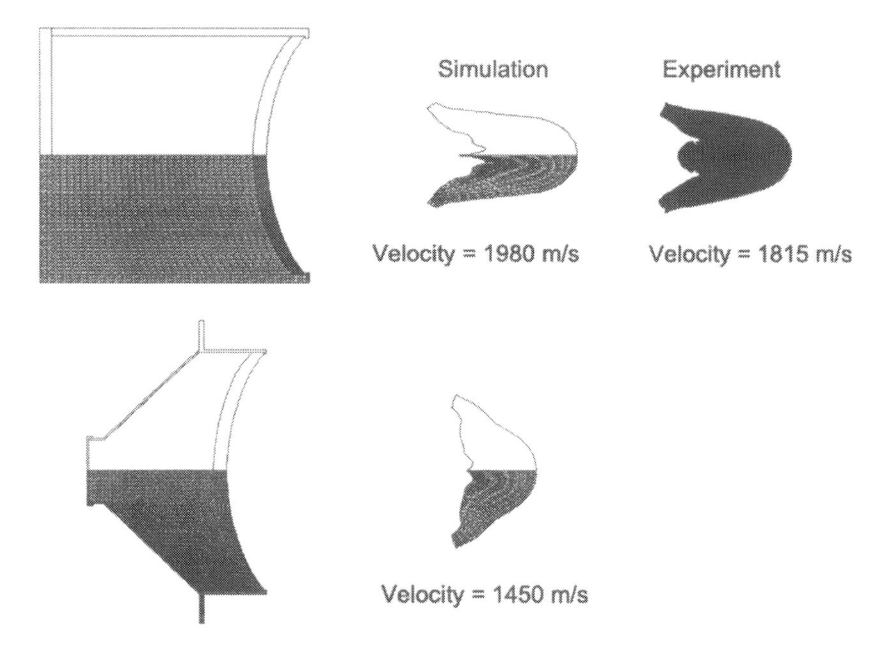

Figure 10.11. Comparison of cylindrical and taper-backed charges.

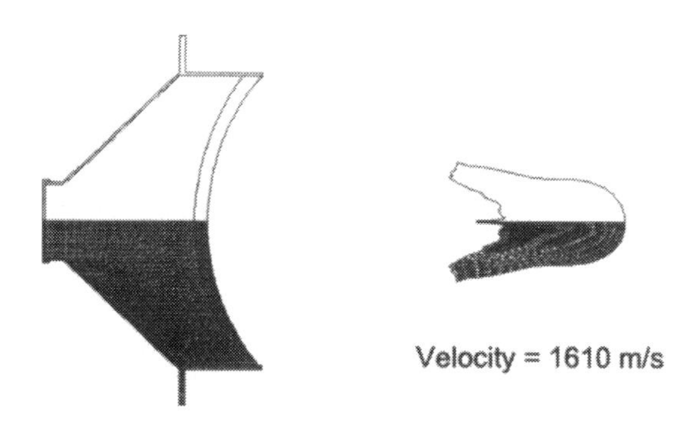

Figure 10.12. Redesigned liner profile.

developed, which use fins on the tail of the projectile instead of the flare [Weimann (1993a)]. This feature reduces the aerodynamic drag on the projectile. A six-finned EFP is shown in Figure 10.15 (center). Aerodynamic performance can be further improved by replacing the blunt nose on the EFP with an ogive nose as shown in Figure 10.15 (bottom) [Weickert and Gallagher (1996)].

Velocity = 1463 m/s

Velocity = 1618 m/s

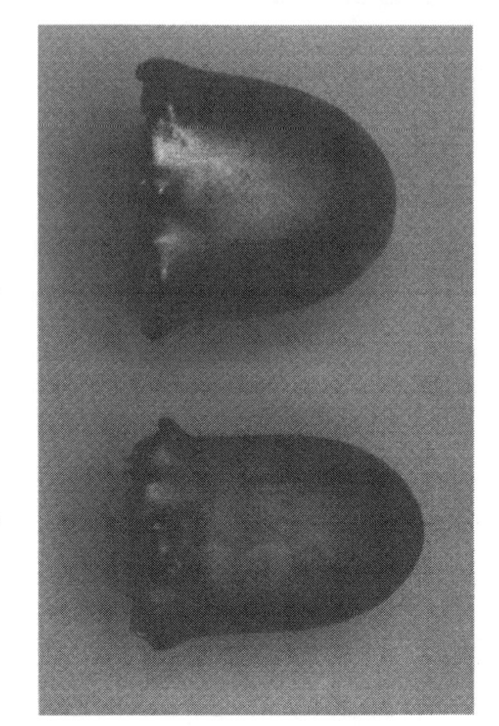

Figure 10.13. Armco iron EFPs. Top: Composition B explosive. Bottom: Octol explosive.

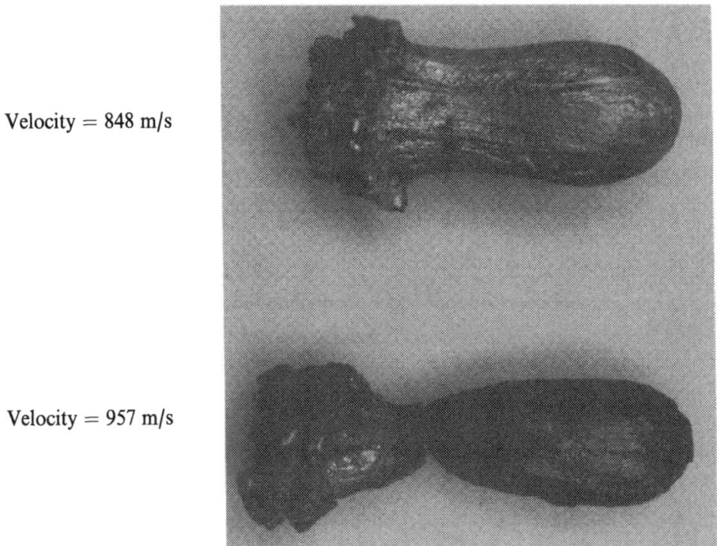

Figure 10.14. Tantalum EFPs. Top: Composition B explosive. Bottom: Octol explosive.

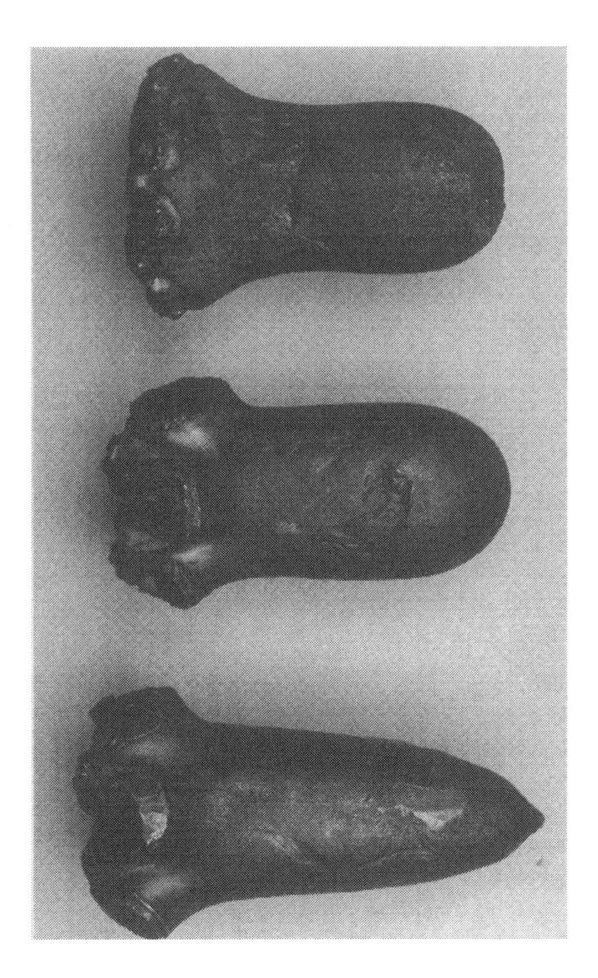

Figure 10.15. Drag stabilized EFPs. Top: Flared tail. Center: Finned tail. Bottom: Finned tail and ogive.

Readers interested in pursuing geometric and material design parameters for EFP devices in more detail or other parameters such as liner material processing or detonation wave shaping should consult Weimann (1993b), Weickert and Gallagher (1992), Weimann et al. (1990), Cullis (1992), Faccini (1992), and Murphy et al. (1992).

10.3. Wall Breaching

There are various applications, both civilian and military, requiring rapid access through the wall of a building or structure. Breaching of walls can

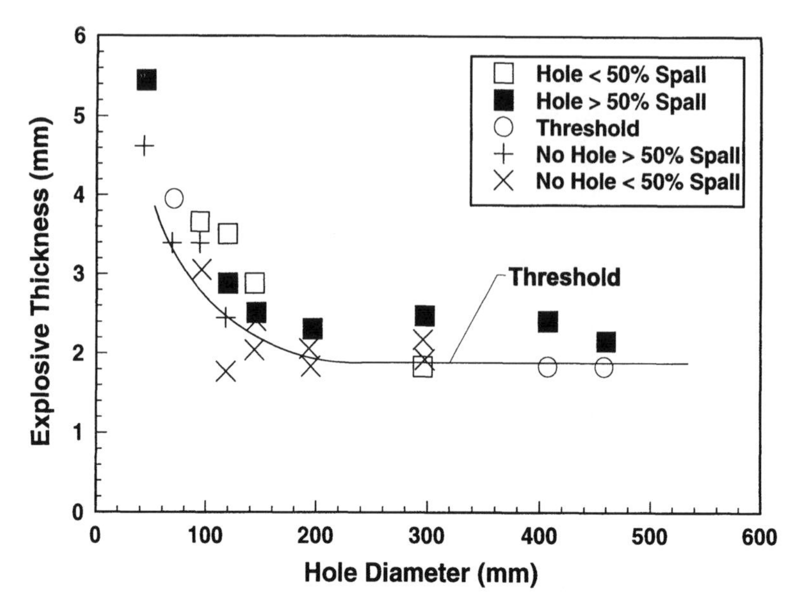

Figure 10.16. Threshold for 6 mm plate [data from Shirey (1980)].

be accomplished using bulk explosives or explosive devices such as shaped charges or explosively formed projectiles. The particular technique selected would depend on the type of structural material for example; steel, masonry or reinforced concrete and the application.

Bulk explosives can be used in sheet or block form for wall breaching. The usual parameter to be determined is the quantity of explosive needed to make the required size of hole in a particular target. For example, Shirey (1980) investigated the response of mild steel plates to breaching by circular disks of sheet explosive in direct contact with the plate. Thresholds of explosive thickness required to produce a breach in steel plates in the thickness range of 6 mm to 25 mm were determined. It was found that for a particular plate thickness, the threshold explosive thickness is a function of the hole diameter (slightly less than the explosive disk diameter) as illustrated in Figure 10.16. For large diameter holes (above a critical value) the explosive thickness is constant. As the hole diameter becomes smaller, edge effects in the detonating explosive become significant and the explosive thickness required to produce a breach is increased. The relationship between plate thickness and threshold explosive thickness (for hole diameters above the critical value) is linear, as illustrated in Figure 10.17.

A series of experiments to determine the charge weight/hole diameter relationship for reinforced concrete targets was conducted by Jonasson and reported by Forsen (1990). In the experiments, hemispheres of explosive were detonated in direct contact with reinforced concrete targets of both one-quarter and full scale in order to investigate scaling behavior. The results of the experiments are given in Figure 10.18. There is a linear relationship

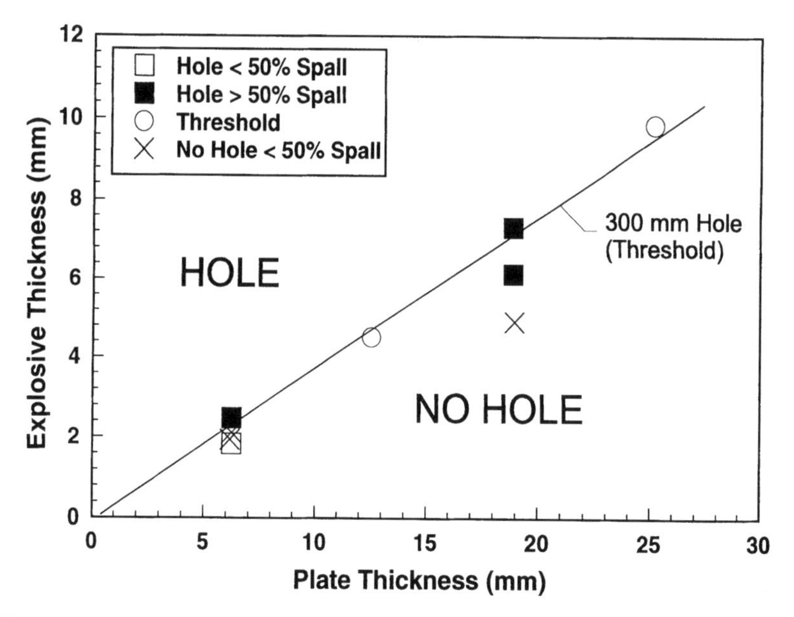

Figure 10.17. Plate thickness versus explosive thickness [data from Shirey (1980)].

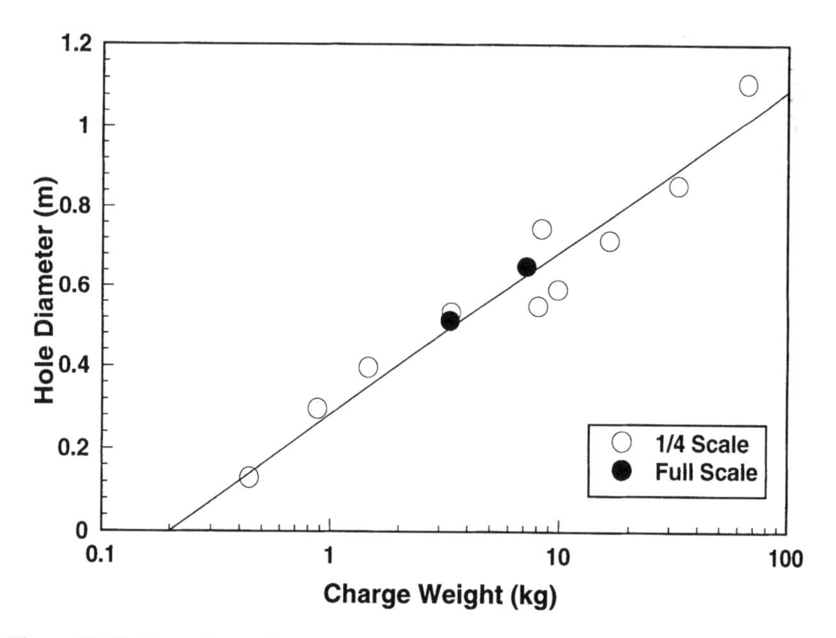

Figure 10.18. Experimental results from contact charges [data from Forsen (1990)].

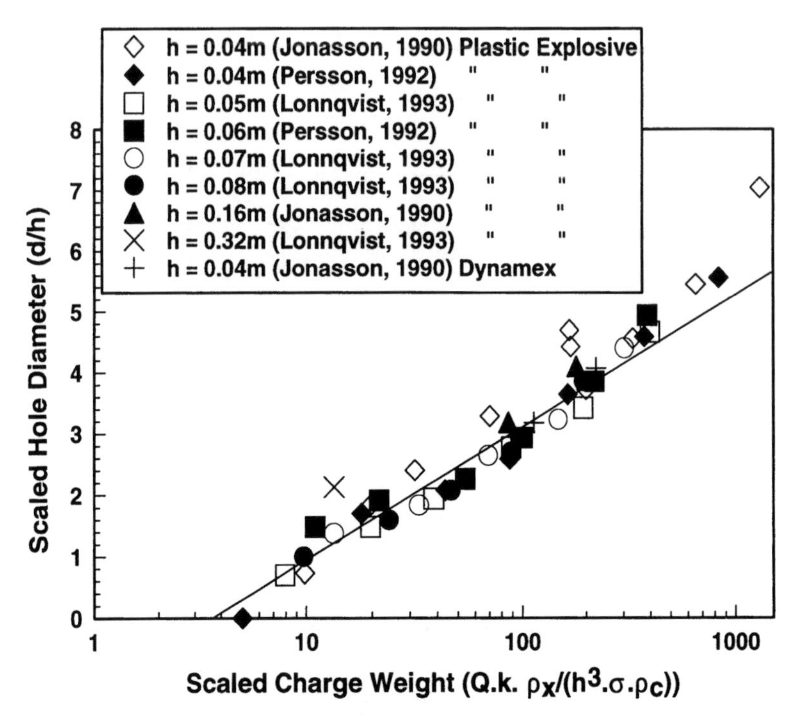

Figure 10.19. Dimensionless charge weight/hole diameter relationship [data from Lonnqvist (1993)].

between the hole diameter and the logarithm of the explosive charge weight. Also shown are the scaled results. Tests were also conducted with spherical, cubical, and cylindrical charges. It was found that these charge configurations gave approximately the same results as for the hemispherical charges.

This research effort was extended by Lonnqvist (1993) to include additional parameters and to develop a dimensionless charge weight/hole diameter relationship. Additional experiments were conducted and the data were combined with previous work by Jonasson and Persson. The results are given in Figure 10.19.

The functional fit to the data is

$$\frac{d}{h} = 0.919 \ln\left(\frac{Qk\rho_x}{h^3\sigma\rho_c}\right),$$

where d = hole diameter (m),
h = slab thickness (m),
Q = explosive charge weight (kg),
k = explosive energy (J/kg),
ρ_x = explosive density (kg/m^3),
σ = concrete strength (Pa),
ρ_c = concrete density (kg/m^3).

Breaching of reinforced concrete walls with bulk explosives can also occur when the explosive charge is not in contact with the structure. This is not a typical situation for demolition of structures, however, it is important for safety considerations in areas in which explosives or explosive devices are handled, for example, an ammunition factory. For this type of situation, the spalling of concrete from the rear surface of the wall is an important consideration. The level of damage to a reinforced concrete wall from an explosive charge is mainly a function of the charge size, distance from the charge to the wall (stand-off), and the wall thickness. Other secondary parameters include concrete strength and reinforcing. The physical processes occurring as the stand-off is decreased are as follows:

(i) small cracks and a surface crater;
(ii) wall deflection and significant cracks;
(iii) spalling on the rear surface of the wall;
(iv) penetration, and
(v) perforation and extreme bending of the reinforcing material.

Typically, these processes are associated with damage categories since it is difficult to quantify the effects of all the parameters involved. Hader (1983) uses three categories; Minor Damage, Spalling, and Perforation. The results from a test series using bare explosive charges are given in Figure 10.20. The log-log plot allows the delineation between the damage categories to be represented by parallel straight lines. In a second series of tests, metal cased charges were used. A comparison of the bare and cased charge damage category results is shown in Figure 10.21. The most important conclusion from this figure is that cased charges result in perforation of

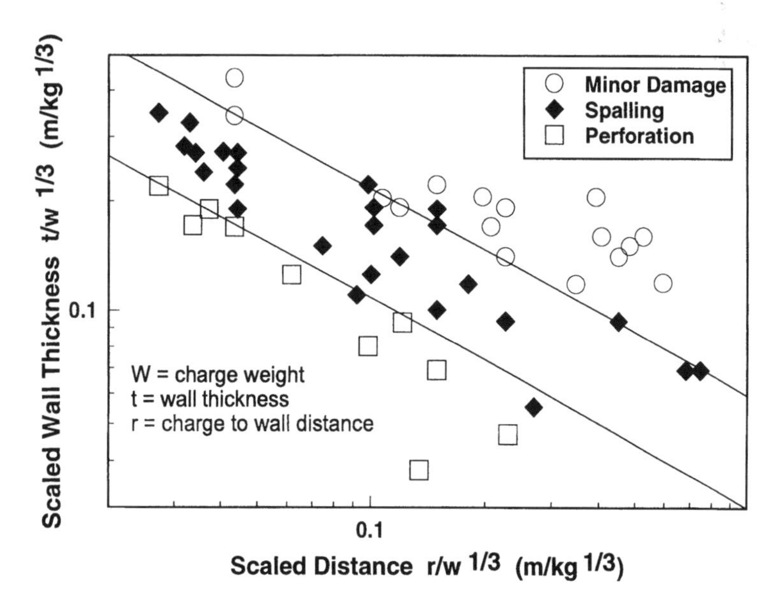

Figure 10.20. Damage to RC walls from uncased charges [data from Hader (1983)].

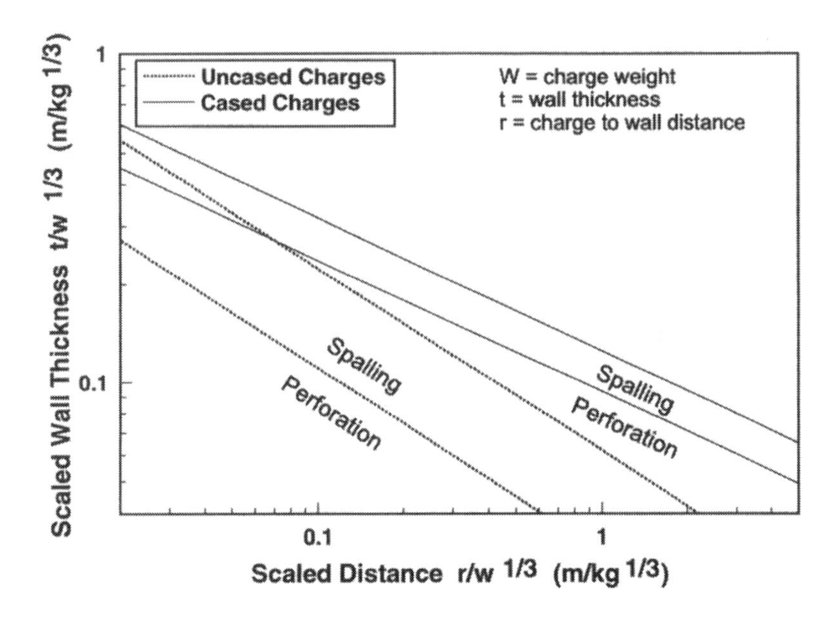

Figure 10.21. Comparison of damage from uncased and cased charges [data from Hader (1983)].

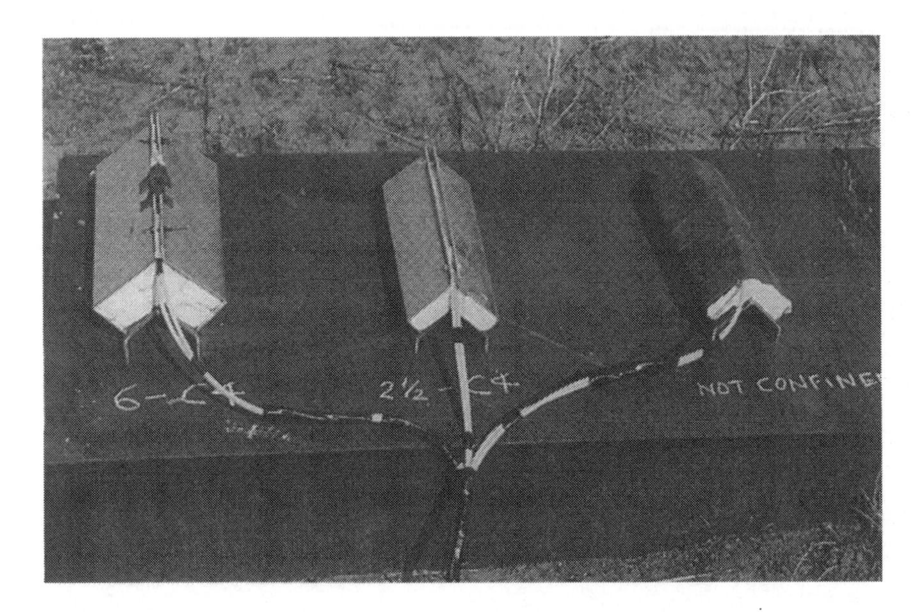

Figure 10.22. Linear shaped charges. Left: 6 blocks C4 explosive. Center: 2.5 blocks C4 explosive. Right: 2.5 blocks C4 explosive. No metal confinement.

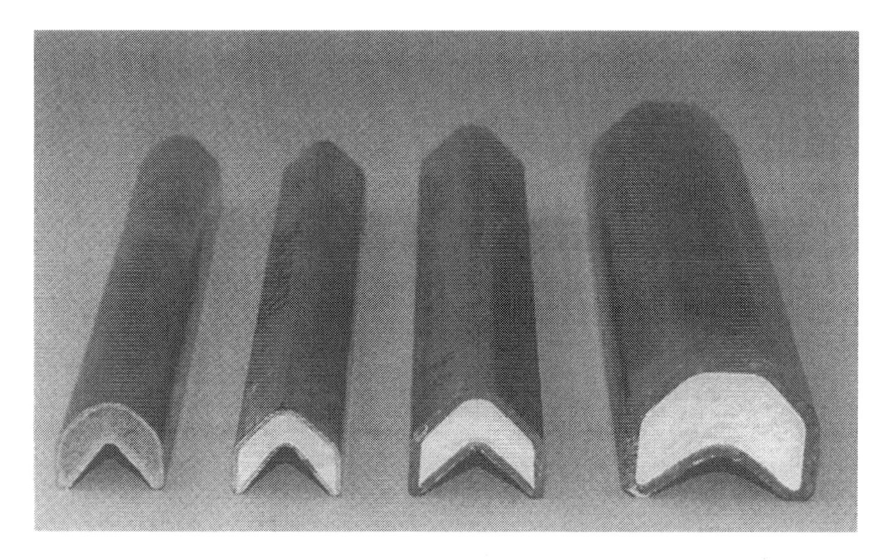

Figure 10.23. Commercial linear shaped charges. Left two charges: Copper case. Right two charges: Lead case.

reinforced concrete walls at distances up to ten times larger than bare charges of the same weight.

Another technique for wall breaching is to use a linear shaped charge. This type of charge produces a long cut in a target typically 1–2 throat widths deep. Linear shaped charges can be constructed using plastic explosive and sheet metal as shown in Figure 10.22, or are available commercially in a wide range of sizes and types (examples in Figure 10.23).

Shaped charges are applied in contact with the wall in a "cookie cutter" arrangement and remove a section of material from the wall. An example of this technique is shown in Figures 10.24 and 10.25. This product, now discontinued, was designed for use by firefighters to vent burning buildings. A wall-breaching device "Breachcase" for military applications is shown in Figures 10.26 and 10.27. Two U-shaped sections of linear shaped charge are housed in a hinged, folding container for storage and transport. When the device is required, it is simply opened, forming a rectangular charge, positioned against a wall and remotely detonated. Nitromethane filled linear shaped charges have also been used for wall breaching as shown in Figures 10.28 and 10.29. The charge (8 cm throat width, six 46 cm long sections) breached a hole approximately 205 cm wide by 128 cm high through 45 cm thick reinforced concrete.

In situations where direct contact charges are not possible, explosively formed projectiles can be used due to their long stand-off capability. Shown in Figure 10.30 is a 300 mm diameter EFP demolition device which produces a backward-fold projectile. This device has significant penetration

Figure 10.24. Jet-Axe.

Figure 10.25. Jet-Axe cut in triple brick wall.

capability against reinforced concrete as shown in Figures 10.31–10.33. The EFP breached the first two concrete walls with sufficient residual energy to topple a third wall (not shown). Similar to shaped charges, a linear charge can be produced. An example of a linear EFP device (8 cm × 8 cm × 25 cm) is shown in Figure 10.34. It consists of a steel casing, copper liner, and an explosive filling of 2.5 kg of Composition B. Test firings of the device with a planar initiation system resulted in a projectile which broke into several pieces as shown in the radiograph (Figure 10.35). Although the projectile broke, considerable damage was done to a 200 mm thick reinforced concrete wall target at an 8 m stand-off (Figure 10.36). To improve the performance of this linear EFP test device, by producing a single projectile, would require either a wave shaper in the explosive or a more complex liner design.

10.4. Bridge Demolition

Bridge demolition techniques depend on many factors, both technical and tactical. In the civilian application the objective would be to remove an old condemned bridge. In this instance, the objective would be to cut the bridge into sections that could be easily removed. An important consideration for explosive demolitions is the collateral damage to nearby structures. The military objective would depend on the particular situation. One example would be to create a gap in the bridge structure large enough so that it could not be crossed with assault bridging equipment, hence denying access to vehicle traffic. The principal technical demolition parameters include: bridge design, construction materials, and the dimensions of key structural members. Bulk explosives, shaped charges, explosively formed projectiles, or combinations of these can be used for bridge demolition.

Bulk explosives can be applied to a bridge demolition using several techniques. Large quantities of explosive can simply be loaded onto the bridge surface and detonated, causing removal of the span or collapse of the bridge. More efficient use of the explosives can be made by targeting key bridge elements. Removal of these elements can cause the bridge to collapse under

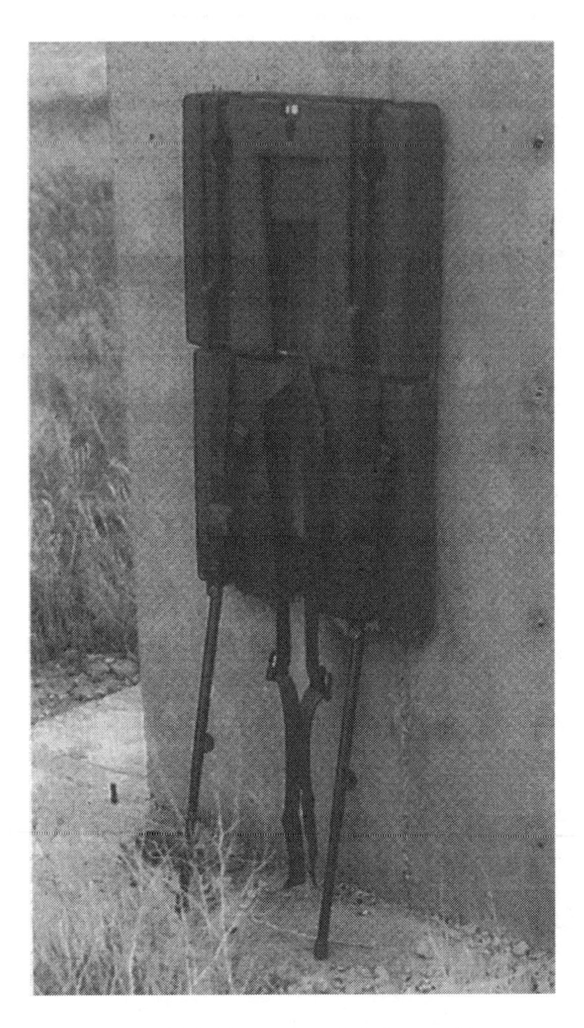

Figure 10.26. Breachcase in firing position.

its own weight. An example of cutting a steel beam using C4 explosive on the web and flanges of the beam is shown in Figures 10.37 and 10.38. Using bulk explosive to destroy a reinforced concrete bridge member typically only strips the concrete from the member, leaving the reinforcing intact. Depending on the bridge design this could be sufficient to cause the bridge to fail under its own weight, otherwise a secondary operation (for example, linear shaped charge) would have to be used to cut the reinforcing. There are various handbook formulas for calculating the amount of explosive required to remove all of the concrete for a given radius from the charge.

Figure 10.27. Results of breachcase on a reinforced concrete wall.

An example by Kraus (1989) is as follows:

$$M = R^3 ac,$$

where M = explosive charge (kg TNT),
R = effective radius (m),
a = tamping factor (4.5 for an untamped charge),
c = resistance factor depending on R.

R	≤1.5 m	≤2.0 m	≤3.5 m	>3.5 m
c	6	5	4	3

Figure 10.28. Nitromethane linear shaped charge (photo courtesy of Golden West Products International).

Figure 10.29. Breach in RC Wall (photo courtesy of Golden West Products International).

Linear shaped charges can also be used for efficient demolition of bridges. Massive linear shaped charges can be placed on top of the bridge deck and when detonated, cut through the deck and beams below. Alternatively, large linear shaped charges can be mounted on the underside of the bridge beams on the flange as shown in Figure 10.39 for a 10 kg shaped charge on a reinforced concrete beam. This beam was turned on its side for the experiment. All of the reinforcing was cut except for the small bars at the top of the beam (Figures 10.40 and 10.41).

The papers by Joachim (1983, 1985) provide useful information on design parameters for linear shaped charges of similar size, suitable for this application. Small shaped charges can also be used in an arrangement similar to the C4 blocks shown previously, where the charges are placed on the web and flanges of the beam. There are many sizes and types of linear

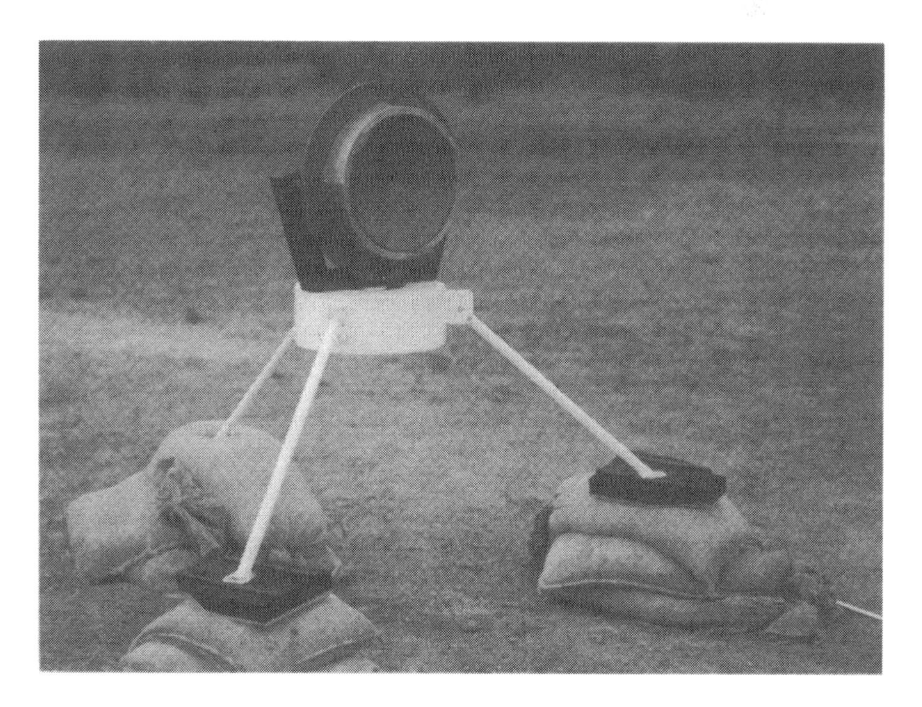

Figure 10.30. 300 mm diameter EFP device (photo courtesy of Defence Science and Technology Organization, Australia).

Figure 10.31. Front view (close-up) of breach through first RC wall.

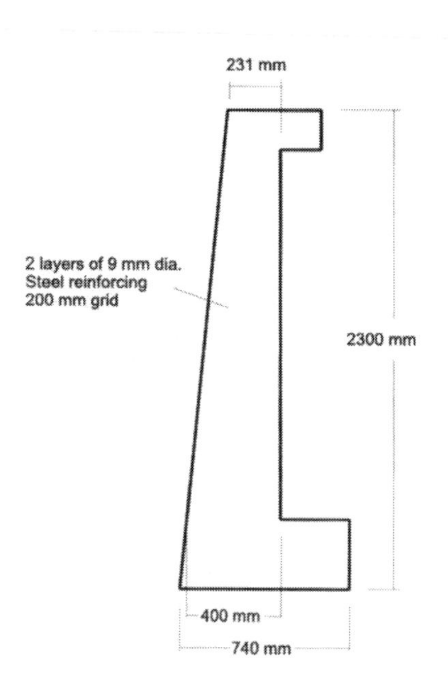

Figure 10.32. Schematic of an RC target wall.

Figure 10.33. Side view of two RC wall targets.

shaped charges, depending on the manufacturer and intended application. Typical penetration performance results are shown in Figure 10.42.

For applications with complex geometry it is desirable to have a linear shaped charge that is flexible and can be contoured to fit the target surface. Lead-sheathed charges in the smaller sizes can be bent by hand, this is not possible with the larger charges. This requirement led to the development of a flexible linear shaped charge called Explosive Cutting Tape (Figure 10.43). The product consists of a copper powder liner in a plastic matrix backed with plastic explosive. The sheathing material is closed-cell foamed-polyethylene [Winter et al. (1993)].

Another method of demolishing steel beams or plates is to use the high pressure generated by colliding shock waves to fracture the metal target. There are several products based on this concept, Shock-wave Refraction Tape (SRT) and Ladder Fracture Tape (LFT). SRT consists of a curved strip of magnet rubber with a small groove on the underside, to promote cracking of the target. Depending on the explosive loading there is from one to four layers of explosive on top of the rubber strip (Figure 10.44). For charges with more than one layer of explosive, a higher velocity of detonation top layer of explosive, in comparison to the lower explosive layer, is used to enhance wave focussing. SRT causes a high-pressure region to be created in the target, followed by relaxation of the pressure, which fractures the target.

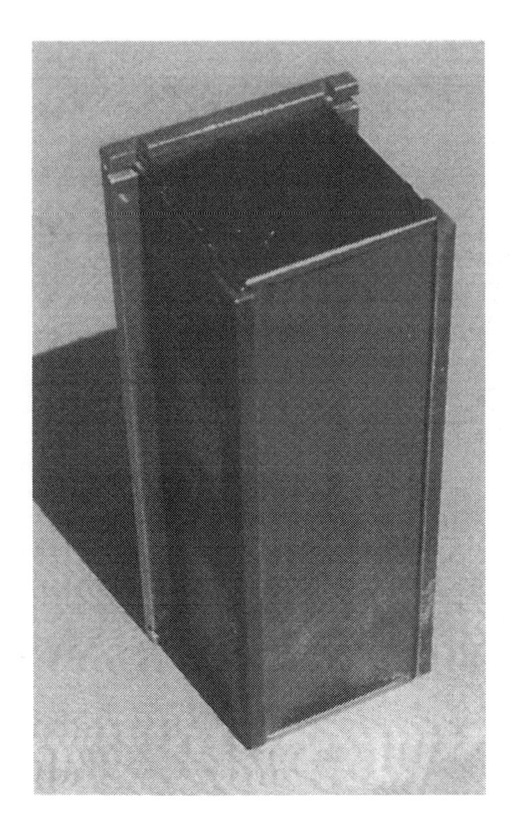

Figure 10.34. Linear EFP.

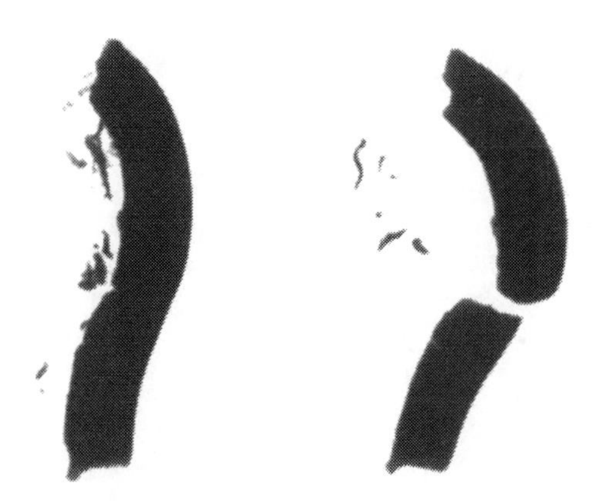

Figure 10.35. Flash radiographs of linear EFP. Left: time = 290 μs; position = 0.6 m. Right: time = 705 μs; position = 1.5 m.

Figure 10.36. Damage to RC wall by linear EFP.

Figure 10.37. C4 explosive attached to steel I beam.

Figure 10.38. Damage to steel I beam.

Figure 10.39. Shaped charge attached to bottom of RC beam.

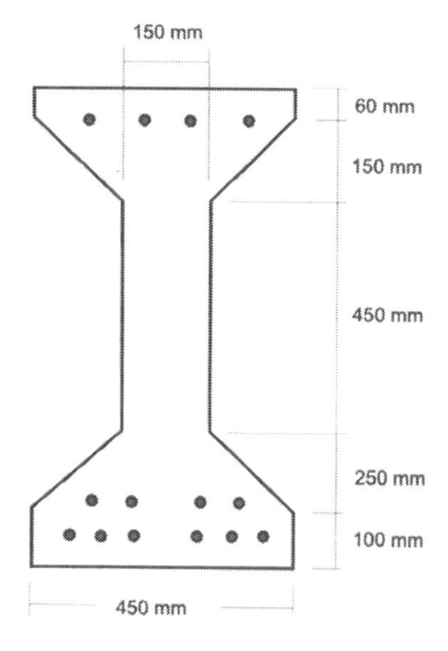

Figure 10.40. Schematic of RC beam section.

Figure 10.41. Damage to RC beam.

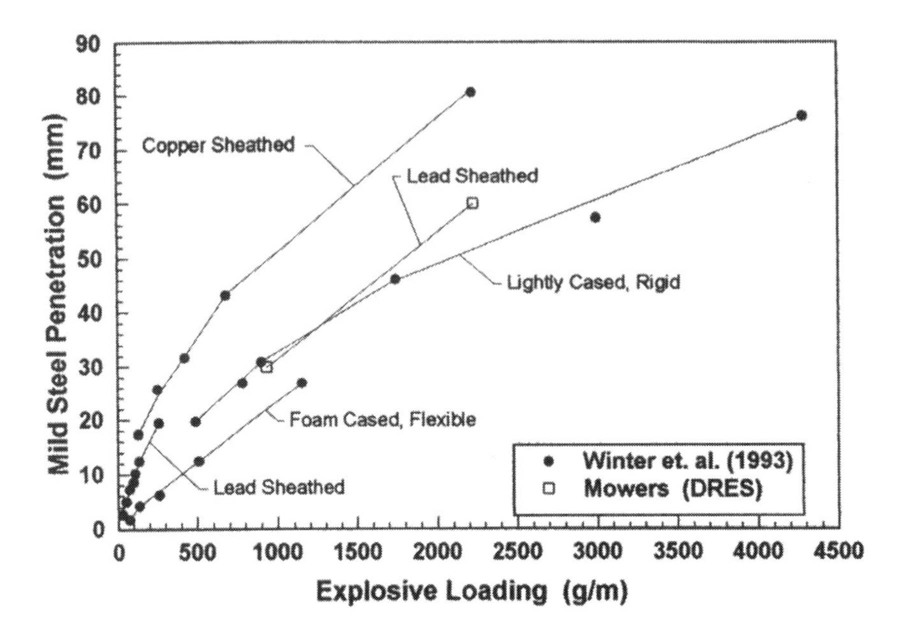

Figure 10.42. Penetration performance of linear shaped charges.

Ladder fracture tape consists of a neoprene rubber moulding with channels which can be filled with plastic explosive [Masinskas and van Leeuwen (1987)]. Once the moulding is filled with explosive, the protruding rubber resembles the rungs and rails on a ladder. Detonation of the explosive from one end of the ladder creates a wave which travels around both ends of the rung barrier and splits into two waves which collide back at the center of the ladder as illustrated in Figure 10.45. This causes a high-pressure region to be formed which continues to the next barrier and then the process repeats.

Figure 10.43. Explosive cutting tape.

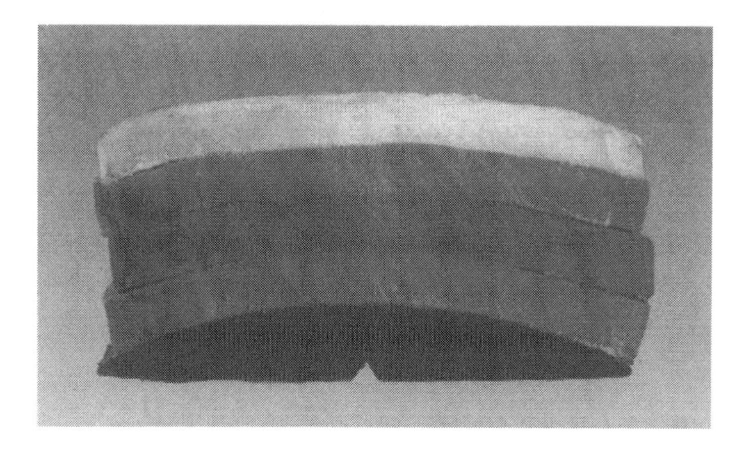

Figure 10.44. Shock-wave refraction tape.

The refection of waves in the target causes a spall surface and fracture of the target plate under the centerline of the LFT.

One final technique for beam demolition using direct contact with the target is to use extrudable explosive which can be applied directly to the web and flanges of the beam as shown in Figures 10.46 and 10.47.

The disadvantage of direct contact charges is that they must be attached to the bridge beam or section to be demolished. This can be particularly

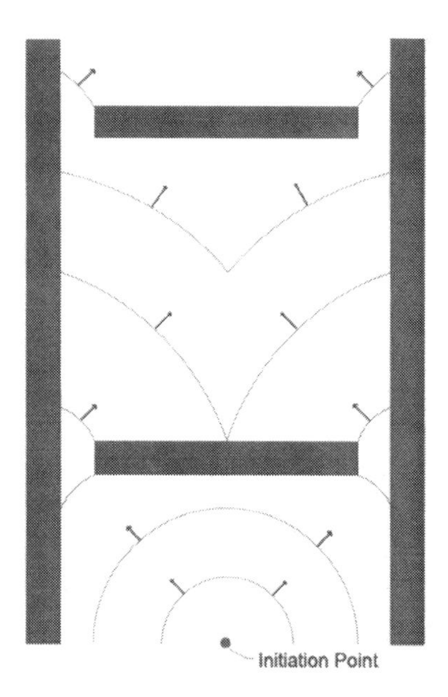

Figure 10.45. Schematic of the functioning of LFT.

Figure 10.46. Extrudable explosive applied to beam.

time-consuming and hazardous to personnel deploying the charges. In a stand-off bridge demolition technique these disadvantages could be eliminated by using explosively formed projectiles fired remotely from the target (tens of meters). The projectile would breach the concrete and cut some of the reinforcing, essentially removing all of the load carrying capacity of the beam. If this was used on all of the main bridge beams, the bridge would collapse under its own weight. This technique was investigated by Worswick et al. (1989) using an EFP device called "Powercone," developed for the mining industry. The devices were found to be effective at breaching large

Figure 10.47. Damage to steel beam.

Figure 10.48. RC T-beam.

cross sections of reinforced concrete. Shown in Figures 10.48–10.50 are the results from firing two Powercone devices into a 1.5 m deep, T-section reinforced concrete beam with a web thickness of 0.35 m.

10.5. Explosive Ordnance Disposal

In the most general sense, the term explosive ordnance disposal (EOD) refers to neutralization of explosive devices. This includes many types of military devices such as bombs, warheads, mortars, and mines, as well as civilian devices such as terrorist bombs. Although some neutralization techniques are applicable to both types, this discussion will be limited to commercially manufactured military devices. In neutralization of a terrorist bomb a technique that can be applied is to disarm the detonator, for example, using an explosively driven liquid [Petrousky et al. (1990)], whereas for military applications it is desirable to neutralize the main explosive charge. The scope is further reduced to include only unexploded ordnance (UXO) and not demilitarization of ordnance. The reason for this is that the techniques for demilitarization of ordnance are completely different from neutralization

Figure 10.49. Powercone test setup.

Figure 10.50. Damage to RC T-beam.

of UXO, for example, using cavitating water jets to wash out the explosive from obsolete munitions [Conn (1986)].

The technique used to neutralize UXO depends on many factors such as size and type of ordnance, fusing mechanism, type of explosive materials contained within the ordnance, if a detonation/deflagration/burning is required, proximity to important installations, and the time allocated to perform the neutralization. Consideration of these factors determines if projectiles, bulk explosives, explosive tapes, shaped charges, or explosively formed projectiles should be used.

The first item to consider is the location of the UXO. If there are no important structures nearby and all personnel can be protected from blast and fragments, then the simplest technique which can be used, provided there are no antidisturbance devices near, attached to, or in the ordnance, is to use a bulk explosive charge (block, sheet, extrudable, or foam) in contact with the ordnance. Detonation of the explosive will cause sympathetic detonation of the explosive filling in the UXO. If the UXO cannot be detonated in situ, for various reasons, then there are two options available. The first is to render the UXO safe by removing or neutralizing the fusing mechanism and transporting the UXO to another location to be neutralized. The second option is to neutralize the UXO in situ but use a technique less violent than detonation.

To render an ordnance device safe the fuse must be removed from the device or made nonfunctional. Removal of the fuse from items such as land mines is extremely difficult since the fuse mechanism is contained within the mine. While neutralization of the fuse in a land mine is possible with a shaped charge, it is difficult to determine the condition of the fuse after firing the shaped charge. For items such as the 105 mm shell, the fuse protrudes from the front of the shell, hence making it simpler to render safe. Shaped charges can be used for this application, but similar to the land mine situation, it is difficult to assess the condition of the fuse after firing of the shaped charge. This was demonstrated by Chick et al. (1994) where a 38 mm diameter conventional shaped charge was fired at the fuse/case junction of a 105 mm shell. The major portion of the fuse was left attached to the shell with explosive fuse components in an unknown condition. Another technique for fuse neutralization is a dearmer device. One of these types of devices uses a 0.50-caliber cartridge to fire a cylindrical projectile at the fuse [Vande Kieft and Bines (1989)]. Other types of dearmers are reviewed by Wyatt (1990). The preferred situation is to totally separate the fuse from the shell, therefore eliminating the possibility of the fuse functioning and detonating the shell. Explosively formed projectiles have been used quite successfully for this purpose [Chick et al. (1994)]. The EFP removes the fuse from the shell without igniting the explosive charge contained in the shell. Since EFPs have a stand-off capability, the render-safe procedure is conducted without having to handle the shell which is a distinct advantage over several other techniques.

Once the shell is rendered safe, then the next step is to remove the main explosive charge. One technique is to use a trepanning tool to cut a hole in the shell casing. This is a rather slow process and as a result various punching charges have been developed to produce a hole in the case of a shell with a wall thickness of up to 12 mm without initiating a reaction of the explosive [de Jong and Kodde (1992)]. Once a hole is cut in the case, then the explosive can be steamed out of the shell and disposed of in a safe manner. This can be a rather lengthy process, several hours, depending on the size of the shell.

The outcome of the technique chosen to neutralize a UXO in situ will depend on many factors as discussed above. Possible outcomes are sensitization, burning, burning transitioning to a deflagration, deflagration, deflagration transitioning to a detonation, and detonation of the explosive charge. Clearly, sensitization of the explosive filling is not a desirable outcome. At the other end of the spectrum, detonation is usually not desired (except for a few particular circumstances) due to the high environmental impact. Burning or deflagration is the desired result. The materials or devices which produce these results are bulk explosives, projectiles, shaped charges, linear shaped charges, explosive tapes, and explosively formed projectiles.

As mentioned previously, bulk explosives can be used to produce sympathetic detonation of UXO. For example, Figure 10.51 shows the use of nitromethane explosive foam applied to neutralization of a land mine. This type of explosive has been used successfully to detonate a variety of bombs and submunitions. Bulk explosive can also be used to cause deflagration in a bomb. Hubbard and Tomlinson (1989b) used stacks of sheet explosive disks

Figure 10.51. Nitromethane foam applied to mine (photo courtesy of Mining Resource Engineering Ltd.).

Figure 10.52. Sheet explosive disks test set-up (photo courtesy of Defence Science and Technology Organization, Australia).

or squares to cause shock-induced deflagration in bombs. This technique breaks the munition into several pieces and large chunks of unreacted explosive remain after the event. The set-up and results of this type of experiment are shown in Figures 10.52 and 10.53. An important aspect of this technique is venting of the gases from the reacting explosive. If the sheet explosive charge is too small, then excess pressure can develop in the munition, resulting in violent deflagration. Venting effects have been studied in detail by Graham (1986). These effects are also important for the other techniques discussed below.

Attack of UXO using projectiles offers another technique of neutralization. Howe et al. (1981) studied the response of confined explosive charges to attack by right circular cylinders and explosively driven flyer plates. Hubbard and

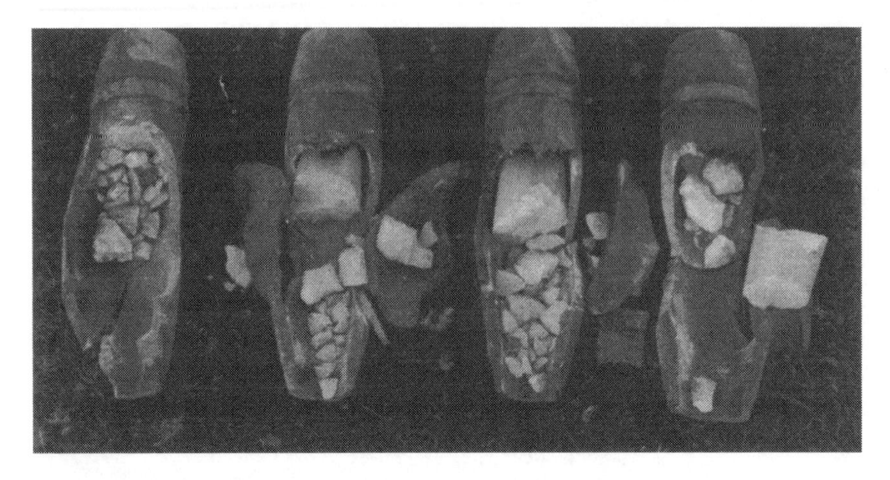

Figure 10.53. Results from experiments using sheet explosive disks (photo courtesy of Defence Science and Technology Organization, Australia).

Tomlinson (1989a) conducted experiments with a range of projectiles from 7.62 mm to 30 mm against a variety of bombs and projectiles. The full range of outcomes, ranging from no reaction to detonation was achieved. Most of the experiments produced deflagrations with varying degrees of violence.

A considerable amount of research has been conducted on the interaction of shaped charge jets with munitions [Held (1987a, b, 1989); Chick and Hatt (1981); Chick et al. (1986a, b, c)]. It is applicable to both vulnerability assessment and EOD. In the work by Chick, it was found that the detonation/failure threshold was very sensitive to the cover or casing thickness. As with projectile impact, the full range of outcomes can be achieved. As a result of this research effort, a hazard assessment protocol has been developed for shaped charge jet attack of munitions containing energetic materials [Chick and Frey (1990)].

Linear shaped charges and explosive tapes (refer to Bridge Demolition, Section 10.4) are available in a wide variety of types and sizes. These devices can be used to slit open a munition and produce a suitable size of opening such that they minimize the risk of the explosive filling reaction rate accelerating due to internal pressure in the munition. Although these types of charges are successful at producing the required opening in the munition, they usually do not ignite the explosive filling, and as a result, a second operation is required to remove the explosive. An elegant one-step solution to this problem was developed by Winter (1996). He used linear shaped charges with a reactive liner material (RLSC). When detonated the RLSC cuts open the munition and ignites the explosive filling. By suitably sizing the RLSC sufficient venting of the explosion products is possible and the explosive filling simply burns. The results from this technique for a 105 mm HESH (High Explosive Squash Head) round are shown in Figures 10.54 and 10.55.

Figure 10.54. RLSC test set-up for 105 mm HESH round (photo courtesy of Defence Science and Technology Organization, Australia).

Figure 10.55. Results from RLSC test (photo courtesy of Defence Science and Technology Organization, Australia).

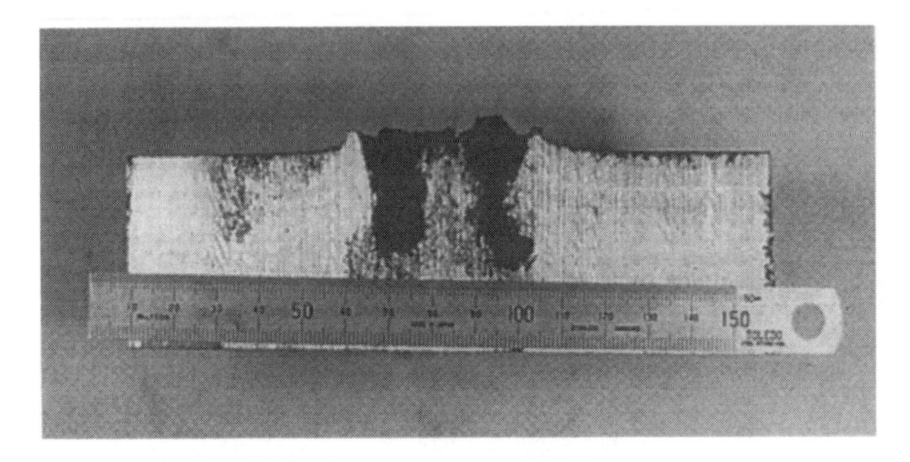

Figure 10.56. Penetration profile in a mild steel block for an EFP (photo courtesy of Defence Science and Technology Organization, Australia).

Explosively formed projectiles have been discussed earlier with respect to removal of the fuse from a UXO. However they can also be used to open the munition case and produce a slow burn or in some cases a deflagration [Chick et al. (1994)]. One of the significant advantages of using an EFP for this purpose is that the stand-off capability of this type of device allows neutralization of the UXO without having to come in contact or disturb the UXO. A novel design of EFP has been demonstrated to be highly successful in cutting out a disk in a munition and inducing a slow burning of the explosive filling in a single operation. The penetration profile in a witness block from this type of EFP is shown in Figure 10.56. These experiments have been conducted at relatively short stand-off distances of two to ten charge diameters. An investigation of EFPs for bomb disposal at a long stand-off distance requirement of 30 m was conducted by van Bree (1995). During this investigation a wide variety of GP bombs were tested with 56 mm and 75 mm caliber EFPs and resulted in a deflagration reaction. Note that the actual stand-off distance used for these experiments was not stated in the paper.

References

——— (1994). *International Conference on Mechanical and Physical Behaviour of Materials Under Dynamic Loading (EURODYMAT 94)*, Oxford, UK.

——— (1995). *Proc. Conf. American Physical Society Topical Group on Shock Compression of Condensed Matter*, August 13–18, Seattle, WA (S.C. Schmidt and W.C. Pao, eds.). Conference Proceedings, No. 37. AIP Press, Woodbury, NY.

Murr, L.E., Staudhammer, K.P., and Meyers, M.A. (eds.) (1995). *Metallurgical and Materials Applications of Shock-Wave and High-Strain-Rate Phenomena (EXPLOMET '95)*, El Paso, TX. Marcel Dekker, New York.

——— (1996). *Proc. 16th International Symposium on Ballistics*, San Francisco, CA. American Defense Preparedness Association; Washington, DC.

Anderson, C.E., Jr. (ed.) (1996). *Hypervelocity Impact. Proceedings of the 1966 Symposium*, Freiburg, Germany. *Internat. J. Impact Engng.*, **20** (1997), 1–6.

Asay, J.R. and Shahinpoor, M. (1993). *High-Pressure Shock Compression of Solids*, Springer-Verlag, New York.

Backofen, J.E., Jr. (1978). The weaponization of shaped charge technology. *Fourth International Symposium on Ballistics*, Monterey, CA.

Baker, W.E. (1973). *Explosions in Air*. University of Texas Press, Austin, TX.

Baker, W.E., Cox, P.A., Westine, P.S., Kulesz, J.J., and Strehlow, R.A. (1983). *Explosion Hazards and Evaluation*. Elsevier Scientific, Amsterdam.

Blazynski, T.Z. (1987). *Materials at High Strain Rates*, Elsevier Applied Science, Amsterdam.

Bushman, A.V., Kanel', G.I., Ni, A.L., and Fortov, V.E. (1993). *Intense Dynamic Loading of Condensed Matter*. Taylor & Francis, London.

Carleone, J. (1993). *Tactical Missile Warheads*. American Institute of Aeronautics and Astronautics.

Chick, M.C. and Hatt, D.J. (1981). Metal jet initiation of bare and covered explosives: summary of the mechanism, empirical model and some applications. DSTO Materials Research Laboratory, Melbourne, Victoria, Report MRL-R-830.

Chick, M., Bussell, T., Frey, R.B., and Boyce, G. (1986a). Initiation of munitions by shaped charge jets. *Proceedings of the Ninth International Symposium on Ballistics*, Shrivenham, UK.

Chick, M., Bussell, T., Frey, R.B., and Boyce, G. (1986b). Assessment of shaped charges for use in explosive ordnance disposal. DSTO Materials Research Laboratory, Melbourne, Victoria, Report MRL-R-1026.

Chick, M.C., Wolfson, M.G., and Learmonth, L.A. (1986c). A calibrated test for the assessment of the sensitivity of explosives to shaped charge jets. DSTO Materials Research Laboratory, Melbourne, Victoria, Report MRL-R-1016.

Chick, M. and Frey, R.B. (1990). Towards a methodology for assessing the terminal effect of a jet strike on munitions. *12th International Symposium on Ballistics*, San Antonio, TX.

Chick, M., Bussell, T., Lam, C.P., McQueen, D., and McVay, L. (1994). An investigation into low environmental impact ordnance disposal methods with ballistic discs. *Twenty-Sixth DOD Explosives Safety Seminar*, Miami, FL.

Conn, A.F. (1986). An automated explosive removal system using cavitating water jets. *Minutes of the Twenty-Second Explosives Safety Seminar*, Anaheim, CA.

Cullis, I.G. (1992). Experiments and modelling in dynamic material properties: Explosively formed projectile research in a European collaborative forum. *13th International Symposium on Ballistics*, Stockholm, Sweden.

de Jong, E.G. and Kodde, H.H. (1992). The development of a punching charge for application as a explosive ordnance disposal technique—period 1986–1990. TNO Prins Maurits Laboratory, The Netherlands, PML1992-100.

Faccini, E.C. (1992). Orbital forging of heavy metal EFP liners. In *High Strain Rate Behaviour or Refractory Metals and Alloys* (Asfahani, R., Chen, E., and Crowson, A., eds.). The Minerals, Metals, and Materials Society.

Forsen, R. (1990). Experiments used for comparison of blast damage to full-scale and one fourth scale reinforced concrete structures. *24th DoD Explosives Safety Seminar*, Saint Louis, MI.

Glasstone, S. and Dolan, P.J. (1977). *The Effects of Nuclear Weapons*, 3rd edn. United States Department of Defense and the Energy Research and Development Administration, Washington, DC.

Graham, K.J. (1986). Explosive response to fragments: Venting studies. Naval Weapons Centre, China Lake, NWC TP 6456.

Hader, H. (1983). Effects of bare and cased explosive charges on reinforced concrete walls. *Symposium Proceedings, The Interaction of Non-Nuclear Munitions with Structures*, U.S. Air Force Academy, CO.

Held, M. (1987a). Experiments of initiation of covered, but unconfined high explosive charges by means of shaped charge jets. *Propellants, Explosives, Pyrotechnics*, **12**.

Held, M. (1987b). Discussion of the experimental findings from the initiation of covered, but unconfined high explosive charges by means of shaped Charge Jets. *Propellants, Explosives, Pyrotechnics*, **12**.

Held, M. (1989). Analysis of shaped charge jet induced reaction of high explosives. *Propellants, Explosives, Pyrotechnics*, **14**.

Howe, P.M., Watson, J.L., and Frey, R.B. (1981). The response of confined explosive charges to fragment attack. *Proceedings of the Seventh Symposium (International) on Detonation*, Annapolis, MD.

Hubbard, P.J. and Tomlinson, R. (1989a). Ballistic attack of a variety of explosive filled munitions. *Proceedings of the 11th International Symposium on Ballistics,* Brussels, Belgium.

Hubbard, P.J. and Tomlinson, R. (1989b). Explosiveness and shock-induced deflagration studies of large confined explosive charges. *Proceedings of the Ninth Symposium (International) on Detonation*, Portland, OR.

Joachim, C.E. (1983). Rapid runway cutting with shaped charges. *Symposium Proceedings, The Interaction of Non-Nuclear Munitions With Structures.* US Air Force Academy, CO.

Joachim, C.E. (1985). Linear shaped charge penetration. *Proceedings of the Second International Symposium on Interaction of Nonnuclear Munitions with Structures*, Panama City Beach, FL.

Kennedy, D.R. (1983). *History of the Shaped Charge Effect—The First 100 Years.* 100th Anniversary of the Discovery of the Shaped Charge Effect By Max Von Foerster. MBB Schrobenhausen, West Germany.

Kinney, G.F. and Graham, K.J. (1985). *Explosive Shocks in Air*, 2nd edn. Springer-Verlag, New York.

Kraus, D. (1989). A new blast method for concrete bridges. *Proceedings of the Fourth International Symposium on Interaction of Nonnuclear Munitions with Structures*, Panama City Beach, FL.

Lonnqvist, L. (1993). The effects of high explosives in contact with reinforced concrete slabs. *Proceedings of the Sixth International Symposium on Interaction of Nonnuclear Munitions with Structures*, Panama City Beach, FL.

Masinskas, J.J. and van Leeuwen, E.H. (1987). Theoretical investigation of the wave shaping process in a section of ladder fracture tape. DSTO Materials Research Laboratory, Melbourne, Victoria, Report MRL-R-1081.

Meyers, M.A. (1994). *Dynamic Behaviour of Materials*. Wiley, New York.

Murphy, M., Weimann, K., Doeringsfeld, K., and Speck, J. (1992). The effect of explosive detonation wave shaping on EFP shape and performance. *13th International Symposium on Ballistics*, Stockholm, Sweden.

Petrousky, J.A., Backofen, J.E., and Butz, D.J. (1990). Shaped charge with explosively driven follow through. United States Patent 4955939.

Shirey, D.L. (1980). Breaching of structural steel plates using explosive disks. *The Shock and Vibration Bulletin.*

van Bree, J.L.M.J. (1995). Disposal of GP bombs by EFP attack. *15th International Symposium on Ballistics*, Jerusalem, Israel.

Vande Kieft, L.J. and Bines, A.L. (1989). Expendable dearmer evaluation. Ballistic Research Laboratory, Aberdeen Proving Ground, MD. Memorandum Report BRL-MR-3772.

Walters, W.P. and Zukas, J.A. (1989). *Fundamentals of Shaped Charges.* Wiley, New York.

Weickert, C.A. (1986). Spin stabilization of self-forging fragments. *Ninth International Symposium on Ballistics*, Shrivenham, UK.

Weickert, C.A. and Gallagher, P.J. (1992). Parametric study of the effect of a confinement ring on the shape of an explosively formed projectile. *13th International Symposium on Ballistics*, Stockholm, Sweden.

Weickert, C.A. and Gallagher, P.J. (1996). Ogive-nosed, finned, explosively formed projectiles. *16th International Symposium on Ballistics*, San Francisco, CA.

Weimann, K., Doeringsfeld, K., Speck, J., and Lopatin, C. (1990). Modelling, testing, and analysis of EFP performance as a function of confinement. *12th International Symposium on Ballistics*, San Antonio, TX.

Weimann, K. (1993a). Flight stability of EFP with star shaped tail. *14th International Symposium on Ballistics*, Quebec, Canada.

Weimann, K. (1993b). Research and development in the area of explosively formed projectiles technology. *Propellants, Explosives, Pyrotechnics*, **18**, 294–298.

Winter, P.L., Thornton, D.M., and Learmonth, L.A. (1993). The application of linear shaped charge technology to explosive demolition of structures. *Dynamic Loading and Manufacturing and Service*, The Institute of Engineers of Australia, Melbourne.

Winter, P.L. (1996). Private Communication.

Worswick, M.J., Mackay, D.J., McQuilkin, A., Weickert, C.A., Storrie, T., and Mowers, S. (1989). Investigation of explosively formed projectiles impacting concrete. *Proceedings of the Fourth International Symposium on Interaction of Nonnuclear Munitions with Structures*, Panama City Beach, FL.

Wyatt, J. (1990). Conventional explosive ordnance disposal. *International Defense Review*, **1**.

Zukas, J.A. (1990). *High Velocity Impact Dynamics.* Wiley, New York.

Zukas, J.A., Nicholas, T., Swift, H.F., Greszczuk, L.B., and Curran, D.R. (1982). *Impact Dynamics.* Wiley, New York.

Index

accelerated ageing (*see* safe life in storage) 319
acceleration (*see also* pick up) 296
accidents 5, 341–343
  Gulf of Tonkin 262
  Palomares, Thule 262
  Texas City 341
adiabatic gamma 61, 80–81, 100–105

Bachmann Process 40
barrier materials for gap tests 292
BDNPA-F (energetic plasticizer) 44
beam cutting 400–401, 403, 405, 410–413
Biot Number 187, 315
blast waves 10–13
blasting caps (*see* detonators)
bomb 413, 416, 418, 420
boundary conditions 184, 185, 188
brisance (*see also* explosives, strength) 139
Bruceton Staircase Method 267, 343
burn 189, 415–416, 420

cgs system of units 25
Chapman–Jouget (C–J)
  density estimating 127
  hypothesis 198
  plane 194
  point 118
  pressure estimating 127
  state 117, 119, 127
characteristic velocities 126
charge diameter effects 130
Charge Hazard Tests 290
Charge Impact Tests 302
  LANL Oblique Impact 303
  LLNL–Pantex Skid 304
  Susan 306
  UK Oblique Impact 302
coherent light (laser) detonators 212
combustion 137, 174
comparatively insensitive explosives 286
composite explosives 165
computer codes 120
conducting composition detonators 209
confinement 274, 308
conservation relations 6–9
conversion factors (systems of units) 25
cook-off testing 314, 346–347, 349, 371–375
critical conditions 180, 181
  criteria (Frank-Kamenetskii Theory) 186, 312–319, 350–352, 362
  energy 204
  power density 204
  temperature 177, 182, 186
critical diameter 131, 133

dearmer 415
deflagration 138, 175, 415–417, 420
deflagration-to-detonation transition 139
deflection angle 236–238
demolition 3, 381–423
  bridge 399–413
dent, equivalent dent 294

detonation 74–77, 137, 139, 175, 415–416
  conservation equations 195
  convergent wave 202
  failure diameter 194
  hydrodynamic theory 194
  jump condition 117
  overdriven 201
  products Hugoniot 128–130
  products isentrope 120, 128
  steady state 115, 194
  theory 189
  transfer 251–254
  transient waves 201
  velocity 117, 122, 124, 125, 138
detonator 139, 382
detonators
  bridgewire/foil 209
  coherent light 212
  conducting composition 209
  DDT (low-voltage) 211
  flash-receptive 206
  flyer plate 211
  slapper 212
  stab-sensitive 206
dimensionless (differential) equations 185
discounting angle 247–249
Drop-Hammer Tests (*see* Impact Tests)

efficiency 233–235
electrical Gurney energy 245
electronic safe and arm unit 216
electrostatic sensitiveness 319, 346
  electrostatic powder testing 320
endothermic binders 44
energetic plasticizers
  (polyNIMMO, FEFO, BDNPA-F, polyglycidyl nitrate, etc.) 44
energy balance 119
Equation of State (EOS) 82–105, 120, 191
  Abel 60, 85–86
  Becker–Kistiakowski–Wilson (BKW) 88
  Davis 95
  Hayes 94–95
  ideal gas 83–84
  intermolecular potentials 88–89
  inverse power potential 86
  Jones–Wilkins–Lee (JWL) 89–91, 121
  linear *U*–*u* 54, 70, 91–93
  mixtures 96–100
  polytropic gas 54–55, 58, 72–73, 84–85
  summary 96
  virial expansion 86–87
  Walsh mirror image 93
  Williamsburg 95
equivalent sphere 186
explosions
  chemical 27
  nuclear 27
  physical 26
explosive cutting tape 405
explosive ordnance disposal 413–420
explosive output 221, 232–235
explosively formed projectile 384–391
  aerodynamic performance 386, 388–389
  blunt nose 389
  charge
    confinement 387–388
    cylindrical 386–391
    linear 399
    taper-backed 388–389
  finned 389
  flared tail 388
  formation
    backward-fold 386–387, 397
    forward-fold 386–387
    W-fold 386–387
  liner
    Armco iron 388
    copper 399
    tantalum 388
  numerical modeling 386–387
  ogive nose 389
  penetration 384, 397
  stabilization 388–389
  stand-off 384
  velocity 384, 388–390
explosiveness 272, 275, 323
explosives
  Act (1875) (UK) 27
  black powder 1, 29, 189, 342
  booster (Tetryl, PETN, RDX, PBXN-7, Comp. A3) 31, 41, 139, 207, 212, 273, 291, 382
  classification 28, 138–139
  cobalt perchlorate 211
  commercial 42, 342
  DINA (liquid) 182
  energetic salts 162–165
    Ammonium Dinitramide (ADN) 165

Ammonium Nitrate (AN) 137, 144, 146, 152, 162, 341–345, 350, 358–375
ANFO 137, 152, 341–345, 359–370
Ammonium Perchlorate (AP) 164
Azides 165
Hydroxylammonium Nitrate (HAN) 165
Mercury Fulminate [$Hg(ONC)_2$] 165
extrudable 411
heterocyclic 152, 160–162
NTO 152, 161
PYX 152, 162
TNAZ 152, 161–162
history 1
liquid oxidizers and explosives 166–168
liquid oxygen explosives (LOX) 166
nitrogen tetroxide (N_2O_4) 167
peroxides 167
TATP 168
military 3, 9–10, 137, 140
new explosives (evolution) 45
nitramenes 156–160, 376–377
CL-20 160
guanadine nitrate 157
HMX 32, 41, 140, 148, 152, 158–159, 341–346, 350
nitramene composites 159
Composition B 40, 141, 159, 388, 399
Composition B-2 159
Composition C-4 159, 400, 403
Cyclotol 159
Octol 159, 273, 343, 388
PBX 9407 159, 273, 291
Torpex-2 159
nitramine 156
nitroguanadine 32, 37, 140, 157
nitrourea 156
RDX 32, 40, 140, 148, 152, 157–159, 206, 341–346, 350, 356, 375
urea nitrate 156
tetryl 32, 38, 160, 291
Nitrate esters 141–146, 376–377
dinitroglycol 141
dynamite 32, 42, 140, 144
guncotton 143
methyl nitrate 141
nitrated sugars 142–143
nitrocellulose 31, 33, 139, 143, 342
nitroglycerine 2, 30, 139–145, 189, 342
nitrostarch 143
PETN 31, 140, 146, 148, 212, 342, 356, 375
smokeless powder 144
triethylene glycol dinitrate 142
nitroalkanes 154–156, 376–377
1,2 dinitroethane 155
1,2 dinitropropane 155
hexanitroethane 155
nitrocubanes 44, 156, 202
nitromethane 154, 397, 416
polynitroethylene 155
tetranitromethane 155
nitroarenes 146–154, 376–377
hexanitroazobenzene 150, 212
hexanitrobenzene 42, 147
hexanitrostilbene 31–32, 41, 148, 212
nitrobenzene 146
picric acid 31, 140, 149, 152
properties 152–153
salts of picric acid 38, 149–150, 152
styphnic acid (trinitroresorcinol) 33, 150, 206
TATB 31, 42, 151, 323
1,3,5-trinitrobenzene 146
2,4,6-trinitrotoluene (TNT) 31, 42, 139–140, 143, 148, 152, 163, 167, 341–343, 357, 375
performance 138–139
physico-chemical properties 34–35
safety 261, 265, 275–339, 341–378
science of 4
sheet 392, 416
strength (*see also* brisance) 139
toxicity 148, 375–378
unconventional 168–171
acetylides 169–170
fuels 170
hydrides 170
metals 168
miscellaneous energetics 170
phosphorus 170
self-igniting materials 170
welding with 17
extinction point 181

flash radiograph 383, 406
Forest Fire model 203
fragments 13–15
  direction 236–238
  velocities 227–228
Friction Tests 286–290, 346
  Powder 286
    ABL anvil 346
    German BAM Friction 288
    UK Mallet 286
    UK Rotary Friction 288
    US Bureau of Mines Pendulum 287
    US Navy Weapons Center Friction Pendulum 288
  Results 289, 290
fuel fire 315
fundamental derivative 61, 64, 100–105
fuze 285
fuzehead 209

gamma estimating 127
gamma law 121
Gap Tests 291
  Instrumented 296
  LANL, US Standard Scale 295
  LANL Small Scale 295
  Picatinny Arsenal 295
  results 298
  sensitiveness 194
  UK Small Scale 86
  Uninstrumented 293
  US Bureau of Mines 293
  US Naval Ordnance Laboratory 293
  US Navy, China Lake 295
gaps, effect of 239
graphite 209
Greek Fire 29
Grüneisen gamma 81, 100–105
gun prematures 273
Gurney
  energy 232–234
  equations 227–228
  laser energy 245–246
  limitations 246–247
  model 221, 224–225
  velocity 227, 235, 238–239
Gurney, R.W. 221

hazard (*see* explosives, safety)
heat of detonation 221, 232–234
hot-spots 182, 188, 203, 207, 314
Hugoniot curve 196
  data 105
  equations 53
  experiments 66–67
hydrodynamics 80

impact interaction 251–252
Impact Tests 277–305, 343–345
  Charge 302
    LANL Oblique Impact 303
    LLNL–Pantex Skid 304
    UK Oblique Impact 302
  Charge Test Results 304, 305
  High-Speed (Susan) 305
  machines 278
    German BAM 283
    Los Alamos Laboratory Scale 281
    Picatinny Arsenal Laboratory Scale 281
    UK Rotter Test 282
    US Bureau of Explosives (NY) Laboratory Scale 283
    US Bureau of Mines 279, 281
    US NOL (Bruceton) 280
    US NOL Laboratory Scale 280
  powder 277
  results 284–285, 344–345
imploding geometries 228–231
impulse estimation 231–232
induction period 177, 202
inelastic collision 250
initiation
  mechanisms 205
  electrical 209
    flash/flame 208
    friction 208
    percussion 208
    stabbing 208
  theories 174
  thermal (heat) 207
initiation trains 214
initiatory explosives 33
insensitive high explosives and propellants 322
insensitive munitions (IM) 263, 273
  IM System or Weapon Tests 326
  IM System or Weapon Test Procedures 335
  Impact Tests 331
    Bullet and Fragment Attack 331
    Shaped-Charge Jet 333
    Shaped-Charge Jet Spall 332

Results of Cook-Off Tests 330
Sympathetic Detonation Test 335
test response criteria 327
Thermal Tests 328
Slow Cook-Off 328
Standard Liquid Fuel Fire 329
interfaces 64–71
Intrusion Tests 309
AWE (UK) Spigot 310
LANL Spigot Intrusion 309
Results 311, 312
US Navy Spigot 311

Judgment in Hazard Assessment 270

Kamlet Φ 238–239

ladder fracture tape 405, 410–411
land mine 416
laser detonators (*see* coherent light detonators)
lenses 18–20
plane wave 19
Low-Amplitude Shock Initiation Test 300
lyddite 38

magnetic flux compression 4
mass balance 119
materials properties of explosives, importance 325
mechanical properties of explosives 273
metal acceleration models 239–245
Method of Minimum Contradictoriness 269
*m*th order reaction (critical conditions) 181
momentum balance 119
munition 415, 417–418

NIMIC (NATO Insensitive Munitions Information Center) 260, 271, 323
Nobel, Alfred and Immanuel 30, 144

Oblique Impact (Skid) Tests 302–304
overpressures 306

particle velocity 198
phase change 77–80
pick-up (*see also* acceleration) 296
piezo crystals 210
platonization 37
powder compaction 4
Powder Hazard Tests
Friction 286
Impact 277
Other 290
Powercone 412–413
pressure-impulse curves 10–13
Probit Analysis 266
propellants
gun (developments) 36
single base 36
double base 36
triple base 37

Rankine–Hugoniot relations 53, 195
rarefaction wave 55–61, 64, 67–69, 202
rate-stick experiments 131
Rayleigh line 117, 196
Rayleigh line slope 118
reaction zone 116, 131, 175, 194
reference frames 61–63
reflection at interface 64–71
reliability testing 265
retonation wave 203
rock blasting 6–9

Safe and Arm Unit (SAFU) 215
safe diameter 318
safe life 318
Safety Certificates 276, 290
scaling 12, 249
sensitive explosives 285
sensitiveness 272
Sensitiveness, Sensitivity, and Explosiveness 271, 323
sensitivity (absolute and relative) 263, 272
Shaped charge 15, 381–384
commercial 397
conical 382
jet 382–383
break-up 382–383
formation 382–383
velocity 382
linear 397, 403–405

Shaped charge (*cont.*)
liner
reactive 418
penetration
conical 15–16, 382–384
linear 397, 405
slug 382–383
standoff 382
Shaped charge modeling 228–231
shell 415
Shellite 38
shock initiation 192
criteria 251–254
Sensitiveness Tests (Gap Tests) 291
tests 291, 298
shock wave 50–55, 64, 137, 139, 175, 189
shock-wave refraction tape 405
shutter 215
side losses 246–249
Skid Test (*see* Oblique Impact Tests) 302
slapper detonator 212
small arms caps 206
sound waves 59
specific impulse 231–232
Spigot Tests 309
spontaneous ignition 176
stab-sensitive detonator 206
standard states 196
stationary state 177, 179, 184
steady thermal state 178, 179
storage profile (accelerated ageing) 318
streaming veloclty 196
subnormal detonation velocity 201
Susan Test 305
symmetric collisions 59
Système International (SI) Units 24

tamping effect 228, 232, 236
Taylor angle 236–238
Taylor wave 118–119
Thermal Explosion Theory 176
Comparison of the Semenov and Frank-Kamenetskii Theories 187
Frank-Kamenetskii Theory 182
differential equations 183, 185
Semenov Theory 176
Thermal Hazard Tests 311
Thermal stability 346–378
case histories 358–378
comparative thermal stabilities 347–349
Abel test 347
Accelerating Rate Calorimeter 347
Henkin Time-to-Explosion 348, 355
Koenen test 349, 355
ODTX 348, 355
SCB test 349
Taliani test 347
quantitative thermal stability 349–354
differential scanning calorimeter 349, 359–361, 364–365, 360–370
differential thermal analyzer 349
ion chromatography 362
thermal gravimetric analyzer 349
thermal analytical tools 354–356
time-to-explosion 356–358
Thermally Initiated (Case) Venting System (TIVS) 275
thermochemical properties (Table 8.12) 320
thermodynamics 81–82
toughness (mechanical) of explosives 274
transmission at interface 64–71

UK Safety Certificate 276
unexploded ordnance 413–420
uninstrumented gap tests 292
units of measurement 24

very sensitive explosives 286
visual observation 374
von Neumann spike 118–119, 201

Walker–Wasley Criterion 204
wall breaching 391–399
Breachcase 397
bulk explosives 392–397
explosively formed projectile 397–399
Jet-Axe 397
nitromethane linear shaped charge 397
reinforced concrete 392–397, 399
spallation 395
steel 392

warming periods 177
wave shaping 18
Wedge Test 299
Woolwich Process 40
Zeldovich–von Neumann–Döring model 200
zero-order reaction 177
ZeuS 387

High-Pressure Shock Compression of Condensed Matter

L.L. Altgilbers, M.D.J. Brown, I. Grishnaev, B.M. Novac, I.R. Smith, I. Tkach, and *Y. Tkach:* Magnetocumulative Generators

T. Antoun, L. Seaman, D.R. Curran, G.I. Kanel, S.V. Razorenov, and *A.V. Utkin:* Spall Fracture

J. Asay and *M. Shahinpoor* (Eds.): High-Pressure Shock Compression of Solids

S.S. Batsanov: Effects of Explosion on Materials: Modification and Synthesis Under High-Pressure Shock Compression

R. Cherét: Detonation of Condensed Explosives

L. Davison, D. Grady, and *M. Shahinpoor* (Eds.): High-Pressure Shock Compression of Solids II

L. Davison, Y. Horie, and *T. Sekine* (Eds.): High-Pressure Shock Compression of Solids V

L. Davison, Y. Horie, and *M. Shahinpoor* (Eds.): High-Pressure Shock Compression of Solids IV

L. Davison and *M. Shahinpoor* (Eds.): High-Pressure Shock Compression of Solids III

A.N. Dremin: Toward Detonation Theory

R. Graham: Solids Under High-Pressure Shock Compression

Y. Horie, L. Davison, and *N.N. Thadhani* (Eds.): High-Pressure Shock Compression of Solids VI

J.N. Johnson and *R. Cherét* (Eds.): Classic Papers in Shock Compression Science

V.F. Nesterenko: Dynamics of Heterogeneous Materials

M. Sućeska: Test Methods for Explosives

J.A. Zukas and *W.P. Walters* (Eds.): Explosive Effects and Applications

Made in the USA
San Bernardino, CA
28 June 2015